Mathematics Education Research: Implications for the 80's

Elizabeth Fennema, Editor

Association for Supervision and Curriculum Development
225 N. Washington Street • Alexandria, Virginia 22314
in cooperation with
National Council of Teachers of Mathematics
1906 Association Drive • Reston, Virginia 22091

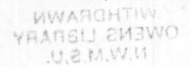

Editing:
Ronald S. Brandt, ASCD Executive Editor
Nancy Carter Modrak, Managing Editor of Booklets

Cover design: Great Incorporated

Stock number: 611-81238
Library of Congress Catalog Card Number: 81-67144
ISBN: 0-87120-107-0

Contents

iii

Foreword

In these days characterized by great emphasis on the so-called basic skills, it is important that we look at where we have been, where we are today, and where it is possible to go in mathematics instruction.

This research-oriented publication is to serve just such purpose. Not that these research reports and their interpretations will give teachers prescriptions for teaching students, but they do give insights into many of the causes of instructional problems in mathematics, thus enabling teachers to plan instruction more responsive to identified needs.

The authors of these chapters have synthesized mathematics research findings for busy teachers. Their realistic observations cover both sexes' feelings about mathematics, the problem-solving processes learners most often use, and the characteristics of interactions between teachers and pupils in the instructional process. For instance, Cooney reports on some interesting research showing that the warm, caring teacher has a positive effect in low socioeconomic status classrooms, but often a negative effect in high SES classes. Nevertheless, overall research findings indicate the effectiveness of a supportive teacher.

The summaries by DeVault and Weaver relative to the uses of computers and calculators, respectively, are particularly timely. Not many educators have dealt effectively with the technology explosion to ensure its contribution to the delivery of instruction.

Dr. Izaak Wirszup, following his comparative study of science and mathematics education in Russia and America, recommended sweeping changes in U.S. education, involving mathematics instruction throughout the K-12 sequence, and in both undergraduate and graduate curriculums.[1]

---

[1] Izaak Wirszup, "The Soviet Challenge," *Educational Leadership* 38 (February 1981): 358-360.

Neither he nor the authors of this book have reported research regarding the effectiveness of the use of textbooks in mathematics instruction. Since a large number of teachers rely on texts for direction, it is reasonable to expect that they are an important variable in the quality of mathematics instruction.

Research as interestingly presented as in this booklet goes a long way toward dispelling myths and stereotypes too often retained by most of us. Our theories and beliefs are revised or changed in the light of new perceptions and experiences. We hope that this much-requested publication will serve to enlighten us and motivate each of us to commit ourselves to recast mathematics education, as we take leadership in developing the curriculums that will educate the citizenry of the new century.

LUCILLE G. JORDAN
President, 1981-82
Association for Supervision and
Curriculum Development

Introduction:
The Value of Mathematics Education Research

Elizabeth Fennema

A president of the National Council of Teachers of Mathematics (NCTM) once said he had never learned anything from research that was of use in teaching. He also quoted a nonreferenced source as saying that "Evaluation in a great number of countries shows that educational research, and research into the field of the pedagogy of mathematics have virtually no influence on school practice" (Egsgard, 1978, p. 554). While some mathematics educators share this strongly negative opinion, others feel just the opposite. Begle and Gibb (another NCTM president) wrote that "research in mathematics education generates an improvement in the teaching of mathematics, and direction for the development of mathematics curricula . . . Research in mathematics education has provided bricks for edifices of cognitive development, skill learning, concept and principle learning, problem solving, individual difference, attitudes, curriculum, instruction, teaching, and teacher education . . . Every mathematics educator (teacher, and teacher of teachers of mathematics) can benefit from these edifices" (Begle and Gibb, 1980, p. 15).

Egsgard, Begle, and Gibb are (or were) highly respected and active participants in the teaching of mathematics. How could they come to such widely divergent beliefs, and which, if either, describes the contribution of research to mathematics education? In all honesty, I believe that neither of these extreme positions is valid. The first underestimates what research can contribute while the other overestimates what is possible at this time.

This chapter, along with the other chapters in this book, is an attempt to realistically show how knowledge of research can contribute to improvement in mathematics education.

Mathematics education research can make contributions to the teaching of mathematics in at least three areas: (1) description of what has been, (2) description of what is, and (3) description of what is possible. In addition, by building theories, research aids in putting the world of mathematics education in broad perspective.

One important contribution of educational research is *to describe what has been,* and to trace the influence of the past on current educational practice. While using different methodologies than experimental or status research, historical study is an important type of educational research. Although it often indicates that we have failed to learn from previous experience, it provides part of the knowledge necessary to make important curricular decisions. Dessart's chapter on curriculum includes research of this type.

Another important role of research is *to find out what exists.* Research provides systematic description of specific situations to see what an objective examination of reality reveals. Surveys, ethnographic, and observational studies are examples of this type of research. Results from such status studies are often surprising and in conflict with widely held beliefs. For example, many people believe that children today are not learning computational skills. The results from the National Assessment of Educational Progress (see Carpenter and others in this book) give information about changes in and the status of children's computational skills. These results indicate that on a nationwide basis, children are doing quite well in computational activities. While many teachers believe they accurately know how they interact with the learners in their classrooms, some studies indicate that teachers are not totally aware of their interaction patterns. What actually occurs in teacher-pupil interactions is reported in the teaching chapter by Grouws and Good. Some processes learners use to solve problems are reported in the problem solving chapter by Kantowski. The chapter on sex-related differences reports girls' and boys' feelings about mathematics.

One specific type of status research is that which deals with *evaluation.* Such studies can serve a variety of purposes. Formative evaluation is a continuous process enabling changes to be made in instructional programs. Summative evaluation determines whether learning has occurred and can be used for a range of programs, from the specific to those that are statewide or nationwide. The chapter by Carpenter and others is an interpretive report of a summative evaluation.

Status research studies are extremely valuable to all who are concerned with mathematics education. They help us do away with faulty thinking based on myths and inaccurate perceptions of reality. They help us know what is possible, and how changes can be made in instruction in order to achieve certain goals.

Another contribution of education research is *to find out what is possible*. Traditional experimental research, which involves precise design and analyses, is carried out in this type of study. For example, conditions of instruction are manipulated in fairly well-defined ways to see if improved learning by specifically described groups of learners follows. The chapters by DeVault and Weaver describe studies in which computers or hand-held calculators were used in a variety of ways in teaching mathematics. Many of these studies were experimental, involving direct comparison of learning by students involved in different types of instruction.

Human beings like to understand what goes on in the world and to put that perceived world into some kind of order. When educational researchers attempt to put an order on what they see happening in relation to education, they are *building theories,* another contribution of educational research. These theories attempt to explain why something has happened and to predict future events. Sometimes they have direct implications for mathematics classrooms. However, many times while a theory might help in understanding the variables at work in a classroom, it has no direct implications for instruction. Perhaps the best known theory that many believed had direct implications for the planning of curriculum is that of cognitive development explicated by Piaget. Hiebert's chapter discusses his perceptions of how at least one portion of this theory has not been particularly helpful in planning mathematics curriculum. He talks about how conservation is apparently a component of some major mathematical ideas, but does not appear to be essential in the learning of these ideas. The chapter by Cooney is an attempt to build a theory about teacher behavior.

Missing from this list of contributions of mathematics education research is any mention of providing information that will tell a mathematics teacher, at any level, what to do in her or his classroom. This is a deliberate omission because I firmly believe research cannot give precise direction to what a specific teacher should do in a particular classroom. This is not to say that research is not helpful to classroom teachers. It is only that research cannot, nor should it even if it could, tell teachers exactly what they should be doing as they plan, conduct, and evaluate instruction.

Research is but one of many ways knowledge is gained by educators (DeVault, in press). Scholarly writing is another way. No one would ever doubt that John Dewey has had a major impact on classrooms and certainly his writing would not be described as research. Teachers also can and must depend on their own experience for knowledge. The wisdom of the classroom teacher should not be negated.

What, then, can research contribute to mathematics education? Research can give us new insight into solution of old and new problems; it can suggest improved classroom procedures; it can make us more objective in our perceptions; and it can make us more thoughtful in all ways as we go about the teaching of mathematics. Research cannot tell us what should be done. Only values determine questions of this type. Research alone cannot determine what a teacher should do on Monday, or on any day of the week. But thoughtful educators will continue to use results of research as one mechanism for the improvement of education for all people.

In the past, many have held unrealistically high expectations of the kind of knowledge research can provide. Researchers may have inadvertently contributed to these high expectations by how they reported and interpreted specific studies. The purpose of this book is to put implications from research in a realistic framework which also considers knowledge gained from other sources.

This book is an attempt to put research in such a form that findings can be readily assimilated by practitioners. The authors were selected because they had worked for a long time in the area in which they were writing. Their task was not to report on individual studies but to synthesize research findings and put them in the context of scholarly writings and classroom teacher wisdom. Each author reported that the task was difficult, but the results reflect their commitment to making research available to nonresearchers in a form that can be used to improve the learning of mathematics by all.

References

Begle, E. G. and Gibb, E. G. "Why Do Research?" In *Research in Mathematics Education,* pp. 3-19. Edited by R. Shumway. Reston, Va.: National Council of Teachers of Mathematics, 1980.

DeVault, M. V. "Future Patterns of School Organization." Lecture given in Tokyo, November 1979, as part of six lectures on technology and the schools of the future. To be published by the National Institute of Educational Research, Tokyo, Japan.

Egsgard, J. C. "President's Report: Problems of the Teacher of Mathematics and Some Solutions." *The Mathematics Teacher* 71 (1978): 550-558.

I. Curriculum

Donald J. Dessart

In the late 1940s, a professor at the University of Wisconsin complained that it was far easier to move a cemetery in mid-January that it was to change the curriculum. Although such a statement may have been true of the curriculum in mathematics before 1950, it certainly is not true of the activity related to the mathematics curriculum after 1950. Since that time, a veritable revolution has taken place.

In this chapter we will review events in research related to this revolution that hold lessons for the curriculum of the future. The next 30 years will offer challenges related to many problems—energy, inflation, declining economics, and international problems of an explosive nature. The role of the mathematics curriculum in this maelstrom will provide demands of monumental proportions for teachers, researchers, and other professionals in mathematics education.

The terms "curriculum" and "mathematics curriculum" are widely used and subject to many different interpretations. *For our purposes, the mathematics curriculum will mean the mathematical content that is to be mastered by the learner and the instructional policies and procedures that are used to organize this content with the intention of promoting effective learning.*

Mathematical Content

The selection of mathematical content for instruction in the schools is subject to many demands and pressures. These demands fall into three categories: psychological, sociological, and structural. In an ideal situation, they play equal and complementary roles in the curricular process;

1

but in reality there is a tendency to stress one demand over another, depending on the pressures of society.

Psychological Demands

From time to time, the selection of mathematical content appears to fall under the influence of two broad psychological theories: (a) the behavoristic, mechanistic theories that view the learner as mastering pieces of mathematical content that collectively produce a whole of learning; and (b) the holistic or field theories that view the learner as comprehending the entirety or gestalt of learning, usually through insight or sudden inspiration.

If we subscribe to the behavioristic point of view, then the goal of instruction is to subdivide the mathematical content into objectives or bits of mathematical knowledge that must be mastered in a "best" sequential order for the learner to succeed. The recent emphasis on behavioral objectives is an example of the results of such an influence.

On the other hand, if we subscribe to the holistic or field theories, then we attempt to provide the fabric that pervades learning and provides a cognitive structure. The more recent emphasis on problem solving as a broad task of understanding the problem, developing a plan and reconsidering the plan, fits into that theory. In the problem-solving process, emphasis is on instilling in the learner the tolerance for manipulating the pieces of the problem in an orderly but not necessarily sequential manner so the learner may arrive at an insight or solution to the problem.

Most teachers subscribe to an eclectic point of view, believing that certain kinds of mathematical content, such as addition or subtraction algorithms, should be organized with a behavioristic theory as the guiding influence; whereas other mathematical content, such as in understanding or finding a geometrical proof, are guided by a holistic theory. In teaching an algorithm, it is important to identify the sequential steps of the process and to have students master these steps. In teaching a geometrical proof, it is more important to provide a varied selection of viewpoints for examining the problem, with the goal that perhaps one of these viewpoints will lead to a solution.

Most teachers are comfortable with a selective point of view that allows one or the other theory to dominate, depending on the content to be learned. But there are "purists" who simply feel that *all* mathematical learning can be described by behavioral objectives or mechanistic descriptions; and there are other purists who feel that *all* mathematical learning should be mastered by a random or incidental organization that leads to insight and solution.

Research studies conducted during the 1930s and 40s attempted to test the effectiveness of these two theories. For example, many studies compared drill instruction with meaningful instruction. In drill instruction the goal was to subdivide the learning experience into parts in which drill or practice was provided for each subdivision. In meaningful instruction an attempt was made to provide a fabric of understanding—a varied set of experiences that would provide a variety of insights into the mathematical content to be mastered. At first it seemed that drill and meaningful instruction were antithetical, but later it became apparent that the two were complementary and that, normally, drill should be preceded by meaningful instruction.

Sociological Demands

Sociological demands for curriculum organization stress individuals' needs for certain mathematical content in order to serve themselves and society. A student must learn arithmetic, algebra, and geometry to satisfy certain vocational demands such as those required of an electrician, an accountant, or an engineer. Consequently, the selection of content must recognize and satisfy these needs. Psychological theories involve academic questions that teachers and curriculum workers can address in the laboratory, while sociological demands are more diverse to identify and to provide for in the curriculum.

There have been various attempts to identify sociological needs. The Commission on Post-War Plans of the National Council of Teachers of Mathematics, meeting after World War II, pointed out the inadequacies of the mathematics curriculum and identified 29 mathematical competencies that should form the minimal or core curriculum. (For instance: can measuring devices be used? Are students skillful in the use of interest tables, income tax tables? Can they construct scale drawings? and so on.) As one reviews these 29 competencies, it is clear that they are socially oriented.

Curriculum makers are alert to the sociological needs of students, and teachers certainly feel the demands, particularly from students who ask, "What is this good for?" Such a question may more truly indicate psychological problems, rather than sociological problems, with the curriculum. Nevertheless, providing for the needs of society is a valid demand placed on teachers and curriculum makers.

Structural Demands

The structure of the discipline of mathematics is often the primary criterion guiding the organization of the curriculum. Certainly, during the modern mathematics movement in the United States, the structure of mathematics was the major force influencing the design of the curriculum. Most modern mathematics projects were concerned with structural questions such as: What is a number? What is the difference between a number and a numeral? What are the laws of arithmetic; such as, the commutative law, the associative law, and so on? What is a variable? What is a function? What is an equation? What is a group? What is a field? All of these questions were aimed at understanding the structure of the discipline.

Interplay of Psychological, Sociological, and Structural Demands

From time to time, one of the three demands seems to influence curriculum makers in mathematics to a greater extent than the others. During times of economic pressures, as during the Great Depression, mathematics teachers lean toward sociological demands as a means of justifying the mathematics curriculum. At other times, such as during the modern mathematics movement, the structural demands seem to be the major factor in designing the curriculum, and yet at other times the psychological demands seem more prevalent, for example, during the recent competency and accountability movements. Obviously, teachers and curriculum makers must attempt to satisfy the current demands of society; but, on the other hand, we would hope that the curriculum would be of a nature that it could satisfy all of the needs of society. Research in this direction or efforts to find suitable models of curriculum design have not been highly successful in the past. Obviously, the complexities of such a model are mind-boggling, and we can readily appreciate that viable models will not be developed quickly or easily. In the meantime, teachers and curriculum makers must be aware of the psychological, sociological, and structural demands, and attempt to provide an amalgamation of the three whenever possible.

Instructional Procedures

The organization of the curriculum depends on the instructional procedures that are to be employed in the classroom. Obviously, the curriculum used in a classroom in which the teacher plays a highly directive role is quite different from a curriculum in which the teacher plays a less directive role. Consequently, curricular organization can be viewed as a function of teacher intervention in the classroom. Three kinds of teacher control will be considered: (a) lecturing or underlining, (b) guided learning, and (c) pure student learning.

Lecturing or Underlining

In lecturing or underlining, the teacher is exerting maximum control of the class activities and the sequencing of events in the classroom. A program designed for this kind of teacher intervention provides a structure or outline of topics to be covered, but leaves the manner of presentation and class activities largely to the discretion of the teacher. Problem sets and exercises are also provided, but the selection of these are left to the discretion of the teacher. This pattern of curricular organization is probably the most predominant pattern used in American schools and colleges and is undoubtedly viewed as the most efficient and practical curricular organization.

Guided Learning

Guided learning may play a significant role in a classroom in which the teacher attempts to engage students in the learning activities. It may also play a large role in programmed instruction or various kinds of individualized instruction. In these procedures, the curriculum must play a more definitive guiding role than in a class in which the teacher lectures. The sequence of events and the manner of presentation must be developed by the curriculum maker rather than solely by the teacher.

Although guided learning is used less frequently in schools than standard lecturing, it does have a more significant role in elementary than in secondary schools. There was a period in the late 1950s and 60s when it appeared that programmed learning would make inroads in the American classroom. But its vitality was short-lived and gave way to more conventional teaching procedures.

Pure Student Learning

The school in which the student develops a curriculum according to his or her own needs is Utopian. Very few elementary or secondary schools employ such procedures, although a few college classes, primarily at the graduate level, have used pure learning methods. In such a procedure, the teacher's role is one of counselor and provider of initial directions. The curriculum, as such, is highly flexible, undeveloped to a large extent prior to the class meeting. Individual freedom and creativity are highly emphasized, and a teacher of extremely mature and learned judgment is needed to function in such a structure.

The Evolving Mathematics Curriculum

The Elementary School Curriculum

Prior to the 1950s, the elementary school curriculum was concerned primarily with developing computational skills with whole numbers, fractions, and decimals. Thorndike dominated much of the thinking. His view was summarized in *The Psychology of Arithmetic* (1924): "We now understand that learning is essentially the formation of connections or bonds between situations and responses . . . and that habit rules in the realm of thought as truly and as fully in the realm of action" (p. vi). The elementary school curriculum under Thorndike's influence consisted largely of identifying specific stimuli (for example, $3 + 4 =$) and specific responses (7) for the skills of computation. Attempts to provide more than developing the "bonds" or "connections" for specific stimuli and their responses were regarded as not only superfluous but a detraction from the main goal of learning.

Reactions to this theory came from mathematics educators like Brownell and Chazal (1935) who wrote an exposition of the drill theory. Brownell, Chazal, and others of this time advocated a "meaning theory" which emphasized the inner relationships of the number system as well as the social uses of arithmetic. Several decades later the curriculum developers of the School Mathematics Study Group (SMSG) relied on the theories of Brownell for justification of their interpretation of "meaning" as understanding the structure of mathematical systems.

Searches for meaning in the elementary school curriculum led naturally to the consideration of other topics. Geometry, which was primarily a domain of the secondary schools, began to make its appearance in the elementary schools largely through the efforts of the modern mathematics innovators. The geometry of the elementary school dealt primarily with informal understandings of concepts to be taught with more precision in later years. For example, points, lines, rays, half-planes, angles, triangles were treated informally in anticipation of more formal definitions in the high school curriculum.

The modern mathematics movement of the 1950s and 1960s emphasized "meaning" also, but primarily from a content point of view. Emphasis was placed on understanding non-decimal numeration systems, the laws of number systems, and the more formalized rules of arithmetic. Drill of computational skills was emphasized to a far lesser degree than in former years and the goal of instruction was to develop an understanding of the number system and its properties. This philosophy held that such formal understandings would make students more flexible and able to apply mathematics to a variety of situations.

The enchantment with modern mathematics began to wane when mathematicians and educators criticized the emphasis on formalism as being too artificial and not consistent with usage in modern life. The decline of students' computational skills was blamed on the modern mathematics curriculums, and the emphasis on skill development under the guise of "basic skills" made its reappearance during the 1970s. By now, the meaning theory was still popular but was coupled with drill as a viable means of developing computational skills.

The basic skills movement led to a concern that today's children were able to compute better than children of former years but were not capable of applying these skills to solving problems. A reaction to this concern was expressed by the NCTM, which recommended that problem solving should be the focus of the curriculum of the 1980s and that basic skills should be broadened to encompass more than merely computational facility.

The Middle and Junior High School Curriculum

Prior to the modern mathematics movement, the middle grades (six, seven, eight) were devoted largely to consolidating computational skills learned in earlier years as well as broadly emphasizing social applications. Students of these years were exposed to applications of arithmetic in banking, installment plan buying, homemaking, and other social uses. This emphasis prompted modern mathematics proponents to regard the junior high school years as a "wasteland" of mathematics education. Beberman was reported to have stated that children of these years were more interested in "the number of angels that danced on the head of a pin" than in social applications.

This opinion motivated modern mathematics curriculum developers to replace social applications with topics of a more highly abstract mathematical nature. The formalism begun in the elementary curriculum was extended to the junior high school years. Understanding equation solving and "proofs" for arithmetic rules were stressed. Greater emphasis on informal geometry and the introduction of new topics, such as probability and statistics, were promoted. During these years, programmed instruction, individualized instruction, and mathematical laboratories were also recommended for the middle years. Many a junior high school housed a special classroom-laboratory where children could pursue special projects related to paper folding, string art, games, and other special topics. One sometimes wonders if these laboratories were actually an "escape" from the more formal material of the modern mathematics programs.

Two events during the 1970s modified the curriculum of those years. The hand-held calculator was becoming more readily available and the metric system was being promoted. Although some teachers deplored the calculator as detracting from students' acquisition of computational skills, the National Advisory Committee on Mathematical Education recommended that a hand-held calculator be made available to each student by the end of the eighth grade. The need for the nation to implement the metric system was recognized, and the junior high school years were seen as an appropriate time to introduce students to this system.

The High School Curriculum

Teaching algebra in grade nine to 15-year-olds had been traditional for many, many years dating back to the late 1800s. Although algebra became somewhat more formal during the modern mathematics movement, its nature did not change radically. Traditional teachers who had emphasized skill development prior to the modern mathematics years were able to continue this emphasis with the modern textbooks by merely ignoring the text material between problem sets.

On the other hand, the geometry curriculum was subject to greater change. Birkoff devised a set of postulates for geometry, making greater use of those properties of the real number system which were unavailable to Euclid. These postulates, published in the *Annals of Mathematics* of April 1932, became the basis for a "new" geometry textbook by Birkhoff and Beatley (1940) of the inauspicious title, *Basic Geometry*.

The influence of Birkhoff's work and that of Hilbert on the SMSG approach to geometry is clearly shown by Brumfiel (1973). In 1980, the SMSG approach dominated the secondary school geometry curriculum in the United States. It is one of the few innovations of SMSG that has avoided most successfully the criticisms leveled at other modern mathematics programs. The major contribution of the SMSG approach was to introduce postulates concerning abstract rulers and protractors into high school geometry. These postulates, which capitalize on properties of the real number system, filled the logical gaps of the Euclidean geometry previously taught in high school.

Although this approach to geometry is the predominant one, it is certainly not the only geometry curriculum used in the United States. NCTM's 36th yearbook, *Geometry in the Mathematics Curriculum* (Henderson, 1973), identifies no less than seven approaches to formal geometry in the senior high school. These approaches include the conventional synthetic Euclidean geometry, approaches using coordinates, a transformational approach, an affine approach, a vector approach, and an approach to satisfy all teachers: an eclectic program in geometry.

Before the 1950s, the high school curriculum beyond geometry usually consisted of a second year of algebra followed by a semester of solid geometry and a semester of trigonometry. This program has been replaced by a unified approach that includes topics from traditional advanced algebra, theory of equations, linear algebra, probability and statistics, and the calculus. The calculus as a separate subject is studied by relatively few students; in 1977-78, only 4 percent of the 17-year-olds in the United States elected it (National Science Foundation, 1980).

Evaluation of Programs

As the modern mathematics programs made inroads into the schools of the United States, administrators, teachers, and parents wanted to know which of the programs—the modern or the traditional—was better for their children. Early attempts at such evaluations might be termed macro-evaluations since they made broad comparisons of classes using traditional curriculums and those using modern programs.

Minnesota National Laboratory Evaluations

In the spring of 1964, the Minnesota National Laboratory reported receiving a grant of $249,000 from the National Science Foundation. The general purpose of the grant was to study the effectiveness of various kinds of mathematics courses, including both conventional courses and new courses developed by the School Mathematics Study Group, the University of Illinois Committee on School Mathematics, the Ball State Teachers College, and the University of Maryland Mathematics Project.

The study was designed to determine differences in achievement between pupils instructed with conventional materials and those instructed with one of the modern programs. Volunteer teachers from a five-state area were invited to participate for a two-year period, teaching a class with conventional materials during the first year, and two classes, one with conventional and one with modern materials, during the second year. Students were tested to determine initial measures of achievement, final achievement, and retention.

In 1968 the results of the evaluations (Rosenbloom and Ryan, 1968) did not show that the modern programs had strongly increased or decreased achievements. There were some exceptions to the overall findings, but even those did not strongly favor the modern curriculums. In subsequent investigations on attitudes, interests, and perceptions of proficiency, no unequivocal differences were found. Dessart and Frandsen (1973, p. 1178) commented on these results:

A conclusion which might have been drawn from these projects was that the experimental materials tested were not worth the vast resources that had gone into their development. The project reporters suggested that such a conclusion should be tempered because of possible lack of validity of the achievement tests for measuring significant objectives of the experimental programs or because of possible lost impact due to poor teacher performance with experimental materials.

New Hampshire Studies

Studies of a smaller scope were also conducted to ascertain advantages or disadvantages of the modern programs. Typical of these was a study performed in New Hampshire which compared groups of students studying from modern, transitional, and traditional materials during 1963-1967 (Austin and Prevost, 1972). In 1965, the pupils studying the modern materials outdistanced the other two groups on the Otis Mental Abilities Test but performed lower on computation testing. By 1967, the modern group scored higher on computation, concepts, and applications tests. Throughout all of these testings, students' abilities to perform arithmetic computations declined.

A pattern was emerging from these evaluations: students in the modern programs seemed to achieve somewhat higher in comprehension of mathematical concept measures but scored lower in measures of computational ability than their counterparts in conventional programs. This led to the humorous comment attributed to Beberman that a modern schoolboy knows that the sum of two natural numbers is a unique natural number, but he doesn't know which one!

National Longitudinal Study of Mathematical Abilities (NLSMA)

The NLSMA was a five-year study that attempted to compare the effectiveness of conventional and modern textbooks as well as the effects of student attitudes and backgrounds of teachers (Begle and Wilson, 1970). Three populations designated X, Y, and Z were studied in the fall and spring of each year. The X population, fourth-grade students in 1962, and the Y population, seventh-graders, were tested for the full five years. The Z population, consisting of tenth-graders, was examined during a three-year period.

The data assembled were enormous. The results related to textbooks indicated that: (1) the variability of means associated with textbook groups decreased as grade levels increased; (2) the SMSG textbook groups performed better than conventional groups on comprehension, analysis, and application levels, but not on the computational level; and (3) some of the modern textbooks produced poor results on all levels from analysis through computation.

The NLSMA revealed that comparing textbooks is at best exceedingly difficult and complex, leading to very few clear generalizations. Many variables must be examined to provide a broad view of the comparisons; otherwise, one may be in the position of comparing textbooks on a single criterion which may not be significant. This lesson was costly to learn as evaluation projects, such as NLSMA, enjoyed the era of federal funding that may never be matched in the future.

Curricular Variable Identification and Study

As we have seen from the evaluations of programs that were conducted during the 1960s and 1970s, the results were not clear-cut and generalizations were very difficult to make. The major obstacle to arriving at meaningful evaluations was the large number of variables that needed consideration. Merely identifying these variables was not sufficient because, in most cases, the variables were not well understood and were not easily measured. During the 1980s, more intensive efforts in variable identification and study seem to be a natural consequence of the research of the previous two decades.

Adjunct Questions and Teaching Problem Solving

One of the difficulties faced by nearly every mathematics teacher is that of teaching "word" problems. An often used but questionable instructional procedure consists of the teacher presenting a model problem to the class and leaving the model on the chalkboard as a pattern for working other problems. Successful students learn to select numbers from the textbook problems to fit the chalkboard model and perform similar operations. Doing enough problems of this kind, students memorize a pattern or algorithm for solving a particular problem. Later, when they meet this type of problem again, it is hoped they will apply the same pattern or algorithm.

Although this instructional procedure has some merit, it probably suffers from two weaknesses: (1) the student is not really "thinking through" the problem but rather is engaging in a "matching-analysis" procedure; and (2) the pattern or algorithm may be forgotten by the student when he or she meets this particular problem in isolation of other similar problems. This instructional procedure does, however, represent an attempt to guide the student through a problem-solving process. Methods of motivating or leading students to engage in problem solving are needed by teachers at nearly all levels. One method that may hold promise is to guide the students by key questions designed to highlight or emphasize the essential elements of a problem and encourage the student to consider these elements. Little research has been done with the usefulness of such questions in mathematics education, but research in other areas may provide some clues for us.

For example, Rothkopf (1966) reported a research study with 159 college students who studied a written passage and were given questions either before or after reading the material. Rothkopf concluded that questions given after reading the material, called "adjunct questions," have both specific and general facilitative effects on post-reading performance.

Mathematics educators have given insufficient attention to the possible uses of adjunct questions—that is, questions that would follow a written paragraph and would be designed to focus the attention of the student on the salient features of that paragraph. For example, if word problems were followed with specific, written questions that would focus the attention of students on the elements of the problem and thus force them to "think through" the problem, would we be more successful in motivating students?

It seems clear that during the 1980s efforts to improve the teaching of word problems will be a priority if the disturbing evidence of the latest NAEP results concerning the success of our students in dealing with word problems influences the directions of teaching (see Chapter II). In such efforts, the usefulness of adjunct questions seems to be an attractive possibility. Furthermore, such questions could be designed to accommodate the Polya (1945) model of understanding the problem, devising a plan, carrying out the plan, and looking back at the problem.

A serious question that arises in the use of adjunct questions is their value in facilitating the questioning abilities of students; that is, will students exposed to adjunct questions initiate questions of their own in attacking word problems or, even further, in a given cultural situation, would students be able to formulate (identify) the problem to be solved? We can hope that such would be the case, but there is a danger that students would become far too dependent on the adjunct questions. Research could provide further insights.

Advance Organizers

Experienced teachers know that before beginning a new topic with a class they must provide an introduction. This introduction probably consists of two phases: (a) a review in which the teacher helps students recall relevant facts, concepts, and principles; and (b) an overview in which the teacher attempts to provide students with general and overarching insights concerning the new material to be studied. This bit of conventional wisdom was given a more precise and theoretical foundation through the work of Ausubel (1968) and others working in the area of advance organizers.

Ausubel pointed out that advance organizers facilitated meaningful learning in three ways: (a) they mobilize relevant anchoring concepts already established in the learner's cognitive structure; (b) at an appropriate level of inclusiveness they provide optional anchorage which promotes initial learning and resistance to later loss, and (c) they render unnecessary much rote memorization students often resort to because they lack sufficient numbers of key-anchoring ideas. Ausubel (1968, p. 148) succinctly summarized the characteristics of advance organizers: "In short, the principal function of the organizer is to bridge the gap between what the learner already knows and what he needs to know before he can successfully learn the task at hand."

Since a good introduction is important to a learning task, it would seem that research on advance organizers should be a productive area for improving teaching. It appears that a relevant question is not whether advance organizers should or should not be used but rather in what ways can advance organizers be constructed to be most effective. Consequently, it seems cogent that any future work on advance organizers should concentrate on their useful characteristics rather than on a testing of their use or disuse.

One of the central issues related to advance organizers is the "goodness of fit" among the elements of the advance organizer, the new material to be learned, and the characteristics of the cognitive structure of the student. The first step in establishing a "good fit" is a matter of identifying the central mathematical concepts and principles of the new material and providing for generalizations of these in the organizer. As difficult as this may be, it is not an insurmountable task. The next step, that of designing the organizer to complement the existing cognitive structure of the student, is an extremely complex task. What is an effective organizer for one student may be utter confusion for another because of the differences in their cognitive structures.

Work by Bloom and his associates is related to the question of designing an organizer to complement the cognitive structure of the student. Bloom (1980, p. 383) identifies "cognitive entry characteristics" as the specific knowledge, abilities, or skills that are necessary prerequisites for a particular learning task. He points out that such prerequisites correlate .70 or greater with measures of achievement of the task.

The critical problem, then, is one of designing effective advance organizers or providing for appropriate cognitive entry characteristics to fit the cognitive structures of the student. But isn't this the crucial problem in teaching? The solution to this problem necessitates studies designed to ascertain the nature of the existing cognitive structure of the student and providing for the further development of these structures; an exercise similar to the one a conscientious teacher employs in tutoring a student over a learning difficulty. The initial step is usually one of attempting to determine the nature of the existing knowledge of the student before beginning the remediation or, in this case, the design of the advance organizer.

Behavioral Objectives

The use of behavioral objectives made a significant contribution to mathematics education because it sharpened the focus upon describing the desired student behavior at the conclusion of a learning experience. As with advanced organizers, the use of behavioral objectives made "pedagogical sense" and they became popular. Research related to the use of behavioral objectives for the most part supported them. As Begle (1979) observed, in over 30 studies completed to determine if the use of behavioral objectives could lead to greater student achievement, half of these studies indicated that student achievement was improved or speeded up. In perhaps only one case did the use of behavioral objectives have a negative effect. In each study the question was essentially one of testing whether students who were exposed to behavioral objectives before a learning experience would achieve higher than those who were not so exposed. The evidence seemed to support their use.

In spite of a promising beginning, the interest in behavioral objectives seemed to decline. As with most "bandwagon" movements, teachers are led initially into the movement by advocates who preach a panacea is at hand. Such was the case with the behavioral objective movement; but when teachers were required to take hours from their busy teaching days not only to write lengthy behavioral objectives for the mathematical topics of the curriculum, but also to monitor their attainment with a class, fatigue soon took its toll. What had initially been regarded as a panacea became a terrific demand on teachers' time.

It soon became evident that the higher cognitive goals of instruction— critical thinking, creating, problem solving—were elusive of capture in the behavioral objective mold. Many of the advocates of the movement recognized this deficiency but felt that, given sufficient time and energy, these goals could be described by behavioral objectives. After all, if we really understand the behavior we desire, then surely we should be capable of describing it.

Learning Hierarchies

Closely related to the notions of advance organizers and behavioral objectives is that of the learning hierarchy. Whereas the advance organizer intends to provide an introduction to the learning experience, and the behavioral objective focuses on the final behavior exhibited by the student at the conclusion of a learning experience, the learning hierarchy is a means of organizing learning tasks to achieve a final goal-objective. The process requires describing the mathematical objective and asking what the student should be able to do in order to achieve that objective. Answering such a question raises the necessity of defining subtasks that should be achieved and subtasks of these subtasks so that a network of tasks is generated. In this network, achieving the final task is dependent on achieving all of the subtasks.

Much of the research conducted with learning hierarchies has been in the form of validation studies. Phillips and Kane (1973) constructed seven different orderings of 11 subtasks for rational number addition. A test was designed to assess mastery at each of the 11 subtasks of the hierarchy. One hundred forty-two students were assigned randomly to the seven treatments corresponding to the seven different orderings of the hierarchy. No particular sequence was found to be consistently superior on achievement, transfer, retention, and time to complete the sequence.

Some Lessons Learned from Research on Advance Organizers, Objectives, and Learning Hierarchies

As we contemplate the efforts that have been expended in research related to advance organizers, behavioral objectives, and learning hierarchies, we quite naturally ask, "What have we learned?" The lessons have been modest, but valuable.

First, one of the most valuable lessons is that there is an organization of the objects of the curriculum called its "psychological organization"; that is, an organization that arises from an analysis and study of the mathematical understandings of children. This organization may be quite different from its axiomatic, logical, mathematical organization. The implication of this lesson is that teachers must understand the psychological organization of the curriculum as well as its mathematical organization. The latter has had the benefit of centuries of the axiomatic method since the time of Euclid, whereas the psychological organization has only been given serious consideration in the last 50 to 100 years. One intuitively feels that all of the psychological research findings related to learning mathematics is waiting for a Euclid to organize them into a viable theory of mathematical learning.

Second, we have learned that there is a danger of attempting to fit all mathematical learning into one model which may turn out to be an inadequate model. The use of the behavioral objective model is an example of this inadequacy. Some view such a shortcoming with alarm when, in fact, it is the natural evolution in any theory; that is, when the theory is inadequate, it must be revised.

Third, it is undoubtedly true that good teachers have known from conventional wisdom the necessity of including in their instructional techniques the basic ideas embodied in advance organizers, behavioral objectives, and learning hierarchies. Researchers have made the valuable contribution of organizing this wisdom into carefully devised theories. Although these theories are often incomplete and inadequate, they do represent a serious beginning of a theory of mathematical learning.

Meaningful Instruction and Drill

Van Engen (1949) discussed three particular theories of meaning: (1) social meaning in which the child understands the mathematics that he or she can observe and use in social situations, (2) structural meaning in which the mathematics becomes meaningful when the child understands the structure of the subject, and (3) the nihilistic theory of meaning which denies meaning to symbols. Brownell (1945) made a statement which is typical of those often quoted, that is, that "Meaning is to be sought in the structure, the organization, and the inner relationships of the subject itself" (p. 481).

In contrast, we might conclude that rote instruction neglects to emphasize the structure, the organization, and the inner relationships of the subject of mathematics and puts emphasis on repetition and fixation of concepts, principles, procedures, or algorithms. In considering a definition for the term "drill," we turn to Sueltz (1953), who discussed drill in the following way: ". . . the words 'drill,' 'practice,' and 'recurring experience' are used to indicate those aspects of learning and teaching that possess elements of similarity and sameness which repeat or recur" (p. 192).

In conclusion, it seems clear that since the research of Brownell and Moser (1949) on meaningful versus mechanical learning, and probably earlier, many mathematics educators subscribed to a statement similar to one by Gibb (1975, p. 59):

> The controversy of drill and practice versus understanding (including the use of hands-on and laboratory types of learning experience) is a long-standing, but unnecessary one. I believe that neither can be considered in isolation from the other. A lot of rote drill and practice in the absence of understanding or useful application does little to promote computational

efficiency. Likewise efforts for developing understanding alone are not effective unless they are tempered with drill and practice to build proficiency in computation, in problem solving, and in thinking logically.

Consequently, it seems reasonable to state that nearly all mathematics educators agree that meaningful instruction and drill go hand in hand, with meaningful instruction simply preceding drill or practice, but that both are necessary for an efficient and profitable learning experience. Willoughby (1970, p. 263) echoed a similar sentiment:

> Virtually everyone who consciously addresses himself to the question of whether it is better for a child to understand mathematical concepts or to commit verbalization of the concepts to memory agrees that understanding is important and desirable. During the recent "revolution" in school mathematics, this point of view became so prevalent that, in some instances, textbooks provided substantial material to help children understand a concept, but virtually no practice or drill work to help them become adept at using it. Available evidence suggests that a child can understand without becoming adept in using the particular skill involved (addition or fractions, for example), but if the skill is one in which he ought to become proficient, practice or drill will be needed. On the other hand, if drill is used without understanding, retention does not seem to be as great, and, of course, the learning of a skill involving the same understanding but different sorts of symbols will be more difficult if the understanding has not been developed.

As noted from Willoughby's comments, drill and practice became unpopular during the modern mathematics movement. The disenchantment with drill actually preceded the modern mathematics revolution. Sueltz (1953, p. 192) observed:

> Twenty-five years ago (about 1928), drill was the common method of learning applied to such school subjects as arithmetic, writing, and spelling. Children were required to write a word 50 times to learn to spell it and the present generation of middle-aged people spent countless minutes in winding up ovals in one direction and then unwinding them in the opposite direction in order to train the muscles to follow the sweeping curve of penmanship. This was drill; it was carried to extremes and became so sterile that during the 10-year period of approximately 1935 to 1945 drill, as a learning procedure, was frowned upon and ridiculed in many educational circles. However, during the same period it remained the dominant pattern employed by many teachers.

The comment by Sueltz that drill remained the dominant pattern employed by many teachers was borne out by research. Milgram (1969) investigated the ways in which elementary teachers used class time in mathematics. A team of observers making twice-weekly observations of 46 intermediate grade teachers in Pennsylvania found the following use of time: (a) going over previous assignment, 25 percent; (b) oral or written drill, 51 percent; (c) introducing new mathematical concepts or developmental activities, 23 percent; and (d) unrelated interruption, 1 percent.

This pattern of instruction has not changed drastically as can be seen from the words of Fey (1979), who analyzed three studies completed by the National Science Foundation (p. 494):

> Despite the difficulty of knowing what teachers understood by the terms "lecture," "discussion," and "individual assignments," the profile of mathematics classes emerging from the survey data is a pattern in which extensive teacher-directed explanation and questioning is followed by student seatwork on paper and pencil assignments. This pattern has been observed in many other recent studies of classroom activity. . . .

Fey cites Welch, one of the NSF investigators, who wrote that the pattern of instruction in all mathematics classes he observed was the same, that is, discussion of the previous assignment, discussion of the new material, followed by seatwork for the students in which they worked the next day's assignment with help from the teacher.

The Textbook

Since the textbook plays such a highly significant role in the life of the student and the teacher, it is surprising that there hasn't been more research related to its effectiveness. Begle (1979, p. 73) summarized the reason for this dearth of research:

> Any two textbooks differ on so many variables that it would be almost impossible to trace the specific variables which cause a specific difference, and without knowing which variables make a difference, we do not know where to start to improve textbooks.

The textbook in many schools is the mathematics curriculum for that grade level. If a topic does not appear in the textbook, it is not taught. This fact of life was certainly used by the curriculum developers of the 1950s and 60s. They knew that in order to change the curriculum, the most effective and efficient method is by changing the contents of textbooks. Since this state of affairs probably will not change in the future, more effort should be expended in research on textbooks. Millions of dollars each year are spent on textbooks, so one can hardly argue the desirability and practicality of such research.

For example, does color make a difference in the effectiveness of learning by children? Most publishers must have concluded that it does make a difference (in sales, at least) as they use color lavishly. Do illustrations, pictures, type style, or the size of the page make a difference? If not, we might question whether funds to develop these features in books could be diverted to other more productive educational uses.

Research on textbooks appears to be a fruitful area of investigation. Some feel that the number of variables is too large; others, on the other hand, feel that research is manageable. For example, Walbesser (1973, p. 76) offered the opinion that all textbooks should contain "(1) performance descriptions of objectives, (2) data on the acquisition of the behaviors described as objectives, and (3) a statement of where the performance list and data are available." He felt so strongly about this viewpoint that he urged a moratorium on the purchase of all textbooks until the author and/or the publisher provided such information!

Walbesser noted that the selection of textbooks is done in a haphazard manner by most selection committees. They depend very heavily on such criteria as endorsement by authorities, bandwagon effects, high identification quotients, little change from previous textbooks, the use of illustrations, the skill of the sales representatives, and the reputations of the authors. Walbesser presented a plan for selecting textbooks that was highly dependent on the use of behavioral objectives and learning hierarchies. This procedure represents a well-defined and rigorous method and provides a promising direction deserving greater attention.

The values of the textbook will probably remain unchanged in the future. Corporations such as IBM, which is a high-technology company at the forefront of innovations, seem to prefer conventional educational practices. Peter Dean (1980), program manager of IBM's Education Developer Services, cited among identifiable trends that "Most managers and employees perceive conventional stand-up classroom instruction as the only 'true' education. Telling tends to be equated with education." Furthermore, he noted that "Most of the self-study material is printed; a modicum of videotape is used" (p. 317).

It seems reasonable for one to conclude that the textbook will continue to be an important tool in the mathematics classroom. More research on its effectiveness and use is certainly in order for the 1980s; in fact, one might place its research in a high priority category.

Summary

In isolating and studying carefully the variables of the curriculum, we may gain a better understanding of the influence of curriculum on learning. Adjunct questions, advance organizers, behavioral objectives, and learning hierarchies offer fruitful areas for investigations. While research on the textbook's effectiveness is difficult, its significance as a learning tool demands that it be studied more carefully and researched more fully.

The late A. S. Barr, a well-known researcher at The University of Wisconsin, often remarked that research is very similar to mining in that much shoveling is required before hitting paydirt! This analogy is so appropriate for research on the mathematics curriculum. Much work lies ahead, but the rewards of a better and more effective education for our children is a reward well worth the efforts.

References

Austin, Gilbert, and Prevost, Fernand. "Longitudinal Evaluation of Mathematical Computation Abilities of New Hampshire's Eighth and Tenth Graders, 1963-1967." *Journal for Research in Mathematics Education* 3 (1972): 59-64.

Ausubel, D. P. *Educational Psychology: A Cognitive View*. New York: Holt, Rinehart, and Winston, 1968.

Begle, E. G. *Critical Variables in Mathematics Education: Findings in a Survey from the Empirical Literature*. Washington, D.C.: Mathematical Association of America and the National Council of Teachers of Mathematics, 1979.

Begle, E. G., and Wilson, J. W. "Evaluation of Mathematics Programs." In *Mathematics Education*. Edited by E. G. Begle. Chicago: National Society for the Study of Education, 1970.

Birkhoff, George D., and Beatley, Ralph. *Basic Geometry*. New York: Chelsea Publishing Co., 1940.

Bloom, Benjamin S. "The New Direction in Educational Research: Alterable Variables." *Phi Delta Kappan* 61 (1980): 382-385.

Brownell, William A. "When is Arithmetic Meaningful?" *Journal of Educational Research* 38 (1945): 481-498.

Brownell, William A., and Chazal, Charlotte B. "The Effects of Premature Drill in Third Grade Arithmetic." *Journal of Educational Research* 29 (1935): 17-28.

Brownell, William A., and Moser, Harold E. *Meaningful vs. Mechanical Learning: A Study in Grade III Subtraction. Duke University Studies in Education, No. 8.* Durham, N.C.: Duke University Press, 1949.

Brumfiel, Charles. "Conventional Approaches Using Synthetic Euclidean Geometry." In *Geometry in the Mathematics Curriculum (36th Yearbook of NCTM)*. Edited by K. B. Henderson. Reston, Va.: National Council of Teachers of Mathematics, 1973.

Dean, Peter M. "Education and Training at IBM." *Phi Delta Kappan* 61 (1980): 317-319.

Dessart, Donald J., and Frandsen, Henry. "Research on Teaching Secondary School Mathematics." In *Second Handbook of Research on Teaching*. Edited by R. M. W. Travers. Chicago: Rand McNally, 1973.

Fey, James T. "Mathematics Teaching Today: Perspectives from Three National Surveys." *Mathematics Teacher* 72 (1979): 490-504.

Gibb, E. Glenadine. "Response to Questions for Discussion at the Conference on Basic Mathematical Skills and Learning." *Volume I: Contributed Position Papers, Conference on Basic Mathematical Skills and Learning, Euclid, Ohio*. Washington, D.C.: U.S. Department of Health, Education and Welfare, Institute of Education, 1975.

Henderson, Kenneth B., ed. *Geometry in the Mathematics Curriculum (36th Yearbook of NCTM)*. Reston, Virginia: National Council of Teachers of Mathematics, 1973.

Milgram, Joel. "Time Utilization in Arithmetic Teaching." *Arithmetic Teacher* 16 (1969): 213-215.

National Science Foundation. *Science Education Databook*. Washington, D.C.: The Foundation, 1980.

Phillips, E. Ray, and Kane, Robert B. "Validating Learning Hierarchies for Sequencing Mathematical Tasks in Elementary School Mathematics." *Journal for*

Research in Mathematics Education 4 (1973): 141-151.

Polya, G. *How To Solve It.* New York: Doubleday, 1945.

Rosenbloom, P. C., and Ryan, J. J. *Secondary Mathematics Evaluation Project: Review of Results.* St. Paul, Minn.: Minnesota National Laboratory, 1968.

Rothkopf, E. Z. "Learning from Written Instructional Materials: An Exploration of the Control of Inspection Behavior by Test-Like Events." *American Educational Research Journal* 3 (1966): 241-249.

Sueltz, Ben A. "Drill-Practice-Recurring Experience." In *The Learning of Mathematics: Its Theory and Practice (21st Yearbook of NCTM).* Edited by Howard Fehr. Washington, D.C.: National Council of Teachers of Mathematics, 1953.

Thorndike, Edward L. *The Psychology of Arithmetic.* New York: Macmillan Company, 1924.

Van Engen, Henry. "An Analysis of Meaning in Arithmetic." *Elementary School Journal* 49 (1949): 321-329, 395-400.

Walbesser, Henry H. "The Textbook as an Instructional Aid." In *Instructional Aids in Mathematics (34th Yearbook of NCTM).* Edited by Emil J. Berger. Washington, D.C.: National Council of Teachers of Mathematics, 1973.

Willoughby, Stephen S. "Issues in the Teaching of Mathematics." In *Mathematics Education (69th Yearbook of the National Society for the Study of Education).* Chicago: University of Chicago Press, 1970.

II. National Assessment

Thomas P. Carpenter
Mary K. Corbitt
Henry S. Kepner
Mary Montgomery Lindquist
and Robert E. Reys

In the last two decades there has been considerable unrest in mathematics education. The modern mathematics reforms of the 1960s were followed by the back-to-basics movement of the 1970s, leaving the mathematics curriculum with many characteristics of two decades earlier. As the 1980s begin, new recommendations for school mathematics are being proposed (NCTM, 1980).

The earlier reforms in the mathematics curriculum have been based on some general principles of learning or student achievement. The modern mathematics movement embraced principles of meaningful learning and discovery learning (Bruner, 1960); and the back-to-basics movement was, in part, a reaction to a perceived decline in achievement test scores (Advisory Panel, 1977). Progress in improving mathematics learning, however, is going to require a much more careful analysis of students' learning and achievement than accompanied previous reforms.

One of the best measures of the achievement of American students is provided by the mathematics assessment of the National Assessment of Educational Progress (NAEP). The second NAEP mathematics assessment was conducted during the 1977-1978 school year. Exercises covering a wide range of objectives were administered to a carefully selected national representative sample of over 70,000 students at ages 9, 13, and 17. Item sampling procedures were used so that between 250 and 450 exercises were administered to each age group with approximately 2,400 students responding to each exercise. Consequently, the results provide an accurate sampling of the knowledge of elementary and secondary students over a broad range of objectives. The assessment also has the advantage of providing analysis of performance on specific exercises. Data on each exercise were analyzed separately to provide a description of students' performance on specific tasks.

The objectives that guided the development of exercises for the assessment were selected by panels of mathematicians, mathematics educators, classroom teachers, and interested lay citizens to reflect important goals of the mathematics curriculum. These groups concluded that the mathematics curriculum should be concerned with a broad range of objectives. Accordingly, the assessment focused on five major content areas: (1) numbers and numeration; (2) variables and relationships; (3) geometry (size, shape, and position); (4) measurement; and (5) other topics, which included probability and statistics, and graphs and tables.

Each content area was assessed at four levels: knowledge, skill, understanding, and application. Knowledge level exercises involved recall of facts and definitions. This included such tasks as ordering numbers; recalling basic addition, subtraction, multiplication, and division facts; identifying geometric figures; and identifying basic measurement units. Skill exercises involved various mathematical manipulations including computation with whole numbers, fractions, decimals, and percents. Also included were making measurements, converting measurement units, reading graphs and tables, and manipulating algebraic expressions. Understanding exercises tested students' knowledge of basic underlying principles such as the concept of a unit covering in measurement. These exercises were constructed so that students could not simply apply a routine algorithm. Application exercises required students to use their own knowledge or skills to solve problems. Both routine textbook problems and nonroutine problems were included in this category.

In addition to these cognitive areas, a number of affective variables were assessed, as well as students' self-reports of the types of activities they engage in during mathematics class. Also, a special set of exercises assessed students' ability to use a calculator to solve various kinds of problems.

The results of the assessment have been summarized elsewhere (Carpenter and others, 1980b, 1980c, 1981b; NAEP, 1979a, b, c, d). Going beyond these results, we have identified several areas in which the National Assessment has provided information about students' knowledge of mathematics that relates directly to the NCTM recommendations for the mathematics curriculum of the 80s. These areas are: (1) the need for a broader definition of basic mathematical skills, (2) the importance of students' understanding of mathematical concepts and processes, (3) the importance of problem solving as the focus of the mathematics curriculum, (4) documentation of the continued development of mathematical skills, (5) implications of calculators for teaching computational skills, (6) the need to increase and extend students' enrollment in mathematics courses, and (7) students' perceptions of their involvement in mathematics class-room activities.

Caution must be observed in interpreting the results from the NAEP mathematics assessment, which was not designed to identify causes of student performance. We have frequently extrapolated beyond the data in drawing conclusions; other authors would possibly reach different conclusions. However, the conclusions presented here are generally supported by a wide range of exercises in addition to the illustrative exercises reported.

A Broader Definition of Basic Skills

The National Council of Teachers of Mathematics (1980) recommendations for school mathematics for the 1980s include the recommendation that basic skills in mathematics should be defined to encompass more than computational facility. Geometry, measurement, probability, and statistics are recognized as important areas of basic skills (National Council of Supervisors of Mathematics, 1978; NCTM, 1980). If students' performance on the second mathematics assessment is a measure of instructional emphasis in the United States, we must conclude that the focus of most mathematics programs is on the development of routine computational skills since students demonstrated a high level of mastery of computational skills, especially those involving whole numbers.

Almost all students demonstrated mastery of basic number facts. About two-thirds of the 9-year-olds could perform simple addition and subtraction computation using algorithms for regrouping. By age 13, almost all students could perform simple computations involving addition, subtraction, and multiplication. Most of the older students were successful with the more difficult calculations such as those summarized in Figure II-1. Students encountered greater difficulty with whole number division and operations with fractions and decimals.

Performance was significantly lower, however, on exercises assessing basic noncomputational skills. In general, the only noncomputational skills for which students demonstrated a high level of mastery were those involving simple intuitive concepts or those concepts or skills they were likely to have encountered and practiced outside of school. This is reflected in students' knowledge of geometric terms. Students were familiar with common everyday terms like square or parallel, but not with terms like tangent and hypotenuse that are used less commonly in everyday vernacular. Over 95 percent of the 13-year-olds could identify squares and parallel lines; but even by age 17, fewer than 60 percent of the students were

Figure II-1. Whole Number Computation

Exercise	Percent Correct	
	Age 13	Age 17

a)   4285
      3273
    +5125
      ?
	85	90

b) Subtract 237 from 504	73	84

c)   671
  × 402
    ?
	66	77

familiar with terms like tangent and hypotenuse. The failure of students to learn basic geometric concepts is illustrated by the fact that only about a fifth of the 13-year-olds and a third of the 17-year-olds could solve a problem involving a simple application of the Pythagorean Theorem.

Performance on measurement exercises generally followed the same pattern as the geometry results. Most students were familiar with measurement concepts and skills that would likely be encountered and practiced outside of school, such as recognizing common units of measure, making simple linear measurements, and telling time. They had a great deal of difficulty, however, with many other basic measurement concepts and skills, especially those involving perimeter, area, and volume. Another basic skill area in which performance was generally low was probability and statistics. Fewer than half the students at any age level demonstrated even a tentative understanding of most basic probability concepts.

The Importance of Understanding

As results in the previous section showed, students failed to master a broad range of basic skills. Further, many of the skills appear to have been learned at a rote, superficial level. Students' performance showed a lack of understanding of basic concepts and processes in many content areas, such as measurement and computation with fractions. For example, almost all students could make simple linear measurements. Over 80 percent of the 9-year-olds and 90 percent of the 13-year-olds could measure the length of a segment to the nearest inch. However, when students were presented with a problem similar to the one illustrated in Figure II-2, 77 percent of the 9-year-olds and 40 percent of the 13-year-olds gave an answer of 5. Thus, although most students would line up the end of the segment when

Figure II-2. How long is this line segment?

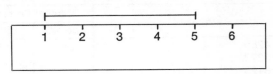

they were measuring it, this change in problem context demonstrated that many of them did not understand the consequences of not doing so.

This superficial understanding was also apparent in many computation exercises. For example, students were relatively successful in multiplying two common fractions, perhaps because multiplying numerators and denominators seems to be a natural way to approach the problem. However, the results from the simple verbal problem shown in Figure II-3 indicate that students had no clear conception of the meaning of fraction multiplication and, therefore, could not apply their skills to solve even a simple problem.

Figure II-3. Fraction Multiplication

Exercise*	Percent Correct	
	Age 13	Age 17
a) 2/3 × 2/5 = ?	70	74
b) Jane lives 2/3 mile from school. When she has walked 2/5 of the way, how far has she walked?	20	21
* Both are similar to unreleased exercises.		

Students' failure to learn basic fraction concepts is also illustrated by performance on several estimation exercises. For example, although 39 percent of the 13-year-olds and 54 percent of the 17-year-olds could calculate the answer to $\frac{7}{15} + \frac{4}{9}$, only 24 and 37 percent, respectively, could make even a reasonable estimate of $\frac{12}{13} + \frac{7}{8}$. Over half of the 13-year-olds and over a third of the 17-year-olds simply tried to add either the numerators or denominators. This response suggests that many students looked for some rote computation rule to apply without even considering the reasonableness of their result.

The importance of understanding may, in part, account for the difference in the level of performance for whole number operations and operations involving fractions and decimals. Most assessment exercises indicate that students had learned the basic concepts underlying whole number computation, and had some notion of the place value concepts involved in the computation algorithms. As a consequence, performance on whole number computation exercises was generally good. As the above examples suggest, however, most students did not have a clear understanding of fraction operations and appear to have operated at a mechanical level. This lack of understanding resulted in relatively poor performance on some fraction computation, and is further highlighted by the serious difficulties encountered in solving simple problems involving fraction operations.

The consequence of focusing on a mechanical application of basic skills is that students become totally dependent on a mechanical algorithm, which is easily forgotten. If students cannot remember a step in the algorithm, they cannot solve even simple problems that might be solved intuitively. For example, the complexity of problems (a) and (b) below appears to have had relatively little effect on their difficulty.

$$\text{(a)} \quad \frac{1}{2} + \frac{1}{3} \qquad\qquad \text{(b)} \quad \frac{7}{15} + \frac{4}{9}$$

In problem (a), students should have been able to find a common denominator almost intuitively. In problem (b), the denominators are not relatively prime and the least common denominator is 45. In spite of the difference in the apparent difficulty levels of the two problems, there was little difference in students' performance on them. Apparently, if students have learned an algorithm, they can apply it successfully in most situations. However, if they have not mastered an algorithm or have forgotten one step, they have difficulty with even simple problems that might be solved intuitively.

The results reported in this section suggest that many students have at best a superficial understanding of many mathematical concepts and processes. Yet around 90 percent of the 13- and 17-year-olds felt that developing understanding was an integral part of mathematics learning, as evidenced by their agreement with the statement "Knowing why an answer is correct is as important as getting the correct answer." Their responses may reflect their actual beliefs, or it may be that the statement was one they had heard from their mathematics teachers and was perceived as the expected response. The second alternative gains credence when considered in contrast to the fact that around 90 percent of both older age groups agreed that "There is always a rule to follow in solving mathematics problems." The students may be concentrating on mastering rules to the extent of ignoring concomitant understanding.

Problem Solving

One of the consequences of learning mathematical skills rotely is that students often cannot apply the learned skills to solve problems. In general, NAEP results show that the majority of students at all age levels had difficulty with any nonroutine problem that required some analysis. It appeared that most students had not learned basic problem-solving skills, and attempted instead to mechanically apply some mathematical calculation to whatever numbers were given in a problem.

A marked discrepancy was found between the results of solution of routine and nonroutine problems. Students generally were successful in solving routine one-step verbal problems such as those often found in textbooks. The results summarized in Figure II-4 are representative of student performance on one-step verbal problems in which the main steps were deciding whether to add, subtract, multiply, or divide and then performing the calculation.

The verbal problem in Figure 4 was presented to 9- and 13-year-old respondents without a calculator and to another group of 9-year-olds who had a calculator available. A third set of respondents was presented the

Figure II-4. Multiple-Choice Subtraction Exercises

Exercise*	Percent Responding		
	Age 9		Age 13
	Without calculator	With calculator	
a) George has 352 arithmetic problems to do for homework. If he has done 178 problems, how many problems does he have left to do?			
174 (correct response)	38	70	82
530	6	8	1
226	6	0	1
Other subtraction error	10	0	4

b)   352
   −178
   ?			

174 (correct response)	50	—	85
530	1	—	0
226	15	—	2
* Both are similar to unreleased exercises.			

same subtraction calculation as a straight computation exercise. As with all exercises on the assessment that required reading, the verbal problems were presented on an audiotape as well as written in the exercise booklet. By age 13, there was very little difference in students' ability to solve the verbal problem and their ability to perform the required calculation. Furthermore, few students at either age level chose the wrong operation.

Although students could successfully solve most simple one-step problems, they had a great deal of difficulty solving nonroutine or multistep problems. In fact, given a problem that required several steps or contained extraneous information, students frequently attempted to apply a single operation to the numbers given in the problem. Students' difficulty with problems that could not be solved with a single operation is illustrated by the results summarized in Figure II-5. In spite of the fact that the problems involved calculations that were well within the students' range of computational skill, many were unable to solve either problem. In both problems, many students simply added or multiplied the numbers without analyzing the problem.

Students have not developed good problem-solving strategies. A basic strategy that helps in analyzing certain types of problems is to draw a pic-

Figure II-5. Multi-Step Problems

Exercise	Percent Responding		
	Age 9	Age 13	Age 17
a) Lemonade costs 95¢ for one 56 ounce bottle. At the school fair, Bob sold cups holding 8 ounces for 20¢ each. How much money did the school make on each bottle?*			
Correct response	—	11	29
Students added, subtracted, or multiplied two of the numbers given in the problem.	—	40	25
b) Mr. Jones put a rectangular fence all the way around his rectangular garden. The garden is ten feet long and six feet wide. How many feet of fencing did he use?			
32 feet (correct response)	9	31	—
16 feet	59	38	—
60 feet	14	21	—
* Similar to an unreleased exercise.			

ture. A problem related to the second exercise in Figure II-5 was given to another set of respondents. In this version, they were given a picture of the rectangle and asked to find the distance around it. This variation in the problem produced a difference of over 30 percentage points in the percent of correct responses for both 9- and 13-year-olds. Since most students could identify a rectangle as shown by results on a simple recognition exercise, many of them apparently did not apply their knowledge by drawing a figure to help them solve the verbal problem. Instead, many of them simply added or multiplied the two numbers given in the problem.

Another basic problem-solving strategy is to look for related problems that one knows how to solve to provide a method for solving the given problem. The results for two closely related problems suggest that students have difficulty transferring a solution method from one problem to even a closely related problem. Although most students could calculate the area of a rectangle, they were unable to recognize that a square simply represented a special case of the rectangle. About half of the 13-year-olds and three-fourths of the 17-year-olds could calculate the area of a rectangle, but only about 10 percent and 40 percent, respectively, could find the area of a square.

NCTM recommends that the development of the ability to solve problems be a major goal of school mathematics in the 1980s. The results of the second NAEP mathematics assessment suggest that we are a long way from achieving that goal. They also suggest a note of warning of how we should *not* approach that goal. Providing more experience with typical textbook verbal problems, while helpful, it not an adequate response to the recommendation. The assessment results indicate that in addition to teaching how to solve simple one-step verbal problems, more emphasis should be placed on nonroutine problems that require more than a simple application of a single arithmetic operation. Part of the cause of students' difficulty with nonroutine problems may result from the fact that their problem-solving experience in school has been limited to one-step problems that can be solved by simply adding, subtracting, multiplying, or dividing. The assessment results indicate that students have relatively little difficulty solving problems that only require them to choose the correct operation. In fact, their difficulties with nonroutine problems seem to result from their interpretation that problem solving simply involves choosing the appropriate arithmetic operation and applying it to the numbers given in the problem.

Instruction that reinforces this simplistic approach to problem solving may contribute to students' difficulty in solving unfamiliar problems. Although it may be argued that children must learn to solve simple one-step problems before they can have any hope of solving more complex problems, an overemphasis on one-step problems may only teach children how to routinely solve this type of problem. It may also teach them that they do not have to think about problems or analyze them in any detail.

Techniques designed to give children success with simple one-step problems that do not generalize to more complex problems may be counter-productive. For example, focusing on key words that are generally associated with a given operation provides a crutch upon which children may come to rely. Such an approach provides no foundation for developing skills for solving unfamiliar problems. Simple one-step problems may provide a basis for developing problem-solving skills, but only if they are approached as true problem-solving situations in which students are asked to think about the problem and develop a plan for solving it based upon the data given in the problem and the unknown they are asked to find.

Students need to learn how to analyze problem situations through instruction that encourages them to think about problems and helps them to develop good problem-solving strategies. Students need ample opportunity to engage in problem-solving activity. If problem solving is regarded as secondary to learning certain basic computational skills, many students are going to be poor problem solvers. Additional discussion of implications of the NAEP results for problem solving at the elementary and secondary levels can be found in Carpenter and others (1980a, 1980d).

Continued Development of Mathematical Skills

Although problem solving and many noncomputational skills clearly require an increased emphasis in the curriculum, we do not deny the importance of computational skills. A reasonable level of computational skill is required for problem solving. We are suggesting, however, that problem solving not be deferred until computational skills are mastered. Problem solving and the learning of more advanced skills reinforce the learning of computational skills and provide meaning for their application.

It is important to recognize that most computational skills are learned over an extended period of time. The results summarized in Figure II-6 suggest that most skills are mastered after their period of primary emphasis in the curriculum. For example, even though a goal of most mathematics problems is that students learn to subtract by age 9, there was significant improvement in performance on subtraction exercises from age 9 to 13 and there was even some improvement between ages 13 and 17. Furthermore, many fundamental errors also disappear as students progress in school. Although over 30 percent of the 9-year-olds subtracted the smaller digit from the larger in a subtraction exercise that required regrouping, only 5 percent of the 13-year-olds and 1 percent of the 17-year-olds committed this error.

Figure II-6. Improvement in Performance by Age

Problem	Percent Correct		
	Age 9	Age 13	Age 17
a) Basic subtraction facts	79	93	95
b) Three-digit subtraction	50	85	92
c) $4/12 + 3/12 = ?$	—	74	90
d) $1/2 + 1/3 = ?$	—	33	66

These results have profound implications for minimum competency programs. Rigid minimum competency programs which hold children back until they have demonstrated mastery of a given set of skills may, in fact, be depriving them of the very experiences that would lead to mastery of the particular skills.

Although some skills will continue to develop through use in other contexts, this is not always the case. The current high school curriculum does not take into account that many basic skills are not well-developed by the time students begin instruction in algebra and geometry. For example, very few 13- and 17-year-olds have mastered percent concepts or skills, but outside of general mathematics classes, there is very little opportunity for high school students to extend or maintain their knowledge of percent.

Implications of Calculators for Teaching Computational Skills

Over 85 percent of the 17-year-olds in the assessment indicated that they had access to a calculator. This availability of calculators would seem to have profound implications for the appropriate level of emphasis that computation should receive and the types of algorithms we should teach.

In spite of the extensive instruction provided on whole number division, only half of the students assessed were reasonably proficient in division by the time they were ready to graduate from high school. With a calculator, however, over 50 percent of the 9-year-olds and over 90 percent of the 17-year-olds could divide accurately. This raises some serious questions as to whether the time spent drilling on division is a productive use of time and effort that might otherwise be devoted to other topics. Certainly, it is clear that our current approach to teaching division is not effective for a substantial number of students.

The division algorithm as well as most of the other algorithms that we teach in school are designed to produce rapid, accurate calculation procedures. Given the widespread availability of hand calculators, it would seem that the continued emphasis on developing facility with computation algorithms should not be as high a priority as it was formerly. Certainly, computation is important; but what is needed are algorithms that students will remember and will be able to generalize to new situations. This brings us back to the issue of understanding. Students are more likely to remember and be able to generalize and apply algorithms if they understand how the algorithms work. Thus, it may be appropriate to begin to shift to computational algorithms that can be more easily understood than the ones currently taught, even if they are less efficient.

The results for the following problem illustrate the potential impact of calculators on our thinking about computation:

> A man has 1,310 baseballs to pack in boxes which hold 24 baseballs each. How many baseballs will be left over after the man has filled as many boxes as he can?

Students had more difficulty solving this problem with a calculator than without using a calculator. Twenty-nine percent of the 13-year-olds correctly solved this problem without a calculator while only 6 percent of the 13-year-olds who had a calculator were successful. Students also had more difficulty comparing and ordering a set of fractions with a calculator than they did without one. They apparently did not understand that fractions can also be thought of as quotients, which allows one to represent them as decimals that are relatively easy to order. These results suggest that students have rigid ways of thinking about numbers and operations. Calculators sometimes require alternative interpretations and require that students have a deeper understanding of numbers and how the operations work.

Calculators also place an increased importance on estimation skills and alertness to reasonableness of results. The results of one exercise given to 13- and 17-year-olds illustrate the importance of this skill and the gross errors that can occur when students using a calculator are oblivious to the reasonableness of a result. Students were asked to divide 7 by 13 using a calculator. About 20 percent of the 13- and 17-year-olds chose the response 5384615 rather than 0.5384615.

Participation in Mathematics Courses

Mathematics learning is a continuous process that encompasses the entire 12 years of elementary and secondary school. Many, if not most, basic skills are not mastered by age 13 and must be reinforced and developed as part of the high school curriculum. Consequently, if we are going to significantly improve the mathematics performance of high school graduates, we must ensure that they continue to take mathematics throughout their high school program. The NCTM (1980) has proposed that at least three years of mathematics should be required of all students in grades 9 through 12. The assessment background data summarized in Figure II-7 indicate that we are currently far short of that goal.

Figure II-7. Mathematics Courses Taken by 17-Year-Olds

Course	Percent having completed at least ½ year
General or Business Mathematics	46
Pre-Algebra	46
Algebra I	72
Geometry	51
Algebra II	37
Trigonometry	13
Pre-Calculus/Calculus	4
Computer Programming	5

Student Perceptions of Mathematics Classes

Among the recommendations of NCTM for the curriculum of the 1980s are several statements that indicate a need for teachers to encourage experimentation and exploration by students as part of the requisite atmosphere that encourages problem solving. Included is a call for teachers to "provide ample opportunities for students to learn communication skills in mathematics" (NCTM, 1980, p. 8) in both reading and talking about mathematics, and a recommendation to teachers to incorporate "diverse instructional strategies, materials, and resources, such as— individual or small group work as well as large group work; . . . the use of manipulatives (where appropriate); . . . the use of materials and references outside the classroom" (pp. 12-13). The implication from these recommendations is that mathematics teachers should provide opportunities for their students to be actively involved in learning and communicating mathematics.

One set of exercises attempted to assess how often students engage in different activities in mathematics classes. Students were presented with a list of classroom activities and asked to rate them in terms of how often they thought the activities occurred in their mathematics class. The activities may be described as student-centered, teacher-centered, classmate-centered, and "other," which included activities requiring active student involvement such as using manipulative objects.

Figure II-8. Ratings of Frequency of Selected Classroom Activities

Activity	Percent Responding			
	Age	Often	Sometimes	Never
Student-centered				
Mathematics tests	9	44	46	9
	13	61	37	1
	17	63	33	3
Mathematics homework	9	43	45	12
	13	67	29	3
	17	57	36	6
Worked mathematics problems alone	9	71	22	7
	13	81	17	9
	17	80	19	1
Worked mathematics problems on the board	9	39	54	7
	13	33	58	9
	17	27	60	13
Used a mathematics textbook	9	75	18	6
	13	81	14	5
	17	87	11	3
Worksheets	9	71	27	2
Teacher-centered				
Listened to the teacher explain a mathematics lesson	9	85	11	3
	13	81	16	2
	17	78	19	2
Watched the teacher work mathematics problems on the board	9	78	19	4
	13	76	21	2
	17	79	18	3
Received individual help from the teacher on mathematics	9	21	67	11
	13	17	71	10
	17	18	70	11

Figure II-8 lists the frequency ratings on the student- and teacher-centered activities. Table headings represent response options for 13- and 17-year-olds to the question "How often have you done these activities in your mathematics classes?"; options for the 9-year-olds were "a lot," "a little," and "never." As the table shows, most students reported that they spent a lot of time listening to and watching the teacher work and explain mathematics problems. They also reported that they spent a lot of time working alone on mathematics problems from the textbook and, for the 9-year-olds, also from worksheets. The percentages of frequency assigned to these activities were the highest for any of the group of activities.

As a group, the classmate-centered and "other" activities received the highest percentages of "never" ratings for all age groups. Among the classmate-centered activities, around half of all age groups said that discussing mathematics in class occurred often; around 60 percent of the 9-year olds and 75 percent of the 13- and 17-year-olds said they sometimes gave help to or received help from their classmates in mathematics. Thirty-five, 44, and 28 percent of the 9-, 13-, and 17-year-olds, respectively, said they never worked mathematics problems with small groups of students.

Most of the older students said they had never made reports or done projects in mathematics classes, and over two-thirds of the 9-year-olds and three-fourths of the older respondents said they had never done mathematics laboratory activities. Further, over half of the 9-year-olds said they had never used objects like counters, rods, or scales in mathematics classes.

These National Assessment results show that students perceive their role in the mathematics classroom to be primarily passive. They are to sit and listen and watch the teacher do the problems; the rest of the time is to be spent working on an individual basis on problems from the text or from worksheets. They feel they have little opportunity to interact with their classmates about the mathematics being studied, to work on exploratory activities, or to work with manipulatives.

An attempt to evaluate the implications of these results for the curriculum of the 1980s leads directly to the issue of the extent of student involvement in the learning process. The results suggest that the current situation, at least from the students' point of view, is one in which mathematics instruction is "show and tell" on the teacher's part, "listen and do" for the students. Students' perception of their involvement is in direct contrast to the recommendations of NCTM. If active student involvement in mathematics learning is as desirable and sought after as the NCTM recommendations imply, then changes in approaches to teaching mathematics that will foster and encourage that involvement must be implemented.

Closing Thoughts

Undoubtedly, it will take at least the entire decade to make significant progress in fully implementing NCTM's recommendations for the 80s. Thus, these recommendations represent goals to strive to achieve by the end of the decade.

The National Assessment results provide one measure of where we are at the beginning of the decade. They also suggest that the development of routine computational skills has been the dominant focus of the school mathematics curriculum, and that the development of problem-solving skills has been inadequate.

Although we are a long way from the kind of program envisioned in the NCTM recommendations, the assessment results provide some basis for cautious optimism. It is probably fair to say that the focus of mathematics instruction has been on computation. There is evidence that students are learning what they are being taught. There is also evidence that curricular reforms can have some impact. On exercises that measured change in performance from the first assessment, there were significant gains of 10 to 20 percentage points on exercises that dealt with metric measurement. These results appear to reflect the increased emphasis on metric measurement in the curriculum over that period of time.

Improved student performance in mathematics is a goal that demands the combined efforts of many people. The results presented here have shown that there is room for much improvement, but there is hope that if we can reorganize the mathematics curriculum to address the NCTM recommendations, students' performance will respond accordingly.

References

Advisory Panel on the Scholastic Aptitude Test Score Decline. *On Further Examination: Report of the Advisory Panel on the Scholastic Aptitude Test Score Decline.* New York: College Entrance Examination Board, 1977.

Bruner, J. S. *The Process of Education.* Cambridge, Mass.: Harvard University Press, 1960.

Carpenter, T. P.; Corbitt, M. K.; Kepner, H. S.; Lindquist, M. M.; and Reys, R. E. "NAEP Note: Problem Solving." *Mathematics Teacher* 73 (September 1980a): 427-433.

Carpenter, T. P.; Corbitt, M. K.; Kepner, H. S.; Lindquist, M. M.; and Reys, R. E. "Implications of the Second NAEP Mathematics Assessment: Elementary School." *Arithmetic Teacher* 27 (April 1980b): 10-12, 44-47.

Carpenter, T. P.; Corbitt, M. K.; Kepner, H. S.; Lindquist, M. M.; and Reys, R. E. "Results of the Second NAEP Mathematics Assessment: Secondary School." *Mathematics Teacher* 73 (May 1980c): 329-338.

Carpenter, T. P.; Corbitt, M. K.; Kepner, H. S.; Lindquist, M. M.; and Reys, R. E. "Solving Verbal Problems: Results and Implications from National Assessment." *Arithmetic Teacher* 28 (September 1980d): 8-12.

Carpenter, T. P.; Corbitt, M. K.; Kepner, H. S.; Lindquist, M. M.; and Reys, R. E. "National Assessment: A Perspective of Students' Mastery of Basic Skills." In *Selected Issues in Mathematics Education.* Edited by M. M. Lindquist. Chicago: National Society for the Study of Education, 1981a.

Carpenter, T. P.; Corbitt, M. K.; Kepner, H. S.; Lindquist, M. M.; and Reys, R. E. *Results of the Second Mathematics Assessment of the National Assessment of Educational Progress.* Reston, Va.: National Council of Teachers of Mathematics, 1981b.

National Assessment of Educational Progress. *Changes in Mathematical Achievement, 1973-78.* Denver: National Assessment of Educational Progress, 1979a.

National Assessment of Educational Progress. *Mathematical Applications.* Denver: National Assessment of Educational Progress, 1979b.

National Assessment of Educational Progress. *Mathematical Skills and Knowledge.* Denver: National Assessment of Educational Progress, 1979c.

National Assessment of Educational Progress. *Mathematical Understandings.* Denver: National Assessment of Educational Progress, 1979d.

National Council of Teachers of Mathematics. *An Agenda for Action: Recommendations for School Mathematics of the 1980s.* Reston, Va.: National Council of Teachers of Mathematics, 1980.

National Council of Supervisors of Mathematics. "Position Paper on Basic Mathematical Skills." *The Mathematics Teacher* 71 (February 1978): 147-152.

Response

Diana Wearne

The National Assessment of Edutional Progress has provided a wealth of information about the mathematics achievement of elementary and secondary students. Carpenter and his colleagues derived a number of significant conclusions from the NAEP data about students' current level of achievement and have offered several noteworthy suggestions for improving school mathematics programs.

Interpreting the results of any test is difficult, particularly when the test is of the magnitude of National Assessment. Several cautions must be exercised in reading reports, some of which are: (1) selections of specific items to interpret may affect the conclusions; (2)

items may not assess the stated objectives; and (3) implications of results may depend on value judgments about the educational importance of individual items.

Whenever sets of items are discussed and it is not possible for the reader to see all of the items (the NAEP tests had between 250 and 430 items at any age level), there exists the possibility that analyzing different sets of items may lead to different conclusions. The authors of the chapter were careful to caution readers as to this possibility.

It is difficult to err in constructing items that assess computation; however, assessing less specific and definable objectives can be difficult. In an effort to assess teaching methods, students were asked to indicate how often they had participated in specific categories of activities, but not the nature or characteristics of the activities. Some 43 percent of the 9-year-olds indicated they often had homework. The homework could have consisted of a page of computation or it could have involved searching for information to record, organize, and use the following day in graphing activities. These are distinctly different types of homework and it is impossible to determine what percentage was of each type.

It also is possible the students' perceptions of how much time was devoted to a given topic or activity were confounded by their attitudes. A 13-year-old who is assigned homework once a week but dislikes it may feel he or she is always having to do homework. A student who has not been in a mathematics course for a year or more may remember only the activities that were especially pleasant or distasteful.

Two-thirds of the 9-year-olds reported they had never participated in mathematics laboratory activities. Some students may have been involved in such activities but did not realize it. For instance, children involved in measurement activities designed to yield numbers for computing may not have considered that a laboratory activity. Half of the 9-year-olds indicated they had never used counters, rods, or scales. They may have used counting manipulatives but called them by other names, such as "popsicle sticks." A child who was involved in an activity for only a portion of the class time and who also completed a workbook page may have responded that textbooks were used always or often.

Another possible misinterpretation of the test results relates to the time at which the tests were administered. Carpenter and others cite the data in Table 20 as evidence that "we are currently far short" of the goal of three years of mathematics for all students in grades 9-12. When the tests were administered in March or April, 72 percent of the 17-year-olds were in the eleventh grade. Some of those students undoubtedly enrolled in a mathematics course in their senior year, increasing the actual percentage of students who took more advanced courses during their high school career.

On another topic, the authors caution against giving too much emphasis to one-step application problems. That may be good advice from a problem-solving point of view, but one-step problems give meaning to mathematical operations where it may otherwise be lacking, as apparently is the case with fractions. Young children understand addition and subtraction within a verbal, one-step problem context before they develop meaning for the symbolic representation.[1] Perhaps a similar link between verbal problem situations and their symbolic representations would help children develop meaning for other operations (multiplication and division) and other kinds of numbers fractions, negative numbers). The one-step problems should force children to think about the meaning of the operations in the verbal context and should not be solved as a series of routine calculations.

[1] Carpenter, T. P. "The Effect of Instruction on First-Grade Children's Initial Solution Processes for Basic Addition and Subtraction Problems." Paper presented at the annual meeting of the American Educational Research Association, Boston, April 1980.

Carpenter and others have painted both a pessimistic and an optimistic picture of mathematics learning in the United States. It is pessimistic because it shows what students are not learning but optimistic because it shows that positive changes have taken place. Citing improved performance in metric measurement because of added emphasis to this topic since the previous assessment, the authors say that pinpointing deficiencies may result in further improvement in student performance. I am not as optimistic.

The pressure to include metric measurement in textbooks came both from inside and outside the mathematics community and change was relatively easy to affect. Improving problem-solving skills is much more complex. Redesigning textbooks so they develop understanding and involve students in problem-solving activities is more than a cosmetic change of including a few additional pages in a book. And needed changes on the part of teachers and administrators are even more difficult to achieve.

One reason for cautious optimism may be the number of well-attended sessions on problem solving at regional and national meetings. Perhaps teachers' interest and continuing emphasis on this important topic by organizations will bring improved problem-solving performance on the next National Assessment.

III. Children's Thinking

James Hiebert

Educators and psychologists have long been interested in children's ability to deal with mathematical tasks. Since the traditional research approach has focused on the outcomes of children's performance, investigations are often designed to determine the effects of certain external conditions on these outcomes. For example, a typical research study might investigate the effects of two different methods of instruction on whole number addition by comparing the number of correct responses on an addition test after instruction. The answers children give are used to infer something about the effectiveness of the external conditions under which the concepts were learned.

In contrast to the concern with conditions outside of the learner and the focus on performance outcomes, several lines of current research are looking directly at the *processes* children use to solve mathematical problems. Consistent with current trends in cognitive psychology, this research focuses on the things that occur inside the child's head. Of course, we cannot actually see inside the mind, so many of the conclusions are based on inference. But the strategies or processes children use provide a window on their thinking. In some areas, where research efforts have been quite intensive in recent years, it is possible to paint a reasonably good picture of children's thinking.

There are two levels at which this type of research on children's mathematical thinking has been carried out. One is an underlying, funda-

Preparation of this chapter was supported in part by a Summer Faculty Research Fellowship, Graduate School, University of Kentucky. The author thanks Thomas P. Carpenter, Elizabeth Fennema, and Thomas A. Romberg for their comments on an earlier draft.

mental level concerned with general principles of thinking and learning. At this level, research focuses on basic cognitive processes that potentially are involved in dealing with a wide range of mathematical situations. Piaget's work is of this kind. The second level deals with processes that are specific to similar types of mathematical problems. Of primary interest are the strategies that children use to solve a given class of problems. Detailed descriptions are given of what children do when they solve linear measurement problems, when they add two whole numbers, when they reduce fractions, or when they solve any other specific type of mathematical problem.

In this chapter we will review and synthesize some of the research at each of these two levels, and look at how the research findings might be used in the design of better instructional programs. Piaget's work and some recent developments in information processing theory provide the basis for the review of general cognitive processes. Several areas of research could be used as examples of the work on more specific mathematical processes. Research on early number concepts and initial arithmetic operations was selected for this review. Children's thinking in this area has received much attention in recent years and the work nicely illustrates this research perspective. It is important to note that this chapter will only sample from the existing research on children's thinking. Readers may also wish to consult other reviews of children's mathematical thinking that are either more extensive, focus on other content, or have been written from a different perspective (for example, Brainerd, 1979; Carpenter, 1976, 1979; Lesh and Mierkiewicz, 1978; Shumway, 1980).

General Cognitive Processes

An abundance of research has been carried out within the past decade in an attempt to uncover relationships between various cognitive processes and performance on mathematical tasks. Two types of cognitive processes have appeared to be most closely related to mathematics learning and have dominated the research in this area. These are the logical reasoning abilities described by Piaget, and information processing capacity, as characterized by recent work in cognitive psychology (Campione and Brown, 1979; Case, 1978a, 1978b).

Logical Reasoning Abilities

In studying how children acquire knowledge and learn about the world, Piaget and associates (Piaget, 1952; Piaget, Inhelder, and Szeminska, 1960) looked very closely at children's thinking. They found many things that surprised them. If they laid out one row of candies and the child laid out a second row with the same number of candies, and then they spread the candies in one of the rows, the child would respond that now the longer row had more candies. Through continued observations of children, Piaget discovered that this response was typical of all young children. Before a certain age (usually around five or six years) children do not conserve number, that is, (they do not recognize that moving the objects in a set has no effect on the number of objects the set contains.) After finding similar nonconservation responses on tasks with other quantities, such as length and area, Piaget concluded that conservation is a hallmark in the development of logical reasoning. It represents a fundamental difference between children's thinking and adults' thinking.

Educators and psychologists have pointed out that the logical reasoning abilities identified by Piaget, such as conservation, may be essential for solving a variety of mathematical problems (Elkind, 1976; Lesh, 1973). An analysis of many mathematical tasks shows that these abilities seem to be logical prerequisities. For example, solving a simple addition or subtraction problem using concrete objects involves moving the objects about and regrouping them in various ways. Many of the strategies involve transformations on objects that logically presuppose the ability to conserve number. A similar analysis, applied to a variety of measurement tasks, suggests that conservation of length may be required to learn foundational concepts of measurement.

While the prerequisite relationships between conservation and learning related mathematical concepts seem quite logical from an adult perspective, they have been difficult to document with children. Recent history has recorded a continuing debate among researchers about the importance of conservation for learning mathematics. Many of the early studies found a general relationship between passing conservation tasks and scoring well on mathematics achievement tests, but some conflicting evidence was also reported. In many ways this early research raised as many questions as it answered (Carpenter, 1980). Recently, investigators have become more sophisticated in their approach to this problem and have resolved many of the previous questions. The success of these studies is due in part to the fact that most of them have focused on specific mathematical concepts rather than on general achievement; some have followed children's progress over a carefully designed instructional sequence; and many have considered the processes children use to solve problems rather than looking only at correct and incorrect responses. These methodological improvements have helped to illuminate the role of conservation in children's ability to learn mathematics.

Arithmetic and measurement have provided the arena for much of the recent research on the relationship between conservation and mathematics learning. Several initial studies reported a significant relationship between number conservation and performance on certain kinds of addition and subtraction problems (LeBlanc, 1971; Sohns, 1974; Steffe, 1970; Steffe and Johnson, 1971). But no clear pattern emerged that might suggest which arithmetic problems depend on conservation and which do not. More recent studies have provided at least a partial answer to this question. Mpiangu and Gentile (1975) taught kindergarten children a variety of simple counting and number skills, and found that nonconservers gained as much from the instruction as conservers. Although the conservers performed more successfully than nonconservers, both before and after instruction, their equal gains led to the conclusion that conservation is not needed to learn arithmetic skills.

A later study by Steffe and others (1976) confirmed that nonconservers can learn simple skills but their results suggest that conservation may be important for solving more complex arithmetic problems. After several months of instruction on various counting strategies, nonconservers had trouble applying them to solve missing addend problems, while conservers were quite successful. Steffe and others argue that conservation is not needed to complete addition and subtraction problems that can be solved with simple counting skills, but it is important for understanding the more complex problems, like those with missing addends.

A study with first-grade children by Hiebert and others (1980) provides further evidence that nonconservers can solve a variety of verbal arithmetic problems. While there were significant differences in the accuracy with which conservers and nonconservers solved some of the problems, there were, for each problem type (addition, subtraction, and missing addend), a number of nonconservers who responded correctly. Furthermore, each kind of solution strategy was used by at least some nonconserving children. In fact, the frequency with which nonconservers applied the more advanced strategies did not differ significantly from their conserving peers. The picture that emerges from these results, along with those of previous studies, is that conservation is not needed to learn elementary counting skills nor to solve simple verbal or symbolic addition and subtraction problems. While it may facilitate performance on missing addend problems there is good reason to believe that it is not a prerequisite.

Research on children's learning of measurement yields results that are similar to those for arithmetic. Several studies have investigated the sequence in which a variety of measurement concepts are acquired. If the ability to conserve is needed to learn certain concepts or skills, then one would expect successful performance on a conservation task to precede mastery of the concept or skill tasks. Apparently this sequence does occur for measurement tasks that assess children's understanding of the inverse relationship between unit size and unit number, that is, the fact that more units are needed to measure a given quantity if they are small than if they are large (Bradbard, 1978; Carpenter, 1975; Hatano and Ito, 1965; Wohlwill, 1970). However, there are many measurement skills that precede the appearance of conservation. This group included: (1) the ability to iterate units, e.g., move a single unit across a surface to measure its length (Bradbard, 1978); (2) proficiency in applying standard measurement techniques, such as using a ruler to measure length (Hatano and Ito, 1965); and (3) the ability to attend to the number of units measured and infer that the quantity which measured the most units is the largest (Carpenter, 1975; Wagman, 1975).

The results of these studies show that nonconservers learn a variety of measurement skills, but they do not indicate what the limits of this learning might be. To obtain this information, Hiebert (1981) instructed length conserving and nonconserving first-grade children on several basic concepts of linear measurement. At least some nonconservers learned each of the concepts and skills except one—using the inverse relationship between unit number and unit size to construct a length. On all tasks but this one, nonconservers used the same kind of solution strategies as their developmentally advanced peers. Apparently conservation is a true prerequisite for this one concept of measurement, but is not needed to master many other measurement concepts.

The argument for using conservation as a readiness measure for instruction is based on the assumption that conservation is a prerequisite for learning various mathematical concepts or skills. Since conservation is not easily taught, and since it presumably represents a fundamental logical reasoning ability, it may be better to postpone instruction on these concepts until the reasoning ability develops. That's the logical argument. Its validity obviously rests with being able to establish empirically that certain mathematical tasks do, in fact, require conservation to solve them.

The research reviewed here focused on the role of conservation in learning initial arithmetic and measurement concepts. The evidence suggests that conservation tasks are of limited value as readiness measures for instruction on these concepts. Except for the concept of the inverse relationship between unit number and unit size in measurement situations, conservation does not seem to be essential for learning to solve school mathematics tasks. There are simply too many children who fail the conservation tasks and perform successfully on the mathematics tasks.

The problem is that even though conservation is a *logical* prerequisite for completing many arithmetic and measurement tasks, children do not seem to use conservation knowledge when they solve the tasks. Children's solution procedures are different than the structural logic of the problem. They move and regroup objects to solve a simple addition problem and do not think to ask the conservation question; they simply count the objects to find the answer. They move a unit to measure the length of an object and do not worry about whether the length of the unit is being conserved. Children seem to focus only on the question at hand and do not recognize that conservation provides an essential logical foundation for the task. Simple skills, such as counting, apparently allow children to bypass the logical structure of many mathematical tasks.

Conservation is not the only reasoning process described by Piaget that potentially affects mathematics learning. Transitive inference and various classification skills are also believed to be fundamental thought processes that support the acquisition of many mathematical concepts (Piaget, 1952; Piaget and others, 1960). However, here too the available research evidence suggests that these abilities are not prerequisites for dealing successfully with logically related mathematics tasks (Hiebert, 1981; Hiebert and others, 1980; Sohns, 1974; Steffe and others, 1976). Researchers may be more successful in establishing relationships between the more advanced, formal reasoning processes in Piaget's theory and the mathematics learning of adolescents (Adi, 1978; Carpenter, 1980). Many of the school mathematics tasks at this level seem to involve directly the abstract reasoning skills measured by the formal reasoning Piagetian tasks. However, so little research has been done at this level that it would be inappropriate to speculate on the nature of these relationships.

Information Processing

The basic notions of information processing theory grew out of a concerted attempt to describe what the learner actually does when solving a problem or acquiring a new skill. The objective is to describe how the learner processes information, and then to use these descriptions to build models of the human information processing system. Many times these models are precise enough so that they can be written in computer language. In this case the validity of the model can be checked by giving the computer and a student the same problem and observing how closely the computer simulates the performance of the student. The value of building these models is that a great deal of thought must be given to detail the processes that are used to solve a particular problem. Computers don't work well unless they are programmed with precision. Consequently, the models help to identify some of the critical points in the thinking process.

Most information processing models include three important features. One is a component labeled working or short-term memory. This is the center of all thinking or information processing. It has a limited capacity, which is usually described in terms of the number of separate pieces of information that can be processed at the same time. A second feature of most models is a description of the processes that could be used to solve a given class of problems. If computer simulations are developed, these processes are described in great detail. A third feature of recent models is a planning and organizing function which serves to oversee the actual processing of information. The so-called meta-cognitive processes include plans and strategies to decide what pieces of information to focus on at any one time, how to organize the information so that it can be processed most efficiently, and which of all available strategies can best solve the problem. While the first feature of the model is concerned primarily with the limits of the system, the second and third have focused more on its capabilities.

Capacity Limitations on Learning. Short-term memory is a critical part of the information processing system because there is a definite limit on the number of information bits that can be handled simultaneously. Try adding 275 and 468 in your head without referring back to the numbers. Even though you know all the rules for completing this simple problem it probably puts some strain on your information processing system. If you made a mistake in computing the answer you can probably put the blame on insufficient short-term memory capacity. Children experience even greater difficulty with these kinds of problems because they have a more restricted capacity. Young children can process only about one-fourth to one-half the number of information pieces that adults can handle.

Some researchers have suggested that children's restricted processing capacity has considerable consequences for the curriculum because it may place severe constraints on children's ability to profit from instruction. Instructional tasks require children to receive, encode, and integrate information. In many cases, children may possess all of the necessary skills for a particular task and still fail the task. The reason for this failure may be children's restricted capacity to deal with all of the information needed to complete the task (Case, 1975).

The research to date has shown that information processing capacity does constrain children's learning to a predictable degree on specially-designed laboratory tasks (Case, 1974). However, it has been more difficult to isolate the effects of this capacity on school mathematics tasks (Hiebert, 1981; Hiebert and others, 1980). Recent work in this area suggests that part of the problem in identifying capacity constraints lies in developing valid and reliable measures of processing capacity (Romberg and Collis, 1980a, 1980b). Some progress is being made in constructing measures but their application to instructional settings is still far from being realized.

Processes and Meta-Processes. Although it is reasonable to think that certain underlying cognitive capacities are needed to learn mathematics, and that an insufficient capacity would limit children's learning, the research on what children *cannot* learn has not been very productive. Recently, a number of investigators have begun looking at the general cognitive processes children do have, and the ones they are capable of learning. These include both the processes that are used to solve problems, and the meta-processes that serve to select and monitor the execution of specific procedures. While short-term memory capacity develops with maturity and cannot be readily improved by specific training, the processes of the system are influenced by instruction. Individuals can be taught strategies that process information more efficiently and push back the limits that might otherwise be imposed by their restricted processing capacity (Brown, 1978; Brown and others, 1981).

It is too early to tell what implications the research in this area might have for the mathematics curriculum. However, several characteristics of this approach are already being used with impressive success to study children's thinking in mathematical situations. One important characteristic of this approach is the emphasis on the proficiencies children have. Children are viewed as capable thinkers who have at least the rudiments of effective problem solving. The assumption is that the things children do make sense to them, and attempts are made to describe in detail, from the children's perspective, the processes they use to deal with information.

A second important characteristic of the recent information processing approach is the concern with careful task analyses. Children may perform quite differently on several tasks measuring the same concept because of differences in task format, the type of response required, or other task variables. Children's real competencies may be hidden by irrelevant task variables. Understanding the task is essential for understanding the processes children use to solve the task. Therefore, a variety of task analysis procedures are used to describe the underlying structure of tasks as well as surface characteristics that may affect performance.

A third significant characteristic of recent research in information processing is the focus on local, rather than global, processes. It has been very difficult to find general principles of learning and thinking that apply to a wide variety of situations. Recognizing this problem, researchers are turning their attention to specific processes that are used on a well-defined set of similar tasks. They believe that it is more productive at this point to describe in detail the particular strategies that are used to solve a homogeneous class of problems than to continue searching for general processes that are involved in a wide range of problems.

Children's Thinking About Number and Arithmetic Concepts

One line of research within the mathematics education community that has successfully applied the general research perspective arising from the information processing approach is the study of number and arithmetic concepts. Recent work in this area has been directed toward describing what young children know about number and arithmetic operations, even before they receive formal instruction; how task variables affect their performance; and what strategies they use to solve different types of arithmetic problems. Although the picture is not yet complete, we are able to describe some important pieces of children's thinking in this area. Consequently, this line of research was selected as an example of current research on children's mathematical thinking. For discussions of research in other areas of mathematics learning and thinking see recent reviews edited by Shumway (1980), Lesh and Mierkiewicz (1978), and Lesh and others (1979).

Development of Early Number Concepts

Children achieve an initial concept of number through counting. Although this conclusion may seem obvious to those who have observed young children answer questions of *how many?,* a series of recent studies has shown how important, and how complex, the counting process is. Fuson (1979, 1980; Fuson and Mierkiewicz, 1980) and Steffe (Steffe and others, 1976; Steffe and Thompson, 1979) trace the development of children's ability to count from when they first verbalize a string of number words to the point where they can use efficient counting techniques to solve a variety of arithmetic problems. An analysis of the counting act shows that a process as simple as finding "how many" objects there are in a set involves the coordination of several separate actions: saying the number word string beginning with one, and pointing to a different object as each number word is spoken.

As children's counting proficiencies continue to develop, two major breakthroughs can be identified. The first occurs when children begin to establish relations among the counting words rather than producing the number word string as a single unit. A symptom of this new facility is that children can now count forward from a number other than one, or count back from a given number. They can give the number that comes just before, or just after, a number without counting from one. A second significant advance occurs when children recognize that the number words themselves can be used as the objects of counting. This new realization substantially increases the power of the counting process in solving problems. For example, children can now count on to find the number that is 5 more than 8 by counting the number words after 8 and stopping at the fifth one. This represents a change in *what* is counted. Concrete objects are no longer needed as counters; number words can serve as the unit items.

The act of counting rests upon several important principles. Described in detail by Gelman and Gallistel (1978), they include the fact that each object to be counted must be assigned one and only one number word, that the same number list must be used every time a set of objects is counted, that the last number word gives the numerosity of the set, and that the order in which the objects are counted does not matter. Gelman (1977, 1978; Gelman and Gallistel, 1978) argues that counting is a natural process for young children, and that before entering school they already understand these principles. According to Gelman, learning how to count is primarily a matter of learning the standard number words (one, two, three . . .) and applying the principles to larger and larger numbers.

Development of Addition and Subtraction Concepts

Along with their counting skills, many preschool children develop some sound, intuitive ideas about arithmetic operations. For example, by four years of age, most children understand that addition increases numerosity and subtraction decreases numerosity, even though they may have trouble calculating the numerical outcome of the increase or decrease (Brush, 1978). If the sets are small enough so young children can count them, many children also seem to recognize that addition and subtraction are inverse operations in the sense that the effect of one cancels the effect of the other, and that, if trying to keep two sets equivalent, adding objects to one set can be compensated for by adding objects to the other set (Gelman and Gallistel, 1978; Gelman and Starkey, 1979). These intuitive notions, together with effective counting skills, provide children with a significant fund of knowledge with which to begin school.

Children's interpretations of arithmetic progress as they receive instruction, but in the first few years this progress appears to be closely tied to the development of their counting abilities. Ginsburg (1977b) believes that counting is so important for children that even after formal instruction "the great majority of young children interpret arithmetic as counting" (p. 13). It is certainly true that before children learn basic addition and subtraction facts they solve arithmetic problems by counting. A series of studies at the Wisconsin Research and Development Center for Individualized Schooling (Carpenter and others, 1981; Carpenter and Moser,

1979) and at the Pittsburgh Learning Research and Development Center (Heller and Greeno, 1979; Riley and Greeno, 1978) have shown that first-grade children, even before receiving instruction, can solve verbal addition and subtraction problems by applying appropriate counting strategies. Furthermore, they do not use the same strategy to solve every problem, but have available a rich repertoire of strategies and use different strategies to solve different types of problems.

One objective of this research has been to determine what factors affect the kinds of strategies that children use to solve different types of verbal addition and subtraction problems. The one factor that consistently stands out as the most significant in this regard is the type of action or relationship between the sets described in the problem (Carpenter and others, 1981a; Riley and Greeno, 1978). The importance of this "semantic structure" is best explained by considering several sample problems. The following problems are all solvable by subtracting the smaller number from the larger: (1) John has 8 apples. He gave 5 apples to Mary. How many apples does John have left? (2) John has 5 apples. Mary gave him some more apples and now he has 8 apples. How many apples did Mary give to John? (3) John has 8 apples. Mary has 5 apples. How many more apples does John have than Mary? If first-grade children are provided with physical objects to be used as counters, and are read these three stories, the majority of children will solve each problem using a different counting strategy. Almost all children who use the counters will solve the first problem by making a set of eight, removing five, and counting the rest. Most children will solve the second problem by making a set of five, adding on additional markers by counting "six, seven, eight," and then counting the number of markers added on. While there is more variation on the third problem, many children will count out a set of five and a set of eight, match the two sets using a one-to-one correspondence, and then count the unmatched markers in the larger set.

It is clear from these examples that many children solve the problems directly by carrying out the action or representing the situation that is described. Although at first glance this may not appear to be a particularly profound conclusion, it carries with it at least two potentially important implications. First, it means that even before receiving instruction, children are sensitive to the critical verbal cues in a story that indicate what action is appropriate to solve the problem. At this point in the learning process, very few children apply the wrong operation to solve a problem. That is, in general they do not add when they should subtract, or subtract when they should add. Not only do they carry out the correct operation, they often match their strategy to the context or semantic structure of the problem.

The fact that initially children use different strategies to solve different types of subtraction problems leads to the second important point. Apparently children perceive these different subtraction situations to be genuinely different problems. Although adults can see the commonality in these problems, and recognize that they can all be solved using the same "subtraction" procedure, it seems that many young children do not possess such a general subtraction concept. They see these as different problems that are solvable by different methods.

Development of an Arithmetic Symbol System

As children proceed in school, they receive instruction in the formal symbolism of arithmetic. They are taught to represent the verbal problems presented earlier as $8 - 5 = \square$. It is at this point that many children experience difficulty. Writing the same number sentence for these three different problems requires children to see them as mathematically similar, an expectation which may go beyond the knowledge and capabilities of first- and second-graders (Gibb, 1956; Vergnaud, 1979).

In addition to the problems children may have with collapsing their many different interpretations of arithmetic situations into a single addition category and a single subtraction category, they also seem to experience difficulty in relating the verbal problem to a symbolic equation, of whatever kind. At the end of first grade, many children can solve verbal addition and subtraction problems, and some can write number sentences which represent these problems. But Carpenter and others (1981b) found that many children view these two processes as being independent. Children in this study often wrote the symbolic equation after, rather than before, finding the solution. The act of writing a number sentence rarely influenced the choice of a solution strategy.

At the heart of this problem is the fact that young children are not always able to give meaning to the formal, arbitrary symbolism of mathematics. Lindvall and Ibarra (1980) report that first- and second-graders have a difficult time demonstrating with concrete objects the meaning of a simple addition or subtraction equation. Grouws (1972) found that even in the third grade, many children did not solve addition and subtraction number sentences when the position of the unknown was somewhere other than by itself, on the right side of the equal sign. The errors were largely noncomputational and indicated that many children did not understand the meaning of the equation. The difficulty with symbolism seems to be pervasive and fundamental.

What happens when primary school children lack the understanding necessary to deal with arithmetic symbols in a meaningful way? The available research suggests that they begin developing their own system of rules to manipulate the symbols and generate answers. Many times they memorize fragments of algorithms or rules and recombine these in unique ways (Davis and McKnight, 1979). Lankford (1972) and Erlwanger (1975) have shown how intricate and "creative" some of these idiosyncratic systems can be. On occasion children's invented, incorrect algorithms are more complex than the correct ones. Apparently children are not incapable of learning complex algorithms and executing them consistently. In fact, Brown and Burton (1978) conclude that even when making errors, elementary school children are generally consistent and systematic. Frequently their errors are the result of methodically following the wrong procedure rather than making random mistakes.

What seems to be missing is a link with reality which might serve as a validating or correcting mechanism. Unable to make sense of the symbol system, many children appear to have no way of knowing whether the processes they are using are correct. While they may be convinced that the procedures they apply are the right ones, this confidence often comes from the belief that they have mastered the rules and tricks of the system, rather than a feeling that the procedures reflect reality (Erlwanger, 1975). A symptom of this problem is the periodic unreasonable responses provided by many elementary school children (see the report of the National Assessment of Educational Progress (NAEP) in Chapter II). Apparently, the formal symbolism of mathematics moves children from their natural, intuitive problem-solving skills that were anchored in real world experiences to rules of symbol manipulation, some of which have lost touch with their reality.

To reiterate using Ginsburg's (1977a) terms, children experience great difficulty translating their informal, experience-rich system into the formal, symbols-and-rules system of school arithmetic. Gaps between these systems begin to develop. Young children often are unable to establish meaningful links between what they know when they get to school and what they soon are asked to do—formalize this knowledge using mathematical symbols. As children lose their intuitive understanding of mathematical problems, or are asked to do mathematics in situations in which they cannot access these intuitive understandings, they begin developing their own unique systems of symbol manipulation, some of which are filled with misconceptions and faulty procedures.

Implications for the Curriculum

The basic assumption of this chapter is that the way in which children think about mathematics and the processes they use to solve mathematical problems must be understood and taken into account by teachers and curriculum builders who design instruction. If instruction is going to build on children's existing knowledge and the problem-solving strategies they have already developed, teachers must be aware of how children think about mathematics. However, an important note of caution should be inserted before discussing the implications of this research for mathematics curriculum and instruction. Research on children's thinking is necessarily *descriptive;* it describes children's behavior in learning or problem-solving situations. Instruction programs are essentially *prescriptive;* they prescribe the conditions that should be set up to facilitate learning. The prescription of programs does not immediately follow from the description of children's thinking (Bruner, 1966; Rohwer, 1970). Children's performance is the outgrowth of their learning experiences, and it is not always clear how a change in these experiences (through a change in curriculum) would influence this performance. Therefore, it is not always possible to prescribe the "best" curriculum from information on children's thinking. But it is possible to suggest several features that can be part of any instructional program.

Although there are many implications that might be drawn from the preceding review of research, two major implications stand out from the rest. One is suggested initially by Piaget's work and deals with the importance of observing children and looking at the world through their eyes. The second grows out of the recent work on children's mathematical thinking and centers on the importance of maintaining a link between children's natural base of experience and the mathematical concepts and symbols they are attempting to learn.

Listening to Children

Children do not think like adults. They view the world from a different perspective; they solve problems by applying qualitatively different forms of thought. A striking example of this is young children's failure on conservation tasks, tasks that seem so "logical" to adults. But the difference in logic does not stop here. The research just reviewed suggests that the failure to understand conservation does not interfere with children's performance on mathematical tasks which, from an adult perspective, seem to depend upon this ability. Nonconservers successfully complete tasks for which conservation seems to be a logical prerequisite. It is clear that children think differently than adults.

Many researchers have pointed out the differences between children and adults, but it was Piaget who most clearly and profoundly demonstrated the nature of these differences. The success of Piaget's work can be attributed in part to his method of research. Piaget observed children as they were solving tasks, questioned them about the reasons for their responses, and tried to understand how they were thinking about the problems. Piaget's work, together with most of the research reviewed in this chapter, makes it clear that children's thinking can only be described by observing children as they are solving tasks. It is difficult, if not impossible, to describe or predict children's thinking by carrying out a rational, adult analysis of the task to be solved. The individual interview used by Piaget (Opper, 1977) suggests itself as a productive way to find out how well children understand basic mathematical concepts and to identify the processes they use to solve mathematical tasks.

The importance of this for instruction is that classroom teachers could apply these individual interview techniques with great benefit. Rather than administering only written tasks, teachers could schedule brief interviews with individual children and observe their performances on a few well-chosen problems. Questions can be asked in an accepting, nonevaluative way to uncover the processes used to solve problems. After identifying these processes, teachers can provide feedback on the appropriateness of solution strategies as well as the correctness of responses. Often children's existing strategies can be modified and built upon to create meaningful, appropriate strategies. Case (1978a) has described the importance of demonstrating to children any inadequacies of their current strategies and guiding them in acquiring more appropriate and efficient ones. In addition, information on children's processes provides teachers with a more fundamental understanding of children's errors.

Research has shown that children's errors are often the result of basic misconceptions rather than random carelessness. The nature of these misconceptions are difficult to diagnose by studying the responses on a paper-and-pencil test. Knowledge of the processes children use provides a deeper level diagnosis of their errors and provides a sound basis from which to prescribe appropriate instructional activities (Romberg, 1977).

Developing Meaning for Mathematical Concepts and Symbols

In 1949, Van Engen pointed out the importance of relating the meaning of real world experiences with the arithmetic concepts and symbols that represent those actions or events. Instruction that emphasizes building these relationships is needed just as much today as it was three decades ago. The only difference is that a little more is known today about the informal knowledge upon which these relationships must be based, and the types of errors which result when the relationships are not built successfully. The implications described below emerge from this recent research base, but their intention could still be summarized by Van Engen's (1949) now-classic call for meaningful arithmetic instruction.

Initial Instruction on Arithmetic Operations. The research on early number and arithmetic concepts indicates that counting processes are critical in children's initial learning. Although all children enter school with some counting skills, not all of them are aware of the more advanced and efficient forms of counting. Rather than having children abandon their natural counting strategies, which have meaning for them, it may be beneficial to include material at the first-grade level directed toward improving these strategies. Children might be shown how to count forward from a given number, count back, and use various heuristic strategies to solve addition and subtraction problems. Two kinds of heuristic strategies that good counters seem to acquire naturally are those which use 10 as an intermediate number fact and those which use doubles (Carpenter and others, 1981a). For example, to solve the problem represented by $6 + 7 = \square$, some children will reason "6 + 4 is 10, and 3 more is 13"; other children will say "6 + 6 is 12, so 6 + 7 is 13." These strategies suggest themselves as likely candidates for instructional content on counting.

An alternative to direct instruction on specific strategies is to provide opportunities for children to develop their own solution processes. Resnick (1980) proposes that instruction should be designed to put learners in the best position to invent or discover appropriate strategies for themselves. There is some evidence that even young children can invent strategies that are more sophisticated than those being taught if they understand the problems and are provided with appropriate aids for solving them (Groen and Resnick, 1977). Therefore, it appears that children may benefit from instructional methods that provide opportunities for them to develop and apply a variety of solution strategies. Strategies that children invent are likely to be strategies that have meaning for them.

Regardless of which approach is used for helping children develop efficient processes for solving addition and subtraction problems, it appears that verbal problems may be a good context in which to introduce these operations. It is frequently assumed that children must first master computational skills before they can apply them to solve problems. However, children develop a variety of counting strategies for solving verbal arithmetic problems before they receive instruction. This suggests that, rather than depending on prior knowledge of computation skills, these problem situations may give meaning to the basic arithmetic operations. In fact, verbal problems may be the most appropriate context in which to introduce addition and subtraction operations. Verbal problems also provide for different interpretations of addition and subtraction, interpretations which children bring with them to the school setting, and which must be eventually integrated for a full understanding of the basic operations.

Initial Instruction on Symbolic Representations. While many first-grade children are quite proficient at solving verbal problems, they often experience substantial difficulty dealing with symbolic expressions. It may be better to postpone formal symbolization until children have had a wide range of experiences with verbal problems and concrete arithmetic situations. There is some evidence which suggests that young children are able to use informal symbol systems (for example, tally marks, pictorial representations) to help them solve problems (Allardice, 1977; Kennedy, 1977). Perhaps these kinds of symbols would provide a more meaningful transition than now exists between children's intuitive understandings of the operations and arithmetic number sentences like $5 - 3 = \square$.

It is clear that when children first encounter the formal symbols of arithmetic, they have a difficult time developing meaning for the symbolic expressions. The emphasis during this instructional period should be on establishing and maintaining a link between the concepts children have already acquired and the symbols that are being introduced. Because of children's well-developed informal knowledge of verbal problem situations, these may provide the best context in which to introduce arithmetic symbols. The meaning children associate with verbal problems could be related to the number sentences that most directly represent the problem situation. To be successful, teachers will need to do more than simply present a related verbal problem alongside the number sentence (Grouws, 1972). Children need a variety of experiences in writing number sentences that represent verbal problems, and writing verbal problems which give meaning to number sentences.

Concrete materials can be used to represent arithmetic concepts and symbols physically. However, to help children see the connection between the physical representation and the symbolic representation, teachers need to structure the concrete activities so that frequent links are made between the physical and symbolic representations. For example, when using base ten blocks in the addition algorithm, symbols should be recorded immediately after the objects have been manipulated in each column (Bell and others, 1976; Merseth, 1978); otherwise the concrete procedure functions as a calculating device which provides the correct answer but which does not facilitate a better understanding of the symbolic process. It is not just the use of concrete materials that improves mathematical understanding, but the explicit construction of a link between meaningful actions on the objects and the related symbol procedures.

In conclusion, the study of children's thinking provides some valuable insights into the processes children use to deal with mathematical situations, and generates some important suggestions for improving instruction. While the focus of this chapter has been on young children's thinking, the implications derived from research are equally appropriate for older children. Understanding what students are thinking as they solve long-division problems, or add two fractions with unlike denominators, is essential for developing instructional activities to help correct students' difficulties. Individual interviews, in which the student is asked to solve a few key tasks, and the teacher asks questions in a noninstructive, nonevaluative manner to clarify the solution procedures, could be employed to uncover the student's processes. Developing meaning for symbols is another instructional task that is equally important at all levels of mathematics learning. For example, teachers might help other children understand the difficult notion that two different fractional symbols can be equivalent (such as $1/2$ and $2/4$) by tying the symbols directly to their concrete and pictorial representations (Ellerbruch and Payne, 1978). Listening to students, and helping them to connect their meaningful base of experience to the symbols of mathematics are critical instructional strategies that can be used with benefit throughout the mathematics curriculum.

References

Adi, H. "Intellectual Development and Reversibility of Thought in Equation Solving." *Journal for Research in Mathematics Education* 9 (1978): 204-213.

Allardice, B. "The Development of Written Representations for Some Mathematical Concepts." *Journal of Children's Mathematical Behavior* 1 (1977): 135-148.

Bell, M. S.; Fuson, K. C.; and Lesh, R. A. *Algebraic and Arithmetic Structures: A Concrete Approach for Elementary School Teachers.* New York: Free Press, 1976.

Bradbard, D. A. "The Concept of Length Measurement in Young Children." Paper presented at the 56th annual meeting of the National Council of Teachers of Mathematics, San Diego, April 1978.

Brainerd, C. J. *The Origins of the Number Concept.* New York: Praeger, 1979.

Brown, A. L. "Knowing When, Where, and How to Remember: A Problem of Metacognition." In *Advances in Instructional Psychology.* Edited by R. Glaser. Hillsdale, N.J.: Lawrence Erlbaum Associates, 1978.

Brown, A. L.; Campione, J. C.; and Day, J. D. "Learning to Learn: On Training Students to Learn from Texts." *Educational Researcher* 10 (1981): 14-21.

Brown, J. S., and Burton, R. R. "Diagnostic Models for Procedural Bugs in Basic Mathematical Skills." *Cognitive Science* 2 (1978): 155-192.

Bruner, J. S. *Toward a Theory of Instruction.* New York: Norton, 1966.

Brush, L. R. "Preschool Children's Knowledge of Addition and Subtraction." *Journal for Research in Mathematics Education* 9 (1978): 44-54.

Campione, J. C., and Brown, A. L. "Toward a Theory of Intelligence: Contributions from Research with Retarded Children." *Intelligence* 2 (1979): 279-304.

Carpenter, T. P. "The Performance of First- and Second-Grade Children on Liquid Conservation and Measurement Problems Employing Equivalence and Order Relations." In *Research on Mathematical Thinking of Young Children.* Edited by L. P. Steffe. Reston, Va.: National Council of Teachers of Mathematics, 1975.

Carpenter, T. P. "Analysis and Synthesis of Existing Research on Measurement." In *Number and Measurement*. Edited by R. A. Lesh. Columbus, Ohio: ERIC/ SMEAC, 1976.

Carpenter, T. P. "Research on Children's Thinking and the Design of Mathematics Instruction." In *Applied Mathematical Problem Solving*. Edited by R. A. Lesh, D. Mierkiewicz, and M. G. Kantowski. Columbus, Ohio: ERIC/SMEAC, 1979.

Carpenter, T. P. "Research in Cognitive Development." In Research in *Mathematics Instruction*. Edited by R. J. Shumway. Reston, Va.; National Council of Teachers of Mathematics, 1980.

Carpenter, T. P., and Moser, J. M. "The Development of Addition and Subtraction Problem Solving Skills." Paper presented at the Wingspread Conference on the Initial Learning of Addition and Subtraction Skills, Racine, Wis., November 1979.

Carpenter, T. P.; Hiebert, J.; and Moser, J. M. "The Effect of Problem Structure on First-Grade Children's Initial Solution Processes for Simple Addition and Subtraction Problems." *Journal for Research in Mathematics Education* 12 (1981a): 27-39.

Carpenter, T. P.; Moser, J. M.; and Hiebert, J. The *Effect of Instruction on First-Grade Children's Solutions of Basic Addition and Subtraction Problems* (Working Paper No. 304). Madison, Wis.: Wisconsin Research and Development Center for Individualized Schooling, 1981b.

Case, R. "Mental Strategies, Mental Capacity, and Instruction: A Neo-Piagetian Investigation." *Journal of Experimental Child Psychology* 18 (1974): 383-397.

Case, R. "Gearing the Demands of Instruction to the Developmental Capacities of the Learner." *Review of Educational Research* 45 (1975): 59-87.

Case, R. "A Developmentally Based Theory and Technology of Instruction." *Review of Educational Research* 48 (1978a): 439-463.

Case, R. "Intellectual Development from Birth to Adulthood: A Neo-Piagetian Interpretation." In *Children's Thinking: What Develops?* Edited by R. S. Siegler. Hillsdale, N.J.: Lawrence Erlbaum Associates, 1978b.

Davis, R., and McKnight, C. "Modeling the Processes of Mathematical Thinking." *Journal of Children's Mathematical Behavior* 2 (1979): 91-113.

Elkind, D. *Child Development and Education: A Piagetian Perspective*. New York: Oxford University Press, 1976.

Ellerbruch, L. W., and Payne, J. N. "A Teaching Sequence from Initial Fraction Concepts Through the Addition of Unlike Fractions." In *Developing Computational Skills: 1978 Yearbook*. Edited by M. N. Suydam. Reston, Va.: National Council of Teachers of Mathematics, 1978.

Erlwanger, S. H. "Case Studies of Children's Conceptions of Mathematics— Part I." *Journal of Children's Mathematical Behavior* 1 (1975): 157-183.

Fuson, K. C. "Counting Solution Procedures in Addition and Subtraction." Paper presented at the Wingspread Conference on the Initial Learning of Addition and Subtraction Skills, Racine, Wis., November 1979.

Fuson, K. C. "The Developing of Counting Words and of the Counting Act." Paper presented at the Fourth International Congress on Mathematical Education, Berkeley, Calif., August 1980.

Fuson, K. C., and Mierkiewicz, D. B. "A Detailed Analysis of the Act of Counting." Paper presented at the annual meeting of the American Educational Research Association, Boston, April 1980.

Gelman, R. "How Young Children Reason About Small Numbers." In *Cognitive Theory (Vol. 2)*. Edited by N. J. Castellan, D. P. Pisoni, and G. R. Potts. Hillsdale, N.J.: Lawrence Erlbaum Associates, 1977.

Gelman, R. "Cognitive Development." *Annual Review of Psychology* 29 (1978): 297-332.

Gelman, R., and Gallistel, C. R. *The Child's Understanding of Number*. Cambridge, Mass.: Harvard University Press, 1978.

Gelman, R., and Starkey, P. "Development of Addition and Subtraction Abilities Prior to Formal Schooling in Arithmetic." Paper presented at the Wingspread Conference on the Initial Learning of Addition and Subtraction Skills, Racine, Wis., November 1979.

Gibb, G. E. "Children's Thinking in the Process of Subtraction." *Journal of Experimental Education* 25 (1956): 71-80.

Ginsburg, H. *Children's Arithmetic: The Learning Process.* New York: Van Nostrand, 1977a.

Ginsburg, H. "The Psychology of Arithmetic Thinking." *Journal of Children's Mathematical Behavior* 1 (1977b): 1-89.

Groen, G., and Resnick, L. B. "Can Preschool Children Invent Addition Algorithms?" *Journal of Educational Psychology* 69 (1977): 645-652.

Grouws, D. A. "Open Sentences: Some Instructional Considerations from Research." *Arithmetic Teacher* 19 (1972): 595-599.

Hatano, G., and Ito, Y. "Development of Length Measuring Behavior." *Japanese Journal of Psychology* 36 (1965): 184-196.

Hiebert, J. "Cognitive Development and Learning Linear Measurement." *Journal for Research in Mathematics Education* 12 (1981): 197-211.

Hiebert, J.; Carpenter, T. P.; and Moser, J. M. *Cognitive Development and Performance on Verbal Addition and Subtraction Problems* (Technical Report No. 560). Madison, Wis.: Wisconsin Research and Development Center for Individualized Schooling, 1980.

Heller, J. I., and Greeno, J. G. "Semantic Processing in Arithmetic Word Problem Solving." Paper presented at the annual meeting of the Midwestern Psychological Association, Chicago, May 1979.

Kennedy, M. L. "Young Children's Use of Written Symbolism to Solve Simple Verbal Addition and Subtraction Problems." *Journal of Children's Mathematical Behavior* 1 (1977): 122-134.

Lankford, F. G. *Some Computational Strategies of Seventh Grade Pupils* (U.S.O.E. No. 2-c-013). Charlottesville, Va.: University of Virginia, 1972.

LeBlanc, J. F. *The Performance of First Grade Children in Four Levels of Conservation of Numerousness and Three I.Q. Groups When Solving Arithmetic Subtraction Problems* (Tech. Rep. No. 171). Madison, Wis.: Wisconsin Research and Development Center for Cognitive Learning, 1971.

Lesh, R. A., ed. *Cognitive Psychology and the Mathematics Laboratory: Papers From a Symposium.* Columbus, Ohio: ERIC/SMEAC, 1973.

Lesh, R., and Mierkiewicz, D., eds. *Recent Research Concerning the Development of Spatial and Geometric Concepts.* Columbus, Ohio: ERIC/SMEAC, 1978.

Lesh, R. A.; Mierkiewicz, D.; and Kantowski, M. G., eds. *Applied Mathematical Problem Solving.* Columbus, Ohio: ERIC/SMEAC, 1979.

Lindvall, C. M., and Ibarra, C. G. "Incorrect Procedures Used by Primary Grade Pupils in Solving Open Addition and Subtraction Sentences." *Journal for Research in Mathematics Education* 11 (1980): 50-62.

Merseth, K. K. "Using Materials and Activities in Teaching Addition and Subtraction Concepts." In *Developing Computational Skills: 1978 Yearbook.* Edited by M. N. Suydam. Reston, Va.: National Council of Teachers of Mathematics, 1978.

Mpiangu, B. D., and Gentile, R. J. "Is Conservation of Number a Necessary Condition for Mathematical Understanding?" *Journal for Research in Mathematics Education* 6 (1975): 179-192.

Opper, S. "Piaget's Clinical Method." *Journal of Children's Mathematical Behavior* 1 (1977): 90-107.

Piaget, J. *The Child's Conception of Number.* London: Routledge & Kegan Paul, 1952.

Piaget, J.; Inhelder, B.; and Szeminska, A. *The Child's Conception of Geometry.* New York: Basic Books, 1960.

Resnick, L. B. "The Role of Invention in the Development of Mathematical Competence." In *Developmental Models of Thinking*. Edited by R. H. Kluwe and H. Spada. New York: Academic Press, 1980.

Riley, M. S., and Greeno, J. G. "Importance of Semantic Structure in the Difficulty of Arithmetic Word Problems." Paper presented at the annual meeting of the Midwestern Psychological Association, Chicago, May 1978.

Rohwer, W. D. "Cognitive Development and Education." In *Carmichael's Manual of Child Psychology (Vol. 1)*. Edited by P. Mussen. New York: Wiley, 1970.

Romberg, T. A. "Second-Level Diagnosis: Addressing Underlying Processes, Not Just Objectives." Paper presented at the Fourth National Conference on Diagnostic and Prescriptive Mathematics, University of Maryland, April 1977.

Romberg, T. A., and Collis, K. F. "Cognitive Level and Performance on Addition and Subtraction Problems." Paper presented at the annual meeting of the International Group for the Psychology of Mathematical Education, Berkeley, August 1980a.

Romberg, T. A., and Collis, K. F. *The Assessment of Children's M-Space* (Tech. Rep. No. 540). Madison, Wis.: Wisconsin Research and Development Center for Individualized Schooling, 1980b.

Shumway, R. J., ed. *Research in Mathematics Education*. Reston, Va.: National Council of Teachers of Mathematics, 1980.

Sohns, M. L. "A Comparison Between Certain Piagetian Logical Thinking Tasks and the Subtraction Ability of First, Second, and Third Grade Children." In *Proceedings of the Fourth Interdisciplinary Seminar: Piagetian Theory and Its Implications for the Helping Profession*. Edited by G. I. Lubin, J. F. Magary, and M. K. Poulsen. Los Angeles: University of Southern California, 1974.

Steffe, L. P. "Differential Performance of First-Grade Children When Solving Arithmetic Addition Problems." *Journal for Research in Mathematics Education* 1 (1970): 144-161.

Steffe, L. P., and Johnson, D. C. "Problem Solving Performances of First-Grade Children." *Journal for Research in Mathematics Education* 2 (1971): 50-64.

Steffe, L. P.; Spikes, W. C.; and Hirstein, J. J. *Summary of Quantitative Comparisons and Class Inclusion as Readiness Variables for Learning First Grade Arithmetical Content*. Athens, Ga.: The Georgia Center for the Study of Learning and Teaching Mathematics, 1976.

Steffe, L. P., and Thompson, P. W. "Children's Counting in Arithmetical Problem Solving." Paper presented at the Wingspread Conference on the Initial Learning of Addition and Subtraction Skills, Racine, Wisconsin, November 1979.

Van Engen, H. "An Analysis of Meaning in Arithmetic." *Elementary School Journal* 49 (1949): 321-329, 395-400.

Vergnaud, G. "The Acquisition of Arithmetical Concepts." *Educational Studies in Mathematics* 10 (1979): 263-274.

Wagman, H. G. "The Child's Conception of Area Measure." In *Children's Mathematical Concepts: Six Piagetian Studies in Mathematics Education*. Edited by M. F. Rosskopf. New York: Teachers College Press, 1975.

Wohlwill, J. F. "The Place of Structured Experience in Early Cognitive Development." *Interchange* 1 (1970): 13-27.

Response

Karen C. Fuson

The Hiebert chapter has identified and described a number of areas of current research on children's thinking. The short space available here permits the discussion of only two of the issues raised in the paper: first, a brief comment on the complex relationship between understanding and skill, and, second, some additional current research findings on ways in which the thinking of young children differs from that of adults and some suggestions to teachers about how to deal with these differences.

Gelman's proposal that preschool children come to understand counting principles and then gradually eliminate their execution errors in counting raises an old educational debate: whether understanding or rote performance comes first. In counting, as probably in other areas, the resolution of this debate is that each comes first in different aspects of counting. For example, the combined research findings of Gelman and Gallistel (1978), Fuson and Mierkiewicz (1980), and Mierkiewicz and Siegler (1980) suggest that children recognize that skipping an object in counting is an error before they are able to eliminate all of

their own object skipping (that is, understanding precedes correct performance). On the other hand, while they rarely say an extra word when pointing to an object, many children do not think that doing so is a counting error (correct performance precedes understanding). Thus, there may be some things that are easier for children to understand than to do and others that are easier simply to do than to understand.

One of the differences between the thinking of children in kindergarten through second (and even perhaps third) grade and that of adults is that children are not nearly as capable as adults are of planning and organizing their activities. That is, there are not only cognitive differences between children and adults; there are also meta-cognitive ones. Children's actions are much more dictated by things or people perceptually present in their immediate environment than by specific goals acting over a relatively long period of time. Children of this age are capable of goal-directed activity, but they also tend to spend much of their time reacting to immediate stimuli in their environment rather than enacting pre-set plans.

Thus, two major functions of a teacher of young children are (1) to provide activities with goals of relatively short duration so that children can maintain their activity in a goal-directed way, and (2) to construct a learning environment containing stimuli to which children can react. The latter suggests the use of concrete objects and concrete situations. Such use has long been proposed for cognitive reasons: Piaget's legacy has been to help us realize not only how concretely bound children's thinking is, but also what capable thinkers young children can be when they are provided with the concrete tools they require for thinking. However, another benefit of the use of concrete objects and situations with children is now evident: a meta-cognitive one. Such perceptually present stimuli in the immediate environment will serve to organize children's behavior and keep it focused on the desired stimuli for learning.

The theoretical work of Vygotsky (1962, 1978; Fuson, 1980) has focused renewed interest on adult-child (or teacher-learner) interaction in learning. This interaction in the learning process can be viewed as one in which the adult and the child engage in a common goal-directed activity with the adult at first carrying out many of the parts of this activity. Gradually the child learns to do these parts and the adult becomes less active, limited to organizing, monitoring, and sup-porting functions. Finally even these functions are eliminated, and the child takes over the whole cycle of behavior. The Soviets call this the movement from the inter-psychological to the intra-psychological plane.

I have outlined and discussed an application of this notion to the mathematics classroom (Fuson, 1979). One of the purposes of such a model is to create a meta-cognitive change in the teacher: to help the teacher step back from the busy acting and reacting that occurs from minute to minute in the classroom and to reflect on these meta-processes inherent in the teaching-learning process. In particular, the goal is to help the teacher think about how to help move children from being nonplanning reactors to actors actively involved in reflecting on and organizing their own learning. Young children do not spontaneously *reflect upon* their actions and thoughts. They can do so, however, with the help or suggestion of the teacher. Repeated help and suggestion will then serve to increase spontaneous reflection.

One way to accomplish this transfer of organization is by using verbalization. Procedures can be verbalized by the teacher; students can then use these verbalizations to regulate their own behavior when they are carrying out the same procedure. This occurs spontaneously in the classroom now: one sees lips moving as a child verbalizes his or her way through an addition solution.

("Write down the four and carry over the one to the tens place.") However, such verbalizations could be used deliberately by teachers.

Children's thinking also differs from that of adults because there is a great deal of information children simply do not know. Thus, one of the primary and obvious functions of a teacher is to help children learn some of the many facts about their world that can then also enhance their reasoning processes.

A related teaching function that is less obvious is to help children recall relevant information they may possess but may not recognize is relevant. Children's knowledge is initially quite context-bound: what they learn is related to the specific context in which they learned it. This is further complicated by the fact that children often notice, encode, and remember surface features of a problem situation rather than the underlying structural dimensions that would be processed by older children and adults. For example, the Soviet researcher Krutetskii (1976) discovered that when children who were very good at mathematics were asked about some story problems they had solved on an earlier occasion, they had encoded and remembered the operation involved in the story. (For example, "Oh, there was one about a boy having some things and another boy giving him some more of them.") Children who were not so good at mathematics had encoded and remembered mathemati-

cally irrelevant features of the stories. ("There was one about trains.")

This difference in the encoding of a situation has now also been found by American researchers, who consider it to be one of the major differences distinguishing experts from novices in a field. Expert chess players, physicists, and so forth notice underlying structural features of situations that novices do not notice. Because primary school children tend to be novices at practically everything, teachers should keep in mind the fact that children are likely to focus on attributes of a situation different from those noted by the expert teacher.

A very important part of the teacher's role, then, is to help children begin to notice structurally important features of a situation (such as numbers of objects) rather than more obvious surface features (such as the color of the objects). Similarly, teachers will explicitly need to point out (or help children discover for themselves) how certain contexts are similar to other contexts the children have experienced, but which they may not recognize are "the same" in some important way. (For instance, 5 + 2 yields the same result as 2 + 5, though the person who started with 2 and got 5 may be much happier than the one who started with 5 and got 2.)

A final difference between the thinking of young school children and adults is that children do not have the same mental operating capacity that adults have, and they cannot assess information as quickly as adults can. For example, in a task in which one is supposed to count the number of objects on a card while remembering the number of objects on previously counted cards, four-year-olds cannot remember the previous count, six-year-olds can remember the count of two cards, and adults can remember the counts of four cards (Case and others, 1979). This means that teachers have to figure out ways to help children use external memory (things written on paper, for instance) rather than overload their mental processes. One example of such an overload involves regrouping in the usual addition algorithm. In the problem 47 + 38, after the 1 is carried over to the tens column, the child must do two mental addition problems, holding the sum for the first one in mind as an addend for the second (1 plus 4 is five, five plus 3 is 8). An alternative procedure (Ames, 1975) using external memory is to add in the 1 to the 4, crossing out the 4 and writing 5 above it. The child then is presented with only the second of the two addition problems (5 plus 3) and does not have to remember anything.

The current overemphasis by researchers on counting to the exclusion of other sources of arithmetic ideas should not be emulated by teachers. The use of measure notions embodied by Cuisenaire rods (seven is a certain length) and of certain figural patterns that can be combined and separated to form sums and differences are also very important, especially for certain children. Providing a rich range of mathematical experiences from which children can choose those most consistent with their particular pattern of thinking is critical.[1] The mathematical world models many kinds of external realities, and the paths to understanding this world are themselves many and varied.

[1] See, for example, Baratta-Lorton (1976); the Nuffield project books for early grades; and Bell, Fuson, and Lesh (1976) for both counting and measure materials for the later grades.

References

Ames, P. Personal communication, April 1975.

Baratta-Lorton, M. *Mathematics Their Way.* Menlo Park: Addison-Wesley, 1976.

Bell, M. S.; Fuson, K. C.; and Lesh, R. A. *Algebraic and Arithmetic Structures.* New York: The Free Press, 1976.

Case, R.; Kurland, M.; and Daneman, M. "Operational Efficiency and the Growth of M-Space." Paper presented at the Biennial Meeting of the Society for Research in Child Development, San Francisco, March 1979.

Fuson, K. C. "Towards a Model for the Teaching of Mathematics as Goal-Directed Activity." In *Explorations in the Modeling of the Learning of Mathematics.* Edited by K. Fuson and W. Geeslin. Columbus, Ohio: ERIC Center, 1979.

Fuson, K. C. "An Explication of Three Theoretical Constructs from Vygotsky." In *Recent Research on Number Learning.* Edited by T. Kieren. A monograph from the Georgia Center for the Study of Learning and Teaching Mathematics. Columbus, Ohio: ERIC/SMEAC, 1980.

Fuson, K. C., and Mierkiewicz, D. "A Detailed Analysis of the Act of Counting." Paper presented at the annual meeting of the American Educational Research Association, Boston, April 1980.

Gelman, R., and Gallistel, C. R. *The Child's Understanding of Number.* Cambridge, Mass.: Harvard University Press, 1978.

Krutetskii, V. A. *The Psychology of Mathematical Abilities in Schoolchildren.* Edited by J. Kilpatrick and I. Wirszup. Translated by J. Teller. Chicago: The University of Chicago Press, 1976.

Mierkiewicz, D., and Siegler, R. S. "Preschoolers' Abilities to Recognize Counting Errors." Paper presented at the Proceedings of the Fourth International Conference for the Psychology of Mathematics Education, Berkeley, August 1980.

Nuffield Mathematics Project Books. New York: John Wiley & Sons, 1967-71.

Vygotsky, L. S. *Thought and Language.* Translated and Edited by E. Hanfmann and G. Vakar. Cambridge, Mass.: MIT Press, 1962.

Vygotsky, L. S. *Mind in Society.* Edited by M. Cole, V. John-Steiner, S. Scribner, and E. Souberman. Cambridge, Mass.: Harvard University Press, 1978.

IV. Teachers' Decision Making

Thomas J. Cooney

Teachers are sometimes likened to actors on stage: they emote and enthuse in order to capture the imagination of the audience. While teachers may need to be good actors, the process of teaching also includes *re*acting. Teaching is an interactive process, one in which the teacher plays off the students and the students play off the teacher. It is a process of gathering information, making a diagnosis, and constructing a response based on that diagnosis. While much of this process may be quite automatic, some situations require conscious decision making. The act of generating and considering alternatives in constructing a response—that is, making an instructional decision—is of paramount importance in teaching.

Shroyer (1978) used the term "critical moments" to denote those moments of classroom teaching when there is an occlusion in the instructional flow. Perhaps a student demonstrates an unanticipated learning problem or gives a particularly insightful response. Such unexpected events cause the teacher to reflect on the interaction and to process certain information in order to construct a reaction. Episodes that depict critical moments are presented later in this chapter to provide a context for considering teaching as a process of decision making.

The Decision-Making Process

Various researchers have studied teachers' decision-making processes (Shavelson, 1976; Peterson and Clark, 1978). Regardless of the theoretical

The author would like to express his appreciation to Dr. Stephen I. Brown of the University of Buffalo for his helpful comments in writing this chapter.

prism through which the processes are viewed and studied, several aspects remain constant: teachers gather and encode information, generate alternatives, and select a course of action.

In Peterson and Clark's study (1978), a scheme consisting of four paths was developed for describing teachers' decision-making processes. The investigators found that Path 1 was most frequently traversed; Part 4 was the second most traversed; student achievement was negatively correlated with Path 3; and Path 4 was positively related to higher learning outcomes. Peterson and Clark's study emphasizes two aspects of teaching central to decision making: (1) the decision-making process is related to educational outcomes, and (2) a critical part of the decision-making process is the generation of alternatives. The generation of alternatives is considered central to viewing the teacher as a decision maker and is deemed essential for a flexible and creative teacher. Peterson and Clark's analysis helps provide a means by which we can consider the role alternatives play in the decision-making process.

Types of Decisions

Teachers make different types of decisions. Some are related to the content, including its selection, and the selection of teaching methods. Other decisions relate to the more interpersonal aspects of teaching, that is, affective concerns. Still other decisions involve management considerations, including the allocation of time. I will use this triadic scheme of classifying decisions as cognitive, affective, or managerial to focus on the various types of decisions that teachers make. I must emphasize, however, that these three categories are not in any way mutually exclusive. Teaching is too complex to permit such a simplistic view. In the real world of the classroom, classification schemes are seldom clearly exhibited. Nevertheless, the classification seems appropriate at least for the purpose of examining factors that influence decisions.

Cognitive Decisions

There are two phases of teaching. The *preactive phase* is what transpires before the teacher begins interacting with students. It typically involves lesson planning. The *interactive phase* involves the classroom interaction between students and the teacher. Content related dcisions, as well as other types of decisions, are made in both the preactive and interactive phases.

Figure IV-1. Scheme for Analyzing Decision Making by Teachers

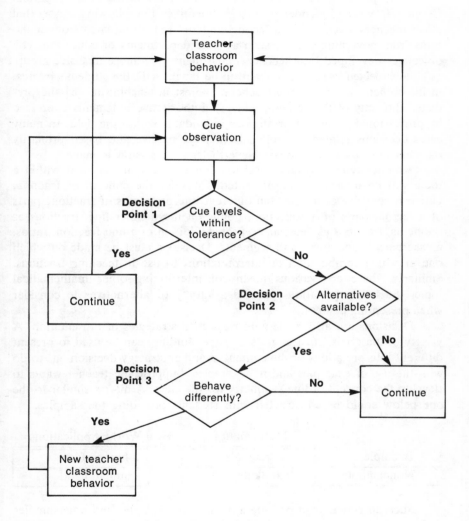

Paths Identified From the Scheme

Decision Points	Path 1	Path 2	Path 3	Path 4
Student Behavior Within				
Tolerance?	Yes	No	No	No
Alternatives Available?	—	No	Yes	Yes
Behave Differently?	—	—	No	Yes

A content decision that occurs in the preactive phase is deciding which content to present and which to exclude from the instructional program. Cooney, Davis, and Henderson (1975) identified the following factors that affect teachers' decisions in selecting content: (a) requirements or regulations from governing bodies, such as state departments of education, (b) objectives developed by a teacher, department, or a more inclusive group, (c) the expected use of the content to be taught, (d) the student's interest in the content as well as the teacher's interest in teaching it, (e) the predicted difficulty of the content, and (f) authoritative judgments expressed by professional groups or prestigious individauls within the field. In many cases decisions related to topic selection are passive and based primarily on what appears in textbooks. Nevertheless, a decision is made.

Another type of content decision concerns how the content within a topic will be interpreted or presented. Consider the concept of fraction. One can conceive of at least ten different interpretations of fraction: parts of a region, parts of a collection, points on the number line, fractions as quotients, fractions as decimals, repeated addition of a unit fraction, ratios, measurement, operators, and segments. Decisions must be made on which one or which combination of interpretations to use in teaching fractions. Similarly, there are various means of interpreting other mathematical topics. Such interpretations provide a variety of alternatives to consider when presenting content.

Decisions are also made with respect to strategies of presentation. A variety of materials, such as rods or paper folding, can be used to present different interpretations of the content. Another strategy decision has to do with the use of examples and nonexamples. Suppose the teacher wants to develop the concept of line symmetry for the class. A matrix similar to the one below could be constructed with students providing the samples.

	Mathematics	Real World Applications
Example	rectangle	a human face
Nonexample	parallelogram	a human hand

Such an activity can provide a mixture of examples and nonexamples and relate the concept to life-like situations. Other cognitive decisions include deciding how to justify theorems, what prerequisite knowledge should be reviewed, or whether to use an expository or a discovery approach.

Cognitive decisions are also made in the interactive phase of teaching. Several classroom episodes are posed below to highlight the nature of these types of decisions.

Episode 1

Mr. Smith's class is learning the Pythagorean Theorem. Students had used unit squares to construct larger squares on the legs of right triangles ABC and DEF.

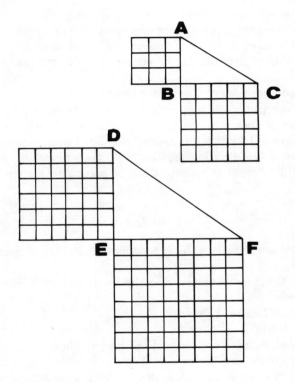

These unit squares were then rearranged to form a larger square of unit squares on the hypotenuse. The following dialogue between the teacher and two students, Billy and Chuck, then transpired.

Teacher: Now consider the right triangle with legs of length a and b and hypotenuse c. *(He draws triangle ABC on the board.)* What does the theorem say about this triangle?

Billy: $a^2 + b^2 = c^2$.

Teacher: Okay. Very good. Now suppose we have a different right triangle with legs of length a and c and hypotenuse b. *(He draws this triangle on the board.)* Now what does the theorem say?

Chuck: The theorem won't work for that triangle. It doesn't apply.

Apparently, Chuck had not grasped the meaning of the theorem. Perhaps he thought of mathematics only in terms of symbols and not in terms of meanings behind the symbols. What alternative actions exist? Possibilities include the following:

    a. Call on another student to state a relationship.

    b. Tell Chuck the theorem does apply and state the correct response.

    c. Ask Chuck to clarify what "doesn't apply" means.

    d. Ask him to state the conditions under which the theorem does or does not apply.

    e. Ask another student if he or she agrees.

The issue is not which alternative is necessarily better for all situations. Rather, the focus should be on the identification of possible alternatives and the decision as to which one seems best suited in a particular context. The making of a wise decision requires the consideration of various alternatives in light of what is known about a particular student in specific situations.

Episode 2

Ms. Jones was reviewing linear functions when the following dialogue occurred.

Teacher: What do we mean, class, by linear function? How would we define it, Mary?

Mary: I don't know. I forgot.

Teacher: Carla?

Carla: Well, it has something to do with a straight line.

Teacher: That's true, but we need more.

Evidently Ms. Jones perceived that students were struggling with the apparent goal of stating a definition. At this point, several alternatives could be considered, including the following:

    a. Call on another student and press for a correct definition.

    b. Provide some sort of a hint on how to "start" the definition and give Carla or another student a chance to state the definition.

    c. Abandon the instructional goal and identify a new goal.

The dialogue continued.

Teacher: Jan?

Jan: Things like $f(x) = 2x + 3$ and $f(x) = 4x - 10$. These are linear functions, aren't they?

Teacher: Yes. That's good. Okay, now let's see how we can graph some linear functions.

The teacher seemed satisfied with the two examples. Was she unclear about the content being taught or at least unclear over the distinction between definitions and examples? Did the teacher make a conscious decision to accept examples rather than a definition? If so, what factors influenced her decision? What was the likely impact of the discussion on the students? Were they confused about what constitutes a definition?

We cannot be sure what cues Ms. Jones attended to when she made her decision to accept the answer, or if she considered any other alternatives. In short, we do not know what information this teacher processed in making the decision. But we do know that for whatever reason a response which was not an answer to the teacher's question was finally accepted. The response may have been accepted as a compromise if Ms. Jones perceived that the task was harder than anticipated (and thus the goal was changed). Or it may have been accepted without Ms. Jones reflecting on the nature of the instructional request.

Consider another situation observed in an elementary school classroom.

Episode 3

Mr. Costa's class was discussing the addition of whole numbers. At one point the discussion focused on a word problem that entailed finding the sum of 1970, 330, and 31. The following dialogue occurred.

Teacher: So what numbers do we need to add?

Sonya: 1970, 330, and 331.

Teacher: Okay. Albert, why don't you show us on the board how to add those numbers? (*Albert goes to the board and writes the following.*)

$$
\begin{array}{r}
1970 \\
330 \\
31 \\
\hline
8370
\end{array}
$$

Albert: The answer is 8370.

Albert's difficulty and misconception are clearly evident. What alternatives exist for the teacher?

    a. Ask another student to come to the board and find the sum.

    b. Show Albert and the class how the numerals should be arranged.

    c. Use the idea of place value to explain briefly how the numbers should be added.

    d. Stop the lesson to review in some detail the process of adding whole numbers.

In this particular case, the teacher decided on option b. The effect seemed to be a continuation of the class discussion in a fairly uninterrupted manner, although an observer might wonder if Albert's confusion had really been resolved.

The following episode highlights the importance of generating alternatives when a lesson goes poorly, and the importance of generating alternative strategies when planning a lesson.

Episode 4

Mrs. Lincoln, a seventh-grade mathematics teacher, was teaching her class how to factor whole numbers into their prime factors. She began by quickly stating the definition of *prime number* and giving two examples of prime numbers. No nonexamples were given. She then presented two demonstrations of how to obtain the prime factorization of a whole number. Students had obvious difficulties, including the above mistakes, as alleged by prime factorizations of the numbers on the left.

$$12 = 4 \times 3$$
$$8 = 5 + 3$$
$$40 = 4 \times 10$$
$$24 = 16 + 8$$

Mrs. Lincoln recognized there was a problem; she repeated the definition, and gave one more demonstration. Students returned to their worksheets, but few corrections were made as they were still quite confused.

Several comments are relevant. First, the students lacked basic prerequisite knowledge with respect to the concepts of prime number and factor. Had the teacher placed greater emphasis on teaching these concepts, particularly through the use of examples and nonexamples of prime numbers and by comparing factors with addends, students would likely have done better.

Second, it seems clear that Mrs. Lincoln had few instructional alternatives to draw on. The role examples and nonexamples can play in designing instructional strategies was mentioned earlier. Kolb (1977) developed a model for predicting the effect of various strategies, including the use of examples and nonexamples for teaching mathematical concepts.

Basically, Kolb's model suggests that examples and nonexamples of concepts produce more learning than presenting characteristics of concepts when students have little prerequisite knowledge. For students with a higher degree of prerequisite knowledge, discussions that focus more on the attributes of a concept, for example, necessary and/or sufficient conditions for concepts, are more effective than focusing on specific examples and nonexamples. The model is complex and involves considerable detail. However, it does highlight the importance of using examples and nonexamples, particularly for students with poor conceptual backgrounds.

In Episode 4, it was clear that many students did not understand the concept of prime number nor of factor. For them an instructional alternative should have been generated which entailed extensive use of examples and nonexamples. Interation of the strategy "define and give one or two examples" was not productive.

Affective Decisions

Teachers need to be sensitive to students and provide ample affective support for them. Instructional decisions involving affective considerations are sometimes based on the teacher's perception of how students are interacting with the content. Comments like "Why do I have to learn this?" are not atypical in mathematics classrooms. The way in which such questions are handled depends on what the teacher perceives to be the reason for such a comment. If the student is asking "How does this content fit with other topics that we have studied or will study?" or "How can the content be applied to help me solve problems in the real world?" then a response dealing with the substance of the discipline is appropriate and, hence, is primarily cognitive in nature. But if the student is really asking "Why am I not doing better in learning this?" then a response oriented toward building the student's confidence appears more appropriate. Thus a teacher is faced with an instructional decision. Within the affective domain in particular, hidden meanings must be attended to as well as the overt context of the remark in order to generate viable alternatives.

Bishop and Whitfield (1972, p. 35) offered the following situation, which suggests the need for an "affective" response:

> If a man can run a mile in four minutes, how far can he run in an hour? A 12-year-old pupil answers: "Fifteen miles." On being questioned about the reasonableness of the answer, he replies: "Well, math is nothing to do with real life, is it?"

Should one expect that a substantive discussion on the relationship or applicability of mathematics to the "real world" would resolve the problem? Perhaps, but it is also conceivable that the student's response has less to do with mathematics per se than it does with an affective problem associated with learning mathematics.

Consider the following two episodes.

Episode 5

Donald has considerable trouble learning mathematics. The current lesson is on solving linear equations of the form $ax + b = c$. Donald is doing rather poorly. The teacher has emphasized that it is important in solving equations to have only one equal sign (=) per line. Donald, along with other students, is sent to the chalkboard to practice solving equations. Donald typically does not do well when performing at the board. The equation to be solved is $2x + 4 = 7$. Donald's solution is:

$$2x + 4 = 7 = 2x + 4 - 4 = 7 - 4 = 2x = 3 = x = 1\frac{1}{2}$$

Episode 6

Pat is a C student in geometry. The class has been studying constructions using a compass and a straightedge. Most students are quite proficient in bisecting a line segment as shown on the left. However, Pat persists in bisecting a segment in the manner indicated on the right.

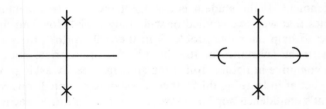

The teacher has continually emphasized to Pat that while her procedure is mathematically correct, it is not the most efficient way and not the method to be used in class. Nevertheless, when asked to find the midpoints of the sides of a triangle, Pat resorts to the second method.

If one were to consider only affective concerns to the exclusion of cognitive ones, then decisions would be easier. But often affective decisions must be tempered with cognitive concerns, as evidenced in Episodes 5 and 6.

In Episode 5, the teacher reinforced Donald with considerable praise for obtaining the correct answer. As a result, Donald felt proud but other students asked if they could solve equations using only "one" line. The teacher seemed intent on emphasizing affective outcomes; desirable affective outcomes were paramount to the teacher. In Pat's case, the teacher was very sharp and critical. Pat probably wouldn't make the same "mistake" again, but at the expense of a loss of enthusiasm for the subject. For this teacher and this situation, cognitive outcomes were evidently of higher priority than affective outcomes.

Many decisions involve striking a balance between cognitive and affective outcomes. Recall Chuck's response concerning the Pythagorean theorem. Chuck had a misconception regarding the theorem. But the teacher might select an alternative action having considerable affective overtones. That is, an alternative might be selected which best ensures Chuck's feelings would not be hurt or best ensures Chuck's continued participation in class discussions. This situation highlights the necessity of considering a number of factors, both cognitive and affective in nature, when making instructional decisions. Artistic teachers are often able to promote both desirable cognitive and affective outcomes. One type of outcome need not be sacrificed for another. But the task of striking a balance is not always easy; it requires careful consideration of several alternatives of action.

Research generally indicates that the warm, supportive teacher is more effective than the critical teacher. Tikunoff and others (1975) conducted an ethnographic study of second- and fifth-grade teachers teaching reading and mathematics in which many teaching variables, affective in nature, were found to be related to achievement. The investigators characterized the significant variables as being related to "those familial interactions in the home which have been attributed traditionally to the successful rearing of children" (p. 22).

Rosenshine and Furst (1971) also suggest that the warm, supportive teacher is more effective than the critical teacher. However, Brophy and Evertson (1976) found that in high socioeconomic status (SES) classrooms praise was negatively related to student learning gains, whereas students in low SES classrooms prospered in warm, supportive classroom atmospheres. This suggests that affective variables may be contextual in nature in terms of how they relate to achievement.

Teachers make continual assessments of students' affective status in the classroom. Although universal quantification is difficult to justify, generally the "familial" variables identified by Tikunoff and others (1975) seem to characterize the effective teacher. But individual instructional decisions may not be unidimensional in value. That is, one may have to strike a balance between cognitive concerns and affective ones when assessing expected payoffs of various teaching behaviors.

Managerial Decisions

Managerial decisions relate to time allocation, organization of classroom activities, and control of disruptive behavior. Some of these decisions can be made in the preactive phase of teaching while others, especially those related to "control" problems, are more specific to the interactive phase of teaching.

Consider Episodes 7 and 8, which involve decisions related to how time is allocated.

Episode 7

A student is subtracting fractions and keeps making mistakes similar to the one below.

$$
\begin{array}{r}
4\ 1/8 = 3\ 11/8 \\
-1\ 7/8 = 1\ \ \ 7/8 \\
\hline
2\ \ \ 4/8 = 2\frac{1}{2}
\end{array}
$$

After the teacher poses several questions, it is clear the student is quite confused.

Episode 8

A geometry teacher is discussing the importance of the parallel postulate in Euclidean geometry. The teacher has emphasized that many theorems in their geometry books are based on the parallel postulate. As an illustration, the teacher argues that the theorem, "The sum of the measures of the angles of a triangle is 180 degrees," follows from the parallel postulate. A bright student asks, "If we didn't have the parallel postulate, does that mean the measures of the angles of a triangle would be different than 180 degrees?"

In Episode 7, should content be reviewed for a single student or for a few students at the risk of "wasting" the time of other students? In Episode 8, should class time be taken to pursue the thought initiated by the bright student? Or should the student be informed that the question was a good one and it would be followed up sometime *after* class? What are the expected results of the two alternatives? The decisions will clearly affect how time is allocated. What is not so clear and is quite value laden is deciding how to strike a balance between discussions of a tangential point for a few students compared with discussions that benefit the remaining students. Given that instructional time is a scarce commodity, allocation of that commodity is critical to determining what is learned.

Some decisions on time allocation occur in the preactive phase of teaching. Ebmeier and Good (1979) found that fourth-grade mathematics teachers could improve achievement by emphasizing six aspects of instruction with tentative time allocations: development (about 20 minutes), homework, emphasis on product questions, seatwork (10 to 15 minutes per day for practice), review/maintenance, and pace (consider the rate of instruction and increase if possible).

Berliner (1978) reported a great deal of variance among teachers in how they allocate time for mathematics instruction, particularly for specific topics, such as fractions, measurement, decimals, or geometry. At the elementary level, the time allocated for mathematics instruction varies considerably from one day to the next because of contextual situations, for instance, if students come back late from a music class or a social studies project takes longer than expected. At the secondary level, the allocated time is more constant, but even within that allocation, a teacher may decide to take care of administrative tasks or attend to other non-mathematical activities. Thus, a decision of one sort or another may significantly affect the amount of time devoted to the study of mathematics.

Another type of decision, which occurs in the interactive phase of teaching, is the decision on how long to wait for students to respond to a question. Rowe (1978) defined two kinds of wait time: (1) the pause following a question by the teacher and (2) the pause following a student's response (usually measured in terms of seconds).

Rowe (1978) found that elementary science teachers typically wait less than one second before commenting on an answer or before asking an additional question. When the two types of wait time were increased, Rowe reported that the length of student responses increased, failures to respond decreased, students' confidence increased, disciplinary problems decreased, slower students participated more and, in general, students were more reflective in their responses.

Consider the likely payoff if wait time of less than one second predominates. Can problem-solving abilities be nurtured and promoted when wait time is consistently less than one second? Not likely. It seems highly desirable for teachers to be explicitly aware of concepts such as "wait time" in order that alternatives can be generated which are consistent with their instructional goals. This is not to claim that awareness of such concepts will yield completely "rational" decisions in the sense that an explicit and highly recognizable decision-making strategy can be readily identified. But it is the belief here that whatever commonsensical decisions are made in the classroom, they can be enhanced by an explicit awareness of alternatives and by having a variety of pedagogical concepts, of which wait time is one, on tap.

Another aspect of managerial decisions involves the ever-present problem of discipline. To deny that teachers are concerned and conscious of potential and actual classroom disruptions is to be oblivious to the realities of classroom teaching. Consider the following episode.

Episode 9

A first-year geometry teacher was discussing the proof of a theorem with the class. In the back of the room a student who was the band's drum major was twirling her baton. After a minute or so, the young teacher noticed her behavior. The teacher's confusion about alternatives was mirrored on his face. Apparently, alternatives did not exist since the teacher avoided the situation. But the impact on the class of the indecisiveness could not be discounted.

Perhaps the response of "do nothing" was the best alternative. But consider an alternative prior to the specific incident. Could the teacher have moved about the room (as was not the actual case) and, as a result, increased his awareness of any potential problems? Did not the decision, determined consciously or unconsciously, to stay in front of the class in a small area inhibit his ability to monitor the student's behavior in the back of the class? Had he decided to move around the room and consciously monitor student behavior, could the embarrassing incident have been avoided? Probably so.

The ability of a teacher to monitor classroom events has been the focus of various investigations. For example, Kounin (1970) studied a number of variables with respect to classroom management and their relationship to achievement. One of the variables identified was called "withitness." This variable dealt with teachers communicating that they know what is going on regarding children's behavior and with their ability to attend to two issues simultaneously. Kounin found withitness to be a strong correlate of achievement. Brophy and Evertson (1976) also found that more successful teachers were more "withit" than less effective teachers. Thus it appears that a teacher's ability to monitor simultaneous classroom events is an important factor in maintaining control and in positively affecting achievement.

There are no explicit directions for solving management problems. But alternatives can be identified for preventing and coping with situations. Perhaps an explicit awareness of possible alternatives can assist teachers in making those difficult decisions and provide greater confidence in themselves for believing they can control classroom events.

Conclusion

Teachers have an immense amount of common sense and good judgment. Many creative teachers have a wealth of alternative methods for dealing with a wide variety of classroom situations. But common sense can be enhanced by an explicit awareness of the importance of generating alternatives and by an explicit knowledge of various pedagogical concepts and principles. Practitioners' maxims and research in concert can play an important role in the generation of alternatives. The art of teaching can be improved by consciously considering alternatives and by expanding the knowledge base for generating alternatives.

Another aspect of improvement can arise from reflecting on why certain alternatives are selected. Value judgments, perceptions about what constitutes the teacher's role, and what constitutes mathematics all provide a sort of filter through which some alternatives pass and others do not. Perhaps a realization of what factors contribute to the selection of alternatives as well as an awareness of the decision-making process itself can provide a basis for several outcomes: additional insights into the teaching process, a richer use of the teacher's knowledge base, and an avenue for teachers' further professional development.

References

Berliner, D. C. "Allocated Time, Engaged Time, and Academic Learning Time in Elementary School Mathematics Instruction." Paper presented at the 56th Annual Meeting of the National Council of Teachers of Mathematics, San Diego, April 1978.

Bishop, A. J., and Whitfield, R. C. *Situations in Teaching*. London: McGraw-Hill Book Company (UK) Limited, 1972.

Brophy, J. E., and Evertson, C. M. *Learning From Teaching: A Developmental Perspective*. Boston: Allyn and Bacon, 1976.

Cooney, T. J.; Davis, E. J.; and Henderson, K. B. *Dynamics of Teaching Secondary School Mathematics*. Boston: Houghton-Mifflin, 1975.

Ebmeier, H., and Good, T. L. "The Effects of Instructing Teachers About Good Teaching on the Mathematics Achievement of Fourth Grade Students." *American Educational Research Journal* 16 (Winter 1979): 1-16.

Kolb, J. R. *A Predictive Model for Teaching Strategies Research. Part I: Derivation of the Model*. Athens, Ga.: The Georgia Center for the Study of Learning and Teaching Mathematics, 1977.

Kounin, J. S. *Discipline and Group Management in Classrooms*. New York: Holt, Rinehart, and Winston, 1970.

Peterson, P. L., and Clark, C. M. "Teachers' Reports of Their Cognitive Processes During Teaching." *American Educational Research Journal* 15 (Fall 1978): 555-565.

Rosenshine, B., and Furst, N. "Research in Teacher Performance Criteria." In *Symposium on Research in Teacher Education*. Edited by B. O. Smith. Englewood Cliffs, N.J.: Prentice-Hall, 1971.

Rowe, M. B. "Wait, Wait, Wait—." *School Science and Mathematics* 78 (March 1978): 207-216.

Shavelson, R. J. "Teachers' Decision Making." In *The Psychology of Teaching Methods (Yearbook of the National Society for the Study of Education)*. Chicago: University of Chicago Press, 1976.

Shroyer, J. C. "Critical Moments in the Teaching of Mathematics." Paper presented at the annual meeting of the American Educational Research Association, Toronto, March 1978.

Tikunoff, W. J.; Berliner, D.C.; and Rist, R. C. *An Ethnographic Study of the Forty Classrooms of the Beginning Teacher Evaluation Study Known Sample* (Tech. Rep. 75-10-5). San Francisco: Far West Laboratory, 1975.

V. Process-Product Research

Thomas L. Good and Douglas A. Grouws

The 1970s produced a considerable amount of substantive research on the teaching of mathematics. Different researchers chose to study the complex processes and interactions associated with mathematics teaching using varied methods. These research approaches may be broadly classified as either qualitative or quantitative. Qualitative approaches include ethnographic studies that draw heavily on the methods typically used by anthropologists and focus on questions concerning what is happening in a classroom and why it is happening. The research on general mathematics teaching by Confrey and Lanier (in press) as well as the research by Easley and his associates (1977) have profitably employed qualitative methodology. Studying mathematics teaching via information-processing approaches also seems to hold potential. Resnick is doing some promising work using this methodology (Resnick and Ford, 1980).

The quantitative approach to studying the teaching of mathematics includes aptitude-treatment interaction studies (Which instructional treatment is best for which student?), process-product studies, and the more traditional treatment-control group type studies (where usually only a few instructional variables are manipulated). The goal of the quantitative studies is to identify what works for specific groups of teachers or specific groups of learners (see, for example, Janicki and Peterson, in press). There is less focus on an individual student or teacher, and more emphasis on groups of learners or teachers.

Both qualitative and quantitative research endeavors are beginning to provide some clear implications for instruction in mathematics. These implications are clear in the sense that they seem logically sound and have also withstood the test of replication or identification in more than one setting. In writing this chapter, we have been asked to focus on the process-product approach to mathematics research and the recent experimental work it has stimulated. This is only a small part of the research being done on the teaching of mathematics, and we are aware of the valuable insights that are being produced by other approaches (see, for example, Davis and McKnight, 1979).

Process-Product Studies

An area that has shown considerable promise in recent years is the study of teacher behaviors. Many of the behaviors that have been associated with effective teaching have been identified from what are commonly called process-product studies. In this type of research, a set of teacher behaviors that seems to hold potential for producing student learning are identified and defined. The frequency and extent of their occurrence are then determined in many classrooms over a fixed period of time. Finally, the correlation between the frequency of occurrence of these teacher behaviors and the average class achievement scores (adjusted for initial differences) during the observation period is computed. A high positive correlation between one of the behaviors and mathematics achievement suggests that effective teachers use this instructional strategy or behavior more often. Replication of the teacher behavior-pupil achievement relationship in subsequent naturalistic studies gives credibility to the finding and suggests the need to examine the variable in field-based experimental studies where cause and effect relationships can be assessed more adequately.

During the past few years a large number of process-product studies of teaching have been conducted (for review, see Brophy, 1979; Good, 1979; Peterson and Walberg, 1979). Several of these studies have specifically examined the teaching of mathematics. In a study of fourth-grade mathematics instruction, Good and Grouws (1977) identified nine effective and nine less effective teachers from a sample of over 100 teachers. Over a three-year period the effective teachers consistently produced better-than-expected mathematics achievement results (residualized gain scores), while the less effective teachers consistently produced lower-than-predicted achievement gains. These differences in outcomes occurred despite the fact that the students taught by relatively effective and ineffective teachers were comparable in ability.

Observational data were collected in 41 classrooms to protect the identity of the relatively effective and ineffective teachers. Approximately equal numbers of observations were made in all classrooms (6-7). Data were collected by two trained observers (both certified teachers) who worked full-time and lived in the target city. Each coder visited all 41 teachers and made about half of the observations obtained in a given classroom. Furthermore, all observations were made without knowledge of the teacher's level of effectiveness.

The data from this study demonstrate an important fact, that patterns of consistent behavior can be identified for both high-effective and less-effective mathematics teachers. Further, there are differences in the patterns for the two groups and these differences suggest behaviors associated with effective and less effective instruction in mathematics, as measured by standardized achievement tests.

Before examining these differential behaviors, two points need to be made. First, many of the teachers in the study did not perform in a consistant fashion. One year they might obtain very good results, and the next year their students might achieve far less than expected. The reasons for these fluctuations are not known and have not been studied. They do suggest, however, that subtle context factors may influence teacher effectiveness (Good, 1979). For example, it may be that if the variability of student ability within a class exceeds a critical point, then the teacher is not successful, even though he or she is behaving and interacting in exactly the ways that had previously produced good results. However, teachers who have inconsistent effects might also vary their behavior from year to year as events in their personal lives allow them more or less time for teaching.

It is highly unlikely that any given behavior in isolation is going to profoundly affect achievement or determine who is an effective teacher. It is far more likely that a number of interrelated behaviors simultaneously (probably under specific conditions) stimulate and enhance student learning in mathematics. Because of the large number of correlations examined in process-product studies, it is also possible that some of the behaviors identified as being associated with effective teaching are not valid. For these reasons it is particularly important in analyzing and interpreting process-product data to look for *clusters* of related behaviors that seem to be associated with effective teaching. With this perspective in mind, let us examine some results.

In the study of fourth-grade mathematics, teacher effectiveness was found to be strongly associated with the following behavioral clusters: (1) general clarity of instruction; (2) task-focused environment; (3) nonevaluative (comparatively little use of praise and criticism) and relatively relaxed learning environment; (4) higher achievement expectations (more homework, faster pace); (5) relatively few behavioral problems; and (6) the class taught as a unit. Teachers who obtained good results were very active (that is, they demonstrated alternative approaches for responding to problems), emphasized the *meaning* of mathematical concepts, and built systematic review procedures into their instructional plans.

In another process-product study focusing on mathematics, Evertson and others (1980) carefully selected a small sample of effective and less effective seventh- and eighth-grade teachers and then systematically observed their teaching. They found that more effective teachers, in contrast to less effective teachers: (1) spent more time on content presentations and discussions and less time on individual seatwork; (2) held higher expectations for their students (assigned homework more frequently, stated concern for academic achievement, and gave academic encouragement more often); and (3) exhibited stronger management skills (minimized inappropriate behavior, made more efficient transitions, and had more student attentiveness).

A number of important relationships exist between the findings of these two studies and earlier research. The conclusion regarding the value of time spent on content presentations coincides very closely with the results of a large number of experimental studies in mathematics (for example, Schuster and Pigge, 1965; Shipp and Deer, 1960; Zahn, 1966; and Dubriel, 1977). These studies have specifically examined the development-practice variable and found without exception that spending more than half of the class period on developing skills and ideas results in higher student achievement. The importance of teacher behaviors that communicate high-achievement expectations in several studies is significant. Similarly, the finding that strong managerial skills and few behavioral problems are positively associated with student achievement seems to be a significant link between the two studies and earlier research (Kounin, 1970).

In a process-product study of the teaching of ninth-grade algebra, Smith (1977) found three interrelated teacher behaviors associated with pupil achievement gains. Smith concluded that these behaviors could probably be associated with a more global variable "involving organization, structuring and clarity of lessons." Here again are findings very similar to those of the studies discussed previously.

Two issues must be kept in mind as generalizations are drawn from process-product studies. First, behaviors identified with effective instruction may not generalize across settings. For example, Evertson and others (1978) found that behaviors that were highly correlated with teacher effectiveness in mathematics were different from teacher behaviors that were associated with the effective teaching of English. No doubt this difference typically is less of a concern when considering teacher behaviors within a subject matter area than when one attempts to generalize across disciplines. Still, the importance of context variables other than subject matter has been illustrated in comparisons of the association between teaching behavior and student achievement in middle-class and working-class schools (Good and others, 1978).

A second caution to be applied when examining specific behaviors associated with effective teaching is that *cause and effect* conclusions should not be stated or implied. Such conclusions are important, but they must be determined by experimental studies.

Experimental Work

A great number of experimental studies have focused on instructional methods in mathematics. Unfortunately, many of them have been isolated studies focusing on only one or two variables, using very small samples. Few have been based on previous process-product research that has comprehensively attempted to determine how more- and less-effective teachers vary in their behaviors.

An exception is the Missouri Mathematics Effectiveness Program (Good and Grouws, 1979a), an experimental study we conducted in fourth-grade classrooms, in which the treatment teachers taught using a system of instruction based on the results of a process-product paradigm and the research work of others. The system of instruction involved the following aspects:

1. Instructional activity was initiated and reviewed in the context of meaning.

2. A substantial portion of each lesson was devoted to content development (the focus was on the teacher actively developing ideas, conveying meaning, giving examples, and so on).

3. Students were prepared for each lesson stage to enhance involvement.

4. The principles of distributed and successful practices were used.

Pre- and post-testing with a standardized achievement test indicated that the performance of students in the experimental group was substantially better than performance of students in control classrooms. End-of-year achievement testing by the school district indicated that experimental classes continued to perform better than control classes three months after the post-testing on the mathematics subtests of a standardized achievement test. Also, experimental students had significantly better attiudes toward mathematics than did control students at the end of the treatment period, as measured on a ten-item attitude scale.

In a follow-up study in sixth-grade classrooms, we experimentally tested a revised and expanded instructional system (Good and Grouws, 1979b). In the fourth-grade study we observed that more learning gains were made in the knowledge and skill areas than in the problem-solving area. Thus the adjusted treatment was designed to improve verbal problem-solving performance without making excessive demands on teachers and without adversely affecting achievement gains in other areas. The additional teaching requests involved teachers daily giving attention to verbal problem solving by using several teaching techniques related to a given problem, such as estimating the answer and writing an open sentence. We also asked teachers to implement the regular instructional program (Good and others, 1977).

The results of the second experimental study showed that the problem-solving performance of students in the treatment group was significantly better than that of students in the control group. However, the achievement gains in other areas (knowledge and skill) were comparable. Although the raw gains of treatment students exceeded those of control students on the general achievement test, these differences were not statistically significant.

These data can be interpreted in two ways. First, it could be that implementing a general instructional program and using explicit strategies for improving verbal problem solving are too much for teachers to do in too short a period of time. An alternative explanation is that control teachers were using many of the strategies called for in the general instructional program. In the previous year teachers in the school district had been exposed to the general instructional program; thus, the information in the general instructional program was not unique, as was the emphasis on verbal problem solving (Grouws and Good, 1978).

Still, in separate field experiments it was possible to affect students' knowledge of mathematics, mathematical skills, and problem-solving abilities. What is less clear is whether or not these three aspects of mathematics instruction can be improved simultaneously or whether better training models would call for teachers to adapt their instruction progressively over time rather than attempting to make comprehensive changes at one point in time. Teachers who are asked to make several complex changes may find the accommodations so major that their instructional system is temporarily disrupted. Much more research is needed along these lines.

We have recently experimented with the instructional model in junior high school settings. Junior high teachers helped modify the program that we used in previous research in elementary schools. Essentially, the instructional program tested in junior high classrooms was very comparable to the one used in the elementary school research. Preliminary analyses suggest that the treatment classrooms out-performed control classrooms, especially in the area of performance on verbal problem solving items. Hence, it seems possible to intervene successfully in both elementary and secondary mathematics classrooms.

Importantly, teachers' reactions to the program have been very positive. This attitude has been reflected in anonymous data that we have collected (Good and Grouws, 1979b) and data that have been produced elsewhere (Keziah, 1980). Apparently, teachers find the requests presented understandable and sufficiently plausible that they are willing to try the program (our implementation data indicate that most teachers who participate in the experiment do use the program as indicated by their classroom behavior), and are willing to continue using the program after the experimental study has been terminated.

Teacher and Student Effects

Peterson (1979) advocates the examination of instructional systems to determine which teaching behaviors best foster the achievement of particular types of students. It also seems reasonable to raise questions about the desirability of a given instructional program for use in changing the particular attitudes and skills of individual teachers. The instructional treatment program that we have been examining in the Missouri Mathematics Effectiveness Project can be described as focusing heavily on *active teaching*. When the effects of the program have been examined in terms of particular student types and particular teacher types (Ebmeier and Good, 1979), it is clear that certain students and certain teachers tend to do better using the treatment than do other combinations of students and teachers. Interestingly, the effects of the program on some of these combinations of student and teacher characteristics have been replicated by Janicki and Peterson (in press). It also seems that the classroom organizational structure (for instance, open-space plans vs. self-contained) also interacts with the treatment program (see, for example, Ebemeier and others, 1980). Without going into a detailed discussion of how student and teacher characteristics interact with the program, it should be noted that in our research context, all experimental groups have done better than all related control groups. However, the magnitude and importance of the differences are more evident for some teacher and student combinations than for others.

It should be evident that there is no single system for presenting mathematics concepts effectively. For example, some of the control teachers in our studies have obtained high levels of student achievement using instructional systems that differ from those presented in the program we have developed. Thus, there are many ways to effectively present mathematics.

However, the instructional program we have developed does seem to be a viable system that teachers are willing to implement. Also, it would seem to be an interesting alternative especially for those teachers who teach in self-contained classrooms and who enjoy an organized approach to instruction. One of the reasons that the program appears to be readily implementable probably is that the teaching strategies were derived from ongoing instructional programs. That is, the program was based upon what relatively effective teachers were already doing in the classroom. Hence, the program appears to have ecological validity and does not demand excessive amounts of teacher time and energy.

Directions For Future Research

Our research approach is only one methodology for attempting to understand, describe, and improve mathematics instruction. One of the chief limitations to this method of studying mathematics is that one can only study teaching practice as it presently exists. Clearly, many exciting ideas for improving mathematics instruction are techniques that have yet to be implemented, and there is much room for creative theorizing about instructional strategies. However, as has been noted elsewhere (Good, 1980), there is considerable variation in teaching behavior in American classrooms. Indeed, one can view the abundant variation in instructional strategies as a rich source of naturally-occurring experiments. The process-product paradigm represents an important research methodology to the extent that teachers' instructional behavior and their effects on students vary in important ways.

At present, most of the process-product research has focused on an examination of teachers who consistently obtain more and less student achievement than do other teachers teaching comparable students. In general, teachers have been selected on the basis of their ability to affect student scores on standardized achievement tests. However, there are many problems with standardized achievement tests—they must be relevant to the instructional goals that teachers are actually pursuing in their classroom instruction. To the extent that this criterion is met, standardized achievement tests represent a reasonable proxy.

It would seem that further use of the process-product paradigm in the study of mathematics instruction should be accompanied by the use of outcome measures other than standardized achievement tests. For example, Confrey (1978) has argued that an important outcome of instruction in mathematics is the conceptual system that students derive from the study of mathematics. It would seem instructive to determine if some teachers consistently help students to develop a more adequate conceptualization of mathematics than do other teachers. Erlwanger (1975) argued that children develop a personal system of beliefs and emotions about mathematics that presumably controls their mathematical behavior in the future. It is important to study how teachers affect students' belief systems, and one potential way to explore this topic is through process-product research. That is, we could attempt to identify teachers who have a distinct impact upon students' beliefs about mathematics. Although the process-product paradigm has been used to explore teaching behavior in terms of its effects on student achievement, we see no reason why the model could not be used to profitably explore students' performance in other areas (problem solving) or other alternative outcomes of mathematics instruction.

We want to emphasize that a process-product approach to the study of mathematics is not the only method appropriate for studying teacher effects, for it has a number of limitations as well as advantages. Our purpose has been to identify some of the useful aspects of this approach and to call for its continued use, along with other methodological strategies, for exploring mathematics instruction. However, if the model is to continue producing positive contributions to theory and research, it would appear necessary to explore other dependent/outcome measures and to integrate the focus on teaching behavior with analyses of student behavior and perceptions (for example, clinical interview strategies). The study of teacher and student variables could be profitably combined with an active examination of the mathematics content being presented. It is likely that different types of instructional and learning strategies would be more or less effective for instruction in particular mathematical concepts or beliefs.

References

Brophy, J. "Teacher Behavior and Its Effects." *Journal of Educational Psychology* 71 (1979): 733-750.

Confrey, J. "Conceptions of Mathematics: An Alternative of Affective and Achievement Outcome Measures." Unpublished, 1978.

Confrey, J., and Lanier, P. "Students' Mathematics Abilities: Improving the Teaching of General Mathematics." *School Science and Mathematics,* in press.

Davis, R., and McKnight, C. *The Conceptualization of Mathematics Learning as a Foundation of Improved Measurement,* (Development Report No. 4.) Urbana: The Curriculum Laboratory, University of Illinois, 1979.

Dubriel, J. "Utilization of Class Time in First Year Algebra." Ph.D. dissertation, University of Missouri, 1977.

Ebmeier, H., and Good, T. "The Effects of Instructing Teachers About Good Teaching on the Mathematic Achievement of Fourth Grade Students." *American Educational Research Journal* 16 (1979): 1-16.

Ebmeier, H.; Good, T.; and Grouws, D. "Comparison of ATI Findings Across Two Large-Scale Experimental Studies in Elementary Education." Paper presented at the American Educational Research Association annual conference, Boston, April 1980.

Easley, J. *On Clinical Studies in Mathematical Education.* Columbus, Ohio: Information Reference Center for Science, Mathematics, and Environmental Education, Ohio State University, 1977.

Erlwanger, S. "Case Studies of Children's Conception of Mathematics—Part I." *Journal of Children's Mathematical Behavior* 1 (1975): 157-283.

Evertson, C.; Anderson, L.; and Brophy, J. *"Texas Junior High School Study: Final Report of Process-Outcome Relationships (Vol. 1, Report No. 4601).* Austin, Texas: Research and Development Center for Teacher Education, University of Texas, 1978.

Evertson, C.; Emmer, E.; and Brophy, J. "Predications of Effective Teaching in Junior High Mathematics Classrooms." *Journal for Research in Mathematics Education* 11 (1980): 167-178.

Good, T. "Teacher Effectiveness in the Elementary School: What We Know About It Now." *Journal of Teacher Education* 30 (1979): 52-64.

Good, T. "Recent Research on Teaching: Some Strategy Guidelines for Conducting Problem Solving Research in Mathematics." Paper presented at the Fourth International Congress on Mathematical Education, Berkeley, August 1980.

Good, T.; Ebmeier, H.; and Beckerman, T. "Teaching Mathematics in High and Low SES Classrooms: An Empirical Comparison." *Journal of Teacher Education* 29 (1978): 85-90.

Good, T., and Grouws, D. "Teaching Effects: A Process-Product Study in Fourth-Grade Mathematics Classrooms." *Journal of Teacher Education* 28 (1977): 49-54.

Good, T., and Grouws, D. "The Missouri Mathematics Effectiveness Project: An Experimental Study in Fourth Grade Classrooms." *Journal of Educational Psychology* 71 (1979a): 355-362.

Good, T., and Grouws, D. "Experimental Study of Mathematics Instruction in Elementary Schools." Final report of the National Institute of Education Grant NIE-G77-003, December 1979(b).

Good, T.; Grouws, D.; Beckerman, T.; Ebmeier, H.; Flatt, L.; and Schneeberger, S. *Teachers Manual: Missouri Mathematics Effectiveness Project* (Technical Report No. 132). Columbia, Mo.: Center for Research in Social Behavior, University of Missouri-Columbia, 1977.

Grouws, D., and Good, T. *Verbal Problem Solving Manual: Missouri Mathematics Effectiveness Project (MMEP)* (Technical Report No. 155). Columbia, Mo.: Center for Research in Social Behavior, University of Missouri-Columbia, 1978.

Janicki, C., and Peterson, P. "Aptitude-Treatment Interaction Effects of Variations in Direct Instruction." *American Educational Research Journal,* XVIII (Spring 1981): 63-81.

Keziah, R. "Implementing Instructional Behaviors that Make a Difference." *Centroid* 6 (1980): 2-4.

Kounin, J. *Discipline and Group Management in Classrooms.* New York: Holt, Rinehart and Winston, 1970.

Peterson, P. "Direct Instruction: Effective For What and For Whom?" *Educational Leadership* 37 (1979): 46-48.

Peterson, P., and Walberg, H., eds. *Research on Teaching: Concepts, Findings and Implications.* Berkeley: McCutchan, 1979.

Resnick, L. B., and Ford, W. W. *The Psychology of Mathematics for Instruction.* Hillsdale, N.J.: Lawrence Erlbaum, 1980.

Shipp, D., and Deer, G. "The Use of Class Time in Arithmetic." *Arithmetic Teacher* 7 (March 1960): 117-121.

Shuster, A., and Pigge, F. "Retention Efficiency of Meaningful Teaching." *Arithmetic Teacher* 12 (January 1965): 24-31.

Smith, L. "Aspects of Teacher Discourse and Student Achievement in Mathematics." *Journal for Research in Mathematics Education* 8 (May 1977): 195-204.

Zahn, K. "Use of Class Time in Eighth Grade." *Arithmetic Teacher* 13 (February 1966): 113-120.

VI. The Sex Factor

Elizabeth Fennema

At the 1978 Annual Meeting of the National Council of Teachers of Mathematics, I concluded a presentation entitled "Sex-Related Differences in Mathematics Achievement: Where and Why" with the following remarks (Fennema, 1978):

> What, then, can be said that is known about sex-related differences in mathematics and factors related to such differences? Certainly when both females and males study the same amount of mathematics, differences in learning mathematics are minimal and perhaps decreasing. Many fewer females elect to study mathematics and therein lies the problem. Factors which appear to contribute to this non-election are females' lesser confidence in learning mathematics and belief that mathematics is a male domain. In addition, differential teacher treatment of males and females is important.
>
> There is nothing inherent which keeps females from learning mathematics at the same level as do males. Intervention programs can and must be designed and implemented within schools which will increase females' participation in mathematics. Such programs must include male students, female students, and their teachers. Only when such intervention programs become effective can true equity in mathematics education be accomplished.

In addition, I had discussed spatial visualization and concluded that it was the only cognitive variable which might be helpful in understanding sex-related differences in mathematics. Most of those statements I still believe. Others I am not so sure about.

Sex-Related Differences in Mathematics

There is a great deal of new information about women and mathematics, plus increasing concern at the action level—the schools. In 1978,

92

the National Institute of Education funded ten major research projects that investigated a number of factors related to the issue. The Women's Educational Equity Act has funded many product development projects, which can be used at all levels of education, to increase females' participation in mathematics. Information about the issue has appeared in a wide variety of publications ranging from the *American Educational Research Journal* to *Chronicle of Higher Education* to *Ms.* magazine. NCTM has had a major task force charged with making recommendations to its Board of Directors. Lead articles about the status of women and mathematics have been in the *Mathematics Teacher* and the *Arithmetic Teacher*. The *Journal for Research in Mathematics Education* has recently published three articles and the editor reports an increasing number of submissions about the topic. Inservice programs designed to increase teachers' awareness are being held nationwide. A major strand at the Fourth International Congress on Mathematics Education was about women and mathematics. The issue is one of the most widely talked about in the mathematics education community since the "New Math." In short, knowledge about the importance of mathematics to females and the inequitable education in mathematics that females have received is easily found. Also available are some intervention programs that have demonstrated effectiveness. What has been the result of this? Is the problem solved? Let's take a look.

The most important place to look to see if change is taking place is in schools themselves. Here exists a major problem. For a number of years, I have been convinced that we cannot talk about what is going on in high schools on a nationwide or statewide basis, or even on a system-wide basis. The analysis of all data I have ever collected, as well as other analyses that I have seen, has led me to the conclusion that sex-related differences in mathematics must be examined on a school-specific basis. Some schools have been remarkably successful in helping females learn mathematics and to feel good about themselves as learners of mathematics. Other schools have not. Some schools have more females than males enrolled in advanced mathematics classes. In many schools, the reverse is true.

Because of the discrepancies which exist among schools in enrollment patterns and also in feelings toward mathematics by females and males, it is somewhat hazardous to generalize. Nevertheless, an examination of the data that do exist is interesting.

Enrollment Patterns

Armstrong (1979) reported a major study whose purpose was to determine the relative importance of selected factors affecting women's participation in mathematics. She collected data from a stratified, random sample of the entire United States. One factor on which she collected data was participation in mathematics classes. She asked twelfth-grade students to check the appropriate boxes if they had taken or were currently enrolled in courses with specific titles. She concluded that few differences exist in course-taking patterns for males and females in the twelfth grade. Her data are shown in Figure VI-1.

Figure VI-1. Sex Differences in Participation in High School Mathematics*

Fall 1978

	Percentages		
	Female	Male	Differences
General Math	88.77	90.12	1.35
Accounting/Business Math	40.21	32.62	−7.59
Consumer Math	25.32	28.78	3.46
Pre-Algebra	64.98	65.08	.10
Algebra I	78.49	81.06	2.57
Geometry	55.29	59.36	4.07
Algebra II	42.15	53.72	11.57
Trigonometry	27.14	30.96	3.82
Probability/Statistics	4.86	9.48	4.62
Computer Programming	13.28	18.16	4.88
Pre-Calculus	18.00	21.49	3.49
Calculus	7.23	8.20	.97

* Armstrong, J., "Women and Mathematics: An Overview of Factors Affecting Women's Participation (Table 3)," paper presented at Research on Women and Education Conference, Washington, D.C., November 1979. (Available from Education Commission of the States, 1860 Lincoln Street, Denver, Colorado 80295.)

More females than males take accounting. In every other category, more males have taken or are taking the class. There appears to be no dramatic difference in course taking. The Second National Assessment of Educational Progress collected data in a manner similar to Armstrong (Fennema and Carpenter, in press). These data indicate approximately the same trend as does the Armstrong sample.

Is the same trend evident when we look at data from a state sample? Wyoming recently completed such a survey, in which mathematics preparation was classified on six levels. Level 6, the least prepared, means students have had only general mathematics. Level 1, the highest level, means students have studied algebra I and II; synthetic and analytic geometry; trigonometry; logarithmic functions (common and natural) and their graphs; mathematical induction; algebra of functions; basic operations on matrices; and limits, continuity, and differentiation of polynomial functions. Figure VI-2 shows the seniors who had attained each level in 1978. At only the two lowest levels did females and males have the same preparation.

California also reports that a greater percentage of boys than girls take four years of mathematics (24 percent male vs. 17 percent female),

Figure VI-2. Percent of College-Bound Seniors Attaining Each of Six Levels of Mathematical Preparedness*

Level	Percent	
	Female	Male
1**	15	30
2	21	36
3	54	64
4	71	78
5	93	94
6	100	100

* Kansky, B., and Olson, M., *Mathematical Preparation Versus Career Aspirations: Study of Wyoming 1978 High School Seniors* (The Science and Mathematics Teaching Center, Box 3992 University Station, Laramie, Wyoming 82071), p. 13.

** Level 1 is the highest level of mathematical preparedness.

while Wisconsin reports a 6:4 ratio of male : female in their most advanced course (Perl, 1980). Overall, these enrollment data are moderately encouraging.

What happens when we look at individual schools rather than compiling across the nation or a state? In 1977, my colleagues and I began developing an intervention program designed to increase high school girls' enrollment in mathematics classes. This project was funded by the Women's Educational Equity Act Program and included a major evaluation component. In order to carry out this evaluation, we wanted to use the program in schools having an imbalance in enrollment by sex in advanced mathematics courses. I assured my colleagues that finding such schools would be no problem since all the literature (including some written by me) said such an imbalance almost always existed. Imagine my surprise (and embarassment) to find that in about one-third of the schools either an equivalent number of females and males or more females than males were enrolled in advanced math classes. I must add that in most of these schools there were not many advanced classes and there were few students of either sex enrolled in them. However, I am convinced now that while enrollment trends may be encouraging on a broad scale, it is only by looking at individual schools that meaningful assessment of females' enrollment in advanced math courses can be made.

Mathematics Achievement

A more critical issue is what current data tell us about whether sex-related differences in achievement exist when the number of mathematics courses boys and girls have studied is held constant. In other words, do girls and boys who report that they are currently enrolled in or have been enrolled in the same mathematics courses achieve equally in mathematics? If these differences in achievement exist, are they large enough so that we should be concerned?

Studies which shed light on mathematics learning by females and males are becoming increasingly sophisticated in at least two ways: (1) mathematics course taking is being considered an important variable to control, and (2) the cognitive complexity of the items used to measure learning is being included. Prior to 1978 studies which considered both of these were basically not available. Now there are four such studies that deal with relatively current data. Two of them reflect information gathered from nationwide samples. The California State Assessment of Mathematics was done in 1978. Students in Grades 6 and 12 were tested on a variety of content areas with items of differing cognitive levels. Comparisons in achievement level were made among groups who reported studying the same number of mathematics courses. A committee was named to evaluate the results and concluded the following about sex-related differences:

> An analysis of the results by sex showed that girls do consistently better than boys in computations with whole numbers, fractions, and decimals. The girls also outperformed boys in simple one-step word problems. However, the committee found that boys typically scored higher on word problems that were either multiple-step problems or required more reasoning ability.
>
> In geometry, the girls scored higher than boys on questions involving recall and identification of geometric shapes, while boys achieved higher than girls on items dealing with spatial relationships and reasoning ability. In measurement, the girls generally scored higher than boys on problems dealing with money; however, boys generally performed better than girls on the other questions.
>
> At the twelfth grade, the relative performance of boys and girls was compared taking into consideration the amount of mathematical preparation of the particular courses that students had completed. The committee noted that the girls continue to outperform boys at the twelfth grade in whole number and decimal computations. However, the girls do not keep up their relative achievement level in fraction computation at the twelfth grade. The committee observed that girls were considerably lower in the

skill areas of measurement, geometry applications, and probability and statistics.

Females tended to achieve higher than males on lower level cognitive tasks, while males scored higher on more complex cognitive tasks.

Armstrong also investigated achievement differences. She compared females and males who had taken the same math courses and concluded: "Twelfth grade males scored significantly higher than females on the problem solving subtest. Thirteen-year-old females scored significantly higher on the lower level mathematical skill of computation." The mathematics Assessment of the Second National Assessment of Educational Progress indicated also that females were somewhat better in computational tasks than were males. Males out-achieved females in higher level cognitive tasks (Fennema and Carpenter, in press).

These four major studies have made me somewhat uncomfortable with the idea that the only thing we have to do is to ensure that females continue to enroll in mathematics during high school. I am still convinced that the majority of females will achieve at the same level as the majority of males if they elect to study the same amount of mathematics. However, differences in achievement on high level cognitive tasks deserve more direct investigation both with large samples and at the individual school level.

Related Variables

In 1978 I reported that females felt less confident in learning mathematics than did males, and they tended to believe that mathematics was less useful to them than to males. These are still two major variables that explain sex-related differences. My colleagues and I are currently involved in a longitudinal study attempting to identify influences on the development of feelings of confidence. Other than to re-emphasize its importance, I have nothing new to report. Belief in the usefulness of mathematics can be changed and later I will talk about how and why.

One major variable that might help explain females' falling behind in learning higher level skills has to do with the practice of such skills. Enrollment in both mathematics-related courses (such as computer science, probability and statistics) and in science courses which use a great deal of mathematics such as chemistry and physics may contribute to the difference. The female : male ratio in such classes is much higher than in traditional mathematics classes. In these related classes, mathematics is used or applied at the same cognitive level at which males are out-achieving females. We learn what we practice; if males tend to participate in higher level skills in other classes, then they will undoubtedly learn them better.

It has long been a hypothesis of mine that teachers expect males to be better problem-solvers than are females and that boys, more than girls, are encouraged to engage in problem-solving activities. If this hypothesis is true, girls would tend to engage in lower level cognitive activities more than do boys. A result of such differential practice would be the different achievement by cognitive level that has been observed.

We do know that teachers are the most important educational influence on students' learning of mathematics. From entry to graduation from school, learners spend thousands of hours in direct contact with teachers. While other educational agents may have influence on educational decisions, it is the day-by-day contact with teachers that is the main influence of the formal educational institution. Part of the teachers' influence is the learners' development of sex role standards. These sex role standards include definitions of acceptable achievement in the various subjects. The differential standards for mathematics achievement is communicated to boys and girls through differential treatment as well as differential expectations of success.

Many studies have indicated teachers treat female and male students differently. In general, males appear to be more salient in the teachers' frame of reference. Teachers interact with males more than with females in both blame and praise contacts (Becker, 1979). More questions are asked of males by teachers. Males are given the opportunity to respond to more high level cognitive questions than are females (Fennema and others, 1980a).

High achieving girls seem particularly vulnerable to teachers' influence. One major study (Good and others, 1973) indicated that high achieving girls received significantly less attention in mathematics classes than high achieving boys. On the other hand, many girls who have been accelerated in mathematics report positive teacher influence (Casserly, 1980) as a cause of their success. This influence was manifest by teachers being 'sex-blind' in the treatment of girls. Teachers treated males and females just alike and had high expectations for females, as well as males.

Another theory that might help in understanding the sex-related differences in mathematics is attribution theory, although I must urge caution in accepting this theory in a simplistic way. Attribution theory has to do with the perceived causes of success and failure experiences. The model that appears to be the most useful to educators as an aid to understanding achievement-related behavior is the one proposed by Weiner (1974). In this model, attributions of success and failure are categorized into the matrix shown in Figure VI-3, with locus of control being one dimension and stability the other.

Figure VI-3. Locus of Causation

		INTERNAL	EXTERNAL
	STABLE	Ability	Task
STABILITY	UNSTABLE	Effort	Luck

Attributions of past successes and failures to certain of the Weiner categories have been shown to be related both to task persistence and task choice (Bar-Tal, 1978). If one attributes success to internal causes, particularly the internal stable cause of ability, then one can expect success in the future and will be more apt to approach or persist at certain tasks. If, however, success is attributed to an external cause, success in the future is not assured and one will avoid the task. A somewhat different situation is true of failure attributions. If one attributes failure to unstable causes, failure can be avoided in the future so the tendency to approach or persist at tasks will be encouraged. Attribution of failure to a stable cause, on the other hand, will lead one to believe that failure can't be avoided.

Although we should be extremely careful of overgeneralizing data and concluding that all males behave one way and all females another way, many studies have reported that females and males tend to exhibit different attributional patterns (Deaux, 1976; Bar-Tal and Frieze, 1977). In a somewhat simplistic summary, males tend to attribute successes to internal causes and failure to external or unstable causes. Females tend to attribute successes to external or unstable causes and failure to internal causes.

These attributional patterns have also been linked to a pattern of behavior called "learned helplessness"—the condition in which failure is viewed as inevitable and insurmountable. This condition results in lowered motivation to persist. Females are more likely than males to display learned helplessness (Dweck and others, 1978).

The principles of attribution theory and learned helplessness can be applied to the problem of math avoidance—the lack of persistence in mathematics related activities. It appears reasonable to hypothesize that if a student attributes successful performance in mathematics to ability, the likelihood of persisting in mathematics is higher than if that success were attributed to an unstable cause such as effort or luck. Conversely, when failure is attributed to ability, lowered persistence will result. The differential in male/female enrollment in mathematics—the lack of persistence among females—might be partially explained by the fact that females, more than males, attribute successful performance to unstable causes and unsuccessful performance to stable ones. A recent study by Wolleat and others (1980) indicated that this is true. Females, when compared to males, exhibited more of the learned helplessness pattern in their attribution of success and failure in mathematics. They were more likely than males to use effort (unstable) and less likely than males to use ability (stable) to explain their successes. When explaining mathematics failures, females invoked the attributions of ability and task difficulty (both stable) more strongly than did males.

Another variable which many believe might help explain both differential course taking and differences in achievement is spatial visualization. Spatial visualization involves visual imagery of objects, movement of the objects or changes in their properties. In other words, objects or their properties must be manipulated in one's 'mind's eye,' or mentally. The relationship between mathematics and spatial visualization is logically evident. Starting at about adolescence, male superiority on tasks involving spatial visualization is found. Many are finding that spatial visualization is related to mathematics achievement differently for males and females (Sherman, 1979).

Currently, my colleagues and I are engaged in gathering data about how mathematics learning is dependent upon spatial visualization. It appears evident that tasks which measure spatial visualization skills have components that can be mathematically analyzed or described. From such an examination, we could hypothesize a direct relationship between mathematics and spatial visualization. An item from the space relations portion of the Differential Aptitude Test (Bennett and others, 1973) requires that a 2-dimensional figure be folded mentally into a 3-dimensional figure. Another spatial visualization test requires that 2- or 3-dimensional rigid figures be rotated and translated to specified locations. The activities required by those tests can be described as mathematical operations. Yet this set of operations is only a minute subset of mathematical ideas which must be learned and, indeed, one could go a long way in the study of mathematics without these specific ideas.

The hypothesis that my colleagues and I are currently investigating is that the critical relationship between mathematics and spatial visualization is not direct, but quite indirect. It involves the translation of words and/or mathematical symbols into a form where spatial visualization skills can be used. For instance, consider the following problem:

A pole 12 feet long has been erected near the bank of a lake. Two and a half feet of the pole have been hammered down into the bottom of the lake; one half foot is above the surface of the water. How deep is the lake?

For children of 11 and 12 years of age, this is a moderately difficult problem. You must add the lengths of two pieces of the post and then subtract that length from the total length, that is, $2\frac{1}{2} + \frac{1}{2} = 3$ and $12 - 3 = 9$. Keeping track of the steps and sequencing them accurately is not easy. Consider the problem from a spatial visualization perspective. If you can visualize in your mind what is involved, the solution of the problem then becomes simpler. An image would enable a person to move the pieces above and below the water together. Then that length could be subtracted from the total length in order to get the correct answer.

Consider a symbolic problem met by children of the same age: $\frac{1}{2} + \frac{1}{3}$. While it can be solved totally with symbols, children of this age, because of their developmental level, often have trouble really understanding the symbolic process involved. If it could be visualized in the mind, spatial visualization skills could be used and the answer found more easily.

We know that females tend to score lower than males on spatial visualization tests. What we do not know is whether females differ from males in their ability to visualize mathematics, that is, in the translation of mathematical ideas and problems into pictures. Neither do we know if good spatial visualizers are better at these translations than are poor spatial visualizers. However, I am increasingly convinced that there is no direct causal relationship between spatial visualization skills and the learning of mathematics in a broad general sense. While I am continuing to investigate the impact of spatial visualization skills, I am less convinced than I once was that spatial visualization is important in helping understand sex-related differences in the studying and learning of mathematics.

In American schools, classrooms don't appear to use mathematical representations which either encourage or require the use of spatial visualization skills. While some primary mathematics programs encourage the use of concrete and pictorial representations of mathematical ideas, by the time children are 10 or 11 years old, symbolic representations are used almost exclusively. Perhaps boys, more than girls, use the concrete representations during primary years and, thus, develop higher skills in using spatial visualization in learning mathematics. As far as I know, however, no one is investigating such a hypothesis.

Interventions

Can schools be changed so that females elect to study more mathematics and learn mathematics as well as do males? All too often, comments are addressed to me that imply that schools alone can't do much. The argument goes like this. Because the studying of mathematics is stereotyped male, and because stereotyping of sex roles is so deeply embedded in society, schools are powerless to improve females' studying of mathematics until society changes. Let me say as emphatically as I can that that argument is fallacious. Schools can increase females' studying of mathematics. Let me cite some evidence that shows strongly that schools can be effective.

Two intervention programs in particular have been intensively evaluated. The first program is called *Multiplying Options and Subtracting Bias*. The rationale used in its development is that merely telling high school females about the importance of mathematics is insufficient. Forces that influence these girls to make their decisions are complex and deeply embedded in societal beliefs about the roles of males and females. Asking females to change their behavior without changing the forces operating on them would place a very heavy burden on their shoulders. What should be done is to change the educational environment of these females so that they are enabled to continue their study of mathematics beyond minimal requirements. This environment is composed of several significant groups of people: mathematics teachers, counselors, parents, male students, and the female students themselves. *Multiplying Options and Subtracting Bias* was designed to change these significant groups' beliefs about women and mathematics as well as to change each group's behavior.

Multiplying Options and Subtracting Bias is composed of four workshops: one each for students, teachers, counselors, and parents. Each workshop is built around a unique version of a videotape designed explicitly for the target audience. Narrated by Marlo Thomas, the tapes use a variety of formats, candid interviews, dramatic vignettes, and expert testimony to describe the problem of mathematics avoidance and some possible solutions. The videotapes and accompanying workshop activities make the target audience aware of the stereotyping of mathematics as a male domain which currently exists, females' feelings of confidence toward mathematics, the usefulness of mathematics for all people, and differential treatment of females as learners of mathematics. Discussed specifically are plans for action by each group. These two-hour workshops are designed to have an impact on a total school.

The program has been evaluated extensively and its use has significantly increased females' enrollment in mathematics courses (Fennema and others, 1980a). Exposure of *Multiplying Options and Subtracting Bias* can substantially influence students' attitudes about mathematics, the stereotyping of mathematics, and students' willingness to take more mathematics courses.

The other intervention program was developed, planned, and implemented by the San Francisco Bay Area Network for Women in Science (now called the Math/Science Network). The Network is a unique cooperative effort of scientists, mathematicians, technicians, and educators from 30 colleges and universities, 15 school districts, and a number of corporations, government agencies, and foundations. The goal of the Network is to increase young women's participation in mathematical studies and to motivate them to enter careers in science and technology.

Seven conferences developed by this network were held in the spring of 1977 and 1978 to increase the entry of women into mathematics- and science-oriented careers. These one-day conferences consisted of a general session with a panel or main speaker, one or two science/math workshops, and one or more career workshops that provided junior and senior high school girls opportunities to interact with women working in math/science-related fields.

The conferences were evaluated in a study involving 2,215 females who had volunteered to attend. Pre- and post-conference questionnaires were administered and responses analyzed. The evaluators concluded that "the conferences (1) increased participants' exposure to women in a variety of technical and scientific fields, (2) increased participants' awareness of the importance of taking mathematics- and science-related courses, and (3) increased participants' plans to take more than two years of high school mathematics" (Cronkite and Perl, 1979).

Evaluations of these intervention programs indicate quite clearly that it is possible to change females' mathematics behavior, and to do so in relatively short periods of time.

Some schools are remarkably more effective than others in persuading females to attempt high achievement in mathematics. Casserly (1980) identified 13 high schools which had an unusually high percentage of females in advanced placement mathematics and science classes. She concluded that the schools had identified these girls as early as fourth grade and the school teachers and peers were supportive of high achievement by the females.

Questions

I would like to conclude with the following questions and answers. Perhaps your answers would be different, but I challenge you to at least think about mine.

Q: Is the sex factor a reality in mathematics education?

A: Yes. Females are receiving an inequitable education in mathematics in many schools. Not only do they elect to study mathematics less than males do, there is some evidence that in a very important part of mathematics learning, they learn less than do males.

Q: Can schools do anything about improving the mathematics education of females?

A: Yes, a great deal. Each school must first find out what its situation is with respect to enrollment and achievement and then plan interventions which specifically address that unique situation.

Q: Can individual teachers do anything?

A: Yes. By becoming truly sex-blind in expectations—by increasing their awareness of all students as individuals who have unique needs which must be met in order to help students achieve at their highest level.

Q: What are the implications of research on sex-related differences for the curriculum of the 80s?

A: Sex-related differences in mathematics can, and should, be eliminated. Equity in mathematics education for females and males is an achievable goal. In order to achieve this goal, each and every school must consider its own specific situation. Much help is available, but the motivation and direction must come from within each school.

References

Armstrong, J. "Women and Mathematics: An Overview of Factors Affecting Women's Participation." Paper presented at Research on Women and Education Conference, Washington, D.C., November 1979.

Bar-Tal, D. "Attributional Analysis of Achievement Related Behavior." *Review of Educational Research* 48 (1978): 259-271.

Bar-Tal, D., and Frieze, J. H. "Achievement Motivation for Males and Females as a Determinant of Attributes for Success and Failure." *Sex Roles* 3 (1977): 303-313.

Becker, J. "A Study of Differential Treatment of Females and Males in Mathematics Classes." Ph.D. dissertation, University of Maryland, 1979.

Bennett, G. K.; Seashore, H. G.; and Ivesman, A. C. *Differential Aptitude Tests Forms S and T.* 4th ed. New York: Psychological Corporation, 1973.

Casserly, P. L. "Factors Affecting Female Participation in Advanced Placement Programs in Mathematics, Chemistry, and Physics." In *Women and the Mathematical Mystique.* Edited by L. H. Fox, L. Brody, and D. Tobin. Baltimore: The Johns Hopkins Press, 1980.

Cronkite, R., and Perl, T. "Evaluating the Impact of an Intervention Program: Math-Science Career Conferences for Young Women." In *Proceedings of the Conference on the Problem of Math Anxiety.* Sponsored by The School of Natural Sciences. California State University, Fresno, 1978.

Deaux, K. "Sex: A Perspective on the Attribution Process." In *New Directions in Attribution Research,* Vol. 1 Edited by J. Harvey, W. Ickes, and R. Kidd. Hillsdale, N.J.: Lawrence Earlbaum Associates, 1976.

Dweck, C. S..; Davidson, W.; Nelson, S.; and Enna, B. "Sex Differences in Learned Helplessness: II. The Contingencies of Evaluative Feedback in the Classroom, and III. An Experimental Analysis." *Developmental Psychology* 14 (1978): 268-276.

Fennema, E. "Sex-Related Differences in Mathematics Achievement: Where and Why?" In *Perspectives on Women and Mathematics.* Edited by J. E. Jacobs. Columbus, Ohio: ERIC Clearinghouse for Science, Mathematics, and Environmental Education, 1978.

Fennema, E., and Carpenter, T. "What the Mathematics Assessment Has to Say About Sex-Related Differences in Mathematics." *The Mathematics Teacher,* in press.

Fennema, E.; Wolleat, P.; and Pedro, J. "Mathematics Attribution Scale." *Journal Supplement Abstract Service* (May 1979), Manuscript No. 1837.

Fennema, E.; Reyes, L. H.; Perl, T. H.; Konsin, M. A.; and Drakenberg, M. "Cognitive and Affective Influences on the Development of Sex-Related Differences in Mathematics." Symposium at the American Educational Research Association, Boston, April 1980a.

Fennema, E.; Wolleat, P.; Pedro, J.; and Becker, A. "Increasing Women's Participation in Mathematics: An Intervention Program." *Journal for Research in Mathematics Education* 12 (November 1980b): 3-14.

Fennema, E.; Wolleat, P.; Becker, A.; and Pedro, J. *Multiplying Options and Subtracting Bias.* Reston, Va.: National Council of Teachers of Mathematics, 1981.

Good, T. L.; Sikes, J. N.; and Brophy, J. E. "Effects of Teacher Sex and Student Sex on Classroom Interaction." *Journal of Educational Psychology* 76 (1973): 74-87.

Kansky, B., and Olson, M. *Mathematical Preparation Versus Career Aspirations: Study of Wyoming 1978 High School Seniors.* Laramie, Wyo.: The Science and Mathematics Teaching Center, 1978.

Perl, T. H. "Discriminating Factors and Sex Differences in Electing Mathematics." Speech before the Mathematics Colloquium at the University of Toledo, Toledo, Ohio, February 1980.

Sherman, J. "Predicting Mathematics Performance in High School Girls and Boys." *Journal of Educational Psychology* 71 (1979): 242-249.

"Student Achievement in California Schools." *1977-78 Annual Report of the Mathematics Section.* California Assessment Program, 1978.

Weiner, B. *Achievement Motivation and Attribution Theory.* Morristown, N.J.: General Learning Press, 1974.

Wolleat, P.; Pedro, J. D.; Becker, A.; and Fennema, E. "Sex Differences in High School Students' Causal Attribution of Performance in Mathematics." *Journal for Research in Mathematics Education* 11 (1980): 356-366.

Response

Grace M. Burton

At NCTM's 1978 annual meeting, the strand "Women and Mathematics" attracted a large cross section of the mathematics education community. Topics ranged from mathematics anxiety to problem-solving ability, from sexism in mathematics textbooks to the distribution of females in leadership roles. Like Fennema, I concluded my presentation with a call for commitment:

The inability of women to succeed

at mathematical endeavors is a bit of hallowed mythology in our folklore, but there are those who seek to change that. Join us. Encourage each mathematically talented Susie you know to excel in mathematical endeavors and to be *proud* of her ability.

I believe we are presently seeing the fruition of that interest in sex-related differences which had re-surfaced in the mid-70s. We are beginning to have the hard data to support—or refute—some of our hunches. Our openness to the findings of research may be tested as some of what we "know" turns out to be not so at all. It is important that we both evaluate the new findings and re-evaluate our beliefs in the light of those findings deemed valid.

One difficulty in evaluating research in the area of sex-related differences is the sheer quantity of the material currently appearing in professional journals. Relevant information is scattered across many disciplines including economics, sociology, neurology, developmental psychology, anthropology, linguistics, philosophy, and biology. Definitions vary from one study to another, even in such basic terms as "feminine," "problem solving," and "spatial ability." Nonsignificant differences tend to be under-reported while differences tend to be over-reported. Much of the data have been collected by self-report. Birth order, race, and socioeconomic class, all of which may be significant factors, are confounded with sex.

Population equivalence cannot be relied upon. Drawing conclusions from older studies presents other difficulties. Extrapolations from animal research, generalizations from the observations of one sex to both sexes, misinterpretations in secondary sources, and societal changes— all have contributed to the present state of the art. At base, though, is the point raised by Fennema—all we can learn from these studies are generalities about males *as a group* or females *as a group*. Traditional research can tell us nothing about what a particular male or female can or will choose to do or be.

It may be that the very vigor of this research activity has given rise to what Fennema calls the new mythology: that is, males and females are basically different in cognitive and psychological make-up. There is now a real danger that this mythology will play a part in educational decision making. Luckily, at least in those institutions receiving federal financial assistance, schools are prohibited from offering single-sex courses or extracurricular activities. Were this not the case, we might see a rash of "Trig for Girls" or "Calculus for Girls" classes. Such efforts to accommodate group differences are shortsighted and inappropriate at best. More often, they are both ineffective and deprecating. As we look back with humor on the 19th century assertions that women actually had significantly different breathing apparatus or nervous systems, we must be careful that we do not subscribe to more modern but possibly equally ludicrous beliefs.

Enrollment Patterns

The study of mathematics is vital to the intellectual development and career progress of both male and female students. I would be among the first to deny that all female students are less likely than all male students to elect to study mathematics when it is no longer a required subject. In some schools, the number of each sex in advanced courses is about equal. I would suspect, although I have no data to back me up, that in those schools where large numbers of young women are continuing to study mathematics beyond the required courses, someone (or more than one someone) is consciously doing something to help this change along.

I firmly believe in the power of the individual to effect change. Each individual has the opportunity to alter the status quo in the direction of greater good. And of course I believe that encouraging each student to actualize his or her intellectual potential to the fullest possible degree is a "good." I repeat without apology, "The place to improve the world is first in one's own heart and head and hands and then to work outward from there" (Pirsig, 1975). Each of us have the obligation to do just that. To accept "More males than females take upper math courses" as an unchalleng-

able assumption is to abdicate that responsibility. Those schools in which female enrollment patterns are different from the norm should be studied, the contributing factors identified and promulgated, and, where possible, modeled in other schools.

Mathematics Achievement

It is hard to accept the result that females tend to achieve higher than males on lower cognitive skills, yet lower than males on higher cognitive skills. Here again, however, we must be cautious about translating that finding into "Girls can't do as well as boys on complex cognitive tasks." I concur with Fennema's statement that the probable causes underlying the reported differences require direct investigation. We must not lose sight of the individual student who has the potential to perform differently from what group norms would lead us to expect. Regardless of the social convention, there have always been women who have delighted in the study of mathematics and who have achieved success despite the fact that cognitive activity was not considered the province of "the gentle sex." Femaleness must not be taken as presumptive evidence of inability to achieve at high mathematical levels.

Related Variables

The strength of variables related to mathematics course-taking such as confidence and belief in the usefulness of mathematics is evident both from research and less structured observation. These factors each school can and should address. Attribution patterns are a newer focus. Building on this current research, those teachers who have long believed in the power of an "Of course you can!" philosophy for themselves and their students may be able to refine their thinking and apply it more effectively.

Interventions

In the concern for well-designed studies and carefully-written reports, it is easy to underestimate the role of the educational practitioner who has the power to speed or retard change in the educational system. If only teachers could be convinced of their power for good! They are *the* powerful influence in the lives of their students for the successful (or unsuccessful) achievement of academic goals, for the development of positive (or negative) attitudes, and for the embracing of (or the escape from) further exploration of mathematics.

Teachers become even more powerful when they know the result of research.

Teachers familiar with research on "wait time" (Rowe, 1978) are unlikely to transmit an "I know your kind can't do it" message by not allowing time for students to answer questions. Teachers who know the Good, Sikes, and Brophy

(1973) findings are more likely to monitor their attending behavior. Teachers who have learned of attribution research will perhaps recognize the subtle differences in those students who blame study patterns for their failure and those who say, "I don't have a math mind." Teachers who have studied the effect of spatial visualization may take special pains to incorporate appealing activities in this dimension in their classes.

I certainly agree that spatial ability is not crucial to the pursuit of mathematics. Now that their diplomas and/or careers are relatively secure, several students and colleagues have told me they never *could* see those rotating shapes in calculus, and that they analyzed their way through projective geometry. On the other hand, certain tasks are facilitated by the ability to mentally rotate or translate figures. If an individual's spatial ability can be developed, develop it we should —not because it will eliminate a sex difference but because it will expand that individual's problem-solving repertoire and enrich his or her life.

Effective intervention programs on all educational levels are needed, and it is heartening that they are being developed, tested, and disseminated. Each and every teacher of mathematics within a school must make a concentrated effort to help each and every student make whatever cognitive and affective strides he or she can. Enlisting the support of counselors, teachers and parents as well as both male and female students is a most promising direction for such programs to take. It recognizes the basic fact that each of us is part of a system, and change in one part of the system effects all the rest of it. It is only when many individuals each make a conscious decision for positive change that such change will occur. Those many individuals must be informed of the existence, extent, and impact of the traditional mythology that women neither can nor should do mathematics.

There are many successful strategies (Menard, 1979). Some appropriate to the local conditions should be chosen, modified to meet the needs of the individual schools, and implemented. Gaining support at the administrative level for these new directions will facilitate the achievement of the desired goals, if only because administrators have the resources to help things happen once they are committed to the value of an idea. In this case, that valuable basic idea is fairness.

Of course, what I am suggesting involves seeing each student as a person first and accepting him or her without comfortable prejudgment and ready-made expectation due to sex, race, social class, sibling performance, or any other factor. It means throwing off any preconceived notion that individuals in any group are by nature logical or illogical, excited or bored by mathematics, ambitious or passive with respect to career. It means remembering there is a vast amount of variation in any group, and that it is an intellectually indefensible act to ascribe characteristics to an individual solely on the basis of group membership.

If each of us in our own spheres of influence demonstrate firm commitment to the importance of the individual, and translate our knowledge from research into action, there may be no need in the future to consider the question, "Is there a sex factor in mathematics education?" We will have provided the best learning environment possible for each student—regardless of sex.

References

Astin, H. S.; Parelman, A.; and Fisher, A. *Sex Roles: A Research Bibliography*. Rockville, Md.: National Institute of Mental Health, 1975.

Burton, G. M. "The Power of the Raised Eyebrow." *The School Counselor* 25 (1977): 116-123.

Burton, G. M. "Regardless of Sex." *The Mathematics Teacher* 72 (1979): 261-270.

Caplan, P. J. "Sex, Age, Behavior and School Subject as Determinants of Report of Learning Problems." *Journal of Learning Disabilities* 10 (1977): 60-62.

Coolidge, M. R. *Why Women Are So*. New York: Henry Holt and Company, 1912.

Golden, G., and Hunter, L. *In All Fairness*. Far West Laboratory for Educational Research and Development, 1974.

Good, T. L.; Sikes, N.; and Brophy, J. E. "Effects of Teacher Sex and Student Sex on Classroom Interaction." *Journal of Educational Psychology* 65 (1973): 74-87.

Gregory, M. K. "Sex Bias in School Referrals." *Journal of School Psychology* 15 (1977): 5-8.

Guttentag, M., and Gray, H. *Undoing Sex Stereotypes.* New York: McGraw-Hill, 1976.

Jacobs, J., ed. *Perspectives on Women and Mathematics.* Columbus, Ohio: ERIC Clearninghouse for Science, Mathematics and Environmental Education, 1978.

Menard, S. *How High the Sky? How Far the Moon?* Washington, D.C.: Office of Education, 1978.

Menard, S. *A Very Special Book.* Washington, D.C.: Women's Educational Equity Act Program, 1979.

Milnar, J. "Sex Stereotypes in Mathematics and Science Textbooks for Elementary and Junior High Schools." In *Sexism and Youth.* Edited by Diane Gersoni-Stavn. New York: R. R. Bowker, 1974.

Pirsig, Robert. *Zen and the Art of Motorcycle Maintenance.* New York: Bantam Books, 1975.

Rowe, M. B. *Teaching Science as Continuous Inquiry.* 2nd ed. New York: McGraw Hill, 1978.

Sadker, M., and Sadker, D. *Beyond Pictures and Pronouns: Sexism in Teacher Education Textbooks.* Washington, D.C.: Office of Education, 1979.

Verheyden-Hilliard, M. E. *A Handbook for Workshops on Sex Equality in Education.* Washington, D.C.: Department of Health, Education and Welfare, 1976.

VII. Problem Solving

Mary Grace Kantowski

The first of NCTM's *Recommendations for School Mathematics of the 1980s* (1980) states, "Problem solving must be the focus of school mathematics in the 1980s." This recommendation not only indicates the importance of problem solving, it also implies that a concerted effort is needed in order to establish problem solving as an integral part of the mathematics curriculum. Before we look ahead to what the 80s hold for problem solving, let us look back to where we were at the beginning of the last decade to see what we have learned about problem solving and where we stand today.

In a comprehensive review of problem-solving research written just prior to the beginning of the 70s, Kilpatrick (1969) noted that "Since the solution of a problem—a mathematics problem in particular—is typically a poor index of the processes used to arrive at the solution, problem-solving processes must be studied by getting subjects to generate observable sequences of behavior." He noted that psychologists had devised numerous techniques for studying problem solving, but that mathematical problems were seldom used in such research. Furthermore, he noted that the larger question of how subjects adapt various heuristic methods to different kinds of problems remains virtually unexplored. The situation described by Kilpatrick a decade ago no longer exists. Mathematics educators *have* been studying problem solving using mathematical problems, and a great deal of effort has been expended in the research community in the study of problem-solving processes.

Until the Kilpatrick report, much of the research and development in problem solving focused on the actual *solution* to the problems or the *answers to the exercises*. Researchers looked at how many problems were

111

solved correctly, without regard for how the solution was attained or how close a student may have come to a correct solution. More recently, research has begun to examine *processes* or *the set of steps students use to find a solution*. In this form of research, the protocols (everything a student says or does as he or she solves a problem) are collected during individual interviews. Although difficult and time consuming, this emphasis on studying how solutions are arrived at has uncovered some interesting regularities common to correct solutions. It has been found, for example, that correct solutions to problems involve setting up a plan, however brief, for the solution (Kantowski, 1977, 1980). Another finding is that different students approach the same problem in a variety of ways, indicating the existence of a style or preference. This would suggest that curriculum developers and textbook authors should consider instruction in problem solving that includes a variety of approaches.

Another change since 1970 has been in the expansion of the meaning of problem solving. At the beginning of the last decade, problem solving to most people meant the solving of verbal or word problems. Although verbal problems remains an area of great interest in the mathematics education community, the term problem solving now includes other problem types such as *nonroutine mathematics problems* and *real (application) problems.*

To many classroom teachers and other educators a *problem* is simply a word problem or an exercise stated in verbal form. Word problems found at the ends of chapters in mathematics books fall into this category. An example of such a verbal problem might be the following:

> Maria bought a hamburger for $.90 and a coke for $.30. If the local sales tax is 5%, how much change should she receive if she gives the clerk $2.00?

Such problems are easily solved by application of algorithms that are a part of standard instruction.

To other educators a problem exists if a situation is nonroutine, that is, if the person attempting the problem has no algorithm at hand that will guarantee a solution. He or she must put together the available knowledge in a new way to find a solution to the problem. Such problems are subjective, that is, what is a nonroutine problem for one person is actually an exercise or a routine problem for another. For most middle school students, the following would be a nonroutine problem:

> Maria has exactly $3.00 and would like to spend it all on her lunch. The menu includes hamburgers at $.90, hot dogs at $.80, onion rings at $.60, french fries @ $.50, and colas @ $.30, $.40, or $.50. The sales tax is 5%. What could Maria have for lunch?

In solving this problem a student has no simple calculation algorithm to follow. The possibilities must be tabulated and some trial and error attempted. Moreover, more than one solution is possible.

In general, a nonroutine problem may be defined as a question which cannot be answered or a situation that cannot be resolved with the knowledge immediately available to a problem solver. In effect, a problem is a situation which differs from an exercise in that the problem solver does not have a procedure or algorithm which will certainly lead to a solution (Kantowski, 1974). That is not to say that such an algorithm does not exist, simply that it is not known to the problem solver at a given point in time. In fact, the solution to a problem may provide a problem solver with algorithms for future exercises.

A third type of problem can be called *applications* or "real problems." Projects such as Unified Science and Mathematics for the Elementary Schools (USMES) deal with real problems, and several curriculum projects, notably Usiskin's *Algebra Through Applications,* have emphasized applications.

Verbal Problems

A comprehensive review of research related to verbal problem solving was undertaken by Sowder and others (1978). Most children in the elementary school are introduced to problem solving through verbal problems. After having been introduced to algorithms in some content area, the next logical step in instruction is to introduce the student to a problem in which the algorithm is being used or applied, to observe if a student is able to use the algorithm correctly, and more importantly, whether he or she is able to select the correct algorithm to use.

If we look at the first verbal problem stated above we see that in this case a student needs to select from and apply several algorithms:

Maria bought a hamburger *and* cola:	
Purchases must be totaled	.90
(Addition must be selected and applied)	.30
	$1.20
Sales tax percent must be changed to a decimal	.05
Amount of sales tax must be found	$ .06
(decimal multiplication must be	
selected and applied)	
Sales tax must be added to price	$1.20
(addition must be selected and applied)	.06
	$1.26

Finally, the total cost must be $2.00
 subtracted from amount given clerk 1.26
(subtraction must be selected and $ .74
 applied)

This kind of problem-solving activity is at a much higher cognitive level than simply performing a computation. There are three steps in the problem-solving sequence for each part of the problem. The student must recognize the structure of the verbal problem, select an appropriate algorithm, then correctly apply the algorithm. Studies give clear evidence that all three steps are necessary for successful problem solving in some measure and that instruction must place emphasis on each of these steps.

Skill in computational processes is necessary for solving problems (Knifong and Holton, 1976; Meyer, 1978). However, having these skills does not guarantee successful problem solving. Results from the second assessment of the National Assessment of Educational Progress support this idea. Although 76 percent of the 9-year-olds and 96 percent of the 13-year-olds could subtract a two-digit number from a two-digit number, only 59 and 87 percent, respectively, could solve a simple application problem using the same subtraction exercise. The results were even more dramatic when the operation was multiplication of fractions. In the case of the 13-year-olds, the percentage of correct responses dropped from 69 on the computation exercise to 20 on the related application problem. Clearly, a factor other than computational skill is involved in solving verbal problems. Results of studies such as those cited point to a need for instructional methods that emphasize something in addition to computation.

Nonroutine Mathematics Problems

In general, nonroutine problems are problems for which a problem solver knows no clear path to the solution and has no algorithm which can be directly applied to guarantee a solution. In the case of the problem stated above, it is not clear from the statement of the problem what the selection of lunch items should be. The student must either use some trial and error to put items together to sum to a given total or organize the data into a table of possible combinations that would give the desired result. Such problems are at a higher cognitive level than simply selection and application of algorithm.

Nonroutine problems are important for several reasons. Experience in solving nonroutine problems can help students transfer methods of problem solving to new situations. Such experiences can also help students grasp the meaning of mathematical structure and develop the ability to see the mathematics in a given situation. Several recent studies give us a good deal of information on the status of solving nonroutine problems. First, it has been found at the elementary, secondary, and postsecondary levels that without specific instruction in techniques for solving nonroutine problems, most students do not know how to approach such problems and do not appear to use strategies in their solution (Lester, 1975; Kantowski, 1974, 1980; Schoenfeld, 1979). In a finding similar to that of Meyer (1978) cited above, Webb (1979) found that conceptual knowledge and heuristic strategy components, among other factors, interact in successful problem solving. This means that it is not simply computational skill and the knowledge of how to apply algorithms that are important in problem solving; it is also important for a student to be able to plan effectively and to use other heuristics such as organizing data into tables and drawing effective diagrams.

Several processes appear to be important in solving nonroutine problems. The solution set-up is the most difficult of the problem solving stages and the most crucial part of the solution (Kulm and Days, 1979). Solution set-up refers to a variety of manipulations of data that could lead to a solution such as organizing data into a table, grouping data into similar sets, or formulating an algebraic equation which would be useful in solving the problem. The ability to set up the problem is related to its successful solution.

Another process, which is closely related to solution set-up and has been emphasized in research dealing with nonroutine problems, is that of planning. In deciding on a plan for solution, a problem solver tries to find a relation to other problems solved previously and decides on a method of solution to try to follow. Plans often precede the solution set-up. Although planning does not always ensure a correct solution, psychologists (for instance, Greeno, Anderson, Rissland), as well as mathematics educators (Webb, 1979; Kantowski, 1977) have found that planning is related to successful problem solving and that most successful solutions of nonroutine problems show some evidence of planning.

Another variable studied in research dealing with the solution of nonroutine problems is that of transfer—memory for and application of methods used in previously solved related problems. Kulm and Days (1979) found that a general-specific sequence of problem presentation— that is, a sequence of problems in which a more general problem is presented initially and followed by a specific case and a similar problem— produced significant transfer of information and that content played a significant role. Both equivalent and similar problems resulted in transfer, although for different type problems. Kantowski, too, in a study of nonroutine problems (1980), found that some reference to related problems existed in a significant number of correct solutions. For example, in one of the instructional sessions, the solution to a problem used in instruction generated the pattern of triangular numbers (1, 3, 6, 10 . . .). Later in the study, many students referred to the pattern of triangular numbers that they had seen generated as they tried to solve problems that looked unrelated on the surface. Some students remembered the pattern; others remembered how the pattern was generated (by adding successive natural numbers to the previous element of the set).

Real Problems

A real problem involves a complex real-life situation that must somehow be resolved. Often there is not an exact solution, but one that is determined to be optimal to fit the conditions. Real problems include traffic flow problems, and problems dealing with financing school functions and effectively utilizing available space. In arriving at a solution to a real problem, students often solve a variety of brief application problems that include substantial computation.

Perhaps the best known study of real problem solving is that done by the evaluators of the USMES program (Shann, Unified Science and Mathematics for the Elementary School, 1976). In this program, students in a class work together in small groups in an effort to solve a "real problem" such as the design of a soft drink that could be served at a party or event that would satisfy most of the group. In solving such problems, students become involved in collecting and compiling data in making decisions about what and how much to purchase and what to charge if the soft drink is to be sold. Computational skills are used in application situations so that the use of the skill becomes meaningful to the student. Among the significant results of the program were the wider repertoire of successful problem-solving behaviors exhibited; larger amounts of time spent in more active, self-directed, and creative behavior; overall higher (although not statistically significant) means on basic skill tests; and significant positive attitudes toward mathematics.

Research on Instruction and Development

Ideas for Instruction

Problem-solving ability develops slowly over a long period (Wilson, 1967; Kantowski, 1974) and grows with experience in solving problems. Therefore, for most students, except perhaps the most gifted in mathematics, systematically planned instruction is an essential factor in the development of problem-solving ability.

A variety of factors seem to affect the ability to be a successful problem solver. In all types of problems with which research has been concerned, the three variables of understanding the problem, planning, and computational skill are important. These three variables constitute the first three phases of Polya's (1973) four phases in the solution of a problem. These phases, as well as Polya's fourth phase, *Looking Back,* will serve as the basis for the suggestions for instruction in problem solving that will follow.

Instruction Must Emphasize Understanding the Problem

One of the most neglected aspects of problem solving is that of *understanding the problem.* As Polya (1973) noted, this aspect of problem solving deals with far more than simply comprehending what is read. Understanding the problem implies grasping the relationships among the conditions of the problem and perceiving what is given mathematically. If a student truly understands a problem, he or she will not only be able to determine what is being sought, but will also recognize if the information given is reasonable or if a solution is impossible with the conditions as given. Reports of many studies indicate that much of the difficulty students have with problem solving stems from their failure to understand the problem.

In reporting the results and implications of the second National Assessment in Mathematics, Carpenter and others (1980b) noted that the multi-step and nonroutine word problems were difficult for all age groups. They noted that "the high levels of incorrect responses seem to indicate that there was little attempt to think through a problem in order to arrive at a reasonable answer" (p. 44). Moreover, in studies undertaken in connection with the Mathematical Problem Solving Project (Lester, 1975) it was found that many students often misread and misinterpreted problems and had difficulty in retaining and coordinating multiple conditions in a problem.

Several examples from the second National Assessment of Educational Progress (Carpenter and others, 1980a) further illustrate the difficulty students have with understanding the problem. One exercise on the assessment required finding the number of buses, each holding a certain number of passengers, that would be needed to carry a given number of people. Thirty-nine percent of the 13-year-olds gave responses indicating they ignored the fact that the number of buses must be a whole number.

In another exercise students were asked how many baseballs would be *left over* if a given number of balls were packed 24 to a box. A large percentage of students gave the quotient rather than the remainder as their response. As the authors of the interpretive report note, "Performance on these and many other exercises indicates that for too many students, problem solving involves little beyond choosing an operation, calculating an answer and reporting an answer" (p. 428). This comment is substantiated by the results of exercises in which extraneous information was given in the problem statements. Large percentages of students simply used all numbers given in some way, indicating a lack of understanding of the problem. When exercises with missing information were given, students also had a great deal of difficulty. In one such exercise, students were asked what else one would need to know in order to solve the problem. Fifty-six percent of the 13-year-olds and 32 percent of the 17-year-olds responded, "I don't know."

As was noted, the assessment results and the results of other studies indicate that many students read a problem and begin to manipulate the numbers in some way—often irrationally. Instruction should include slow-down mechanisms to motivate students to make a concerted effort to *understand* what is being asked. One such procedure is reported by Kalmykova (1975). Students read problems with inflection and try to convey meaning to one another. Then, students set up very simple problems to develop early habits of trying to understand a problem being presented.

Understanding a problem often involves translating it into some other form—a diagram, an equation, a matrix or a model. The important emphasis in instruction should be that the translation follows understanding, or leads to deeper understanding of the problem and not simply be a rote behavior.

Students in the elementary school are often taught to change "word sentences" into "number sentences." For example, in the problem

> The two fourth-grade classes at Longwood School are saving boxtops for athletic equipment. The need 10,000 boxtops to get a jungle-gym set. One class has collected 3,871 tops and the other class has 4,106 tops. How many more tops do they need?

One possible translation might be:

NEED	=	10,000	−	HAVE
HAVE	=	3,871	+	4,106
HAVE	=	7,977		
NEED	=	10,000	−	7,977 tops

To be able to make the translation, the student must first understand that what is needed is the difference between the total both classes have and the number needed, and will set up the translation accordingly. Students who do translations for problems such as this one by rote will often focus on the words "how many more" and set up a translation such as 4,106 − 3,871 = □.

Students should also be aware of the importance of understanding. In another of the Soviet studies, Gurova (1969) found that pupils' awareness of their own processes while solving problems had a positive effect on problem solving. One technique, which could be used in instruction to help students understand a problem, is to pose problems with missing information. This makes the problem impossible to solve. Another technique is to assign problems with redundant or contradictory information.

Instruction in Planning

A second suggestion for instruction deals with helping students to construct *plans* for approaching problems before they begin to "attack" a problem. It is becoming increasingly clear that planning strategies are essential to arriving at correct solutions. Many researchers (such as Lester, 1975; Kantowski, 1975) have found that prior to instruction, many students do not appear to use any strategies in problem solving or use some form of random trial and error. However, instruction in planning techniques appears to have positive effect on the use of planning strategies and, consequently, on improvement in problem solving. Moreover, a variety of plans for the same problem should be suggested to emphasize to students that there is more than one way to solve a problem.

Emphasis on a Variety of Solutions

Several studies, on both the elementary and secondary levels, indicate that students exhibit a great variety of problem-solving styles and preferences for certain ways to go about solving a problem. Moser (1979) studied children's representations of certain addition and subtraction problems with some emphasis on the students' interpretation of the problem structure and their imposition of structure on the given problems. One of the interesting features of the research findings was that even children in the first grade have a rich repertoire of strategies that they apply to a problem based on their interpretation of its meaning to them. For example, children who have not yet been taught an algorithm for subtraction approach simple "take away" problems in a variety of ways. Some count to the higher number on their fingers, then move backwards the desired number of places to get the result; others use a "counting up" procedure, beginning with the smaller given number. If concrete aids are available, some children set up a 1 : 1 correspondence and count the unmatched items to find the answer, while others use an "add on" model, starting with the smaller given number and finding the number needed to be added on to give the desired result.

In a study of nonroutine problem solving among secondary school students, Kantowski (1980) found evidence of a variety of styles in a given group. In the instruction used in the study, several solutions were given for each problem, some more elegant and efficient than others. The solution paths selected by the students varied widely. They did not always select what would be considered efficient solution paths, but those they followed suited their styles. Often, as problem-solving ability developed, students moved to more elegant solutions for problems similar to those they had solved earlier. Instruction should include a variety of solutions to problems to appeal to the variety of styles that might be present in a given group.

Hints or Cues for Solution

One of the most frustrating aspects of problem solving for teachers and students alike is its all-or-nothing aspect. There is nothing more annoying than being unable to find the *one step,* formula, or piece of information that would unlock the solution to a problem. Students can be helped to move closer to a solution even when they are unable to find one piece of information. The Soviets have an interesting concept in the "zone of proximal development." It is a zone in which a student is able to operate *with assistance,* but in which he or she is not able to operate alone. The application of this concept to instruction in problem solving could be valuable. Students can be provided with hints or cues that could be used or ignored during the solution of a problem. Such cues could be useful to students who are at a dead end and for whom the solution would "open up" if a cue or hint were provided. In the problem involving boxtops stated above one "cue" might be the question:

What is the total number of boxtops *both* fourth grades have?

Another "cue" could be the first translation:

$$10,000 - \boxed{\text{HAVE}} = \boxed{\text{NEED}}$$

and the question:

What is the value of $\boxed{\text{HAVE}}$?

For more complex problems, hints might include formulas (such as those for area or perimeter) or strategies that might be useful (such as guessing using a small number). Such "cues," or "hints" can also aid the students in understanding the problem. Students might be motivated to persevere, which could also result in more positive attitudes.

Relationship of Problem Solving to Proficiency in Basic Computational Skill

In recent testimony before the U.S. House of Representatives, Ed Esty observed that "in general, it appears that what is being taught is being learned. We see this on the satisfactory performance on the lower level skills and, unfortunately, in the drop in performance on the high level problem solving skills" (*NAEP Newsletter,* February 1980). Furthermore, too much emphasis on computational drill may be counterproductive to development of the flexibility needed for problem solving. How can this demand for emphasis on basic skills be reconciled with the need for development of problem-solving ability in the limited time available for mathematics instruction? Results from the USMES evaluation indicate that students who engage in problem-solving experiences do not suffer in basic skill development. Students in experimental groups did at least as well on tests of basic skill (Shann, 1976). The Dutch look at basic skill from an interesting point of view. They distinguish between skill and mechanical practice and contend that "anyone who can execute standard routines but is incapable of solving a new problem *has no skill"* (Van Dormolen, 1976). Problem solving is a basic skill.

Applications for use of algorithms must be included when the algorithms are taught. *Selection* of an algorithm is one of the difficulties hampering effective problem solving (Kulm and Days, 1979). If applications were taught along with algorithms, this difficulty might be alleviated.

Need for Changes in Evaluation Practices

Students are very *product*-oriented as is obvious from many of the studies reviewed. They are concerned with the *answer* and whether that answer is right or wrong. In several cases, it was seen that *understanding, setting up the problem,* and *planning* were factors in successful problem solving. Evaluation procedures should take into account these processes in order to motivate students to develop them. The curriculum must include some emphasis on understanding and planning. Exercises should reflect these emphases and teacher-made tests and grading procedures should give credit for using these processes. Questions such as "What do you need to be able to continue to solve this problem?" would make students think about what to do next and would assure students partial credit for understanding what to do instead of not giving them any credit at all if a computational error was made somewhere along the way. Students often work for grades, therefore grading practices should reflect all aspects of problem solving, including use of correct processes during solution.

What the Curriculum Holds for the 80s in Problem Solving

Thus far, we have looked at some implications of research in problem solving for the teacher who is trying to translate some of what has been found to classroom practice. But research, both educational and technological, has far-reaching implications for curriculum developers in addition to the suggestions for instruction cited.

Conspicuous by its absence in the discussion to this point has been the role of the calculator and, more dramatically perhaps, the role of the microcomputer in the problem-solving curriculum of the 80s. One reason for this, of course, is that it is too soon to see many published research studies dealing with the microcomputer. That does not, however, mean that work is not being done in problem solving using the computer. Of those 61 presentations in problem solving at NCTM's 1979 National Convention, several included instructional techniques using the microcomputer.

Teaching for problem solving is one of the most difficult tasks facing the teacher at any level. Let us consider for a moment some reasons for this difficulty.

(1) Often there are no new concepts to introduce or algorithmic skills to teach—the object of the instruction is to have students put together knowledge they have already acquired to solve the given problems, and techniques to do this are not readily available.

(2) All students in a given group are not familiar with the necessary content or the algorithms needed to solve some of the problems encountered, and the teacher is then faced with the problem of how to handle the diversity of backgrounds.

(3) That students work at different rates is especially true in the case of problem solving. Some students need much more time than others to understand a problem and to find what is being sought.

(4) There are many problem-solving styles resulting in different paths to the solution of problems, particularly those that are nonroutine (Moser, 1979; Kantowski, 1980). It is difficult for a single teacher to take into account the variety of styles that occur.

(5) Teachers are faced with increasing requirements in the curriculum and are pressured to emphasize computational skills and so have little time to assist students who are having difficulty in problem solving experiences.

(6) There is a lack of good sequences of related problems to use in instruction.

Many of these problems can be partially resolved through the use of computers. The computer brings the potential for radical change in the mathematics curriculum and for great support for teachers, bringing some relief in overcoming the difficulties in problem solving instruction outlined above. The microcomputer is not a panacea. It cannot resolve all the difficulties encountered in an effort to teach for problem solving, but it can definitely provide support in many of the problem areas. Specifically, a student interacting with a microcomputer can work at his or her own pace and take as little or as much time as needed on a given problem. The branching capability of the computer is perhaps the single aspect of this machine that makes it such an invaluable tool in teaching for problem solving. It can enable a student to request hints or cues or other information such as an algorithm that may have been forgotten or some instruction in content that was, perhaps, never learned. More important, possibly, this branching capability can take into account the many preferences or styles encountered in any group, by permitting a student to select a desired path to solution and to request cues to aid in finding a solution in his or her own preferred style. The capacity of the computer to provide for differences in educational backgrounds and preferred styles enables the teacher to deal effectively with what could otherwise be an unmanageable situation. Moreover, the capability of the computer to collect and process data can be used to provide feedback to a teacher on many aspects of a student's problem-solving experience. For example, a record of paths followed, hints or cues selected, as well as information needed (algorithms, instruction in content) can give a teacher a valuable profile of a student. Software to handle such demands of instruction for problem solving will be an important demand in curriculum development for the 80s.

The graphics mode of the computer cannot only provide excellent color diagrams related to problems, it can also simulate motion in a way not possible on the printed page or even in other media. This capability of the computer has implications for curriculum development as well as for further research related to spatial abilities. It has been observed in working with students (Kantowski, 1980) that the complexity of a figure will often keep a student from finding a solution to a problem. If the figures can be seen as they are generated, perhaps their complexity will be less overwhelming, and a solution more readily found.

The availability of the calculator mode of the computer also makes it an invaluable tool in teaching for problem solving. Conceptually simple problems previously unsuitable for widespread use because of tedious and time-consuming calculations may now be included at relatively low levels of instruction. Moreover, real problems that could often not be studied because of the tedious calculations required can and should be included in the curriculum at all levels. For example, in a recent exploratory study, I found that students were able to solve very complex problems in number theory with the aid of a microcomputer. In subsequent interviews with students after the problem-solving sessions, many admitted that they never would have attempted to solve some of the problems without the aid of the computer (Kantowski, 1980).

Need for Problems to Use in Instruction

Because students learn by solving similar kinds of problems (Polya, 1973; Cambridge Conference, 1963) sets of related problems need to be developed. Such sets could include problems of similar mathematical structure, problems involving similar content, or problems for which similar solution techniques would be useful. This reiterates the call for sequences of problems made by the participants of the Cambridge Conference. Although almost two decades have passed since the publication of their report, their belief that "the composition of problem sequences is one of the largest and one of the most urgent tasks of curricular development" (p. 28) is still relevant today.

In summary, the future looks very bright for the teaching of problem solving in the curriculum of the 80s. During the last decade, we have begun to see new trends for further research and directions for curriculum development suggested by the studies of the last few years. The curriculum promises to be more child-centered if new technology can be used to advantage in dealing with the diversity of problem solving styles. And, after all, the learner is what we are all about. Success in problem solving is a very rewarding experience—finding clues to so much lack of success in the past could be the breakthrough we've been seeking. The hope for the curriculum of the 80s is that the problem-solving component will present a new perspective to teachers and students alike. The prospect of new and more challenging problems and the applications promised by the advent of computer technology give us assurances of an exciting decade.

References

Cambridge Conference on School Mathematics. *Goals for School Mathematics.* Boston: Houghton Mifflin Co., 1963.

Carpenter, T. P.; Corbitt, M. K.; Kepner, H. S.; Lindquist, M. M.; and Reys, R. E. "Problem Solving." *Mathematics Teacher* 73 (1980a): 427-433.

Carpenter, T. P.; Corbitt, M. K.; Kepner, H. S.; Lindquist, M. M.; and Reys, R. E. "Results and Implications of the Second NAEP Mathematics Assessment: Elementary School." *Arithmetic Teacher* 27·(1980b): 44-47.

Carpenter, T. P.; Hiebert, J.; and Moser, J. *The Effect of Problem Structure on First Graders' Initial Solution Processes for Simple Addition and Subtraction Problems* (Technical Report No. 576). Madison, Wis.: Wisconsin Research and Development Center for Individualized Schooling, 1979.

Conference of Basic Mathematical Skills and Learning. *Vol. II: Working Group Reports.* Washington, D.C.: National Institute of Education, 1975.

Conference Board of the Mathematical Sciences, Committee on Computer Education. *Recommendation Regarding Computers in High School Education, 1972.*

Days, H. C.; Wheatley, G. H.; and Kulm, G. "Problem Structure, Cognitive Level and Problem Solving Performance." *Journal for Research in Mathematics Education* 10 (1979): 135-146.

Deringer, D., and Molner, A., eds. *The Next Ten Years, Perspectives and Recommendations.* Washington, D.C.: National Science Foundation, 1979.

Driscoll, M. J. "Mathematical Problem Solving: Not Just a Matter of Words. Research Within Reach." St. Louis, Mo.: CEMREL, 1979.

Ehr, C. "Cognitive Style and Information Selection During the Solution of Mathematics Word Problems." Ph.D. dissertation, University of Georgia, 1979.

Gurova, L. L. "Schoolchildren's Awareness of Their Own Mental Operations in Solving Arithmetic Problems." In *Problem Solving in Arithmetic and Algebra. Soviet Studies in the Psychology of Learning and Teaching Mathematics, Vol. 3.* Edited by J. Kilpatrick and I. Wirszup. Stanford, Calif.: School Mathematics Study Group, 1969.

Kalmykova, Z. I. "Analysis and Synthesis as Problem Solving Methods." In *Soviet Studies in the Psychology of Learning and Teaching Mathematics, Vol. 2.* Edited by M. G. Kantowski, J. Kilpatrick, E. G. Begel, I. Wirszup, and J. W. Wilson, Stanford, Calif.: School Mathematics Study Group, 1975.

Kantowski, M. G. "Processes Involved in Mathematical Problem Solving." Ph.D. dissertation, University of Georgia, 1974.

Kantowski, M. G. "Processes Involved in Mathematical Problem Solving." *Journal for Research in Mathematics Education* 8 (1977): 163-180.

Kantowski, M. G. *The Use of Heuristics in Problem Solving: An Exploratory Study.* Final Report, National Science Foundation Project SED 77-18543, 1980.

Kilpatrick, Jeremy. "Problem Solving and Creative Behavior in Mathematics." In *Studies in Mathematics, Vol. 195. Reviews of Recent Research in Mathematics Education.* Edited by J. W. Wilson and L. Ray Carry. Stanford, Calif.: School Mathematics Study Group, 1969.

Knifong, J. D., and Holton, B. "An Analysis of Children's Solutions to Word Problems." *Journal for Research in Mathematics Education* 7 (1976): 106-112.

Kulm, G., and Days, H. "Information Transfer in Solving Problems." *Journal for Research in Mathematics Education* 10 (1979): 94-102.

Lester, F. K. "Mathematical Problem Solving in the Elementary School: Some Educational and Psychological Considerations." In *Mathematical Problem Solving.* Edited by L. Hatfield. Columbus, Ohio: ERIC, 1975.

Licklider, J. C. R. "Impact of Information and Technology on Education in Science and Technology." In *The Next Ten Years, Perspectives and Recommendations.* Edited by D. Deringer and A. Molner. Washington, D.C.: National Science Foundation, 1979.

Meyer, R. A. "Mathematical Problem Solving Performance and Intellectual Abilities of Fourth-Grade Children." *Journal for Research in Mathematics Education* 9 (1978): 334-348.

Moser, J. M. *Young Children's Representation of Addition and Subtraction Problems* (Technical Paper No. 74). Madison, Wis.: Wisconsin Research and Development Center for Individualized Schooling, 1979.

National Assessment of Educational Progress. "Legislators Study Decline in Math Skills." *NAEP Newsletter* 13 (1980): 1-2.

National Council of Teachers of Mathematics. *An Agenda for Action: Recommendation for School Mathematics of the 1980's.* Reston, Va.: National Council for Teachers of Mathematics, 1980.

Polya, G. *How to Solve It.* 2nd ed. New York: Doubleday, 1973.

Polya, G., and Kilpatrick, J. *The Stanford Mathematics Problem Book.* New York: Teachers College Press, 1974.

Report of the Conference of Computer Oriented Mathematics and the Secondary School. Washington, D.C.: National Council of Teachers of Mathematics, 1963.

Schaaf, O. *Problem Solving in Mathematics, Lane County Mathematics Project Technical Report.* Eugene, Ore.: Lane County, 1980.

Schoenfeld, A. H. "Explicit Heuristic Training as a Variable in Problem Solving Performance." *Journal for Research in Mathematics Education* 10 (1979): 173-187.

Shann, M. H. *Measuring Problem Solving Skills and Processes in Elementary School Children.* Boston: Boston University, 1976.

Shann, M.; Reali, N.; and Bender, H. "Student Effects of an Interdisciplinary Curriculum for Real Problem Solving." Final Report, National Science Foundation Project PES 74-00542 A01, *Unified Science and Mathematics for the Elementary School.* Boston: Boston University, 1975.

Silver, E. A. "Student Perceptions of Relationships Among Mathematical Verbal Problems." *Journal for Research in Mathematics Education* 10 (1979): 195-210.

Sowder, L.; Barnett, J.; and Vos, K. *A Review of Selected Literature in Applied Problem Solving.* Final Technical Report, National Science Foundation Grant No. SED 77-19157. Northern Illinois University, 1978.

U.S. Congress. House. Committee on Science and Technology. *Computers and the Learning Society. Hearings Before the Subcommittee on Domestic and International Scientific Planning, Analysis and Cooperation.* Washington, D.C.: U.S. Government Printing Office, 1978.

Usiskin, Z. *Algebra Through Applications.* Reston, Va.: National Council of Teachers of Mathematics, 1980.

Van Dormolen, J. *Mathematical Skills.* Groninger, The Netherlands: Walters-Noordhoff, 1976.

Webb, N. "Processes, Conceptual Knowledge and Mathematical Problem Solving Ability." *Journal for Research in Mathematics Education* 10 (1979): 83-93.

Wilson, J. "Generality of Heuristics as an Instructional Variable." Ph.D. dissertation, Stanford University, 1967.

Response

Larry K. Sowder

Kantowski's chapter is especially important because of the likely forthcoming re-emphasis on problem solving in the mathematics curriculum. She has done an excellent job of clarifying how researchers use the term "problem," of describing research trends, of identifying problem-solving components, and of suggesting some research implications for instruction and curriculum development. I heartily endorse the spirit of her remarks.

The Problem-Solving Components

Understanding the Problem. During this first phase of solving a problem, a student should be able to call on a firm conceptual basis. For example, consider a student whose "understanding" of fractions is confined to rules of computation. The student is certain to be handicapped in dealing with a verbal problem in which a fraction appears, unless some rotely-learned procedure happens to fit the context of the problem. Studies which have manipulated the amount of time devoted to concept development have found that *at least 50 percent of class time should be spent on concept development* (Shipp and Deer, 1960; Shuster and Pigge, 1965; Zahn, 1966). In our example, a fraction should not be allowed to be only two numerals separated by a bar. Nor should a student's concept be limited to only one kind of model, like pie shapes. Fractional number concepts must be firmly founded in several models like folded papers, pieces of string, colored rods or strips, and sets of objects, for example, as well as work with circular and rectangular regions and number lines. The student would then have a richer conceptual basis from which to draw when confronted with "¾ of the

distance" or "¾ of the children" or "¾ of the amount" in a verbal problem.

In addition to the anecdotal evidence cited by Kantowski, some research studies suggest that students can profit from reflecting about problems. For example, middle schoolers perform better on problem solving if allowed to suggest possible ways to solve a problem, to discuss these ways, and then to arrive at a consensus (Blomstedt, 1974). Such a procedure contrasts sharply with the common modeling practice in which the teacher "shows" the students how to solve a problem. In the same vein, Rowe (1969) reported that teachers wait only about one second after asking a question before calling on a student, and then again only about one second after the student response before proceeding to the next question or remark. How much time does that allow for reflection? She found that the quality and quantity of student responses increased significantly when teachers sought to lengthen their wait-times to five seconds. Surely such results have implications for our questioning during problem solving! Finally, at least two studies (Graham, 1978; Keil, 1964) suggest that students can profit from making up their own problems and solving them. It seems clear that teaching procedures other than "monkey see, monkey do" are called for if our students are to develop a deeper understanding of a problem.

Planning. For the routine elementary school verbal problem, planning often comes down to the selection of the proper operation(s). The points in the previous section, especially the desirability of a strong conceptual basis, are directly related to one's ability to decide whether to add, subtract, multiply, or divide.

Kantowski mentions the use of cues or hints for solution of a problem, but this advice may be misinterpreted. For example, some teachers tell students to look for "key" or "clue" words which can suggest what operation to perform. Such advice is well-intended and sound insofar as the student then thinks about how the variables in the problem are related. The advice, however, can be misapplied by children. Consider a student who has been told that the word "gave" signals subtraction. If the student thoughtlessly applies that "rule" to this problem: "Pat had 50¢. Then Grandma gave Pat 35¢. How much did Pat have then?"—the student will not get the correct solution. Note that this thoughtless application is possible if the understanding-the-problem component is bypassed. Cues and hints for solution must not be used to circumvent thinking about the problem. Even first-graders blindly apply the keyword approach if they have been taught it (Nesher and Teubal, 1975).

Carrying Out the Plan. The solutions of most routine verbal problems and of many nonroutine problems involve computations. A quote from the NCTM agenda for the 1980s is appropriate:

> It is recognized that a significant portion of instruction in the early grades must be devoted to the direct acquisition of number concepts and skills without the use of calculators. However, when the burden of lengthy computations outweighs the educational contribution of the process, the calculator should become readily available (p. 8).

We must not delude ourselves by thinking that inserting longer or more frequent lists of story problems into a textbook is in itself the way to insure a focus on problem solving, if the lists do not require understanding the problem and planning. If problem solving *is* to be the focus, then these components, rather than computation, should receive the emphasis during work on story problems. It is consistent with such an emphasis to leave at least some of the computation connected with verbal problems to calculators.

Looking Back. Looking back—reviewing the solution of a problem just solved, seeking other possible solutions, and thinking of other problems that could be solved in the same way—is a component of problem solving that has not been subjected to much research even though it often appears in lists of problem-solving advice (Polya,

1971). Looking back would offer another chance for students to reflect on the problem and its solution. Despite the urgency often felt in the classroom to move on to the next problem immediately after solving one, a minute or two spent on looking back could allow attending to the processes involved rather than leaving the impression that the sole concern is the answer.

Research and Development

The NCTM agenda at least reminds us that the horse belongs before the cart. Whether the horse remains obscured by the cart remains to be seen. A great challenge to curriculum developers will be to make it possible for busy teachers of the 1980s to focus on problem solving. Supervisors will also have a great challenge: convincing some teachers that such a hard-to-teach emphasis is the proper focus. Kantowski has been perhaps too sanguine in making a case for microcomputers as a prime problem-solving vehicle. Certainly microcomputers in the schools are easily justified on a computer literacy basis; software to accomplish the many things that Kantowski envisions, however, is in an infant state. (Indeed, Kantowski is engaged in pioneering work using the microcomputer as an aid in problem-solving instruction.)

One difficult area of research deserves more attention: affective factors in problem solving. We have all observed even young children work concentratedly on an occasional task which has intrigued them. It would seem natural, then, to try to identify intriguing tasks and to isolate what makes them intriguing. Researchers have tried to tie verbal problems to student interests but by-and-large have not found that approach to yield better performance (Cohen, 1976; Travers, 1967). Perhaps having students write their own problems gave the better results (cited above) because of some affective-cognitive interplay.

It *is* encouraging that so many researchers are giving attention to problem solving, routine and nonroutine. With one and a half million "mini-laboratories" in operation every school day, with teachers trying different things, often on a hunch basis, one can still wish for a coherent, concerted research approach. If every school district were to seek out or develop a study of some aspect of problem solving, perhaps our teaching of problem solving could proceed on a more scientific basis.

References

Blomstedt, R. K. "The Effects of Consensus on Verbal Problem Solving in Middle School Mathematics." Ph.D. dissertation, University of Texas at Austin, 1974.

Cohen, M. P. "Interest and Its Relationship to Problem-solving Ability among Secondary School Mathematics Students." Ph.D. dissertation, University of Texas at Austin, 1976.

Graham, V. G. "The Effect of Incorporating Sequential Steps and Pupil-constructed Problems on Performance and Attitude in Solving One-step Verbal Problems Involving Whole Numbers." Ph.D. dissertation, Catholic University of America, 1978.

Keil, G. E. "Writing and Solving Original Problems as a Means of Improving Verbal Arithmetic Problem Solving Ability." Ph.D. dissertation, Indiana University, 1964.

National Council of Teachers of Mathematics. *An Agenda for Action.* Reston, Va.: National Council of Teachers of Mathematics, 1980.

Nesher, P., and Teubal, E. "Verbal Cues as an Interfering Factor in Verbal Problem Solving." *Educational Studies in Mathematics* 6 (March 1975): 41-51.

Polya, G. *How to Solve It.* 2nd ed. Princeton, N.J.: Princeton University Press, 1971.

Rowe, M. B. "Science, Silence, and Sanctions." *Science and Children* 6 (March 1969): 11-13.

Shipp, D. E., and Deer, G. H. "The Use of Class Time in Arithmetic." *The Arithmetic Teacher* 7 (March 1960): 117-121.

Shuster, A. H., and Pigge, F. L. "Retention Efficiency of Meaningful Teaching." *The Arithmetic Teacher* 12 (January 1965): 24-31.

Travers, K. J. "A Test of Pupil Preference for Problem-solving Situations in Junior High School Mathematics." *Journal of Experimental Education* 35 (Summer 1967): 9-18.

Zahn, K. "Use of Class Time in Eighth-grade Arithmetic." *The Arithmetic Teacher* 13 (February 1966): 113-120.

VIII. Computers

M. Vere DeVault

Mathematics instruction in elementary and secondary schools is frequently perceived to be more amenable to the use of computers than are other areas of the school curriculum. This is based on the perception of mathematics as a subject with clearly defined objectives and outcomes that can be reliably measured by devices readily at hand or easily constructed by teachers or researchers.

Because the purpose of this book is to provide implications of research evidence rather than a historical review of research, the studies include those that have been undertaken, completed, or published during the decade of the 70s. Vinsonhaler and Boss (1972) summarize ten major studies of CAI drill and practice that were completed prior to those reviewed in this chapter.

As with other chapters in this book, the literature reviewed here is limited to that which is clearly research. Some related topics otherwise of much interest to teachers and curriculum developers are omitted. Included, for instance, are few descriptions of current CAI programs, little identification of problem-solving modes for which computers are being used, and no descriptions of individualized instruction programs that are being managed by computers. A number of resources are available elsewhere for the interested reader. Wang (1978) lists nearly 3,000 programs being used throughout the United States. Bukoshi and Korotkin (1976), reporting on computer activities in secondary education, indicate that in 1975, 58 percent of secondary schools used computers for administrative or instructional purposes. This is an increase of 34 percent since 1970. Forty-three percent of the 1,459 computer-based courses were in mathematics.

Even though schools seem to be organized as though learning occurs

131

in compartmentalized fashion, we know that *even* with mathematics, a learner's self-concept, ability to read, and interest and functioning in the social and scientific world that surrounds us not only influence learning, but provide an integral part of growing up during the elementary and secondary school years. Much of what we continue to learn about reading, for instance, has implications for mathematics instruction; what we know about the use of computers with adults, as in military research, must have implications for mathematics; and the vast literature on the technology of mass media must also have implications for mathematics. Yet all of these sources of information are to go untapped in this chapter.

Computers

Computer technology used in mathematics education can be divided into two categories. The first of these is computer-assisted instruction (CAI), which, in turn, can be divided into drill-and-practice programs, instruction in mathematical concepts, problem solving, and computer programming. The second use of computer technology is in computer-managed instruction (CMI).

Computer-Assisted Instruction

The Computer Curriculum Corporation (CCC), under the direction of Patrick Suppes, and the mathematics instruction program included in Programmed Logic for Automated Teaching Operations (PLATO), under the direction of Robert Davis, represent the most extensive CAI programming and research efforts in mathematics education underway during the 70s. There are some interesting and significant common characteristics of these two comprehensive and visible programs. Both programs represent long-term developmental efforts by their respective directors. Suppes and Davis were mathematicians who turned their attention to school mathematics instruction early in the 60s; they were heavily involved in the development, implementation, and dissemination of mathematics programs for the schools during the period of the new math. Both worked extensively with the schools and directly with children within those schools. Both developed school program materials that were supported initially by research and development funds, with later editions of the materials made available to schools through commercial publishers. Suppes and Davis drew heavily on these earlier experiences and materials in the development of the courseware that has become central to their CAI efforts. A central point that must be made about these two programs, then, is that the content or substance of these CAI efforts has a substantial developmental history.

There are a number of other important points to be made about the two programs. They function on two of the largest computer systems committed to instructional use. Suppes has made interactive satellite transmissions of programs to South America and to Eastern states of the U.S.A. from his center in Palo Alto, California. The work Davis has done at Illinois uses only a small portion of the PLATO system for his elementary mathematics program. Suppes and Davis have both accompanied their instructional efforts with research efforts designed to monitor the effectiveness of their programs. Suppes has been at work with computers longer than Davis and, therefore, has a much larger body of research support for his efforts. In addition, Suppes' evidence includes data from a wider variety of populations than does the Davis research. Although both programs are intended to supplement the work of the regular classroom teacher, Suppes has focused largely on drill and practice, whereas Davis has included drill and practice along with other CAI instruction.

The Computer Curriculum Corporation (CCC)

Suppes (1979) has reported that CCC courses include the largest number of students using CAI in this country. By 1978, that number was in excess of 150,000 students in 24 states. Most of these students were disadvantaged or handicapped. A description of these courses and the strands strategy they represent is contained in Suppes (1979) and in Macken and Suppes (1976). The content strands include number concepts, horizontal addition, horizontal subtraction, vertical addition, vertical subtraction, equations, measurement, horizontal multiplication, laws of arithmetic, vertical multiplication, division, fractions, decimals, and negative numbers.

The evaluation of the effectiveness of CCC mathematics instruction as implemented in 21 different sites is reported by Macken and Suppes (1976) and by Poulsen and Macken (1978). Most populations were largely disadvantaged or handicapped youth. Many were Title I students. Others were either deaf, low IQ, or minority students whose opportunities for previous schooling had been severely limited.

In general, the data in these several studies were analyzed to answer three questions: (1) How was time on the CAI terminal related to achievement gains? (2) How was gain in achievement related to expected gains? (3) How was CAI placement related to standardized test placement of students? Additionally, there was considerable evidence gathered concerning the satisfaction of parents, teachers, and students with their work in CAI.

There is consistent evidence to support the claim that achievement gains are related to the amount of time students spend in CAI. (Program developers insist that ten minutes per day, five days per week, for a total of 1,500 minutes per year is the optimum time students should spend in CAI. Though planning for that amount of time is part of the conditions for participation in the CCC program, there is much evidence that it is a seldom-achieved goal.)

The most extensive CCC study concerning the relation of time at the CAI terminal to achievement (Poulsen and Macken, 1978) included data from a number of schools throughout the southern half of California. The subjects were largely disadvantaged students in grades three through nine. The authors report that ". . . no group received more than 75% of the recommended time, and most groups received considerably less, even though most students were scheduled to receive ten minutes of CAI per day . . . " (p. 3). The school day schedule is continuously altered with extra activities that occupy children's time and interests. Time taken away from regularly scheduled class activities is as likely to be taken from CAI as it is from any other scheduled activities in the school day. The proportion of the 1,500 recommended minutes per year actually spent in CAI ranged from 23 to 75 percent among the 20 groups reported. Within grade groups from three through nine in each of seven schools, the correlations between time in CAI and grade placement gain ranged from .53 to .99. The average correlation for all schools was .86. Such a correlation suggests a very strong relationship between time in CAI and grade-placement or achievement in mathematics.

The ratio of actual to expected gain in mathematics achievement as measured by the standardized test then in use in each district was consistently high throughout the several studies reported (Poulsen and Macken, 1978). In Freeport, New York, the ratio of actual gain to expected gain was 1.54 for 142 students who were initially one year below grade level and 1.91 for 20 Hispanic students who had experienced little previous schooling. In a study in Isleta, New Mexico, 96 students were in CAI for a period of seven months. In each of the four classes, the average gain was more than one month for each month in CAI; and the average growth rate was 1.33 months per month. The mean gains for the year made by Title I students in CAI at Shawnee Mission, Kansas, as measured by pretests and post-tests with the Key Math Diagnostic Arithmetic Test, ranged from .99 in grade two to 1.77 in grade five, with an average gain of 1.41 across grades one to six. There is, then, in these and other reported studies (Macken and Suppes, 1976; Poulsen and Macken, 1978), ample evidence that the CAI program developed and implemented through CCC consistently obtained results that surpassed expected gains for students least expected to succeed in mathematics as a result of regular classroom instruction.

Of specific relevance to the CCC mathematics curriculum is the question of the relationship between CAI placement, as determined by the student's performance in the instructional program, and standardized test placement. These studies reported consistently high correlations between these two measures. Among the 20 groups reported in the southern California study, the correlations between CAI final grade placement and the California Test of Basic Skills (CTBS), placement ranged from .28 to .87. It should be noted, however, that the next lowest correlation after .28 was .53 and the average of all correlations was .74. In the Ft. Worth study, correlations between CAI placement and the Stanford Achievement Test (SAT) mathematics computation grade placement were .62, .51, and .56, respectively, for grades three, four, and five (Macken and Suppes, 1976). The correlations between CAI placement and SAT mathematics applications grade placements were .35, .52, and .59 for grades three, four, and five. There does, then, seem to be a relatively high correlation between CAI placement and placement on standardized measures. These findings indicate that achievement on CAI mathematics as defined and developed by CCC is an appropriate curriculum for the development of skills and concepts measured by standardized instruments such as the Stanford and California achievement tests.

Though there has been a less direct attempt to measure the affective impact of CAI in the CCC program, a number of reported comments seem pertinent. Crandall (1977) reports that the CCC program in Los Nietos, California seemed to reduce truancy and vandalism in his school. St. Aubin, reporting on the Dolton, Illinois CAI program for the handicapped (see Macken and Suppes, 1976), claims that using the computer for individual work resulted in students' improved perceptions of themselves and their school. He reports that after some initial hesitancy, teacher response to CAI was positive and exciting as they saw improvement in the children's self-image.

To summarize the research findings of the CCC program, we have positive results for each of the questions for which answers were sought.

(1) Time children spent at CAI terminals was positively related to their achievement.

(2) There was substantial evidence that actual achievement gains exceeded expected gains based on previous experience of the subjects.

(3) There was evidence that grade placement as determined by the CAI program was highly correlated with grade placement on standardized tests.

(4) There was much subjective evidence to support the claim that attitudes of students and teachers toward CAI were positive.

PLATO and Elementary School Mathematics

Another CAI program designed for use in the elementary school (grades four through six) is the elementary mathematics program developed over a number of years by Davis and modified for use with the PLATO computer on the University of Illinois campus at Champaign-Urbana. Sixty interactive PLATO terminals are dedicated to elementary school mathematics. Four of these terminals are present in each participating classroom. "Each student received ½ hour of mathematics lessons, via computer, each school day, plus whatever instruction the teacher chose to provide. In fact, each teacher continued the 'regular' math curriculum from pre-PLATO years, except that a few teachers made adjustments to help relate the 'regular' curriculum to the PLATO curriculum" (Davis, 1980b).

The four content strands of the PLATO program were derived from Davis' earlier work. The first three strands are viewed by Davis as representing, respectively, content that is usually taught successfully in the schools (whole numbers); content that is not so successfully taught in the schools (fractions); and content that is not usually taught in grades four through six (graphs and functions). The fourth strand, concerned with programming computers, provides an option for children as a fringe benefit, but is not viewed as a part of the demonstration-research project.

It should be recognized that the PLATO project is not one that follows a typical programmed instruction format. Rather, PLATO mathematics is presented via terminals that include both the teletype and an audiovisual interface between computer and student, making it possible to follow the general instructional formats used in the earlier Madison Project materials. Davis calls the Madison Project strategy *paradigmatic learning experiences.*

The major research to determine the effectiveness of this instruction was undertaken during the 1975-76 school year by Swinton, Amarel, and Morgan (1978). Students in a dozen classrooms using the computer program were matched with students in classrooms that did not use the computer. At every grade level, children using the computer made significantly greater achievement gains than children in the control group on measures associated with the program as well as on computation and applications subtests of the California Test of Basic Skills. Of importance also were the positive attitudes exhibited by children.

Paraphrasing the final report to NSF (Slottow and others, 1977), Davis (1980b, p. 9) speaks of the strongly positive reactions of students and teachers to the PLATO mathematics program:

On every single attitude question used, differences strongly favorable to PLATO were observed. Pupils were enthusiastic about the mathematics lessons which the computer presented on the TV-like screens, may students sought *extra* sessions, their attitudes toward mathematics improved (as measured by a questionnaire), and so did their attitudes toward their own ability to deal with mathematics. Teacher assessments, though inevitably subjective, were very strongly positive, including even reports that PLATO had decreased anti-social behavior.

A few children's quotes taken from Stake's report (1978) of PLATO and fourth-grade mathematics illustrate the informal relationship children have with their computer teacher:

Dear Plato,
    Why does PLATO get messed up a lot?
                                        From Cool Cat

Dear Plato,
    This was a very nice session. Not too hard or too easy.
I am glad someone was able to invent you.
                                        Sally R.

Dear Plato,
    I like the games you play. But now I have to go.
                                        Kitty N.

Swinton, Amarel, and Morgan (1978, p. 24) report:

A particularly important outcome was revealed in positive effects on in- struments designed to measure students' understandings of any ability to represent concepts and operations, beyond mere facility in manipulation of symbols. The PLATO system here demonstrated that it was capable of *teaching*, as well as of providing drill and practice of concepts already introduced by classroom teachers.

That teachers and the PLATO program are important complements to one another cannot be denied. Swinton and others (1978, p. 25) con- clude that "Teacher effects are real, large, and idiosyncratic." The PLATO mathematics program is not teacher proof; it is not independent of the decisions and actions of individual teachers. Rather, the PLATO system is experienced differently by children in classrooms of different teachers. The authors report that teachers perform most effectively when they are given control over the curriculum. Though this is the case in many uses of computers in mathematics instruction, it is more apparent in PLATO than in Suppes' CCC program, where children are scheduled at CAI terminals out of the classroom and management diagnosis and prescription decisions are designed in the program.

PLATO mathematics was being developed in a number of dimen- sions at the same time, and Swinton and others (1978, p. 25) warn that

". . . simultaneous system and curriculum development is hazardous. . . ." They make a point of suggesting that more attention needs to be given to the development of courseware prior to research efforts than is frequently the case, and that it probably does make a difference who is involved in the development work. These researchers express a preference for those persons deeply involved in the subject matter, with extensive teaching experience and with a proven track record in curriculum development work, over those persons whose first interest is in the computer and who then seek a subject matter in which to make an application of computer expertise. The authors conclude with an expression of support for continued developmental efforts in the PLATO project and describe the system as having ". . . demonstrated its potential as a curriculum test bed . . ." (p. 26).

Comparative Research Studies

In addition to the major research and development efforts of Suppes and Davis, there are a number of other projects that have been reported in the literature during the 70s.

Nine studies comparing achievement and attitudes of students using computers with noncomputer students are summarized in Figure VIII-1. No researcher is represented more than once in this list, indicating that research efforts appear to be isolated. A study of student attitudes by Hess and Tenezakis (1973) is an interesting example included among these studies. In the study, there were 189 seventh to ninth grade subjects, 50 of whom had taken the CAI drill-and-practice program with the Suppes CCC materials for a period of one or two years. The 139 non-CAI comparison students were, on the average, performing better than the 50 CAI subjects, though among the CAI students were several who had been in the program because of their need for remediation.

Subjects were asked to compare their perceptions of computers with their perceptions of teachers and of textbooks. The CAI group indicated that computers had some real advantages over the classroom teacher. They viewed computers as fairer, easier, clearer, bigger, more likeable, and better than the teacher. The CAI group perceived the computer as having more information and making fewer mistakes than the teacher. According to the authors, students perceived computers as being more "charismatic" than teachers, with greater endurance of work, greater infallibility, and greater capability to help a student improve grades in mathematics. Non-CAI students viewed the computer even more favorably, perhaps reflecting the mystique of the computer that is so prevalent in society. In a further discussion of the effects of CAI instruction on students' perceptions of

Figure VIII-1. Studies Comparing Achievement and/or Attitudes of Students Using Computers with Noncomputer Students

Name of researchers	Year of Report	Grade level	Program type	Achieve-ment	Attitude
Crawford	1970	7	Drill & Practice	ns	
Hatfield and Kieren	1972	7,11	Programming	+	+
Street	1972	3-7	Drill & Practice	ns	−s
Martin	1973	3,4	Drill & Practice	+*	+
Hess and Tenezakis	1973	7-9	Drill & Practice		+
Milner	1973	5	Programming	ns	+
Smith	1973	7-9	Drill & Practice		+
Robitaille and Sherrill	1977	9	Programming	−s	+
Morgan	1977	3-6	Drill & Practice	s	+

   s = differences significantly in favor of computer students
−s = differences significantly in favor of noncomputer students
  + = differences in favor of computer students
  − = differences in favor of noncomputer students
 ns = differences not significant
  * = fourth-grade boys and low-ability students achieved more than others.

teachers, Brod (1972) reports that especially during the first year of CAI instruction, involvement reduced students' dependence on the teacher for task-specific resources. This undermining of the teacher's authority, he suggests, represents an unanticipated and undesirable consequence of CAI instruction.

Thirty-two dissertation studies (1969-1979) reporting on the comparative effectiveness of CAI and regular classroom instruction in changing achievement and attitudes of students were reviewed. In most of the studies, CAI was used for drill and practice, although in five, the computer served as tutor; in seven, the emphasis was on programming; and in two, problem solving. In comparison with the research studies reported in Figure VIII-1, results among the dissertation studies appear to be less consistently positive both for achievement and for attitudes. Of the 30 studies that compared achievement differences, 18 reported no significant differences; whereas 12 did report some significant differences that favored the

CAI groups. Of the 13 studies that compared attitudes, eight reported no significant differences and five reported significant differences (four for the CAI group and one for the non-CAI group). These dissertation studies as a group fail to generate support for a relationship between computer-assisted instruction and the attitudes of students. Though the achievement picture was varied, nonsignificant studies outnumbered those that significantly favored the computer group by 18 to 12.

Clearly, there are many problems with any attempt to summarize the findings of comparative studies of CAI. The unknown quality of the CAI instructional components (courseware) and the certain unevenness of that quality raises many questions about individual studies and about the results as a group. The hardware/software configurations are also varied from study to study. Indeed, there are a number of instances in which the still-developing state of either the hardware, the software, or both created problems that clearly had an impact on the results of these studies. The nature and extent of teacher involvement in planning for the supplementary CAI instruction differed from study to study and may have been quite minimal in most or all of the studies in this group. The lack of teacher involvement early in planning and implementation of computer use in classrooms would seem to be a major problem in many current computer applications. Nonetheless, there seems to be only minimal evidence from these studies that one could confidently proceed with such CAI programs in contexts, such as those used for these students with the expectation that achievement will be improved.

Additional Studies of CAI

A few studies have been reported that have investigated specific aspects of CAI, but have not been concerned with the comparative results between students with and without CAI instruction as in the case of studies included in the previous section.

Taylor (1975) reports on adaptive mastery, typical mastery, and traditional nonmastery models which employed different criteria for terminating CAI practice. All seventh-grade subjects received instruction in basic arithmetic skills. The adaptive mastery model differed from the typical mastery model in that it provided variable amounts of practice depending on feedback rather than a fixed amount of practice. Though the adaptive mastery model required less time, fewer practice items, and minimized overpractice, students in this group reached the same level of performance on post-tests and on delayed retention tests as did students in the other two models.

Keats and Hansen (1972) report a study in which they investigated the effectiveness of different kinds of CAI feedback. The area of study for 45 ninth-graders was proofs in mathematics. Feedback for the three groups of 15 students each included verbal definitions, numerical examples, or a combination of both types. The latter was thought to be more like that of typical classroom instruction. Though there were no significant differences in post-test scores, error analysis by groups over the 11 exercises* revealed that feedback in the form of verbal definition was more helpful than providing the learners with numerical examples or with a combination of the two. The authors conclude that this finding supports previous research, and especially is in keeping with Ausubel's (1961) support for providing the learner with a verbal explanation of underlying principles.

In a study by Herceg (1973), top track and middle track algebra II students were assigned to three computer treatment groups: individual rate setting with formally presented objectives, traditional classroom setting with formally presented objectives, and traditional classroom setting without formally presented objectives. Top track students did not achieve significantly higher when they were provided objectives for the unit. Middle track students, however, did achieve significantly higher when they were made aware of the objectives of the unit, although those in the individual rate setting treatment achieved significantly lower than students in the traditional classroom setting.

Dienes (1972) investigated the pacing question as it relates to drill-and-practice computer applications with sixth–grade students. His study was divided into two parts. In the first part, all 167 students completed a part of the computer program at their own pace. On the basis of this self-paced experience, students were assigned to treatment groups for which the pace of inaccurate students was decreased while it was increased for slow and accurate students. Fast and accurate responders were assigned to a "task-mean" treatment. Control groups continued to proceed at their own pace. There were significant differences in the achievement of treatment and control groups. Such external pacing assistance was beneficial to those students who did not adopt appropriate pacing habits.

Alspaugh (1971) reports that high school students learned FORTRAN programming language as well as college students, although they required twice the number of hours of instruction. It was suggested that "the grade placement for beginning FORTRAN courses can be lowered from grades 15-16 to grades 11-12 with comparable achievement . . ." (p. 47).

The computer has been used frequently as a tool to investigate learning and teaching strategies. Such a study is that by Kraus (1980) in which he investigated the heuristics of problem solving as subjects played a computer version of the game of NIM.

Computer-Managed Instruction

The 70s witnessed considerable activity in the development of computer-managed instruction (CMI) programs, though little research has come from those efforts. CMI systems provide a means for keeping information concerning available learning resources and the learning progress of individual students. In individualized instruction programs, CMI systems assist with the diagnosis and prescription of learning activities. Baker (1978) has provided an excellent review of such systems and the status of CMI by the mid-70s.

Several dissertation studies have compared the achievement and/or the attitudes of students who have experienced CMI with students in traditional classrooms. Of the six studies reported in Figure VIII-2, most report positive results that favor computer students over non-CMI students, though these differences are seldom significant.

Figure VIII-2. Dissertation Studies Comparing Achievement and/or Attitudes of CMI Students with Non-CMI Students

Name of researcher	Year of study	Grade level	Achievement	Attitude
Miller, Daniel	1970	6	s	
Miller, Donald	1970	6		s
Lee	1972	5		+
Akkerhuis	1972	6	s	ns
Wilkins	1975	8	+	
Chanoine	1977	4-6	+	

+ = differences in favor of computer students
s = differences significantly in favor of computer students
ns = differences not significant

Implications from Research in Computer Applications

The research of the past decade has been conducted on programs using the facilities of large-scale computers. As a new decade begins, the microcomputer is clearly seen as the way of the future. The flexibility of these computers, the control of these microcomputers at the local level (even at the classroom level), and the potential for involving students and teachers in a wider range of computer problems and technologies place issues raised during the past decade into new contexts. In these new contexts, new questions will be asked, many of which differ substantially from those raised in the immediate past. Nonetheless, a number of implications can be drawn for the future from the research that has been identified here. On the one hand, it appears that studies investigating computer technologies in mathematics instruction are too few, too piecemeal, and too unclear in their results to provide certain direction for elementary and secondary education. On the other hand, however, there does seem to be evidence emanating from specific centers that provides positive expectation from technology for the decade of the 80s.

Consistently positive results appear to come from centers that have a long, dedicated history of effort directed toward mathematics instruction. Results from the Stanford work of Suppes and that of Davis with PLATO at Illinois provide substantial evidence that CAI can be consistently effective in mathematics instruction and that such instruction even now is within reasonable cost bounds. During the 80s application of microcomputers to instructional tasks may be expected to greatly reduce costs compared to those incurred by the use of larger system computers that have been available during the 70s, and on which most or all research reported here was conducted.

Inconclusive results tend to accompany those projects which are associated with short-term efforts or are in their first two to four years of operation. Most of the dissertation studies appear to be a part of this set of short-term effort. No doubt many of these dissertation studies will provide the experience and insights on which the trends in the late 80s will be based, but few are sources of research evidence that can be used to support technological applications in mathematics instruction at the current time.

Mathematics learning by disadvantaged youth can be improved through certain computer applications of CAI. Studies that have investigated the impact of CAI on the mathematics learning of Native Americans, the deaf, inner-city Blacks, and bilingual Spanish-speaking youth have shown that mathematics achievement can substantially exceed expectations based on previous experiences. There is currently less evidence that "average" or "above average" learners can be helped as much. Federal, state, and foundation funding has been much more available for disadvantaged subgroups of children than for the general population. This has encouraged researchers to seek funding for hardware, software, and courseware development for applications to these special groups; hence, our current evidence provides more information about such learners than we have for others.

The history of educational innovation during the past two decades has been one that has taught us the importance of homegrown products. However, with the cost and time required for technological innovation, *we may be at a time when local districts will find it necessary to draw more heavily on the successful products developed and demonstrated effective elsewhere.*

Clearly, the development of computer-assisted instruction programs for use in the schools is in its infancy. The integration of computer activity with other activities underway in the classroom or with other activities that are particularly feasible because of the computer technology is only now beginning to be developed. *Needed are CAI programs that lead learners away from the computer and make it possible because of the computer to explore the world in ways not otherwise possible.* The addiction that accompanies those who pursue computer programming professionally must not be encouraged by the manner in which computers are used for instruction in the decade of the 80s.

There is some evidence that students and teachers react favorably to the use of CAI or CMI. Attitudes can appear positive even in instances where the results of achievement data are indifferent. It would seem, therefore, a fair warning that *attitudes alone may not be an adequate measure of the effectiveness of computerized classroom instruction.* The minimal evidence suggesting that the introduction of computers in the classroom, at least initially, reduces students' perceptions of the teacher's authority has considerable implication for the preservice and inservice preparation of teachers. *High technology introduced into the classroom must be brought by the teacher, not by an outsider to whom the teacher must turn each time a question arises concerning the use of that technology.* With the advent of microcomputers, teachers need relatively minimal preparation to take charge of those computers in the classroom. As an extension of the teacher who is competent to operate the computer and who understands how it may be best integrated into the classroom curriculum, technology can be used to sustain or boost the authority of and respect for the teacher in the classroom. When teacher and technology are seén as separate components of the classroom environment, humanism and technology also become separate environments. DeVault and Chapin (1980) have pointed to the need for a balance between supply and demand as the supply of microcomputer technology becomes more readily available through technical advances and cost reductions. The demand side of the supply and demand equation rests in the hands of the teacher, and technology must respond to teachers' perceptions of classroom instruction. Throughout development and implementation efforts, teachers must have roles that are comparable in importance and impact with those of the technologists if the introduction of this technology is to succeed in the classroom.

Certainly much of the adverse criticism of technology in the schools comes from fear that many humanistic characteristics of elementary and secondary schools will be sacrificed for the perceived or actual efficiency and effectiveness of technology. Much of our concern for humanism is centered around the search for better ways to meet individual differences and to assist students in reaching their full potentials.

Suppes (1979) alluded to these humanistic values:

> We do not yet realize the full potential of each individual in our society, but it is my own firm conviction that one of the best uses we can make of high technology in the coming decades is to reduce the personal tyranny of one individual over another, especially wherever that tyranny depends upon ignorance.

His own research and that of others investigating the effectiveness of his CAI programs indicate that those learners who may be expected to be low achievers in terms of their past histories can be helped through CAI. Are there values that one must sacrifice to attain these levels of achievement? What are they? How can we protect these values in the schools as technology becomes increasingly prevalent in the next decades?

Hoban (1977) placed the highest priority on human values over technology:

> Explicity, the major theme is that a symbiotic relationship exists between educational technology and human values and that in this symbiosis, human values are or should be invariantly transcendent.

These positions may represent extremes among the many currently in the literature expressing concern about the relationship between human values and the role of technology. Though much has been written about this concern, less has been done to clarify what these concerns mean for classroom practice. In the decades immediately ahead, teachers and others responsible for classroom environments must address these complex questions. The answers such practitioners provide will do much to shape the nature of instruction in the 21st century.

References

Akkerhuis, G. "A Comparison of Pupil Achievement and Pupil Attitudes With and Without the Assistance of Batch Computer-Supported Instruction." Ph.D. dissertation, University of Southern California, 1972.

Alspaugh, J. W. "The Relationship of Grade Placement to Programming Aptitude and FORTRAN Programming Achievement." *Journal for Research in Mathematics Education* 2 (1972): 44-48.

Andreoli, T. F. "The Effects of a Programmed Course in Computer Programming with Different Feedback Procedures on Mathematical Reasoning Ability." Ph.D. dissertation, University of Connecticut, 1976.

Ausubel, D. P. "Learning by Discovery: Rationale and Mystique." *The Bulletin of the National Association of Secondary School Principals* 45 (1961): 18-58.

Baker, F. B. "Computer-Based Instructional Management Systems: A First Look." *Review of Educational Research* 41 (1971): 51-70.

Baker, F. B. *Computer Managed Instruction: Theory and Practice.* Englewood Cliffs, N. J.: Educational Technology Publications, 1978.

Baltz, B. L. "Computer Graphics as an Aid to Teaching Mathematics." Ph.D. dissertation, Ohio State University, 1977.

Barnes, O. D. "The Effect of Learner Controlled Computer Assisted Instruction on Performance in Multiplication Skills." Ph.D. dissertation, University of Southern California, 1970.

Boyd, A. L. "Computer Aided Mathematics Instruction for Low-Achieving Students." Ph.D. dissertation, University of Illinois at Champaign-Urbana, 1974.

Brod, R. L. "The Computer as an Authority Figure: Some Effects of CAI on Student Reception of Teacher Authority." Ph.D. dissertation, Stanford University, 1972.

Bukoshi, W. J., and Korotkin, A. L. "Computing Activities in Secondary Education." *Educational Technology* 16 (1976): 9-23.

Casner, J. L. "A Study of Attitudes Toward Mathematics of Eighth Grade Students Receiving Computer-Assisted Instruction and Students Receiving Conventional Classroom Instruction." Ph.D. dissertation, University of Kansas, 1977.

Chanoine, J. R. "Learning of Elementary Students in an Individualized Mathematics Program with a Computer-Assisted Management System." Ph.D. dissertation, Wayne State University, 1977.

Cole, W. L. "The Evaluation of a One-Semester Senior High School Mathematics Course Designed for Acquiring Basic Mathematics Skills Using Computer-Assisted Instruction." Ph.D. dissertation, Wayne State University, 1971.

Confer, R. W. "The Effect of One Style of Computer-Assisted Instruction on the Achievement of Students Who are Repeating General Mathematics." Ph.D. dissertation, University of Pittsburgh, 1971.

Crandall, N. "CAI Gets Credit for Dramatic Achievement Gain for Minorities." *Phi Delta Kappan* 59 (1977): 290.

Crawford, A. N. "A Pilot Study of Computer-Assisted Drill and Practice in Seventh Grade Remedial Mathematics." *California Journal of Educational Research* 21 (1970): 170-181.

Davies, T. P. "An Evaluation of Computer-Assisted Instruction Using a Drill-and-Practice Program in Mathematics." Ph.D. disertation, United States International University, 1972.

Davis, R. B. *Discovery in Mathematics: A Text for Teachers.* New Rochelle, New York: Cuisenaire Company of America, 1980a.

Davis, R. B. "Alternative Uses of Computer in Schools: Cognition vs. Natural Language Statements." Paper presented at the annual meeting of the American Educational Research Association, Boston, April 1980b.

DeVault, M. V., and Chapin, J. "The Teacher Factor in the Supply and Demand Curve for Technology in the Schools." *Educational Technology* XXI (1980): 7-14.

Dienes, Z. B. "The Time Factor in Computer-Assisted Instruction." Ph.D. dissertation, University of Toronto, 1972.

Hansen, D. N. "Empirical Investigations Versus Anecdotal Observations Concerning Anxiety in Computer-Assisted Instruction." *Journal of School Psychology* 8 (1970): 315-316.

Harris, R. T. "An Evaluation of Computer-Assisted Instruction in Mathematics Using Test-and-Practice Method for Third and Sixth Grade Students." Ph.D. dissertation, United States International University, 1976.

Hatfield, L. L. "Computer-Assisted Mathematics: An Investigation of the Effectiveness of the Computer Used as a Tool to Learn Mathematics." Ph.D. dissertation, University of Minnesota, 1969.

Hatfield, L. L., and Kieren, T. E. "Computer-Assisted Problem Solving in School Mathematics." *Journal for Research in Mathematics Education* 3 (1972): 99-112.

Herceg, J. "A Study of the Coordinator's Role in the Introduction of Formally Presented Objectives and Individualized Learning Rates in Computer-Assisted Mathematics." Ph.D. dissertation, University of Pittsburgh, 1972.

Hess, R. D., and Tenezakis, M.D. "Selected Findings." *AV Communication Review* 3 (1973): 311-325.

Hoban, C. F. "Educational Technology and Human Values." *AV Communication Review* 25 (1977): 221-242.

Jamison, D.; Suppes, P.; and Wells, S. "The Effectiveness of Alternative Instructional Media: A Survey." *Review of Educational Research* 44 (1974): 1-67.

Katz, S. M. "A Comparison of the Effects of Two Computer Augmented Methods of Instruction with Traditional Methods upon Achievement of Algebra II Students in a Comprehensive High School." Ph.D. dissertation, Temple University, 1971.

Keats, J. B., and Hansen, D. N. "Definitions and Examples as Feedback in a CAI Stimulus-Centered Mathematics Program." *Journal for Research in Mathematics Education* 3 (1972): 113-122.

Kraus, W. H. "An Exploratory Study of the Use of Problem Solving Heuristics in the Playing of Games Involving Mathematics." Ph.D. Dissertation, University of Wisconsin-Madison, 1980.

Kriewall, T. E. "Applications of Information Theory and Acceptance Sampling Principles to the Management of Mathematics Instruction." Ph.D. dissertation, University of Wisconsin, 1969.

Lamb, R. L. "A Study on the Coordination of Graph Theory and Computer Science at the Secondary Level." Ph.D. dissertation, Georgia State University, 1976.

Lee, K. "A Study of the Effectiveness of Computer-Assisted Reporting of Fifth Grade Pupils' Mathematical Progress as Perceived by Parents and Pupils." Ph.D. dissertation, University of Iowa, 1972.

Macken, E., and Suppes, P. "Evaluation Studies of CCC Elementary School Curriculums." In *CCC Educational Studies*. Palo Alto: Computer Curriculum Corporation, 1976.

McLean, R. F. "A Comparison of Three Methods of Presenting Instruction in Introductory Multiplication to Elementary School Children." Ph.D. dissertation, Temple University, 1974.

Martin, G. R. *TIES Research Project Report: The 1972-73 Drill and Practice Study*. St. Paul, Minn.: Total Information Educational Systems, 1973.

Miller, D. D. "The Effect of Automated Marking on Sixth Graders' Mathematics Achievement." Ph.D. dissertation, Arizona State University, 1970.

Miller, D. D. "The Affective Influence of a Marking Machine on an Elementary Classroom." Ph.D. dissertation, Arizona State University, 1970.

Milner, S. "The Effects of Computer Programming on Performance in Mathematics." Paper presented at the annual meeting of the American Educational Research Association, New Orleans, February 1973.

Morgan, C. E., and others. *Evaluation of Computer-Assisted Instruction, 1975-76*. Rockville, Md.: Montgomery County Public Schools, 1977.

Nagel, T. S. "A Descriptive Study of Cognitive and Affective Variables Associated with Achievement in a Computer-Assisted Instruction Learning Situation." Ph.D. dissertation, Michigan State University, 1969.

National Council of Teachers of Mathematics. *Computer-Oriented Mathematics: An Introduction for Teachers*. Washington, D.C.: National Council of Teachers of Mathematics, 1963.

National Council of Teachers of Mathematics. *Computer-Assisted Instruction and the Teaching of Mathematics*. Washington, D.C.: National Council of Teachers of Mathematics, 1969.

O'Connell, W. B., Jr. "An Investigation of the Value of Exposing Slow-Learner Ninth Year Mathematics Pupils to a Relatively Short Computer Experience." Ph.D. dissertation, University of Rochester, 1973.

Ostheller, K. O. "The Feasibility of Using Computer-Assisted Instruction to Teach Mathematics in the Senior High School." Ph.D. dissertation, Washington State University, 1970.

Pachter, S. N. "A Computer-Assisted Tutorial Module for Teaching the Factoring of Second Degree Polynomials to Regents Level Ninth Year Mathematics Students." Ph.D. dissertation, Teachers College, 1979.

Philips, C. A. "A Study of the Effect of Computer Assistance Upon the Achievement of Students Utilizing a Curriculum Based on Individually Prescribed Instruction Mathematics." Ph.D. dissertation, Temple University, 1973.

Poulsen, G., and Macken, E. "Evaluation Studies of CCC Elementary Curriculums, 1975-1977." In *CCC Educational Studies*. Palo Alto: Computer Curriculum Corporation, 1978.

Robitaille, D. F., and others. "The Effect of Computer Utilization on the Achievement and Attitudes of Ninth-Grade Mathematics Students." *Journal for Research in Mathematics Education* 8 (1977): 26-32.

Ronan, F. D. "Study of the Effectiveness of a Computer When Used as a Teaching and Learning Tool in High School Mathematics." Ph.D. dissertation, University of Michigan, 1971.

Saunders, J. "The Effects of Using Computer-Enhanced Resource Materials on Achievement and Attitudes in Second Year Algebra." Ph.D. dissertation, University of Pittsburgh, 1978.

Slottow, D.; Davis, R. B.; and others. *Final Report: Demonstration of the PLATO IV Computer-Based Education System, January 1, 1972—June 30, 1976.* (Final report of NSF Contract No. C-723.) Urbana, Ill.: Computer-Based Education Research Laboratory, University of Illinois, 1977.

Smith, I. D. "The Effects of Computer-Assisted Instruction on Student Self-Concept, Locus of Control, and Level of Aspiration." Ph.D. dissertation, Stanford University, 1971.

Smith, I. D. "Impact of Computer-Assisted Instruction on Student Attitudes." *Journal of Educational Psychology* 64 (1973): 366-372.

Stake, B. E. "PLATO and Fourth Grade Mathematics." Paper presented at the annual meeting of the American Educational Research Association, Toronto, March 1978.

Street, W. P. *Computerized Instruction in Mathematics Versus Other Methods of Mathematics Instruction Under ESEA Title I Programs in Kentucky.* Lexington, Ky.: University of Kentucky, 1972.

Suppes, P.; Jerman, J.; and Brian, D. *Computer-Assisted Instruction: Stanford's 1965-66 Arithmetic Program.* New York: Academic Press, 1968.

Suppes, P. "Current Trends in Computer-Assisted Instruction." *Advances in Computers* 18 (1979): 173-229.

Swinton, S. S.; Amarel, M.; and Morgan, J. A. *The PLATO Elementary Demonstration, Educational Outcome Evaluation Final Report: Summary and Conclusions.* Princeton, N.J.: Educational Testing Service, 1978.

Taylor, S. S. "The Effects of Mastery, Adaptive Mastery, and Non-Mastery Models on the Learning of a Mathematical Task." Paper presented at the annual meeting of the American Educational Research Association, Washington, D.C., March-April 1975.

Vinsonhaler, J., and Bass, R. "A Summary of Ten Major Studies of CAI Drill and Practice." *Educational Technology* 12 (1972): 29-32.

Wang, A. C., ed. *Index to Computer-Based Learning.* Milwaukee: Instructional Media Lab, University of Wisconsin, 1978.

Washburne, R. M. "CEMP: A Computer Enriched Mathematics Program." Ph.D. dissertation, Cornell University, 1969.

Whelchel, B. R. "Computer-Assisted Instruction as a Determinant for Scholastic Achievement." Ph.D. dissertation, Stanford University, 1974.

Wilkins, P. W. "The Effects of Computer Assisted Classroom Management on the Achievement and Attitudes of Eighth Grade Mathematics Students." Ph.D. dissertation, Arizona State University, 1975.

Wilkinson, A. "An Analysis of the Effect of Instruction in Electronic Computer Programming Logic on Mathematical Reasoning Ability." Ph.D. dissertation, Lehigh University, 1972.

Wright, E. B. "Investigation of Selected Decision-Making Processes for Aspects of a Computer-Assisted and Mastery Learning Model in Basic Mathematics." Ph.D. dissertation, Pennsylvania State University, 1977.

Response

Robert B. Davis

How we use computers in education may well shape the future of education. Though it won't be the only influence, it is likely to be an important one. This is alarming because decisions about the educational uses of computers are not being made in the thoughtful, careful way that is called for. DeVault's excellent review is entirely correct (and stands virtually alone!) in looking at the pre-computer practices from which computer uses have grown. From the use of flash cards to teach "addition facts" like 3 + 2 = ?" there have grown computer programs that ask "3 + 2 = ?" From memorizing verbal definitions in pre-computer mathematics lessons, there have grown computer programs that ask one to type in (or select) verbal definitions.

In the case of our own Madison Project work, in pre-computer days we became convinced (from observing students) that verbal definitions don't work well with most children. They know a cat when they see one, but they cannot give you a verbal definition of "cat," and they can't learn from verbal definitions of this type. Hence, we developed the *paradigm teaching strategy* (Davis and others, 1978; Davis 1980a, 1980b, 1981) to give children experience with a mathematical concept without trying to use words to tell them about it. (Would you like to use words to tell children about an elephant if they had never seen one? Or would you rather take them to the zoo and show them an elephant?) We have created mathematics lessons for computer delivery built on this idea of paradigm teaching strategy: the computer program gives children experience with fractions, functions, and negative numbers, and does not try to introduce these ideas by purely verbal statements.

Now, the Madison Project approach here may be wise, or it may be foolish. In this short note one cannot argue the ultimate merits of any particular approach. But DeVault's point—a major one!—is that the Madison Project computer lessons grew out of the pre-computer Madison Project teaching practices. This is not peculiar to the Madison Project. The same situation exists for all computer-delivered lessons (so-called "courseware"). Whatever teacher or other specialist developed the lessons, he or she was building on pre-computer practices and expectations. I do not deplore this—on the contrary, it is inevitable, at least at first. But it means we must ask: how good were the pre-computer lessons from which the computer-administered versions have grown?

Computers Sharpen Choices

There is a reason why this has suddenly become critical. Perhaps above all else, computers compel us to make commitments. When a student is practicing, say, factoring polynomials under the guidance of a sympathetic teacher, there can be so much going on—so many transactions between the participants—

that it may be hard to say exactly what is happening. Sometimes for better, and sometimes for worse, subtlety and ambiguity rule the day. But put that same activity on a computer-administered lesson ("CAI mode"), and it quickly takes a more definite shape. On the computer, it becomes definitely drill, or definitely a game, or definitely a demanding lesson in reading comprehension, or definitely some other thing. Much of the ambiguity is gone, and we are faced with questions such as: Do we really *want* this much drill? Do we really *want* this many games? Should factoring polynomials be presented in a game-like atmosphere, anyhow? For that matter, should factoring polynomials be treated like drill? Before computers, these choices were less sharp. (One of my teachers in junior high argued that because of typewriters, spelling had become more important—one could not hide misspelling under a cloak of illegibility. Similarly, the definiteness of computer lessons precludes hiding uncertainty under a cloak of ambiguity and subtlety.)

Bases for Decisions

How, then, is one to choose among different possibilities for computer CAI lessons? At least four methods must be considered: (a) use of paper-and-pencil tests (especially multiple-choice tests), (b) use of methods for revealing the performance and present status of individual students, (c) use of task-based interviews, (d) direct examination of the computer lessons themselves.

There is abundant evidence that method (a), despite its unfortunate appeal (and its resultant popularity) is in fact the least satisfactory. Multiple-choice tests appear to produce "hard data." This appearance is deceiving. Such tests produce *numbers,* but do not give adequate descriptions of how students are *thinking* about mathematical problems. Erlwanger (1973) found students who seemed, on test scores, to be making satisfactory progress, but for whom many mathematical symbols were meaningless—for example, students who had no idea of the size of decimals and fractions, who did not know whether .7 was larger than 6 or smaller than 1. Alderman and others (1979) confirmed this, finding students who believed that 3/10 was equal to 3.10. Alderman and others also found that 50 percent of the advantage of one curriculum over another in one comparison study was due to the specific format in which questions were posed. Change the format, but not the content, of the questions, and half the advantage of the curriculum disappeared. Porter (1980) and his colleagues, in a group of careful studies, found very little commonality between what was presented in textbooks, what was taught in class, and what was covered on the best-selling tests (at the level of fourth-grade mathematics). *We have not been testing what we have*

thought we were teaching. Other fundamental reasons for fearing that test results can lead us in wrong directions are presented in Houts (1977), Tyler and White (1979), and elsewhere.

Indeed, my own main concern about computers is not about computers themselves. It is that test scores, because they are easily obtained and erroneously believed to be "scientific," will lead us into making incorrect choices, and thus into misapplying the promise of computers.

Measurements cannot, by their nature, resolve fundamental questions (see, for example, Kuhn, 1962). Results can always be interpreted in different ways, *if really fundamental uncertainties are involved.* Suppose we suspected that Curriculum A was sexist. Would we be satisfied by comparing test or questionnaire results of students in Curriculum A with those of students in a control group? Surely not; if no differences were found, there would remain the possibilities that our test or questionnaire was not sensitive enough, that the effect on students developed slowly and required a longer period of time to produce effects, or that the control curriculum itself was sexist. (This is not fanciful. Alderman and others found that students in a CAI curriculum had serious misconceptions about mathematics, *but so did the students in the control group.* One cannot defend ineffective curriculums by arguing that they are no worse than other ineffective curriculums!)

The large-scale introduction of computers into education is likely to be comparable to the large-scale introduction of gasoline-powered vehicles. Automobiles (which were not our only possible choice!) have facilitated suburban living (which can be pleasant) and thus contributed to the decline of our central cities (which had been based on proximity). They made the United States vulnerable to the political demands of the OPEC nations; they have contributed to our unfavorable balance of payments in international trade; they played a role in the destruction of the urban transit system in Los Angeles; and they have proved severely harmful to the environment. Had we based our early decisions about motor vehicles on an unthinking reliance on measurements, would we have measured the right things?

In the crucial decisions concerning computers, there can be no substitute for careful analytical thought, especially thought about our fundamental goals and fundamental values. This kind of analysis is NOT presently taking place.

The references listed below pursue further the problem of making wise decisions about the use of computers, and suggest a variety of alternative ways of using computers in education, some of which have not yet received the attention they deserve.

References

Alderman, Donald L.; Swinton, Spencer S.; and Braswell, James S. "Assessing Basic Arithmetic Skills and Understanding Across Curricula: Computer-Assisted Instruction and Compensatory Education." *Journal of Children's Mathematical Behavior* 2 (Spring 1979): 3-28.

Davis, Robert B. "Alternative Uses of Computers in Schools: Cognition vs. Natural Language Statements." Paper presented at the annual meeting of the American Educational Research Association, Boston, April 1980a.

Davis, Robert B. *Discovery in Mathematics: A Text for Teachers.* New Rochelle, N.Y.: Cuisenaire Company of America, 1980b.

Davis, Robert B. "What Classroom Role Should the PLATO Computer System Play?" *Proceedings of the 1974 National Computer Conference,* Chicago, May 1974.

Davis, Robert B. *Learning and Thinking in Mathematics.* London, England: Croom Helm, 1981.

Davis, Robert B.; Jockusch, Elizabeth; and McKnight, Curtis. "Cognitive Processes in Learning Algebra." *Journal of Children's Mathematical Behavior* 2 (1978): 10-320.

Erlwanger, Stanley H. "Benny's Conception of Rules and Answers in IPI Mathematics." *Journal of Children's Mathematical Behavior* 2 (Autumn 1973): 7-28.

Houts, Paul L. *The Myth of Measurability.* New York: Hart Publishing Co., 1977.

Kuhn, Thomas S. *The Structure of Scientific Revolutions.* Chicago: The University of Chicago Press, 1962.

Papert, Seymour. *Computer-Based MicroWorlds as Incubators for Powerful Ideas.* Cambridge: Massachusetts Institute of Technology, Artificial Intellgence Laboratory, 1978.

Papert, Seymour. *Mindstorms: Children, Schools and Powerful Ideas.* New York: Basic Books, 1980.

Papert, Seymour, and Solomon, Cynthia. "Twenty Things To Do With a

Computer." *Educational Technology* 12 (Spring 1972): 9-18.

Porter, Andrew. "The Importance of Defining the Content of School Instruction from Multiple Perspectives." Paper presented at the annual meeting of the American Educational Research Association, Boston, April 1980.

Tyler, Ralph W., and White, Sheldon H. *Testing, Teaching and Learning.* Washington, D.C.: The National Institute of Education, 1979.

IX. Calculators

J. Fred Weaver

Rarely, if ever before, has there been as much exploration and investigation regarding a particular aspect of mathematics instruction, over so wide an educational range, within so short a period of time, as has been the case concerning use of the electronic calculator. In the second of her state-of-the-art reviews, Suydam (1979a) asserted that "Almost 100 studies on the effect of calculator use have been conducted during the past four or five years. This is more investigations than on almost any other topic or tool or technique for mathematics instruction during this century" (p. 3). Reports of calculator use in school settings continued to be released, more or less unabated, during 1979 and 1980. In this chapter attention is given only to studies at the precollege level, grades K-12 (although many postsecondary investigations have been conducted and reported).

Delimitation by Exclusion

There are several things this chapter does *not* purport to be. It is not a comprehensive or definitive listing of research on calculator use in school settings. Such a listing would do no more than duplicate material found elsewhere (for example, Suydam, 1979b). This chapter will not summarize extensively and review critically any particular collection of investigations on calculator use in school settings. Such summaries and reviews also are readily available elsewhere (Suydam, 1979c, 1979d; Roberts, 1980). And this chapter makes no attempt to systematically and formally integrate or synthesize research on calculator use, whether by the commonly used voting method or by a more sophisticated meta-analysis.

154

What follows is a somewhat subjective distillation of the essence of consequential research findings to date on calculator use in school settings, the implications of such findings for classroom instruction, and some indication of research directions that need to be taken during the 1980s.

A Diverse Domain

The domain of research pertaining to calculator use in school settings has *diversity* as one of its principal attributes. Consider these illustrations.

At one extreme we find reports of things that may be termed "informal explorations" or "feasibility studies," which were limited in one way or another. For instance: in one case a sample of only *three* pupils was involved; in another case the exploration time consisted of *two class periods;* and some published reports described or illustrated ways in which calculators were used, and identified certain findings or conclusions, but without any supporting objective evidence or data.

At the other extreme we find reports of *experimental investigations* that were substantial in scope: in some instances the treatments extended over an entire school year; and in one instance the sample involved pupils from 50 classes, grades two through six, from five Midwestern states.

Between the two extremes we find a broad spectrum of investigations that vary markedly: in quality; in the ways in which, and the extent to which, calculators were used; in the class environments in which calculators were used; in the nature and scope of content involved; and in the effects considered (cognitive and affective). Typically included in these investigations were the following:

- drill on basic multiplication facts having factors of 7, 8 or 9;
- development of the "concept and skill of long division" at the fifth-grade level;
- a potpourri of work with rational numbers (in common- and/or decimal-fraction form), with percents, and with ratios and/or proportions;
- work within algebraic and trigonometric contexts, with varying degrees of content coverage as in other instances that follow;
- classes in general mathematics and consumer mathematics;
- work with remedial and/or low-achieving students, including the mildly handicapped;
- specialized-content classes or courses such as business arithmetic and chemistry;
- work that focused on particular properties such as "doing/undoing," and on problem-solving processes and strategies;

- consideration of student performance associated with a particular calculator *type* (for example, RPN—reverse Polish notation); and
- surveys pertaining to opinions and practices regarding calculator use in school settings.

The preceding categories are illustrative rather than exhaustive or definitive. Suydam's (1979b) listing should be consulted for more extensive classifications and for explicit references.

Freedom from Fear

There is no doubt that many calculator investigations have been prompted by a frequently expressed *fear* on the part of those persons who believe that "The principal objectives of mathematics instruction (at least in K-9) are that children learn the basic facts and pencil-and-paper algorithms. Such learning will not occur if hand-held calculators are made available in the schools" (Shumway, 1976, p. 572).[1]

Based on research evidence, is there valid cause for such fear (which is more intense the lower the grade level)? In seeking to answer that question I have drawn heavily on conclusions from several research reviews, each of which was fully cognizant of limitations inherent in certain of the investigations involved.

First, Suydam (1977) summarized 40 findings from 21 experimental and several action or preliminary investigations in which instructional effects of one kind or another were compared for *calculator* and *noncalculator* groups: "In 19 cases the Calculator group achieved significantly higher on pencil-and-paper tests (with which the calculator was not used). No significant differences were found in 18 instances. *In only three instances was achievement significantly higher for the Noncalculator group.*" Suydam concluded, "Such gross tabulations provide some support for the belief that calculators can be used to promote achievement" (p. 1, italics added).

Next, in a subsequent review involving a more extensive research base, Suydam (1978) indicated that "In most of the studies at the elementary school level, the data were collected to provide an answer (to parents and school boards, as well as to teachers) to the question, *'Will the use of calculators hurt mathematical achievement?'* The answer appears to be

---

[1] This quotation was Shumway's way of summarizing the argument of those opposed to calculator use in schools. The statement should *not* be construed to reflect Shumway's own view, or mine, regarding "the principal objectives of mathematics instruction" and the role of calculator use in relation thereto.

'*No.*' . . . What we do know is that the calculator, in general, facilitates mathematical achievement across a wide variety of topics, and this finding is verified at both elementary and secondary levels" (p. 7, italics added).

And a year later, in discussing the investigations upon which her second state-of-the-art review was based, Suydam (1979a) stated, "Many of these studies had one goal: to ascertain whether or not the use of calculators would harm students' mathematical achievement. The answer continues to be 'No.' The calculator does not appear to affect achievement adversely. In all but a few instances, achievement scores are *as high or higher* when calculators are used for mathematics instruction (but not on tests) than when they are not used for instruction" (p. 3).

Roberts (1980) reached similar conclusions, along with some additional ones also cited by Suydam, in his critical review of 11 elementary- and 13 secondary-level investigations:

> The majority of the studies completed at the elementary level showed computational advantages (6 of the 11) from the introduction of calculator usage into the mathematics instruction, . . . However, in only one study of the five investigating concepts were there conceptual benefits due to calculator usage and in only one study of the four investigating attitudes were there attitudinal benefits (p. 76).
>
> A majority of the secondary-level studies (6 of the 11 computation studies) found computational benefits due to calculator use. However, as was the case in the elementary studies, very little support was found for the hypothesis that calculator benefits transfer to the more conceptual (1 of the 8 concept studies) and affective areas (2 of the 9 attitudinal studies) (pp. 79-80).
>
> There seems to be little doubt about the computational value associated with calculator use. . . . However, [with respect to] conceptual and attitudinal impacts due to calculator use, there is less consensus as to what facts can be gleaned from the research literature (p. 94).

Since the review by Roberts and those by Suydam, two additional comprehensive investigations regarding the effects of calculator use vs. nonuse at the elementary-school level have been reported.[2]

Findings from a year-long investigation involving two different instructional programs led Moser (1979) to conclude that "Use of calculators with ongoing curricula at the second- and third-grade levels had no harmful effect upon arithmetic achievement" (p. xiii).

Two reports of a study—funded by the National Science Foundation and conducted in grades two through six (50 classes from five Midwestern

---

[2] These two additional investigations were included in Suydam's most recent state-of-the-art review (August 1980) which had not been released at the time this chapter was prepared.

states, with treatments that were in effect for 67 school days within an 18-week period)—cited these observations:

"The results . . . show no evidence of a decline in mathematics learning in classes that used calculators and there was some evidence that children in the primary grades benefit from using calculators in the study of mathematics" (Wheatley and others, 1979, p. 21).

"Children grow significantly on basic fact and mathematics achievement tests taken without the use of calculators regardless of whether or not calculators were used during instruction. . . . Children . . . did not develop any of the feared debilitations when tested without calculators because of calculator use for instruction" (Shumway and others, 1981, pp. 139, 140).

What, then, can be concluded about the *fear* that calculators may inhibit mathematics learning in schools? All in all, when calculators were used in the variety of ways investigated to date across a rather wide range of grade levels and content areas, evidence suggests that we have *no cause for alarm or concern* about potentially harmful effects associated with calculator use. This is particularly true with respect to computational performance, for which a nontrivial amount of evidence of *facilitating* effects has been reported. Even in the case of conceptual and affective aspects of mathematical learning, there is no extensive or strong body of evidence that suggests any pronounced inhibitory effects associated with calculator use. Seldom is the research literature so clear as it is in this respect.

An Implication for Classrooms at the Outset of the 1980s

I am convinced that we can embark on school mathematics instruction at the outset of the 80s with *freedom from fear*—freedom from fear that calculator use will have harmful or debilitating effects on students' mathematical achievement. Fear that calculator use will have marked negative cognitive or affective influences on students can no longer be used as a reason, or an *excuse,* for not welcoming and including calculators among the instructional aids and materials that have potential contributions to make in connection with school mathematics programs. The extent or fullness of that potential, however, remains to be ascertained.

How Calculators Are Used

Several surveys have been conducted to find out how calculators are used by classroom teachers in connection with their mathematics instruction. Suydam (1978) indicated:

At the elementary school level, four types of uses are predominant:

(1) Checking computational work done with pencil and paper.

(2) Games, which may or may not have much to do with furthering the mathematical content, but do provide motivation.

(3) Calculation: when numbers are to be operated with, the calculator is used with the regular textbooks or program.

(4) Exploratory activities, leading to the development of calculator-specific activities where the calculator is used to teach mathematical ideas.

At the secondary school level, the emphasis varies:

(1) Calculation, used whenever numbers must be operated with.

(2) Recreations and games.

(3) Exploration: because secondary school mathematics teachers' backgrounds are generally good, there is much more of this type of activity than at the elementary school level. In addition, the students who continue in higher-level courses are often intrigued to explore.

(4) Use of calculator-specific materials. There is at least one text integrating the use of calculators, with several others being field-tested (p. 4).

And from their survey of calculator use in grades 1, 3, 5, and 7, Graeber and others (1977) reported, "In the first grade, calculators were used most frequently for drill; the next three most frequent usages were for checking, motivation, and remediation. Use of the calculator for drill decreased with grade level. Above first grade the most frequent usage was for checking. Motivation and word problems were the next most frequently reported uses for calculators at the higher grade levels" (quoted by Suydam, 1978, p. 5).

It really is not surprising to find that the most common uses of calculators are relatively pedestrian ones. This should change as teachers learn about and personally explore more significant roles for calculators.

Looking Ahead

Although I can safely conclude that students' use of calculators will not inhibit their mathematical learning, *new research directions* and *imaginative curriculum development* are pre-eminent among things that are needed during the 1980s. The close relationship between them has been emphasized in a National Institute of Education and National Science Foundation (NIE/NSF) document (n.d.), *Report of the Conference on Needed Research and Development on Hand-Held Calculators in School Mathematics:* "Research must go hand in hand with development. . . . Developers should review relevant research in designing their curriculums, and researchers should investigate existing curriculum materials in choosing suitable contexts for their investigations" (p. 9).

In view of this reciprocating relationship between research and development, one is not necessarily distinguished from the other in some of the material that follows.

A Broad Research Need

Begle's examination of reports of calculator investigations led him to conclude that "In almost all these studies, the calculator was used merely as a supplement to a regular course. We have yet to see the results of evaluation of instructional programs which explicitly make use of the special capabilities of calculators" (p. 114).

In a similar vein, Roberts (1980), for instance, contended that a "crucial" consideration of future research "will be the necessity to develop treatments that utilize unique capabilities inherent to calculators. So far, most studies have not adequately integrated calculator use into the instructional process" (p. 95).

In Defense of the Past

I certainly concur with Begle and Roberts in their needs assessment. But it is important to recognize that calculator research to date has been essentially the first phase of an evolutionary process that is more or less natural, and not at all undesirable.

For one thing, some *feasibility* investigations—limited in scope, duration, and control—were necessary. We had to know whether certain things were even plausible—whether certain expectations were at all realistic—before more substantive studies could be considered at all sensibly. If young children, for instance, were prone to make many errors in using a calculator keyboard and in reading a calculator display (which we now know is *not* commonly the case), certain subsequent investigations involving young children would have been pointless. And knowledge of whether pupils are sufficiently sensitive to the use of various technical features of a calculator (such as automatic constants, memories of one kind or another, and logic systems involved) would be essential to deciding whether to even attempt to investigate certain calculator uses, treatments, or algorithms.

For another thing, it is not at all surprising or undesirable that many studies have been tied rather closely to existing curriculums. Marked changes in curriculums cannot be effected suddenly, desirable though some changes may be. If teachers and students were to use calculators at all, such use had to be first within the context of present curricular content. Moser (1979) was of the conviction that "research with an existing curriculum is judged to be a necessary prerequisite for future research" (p. 14).

The extensive suggestions he gave to second- and third-grade teachers in detailed day-by-day written form, and the ways in which researchers were available daily to consult with teachers and pupils in the "five-states study" (Wheatley and others, 1979; Shuway and others, 1981) represent exemplary efforts to effect *calculator-assisted* instruction *within existing curriculums* at the elementary-school level. In instances in which similar efforts were undertaken at middle- and secondary-school levels, work generally was of shorter durations of time with selected topics or pieces of content.

Curricular Changes

If programs of school mathematics instruction (and research pertaining to it) are to take full advantage of the "special" and "unique capabilities" associated with calculators, some curricular changes—substantial ones, in certain respects—must be effected. One view of this at the pre-algebra level has been suggested by the National Advisory Committee on Mathematical Education (1975):

> The challenge to traditional instructional priorities is clear and present. . . . First, the elementary school curriculum will be restructured to include much earlier introduction and greater emphasis on decimal fractions, with corresponding delay and de-emphasis on common fraction notation and algorithms. . . .
> Second, while students will quickly discover decimals as they experiment with calculators, they will also encounter concepts and operations involving negative integers, exponents, square roots, scientific notation and large numbers—all commonly topics of junior high school instruction. . . .
> Third, arithmetic proficiency has commonly been assumed as an unavoidable prerequisite to conceptual study and application of mathematical ideas. This practice has condemned many low achieving students to a succession of general mathematics courses that begin with and seldom progress beyond drill in arithmetic skills. Providing these students with calculators has the potential to open a rich new supply of important mathematical ideas for these students—including probability, statistics, functions, graphs, and co-ordinate geometry—at the same time breaking down self-defeating negative attitudes acquired through years of arithmetic failure (pp. 41, 42).

Less marked changes at higher instructional levels have been proposed. For instance, the NIE/NSF document (n.d.) suggested that "Use of calculators will require less revision of some current courses, such as high school algebra, geometry, and elementary functions and analysis. In all these courses, however, some parts need to be revised to include more applications that exploit the full potential of calculators" (pp. 11-12). Jewell's (1979) textbook analyses led him to conclude that approximately one-half of the content of algebra, geometry, and elementary functions texts and one-eighth of an algebra-trigonometry text could be appropriate for meaningful calculator application.

These illustrations indicate that across grades K-12 high priority must be given to research and development efforts that will (1) generate and evaluate instructional programs that make explicit use of special capabilities of calculators, (2) generate and evaluate treatments that use unique capabilities inherent in calculators, and (3) adequately integrate calculator use into the instructional process so that the calculator's presumed potential to facilitate and enhance the teaching and learning of school mathematics may be suitably assessed. Such assessment should give particular attention to the development and acquisition of problem-solving skills, mathematical concepts, and algorithmic processes.[3] I share Moser's (1979) belief that "the full benefit of calculator use in schools will never be realized until existing curricula are modified or new ones are developed that take advantage of a calculator's distinct features" (p. 14).

More than Mathematical Content

Research and development efforts associated with curricular change should involve more than content considerations per se. I wish to illustrate this by citing the possibility of a somewhat unconventional instructional sequence that would seriously challenge a position commonly held by a good many persons—a position expressed in the following way by Judd (1975):

> "Students must have a good background in manipulative math experiences before they can understand the inputs and outputs of the calculator. . . . Don't in short, put a calculator in the hands of a student before he . . . . understands the nature of the processes basic to arithmetic. Only after the students understand the meaning of the function they are performing should they be given a magic box to carry them to completion" (p. 48).

Now consider examples of the forms:

$$a + b = n \qquad a \times b = n$$
$$a - b = n \qquad a \div b = n$$

---

[3] In relation to both classroom instruction and research, calculators give rise to a massive set of issues and problems associated with *testing*—a veritable "can of worms" much too intertwined to be considered within the space limitations imposed upon this chapter. One inkling of this is glimpsed in the NACOME Report (1975): "Present standards of mathematical achievement will most certainly be invalidated in 'calculator classes.' An exploratory study in the Berkeley, California public schools [Kelley and Lansing, 1975] indicated that performance of low achieving junior high school students on the Comprehensive Tests of Basic Skills improved by 1.6 grade levels simply by permitting use of calculators" (p. 42).

where a and b are particular integers that are given, and n is to be calculated. It is assumed that students have been introduced to the integer *concept* (and referents for it), but not to operations on integers. Very deliverately let the calculator play a "magic box" role. For each operation in turn, let students select particular integers a and b (with some possible teacher guidance or suggestion in order to sample the domain fully) and use calculators to generate corresponding sums, or differences, or products, or quotients;[4] and then use a particular set of assignments as the basis for intuiting the assignment or operational rule. Only *after* that would attention be directed to uses for and applications of an operation on integers and the calculation rule associated therewith—along with instructional activities that provide justification for a rule and its "sensibleness."

Such a procedure is not at all out of line with Wittrock's (1974) hypothesis which may be "succinctly, but abstractly stated, . . . that human learning with understanding is a generative process involving the construction of (a) organizational structures for storing and retrieving information, and (b) processes for relating new information to the stored information. Stated more directly, all learning that involves understanding is discovery learning" (p. 182).

It was recommended in the NIE/NSF (n.d.) document that due consideration be given to *"current psychological, behavioral, and learning theory models"* (p. 19). Not only is the suggested approach to operations on integers in keeping with that recommendation, but early in the last decade Fennema's (1972) research raised some serious questions about the commonly held belief that children's learning should invariably proceed from the concrete/manipulative to the abstract/symbolic. Her finding, that in second-graders' introductory work with whole-number multiplication certain advantages accrued from using a symbolic referent in contrast with a concrete referent, is of considerable potential significance and import. Calculators provide an excellent means of investigating this phenomenon further and in a wide variety of mathematical contexts.

And when such investigations also take into consideration work being done in areas such as information processing and cognitive psychology, in cognitive or learning styles, and in "right brain" vs. "left brain" functions, we very well may find that a concrete/manipulative-to-abstract/symbolic instructional sequence is not as sacrosanct as we seem to believe it to be. For some students at certain times it may be desirable, even preferable, to introduce new mathematical content in a symbolic rather than

---

[4] The fact that the set of integers is not closed under division poses problems in the selection of values for a and b when working with examples of the form $a \div b = n$.

a concrete mode; and to change, even reverse, other facets of instructional sequences. During the 1980s our research and development efforts pertaining to calculators certainly should be planned to include specific tests of some of our cherished tenets regarding mathematics learning and teaching.

Other "Time-ly" Issues

There is a sense in which the immediately preceding consideration involved a *time* factor—the time when certain things occur in an instructional sequence. I now wish to direct attention briefly to three other issues in which *time* is of consequence.

First: Although far from easy to implement and control, *longitudinal* studies of effects of calculator use are essential. It is far from sufficient to know what may, or may not, be expected from relatively short-term uses of calculators, even over an entire school year. Effects over time, over a period of several school years, must be assessed before we have sufficient knowledge to answer some of the questions that have arisen, and will arise, regarding the use of calculators. Often it is not simply time per se that is of consequence, but the *cumulative* effects over time. Cross-sectional investigations have a contribution to make, but they cannot give us precisely the same kind of information to be gleaned from longitudinal studies.

Second: We already know that calculators make it possible for students to work more exercises and to deal with more applications and problem situations (in which computations are necessary) within a given unit of time. But is "do more of the same thing," particularly in the case of exercises, using time to best advantage? Likely not, and we should assess the cumulative effects of introducing and extending additional content within the time "gained," as opposed to doing more of the same exercises or whatever. It is rather commonly believed that much is to be gained by expanding curricular content; but will that in reality be the case—and if so, in what way(s)?

As we seek to broaden the scope of curricular content (presumably through time to be gained by calculator use for routine computations), it is my hope that we will not repeat a serious mistake reflected in the report of the Cambridge Conference on School Mathematics (1963) in which some wholly unrealistic content expansions were suggested, especially for grades K-8. It is easy to become literally "too mathematical" for the rank and file of students within those grades, and it behooves us to be realistic and sensible regarding the development of a broader content base for calculator use and exploitation in grades K-8, and the planning of investigations related thereto.

Third: As I suggested in an earlier source (Weaver, 1976), we need to investigate the long-term effects of (1) *introducing* certain pencil-and-paper algorithms *after* rather than before students have worked with numbers of certain magnitudes and domains using calculators, and (2) then emphasizing acquaintance and reasonable skill with such algorithms, rather than the attainment of a high degree of mastery coupled with a highly efficient level of performance.

Research and development activity should go even further and direct attention to the feasibility of an organizational pattern shown in Figure IX-1 for suitable phases of mathematics programs at elementary- and middle- or junior high-school levels.

A student's progress along each pathway is relatively independent of progress along the other. Calculators are used principally in connection with *A;* and when used in connection with *B,* their role is a different one than when used in connection with *A.* As computational proficiency is attained in an area within *B,* it may be used as needed and desired within *A.* But *lack of computational proficiency would never impede a student's progress in connection with A.*

The reorganized pattern suggested by Figure IX-1 rightfully makes *A* (rather than *B*) the principal focus of instruction.

Figure IX-1. Calculator-Influenced Reorganized Instructional Pattern

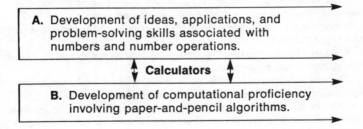

Using Research-Based Information in the Classroom

I am confident that during the early years of the 1980s we will see evidence of much-needed efforts to implement the NIE/NSF (n.d.) recommendation that "New means should be explored to rapidly communicate results of experiments with calculators and proposals for their use to the teaching profession, especially at the elementary school level" (p. 20). But it is not sufficient simply to disseminate information, regardless of how widely and innovatively that may be done. It is essential that disseminated information be *used,* in one way or another, in connection with classroom instruction.

Within a broader context Suydam and Weaver (1975) made the following suggestions which are fully applicable within the present context of calculator-research concerns:

> Teachers should test research findings [and treatments] in their own classrooms. Remember that just because research says something was best for a *group* of teachers in a *variety* of classrooms, doesn't necessarily mean that it would be best for *you* as an individual teacher in your *particular* classroom. . . . Teachers have individual differences as well as pupils! . . .
> Teachers must be careful not to let prior judgments influence their willingness to try out and explore: open-mindedness is important. . . . Be willing to investigate. . . .
> Research is not an end in itself—it should lead to some kind of action. You decide to change, or not to change; you will accept something, you will reject something. . . . *Do* something as a result of research: incorporate the conclusions of research [as tempered by unique attributes of your own situation and circumstances] into your daily teaching (p. 6).

In Conclusion

NCTM (1980) has recommended that "mathematics programs take full advantage of the power of calculators and computers at all grade levels" (p. 1), and has made the following related recommendations for *action:*

> All students should have access to calculators and increasingly to computers throughout their school mathematics program.
> Schools should provide calculators and computers for use in elementary and secondary school classrooms.
> Schools should provide budgets sufficient for calculator and computer maintenance and replacement costs.
> *The use of electronic tools such as calculators and computers should be integrated into the core mathematics curriculum.*
> Calculators should be available for appropriate use in all mathematics classrooms, and instructional objectives should include the ability to determine sensible and appropriate uses.
> Calculators and computers should be used in imaginative ways for exploring, discovering, and developing mathematical concepts and not merely for checking computational values or for drill and practice.
> *Curriculum materials that integrate and require the use of the calculator and computer in diverse and imaginative ways should be developed and made available.*
> Schools should insist that materials truly take full advantage of the immense and vastly diverse potential of the new media (p. 9).

Let us not only begin the 1980s with *freedom from fear,* but also progress through the 1980s with ever-increasing assurance of ways in which calculator use can facilitate school mathematics instruction, K-12, and enhance its quality. Forty-five years ago W. A. Brownell (1935) championed a change in school mathematics programs, formulating that which he termed the "meaning theory" in which "The basic tenet in the proposed instructional reorganization is to make arithmetic less a challenge to the pupil's memory and more a challenge to his intelligence" (p. 32).

Now, as we enter the 1980s, we are in a position to reformulate school mathematics programs in a manner that will free them from the shackles of the attainment of computational skills with pencil-and-paper algorithms as the basis upon which instruction is initiated, organized, and sequenced at the pre-secondary level; and will have analogous reorganizational implications for programs at the secondary level. *The calculator is the key.* Now is the time to turn that key in all earnestness.

References

Begle, E. G. *Critical Variables in Mathematics Education: Findings from a Survey of the Empirical Literature.* Washington, D.C.: Mathematical Association of America and National Council of Teachers of Mathematics, 1979.

Brownell, W. R. "Psychological Considerations in the Learning and the Teaching of Arithmetic." In *The Teaching of Arithmetic (10th Yearbook of NCTM).* Edited by W. D. Reeve. New York: Bureau of Publications, Teachers College, Columbia University, 1935.

Cambridge Conference on School Mathematics. *Goals for School Mathematics: The Report of the Cambridge Conference on School Mathematics.* Boston: Houghton Mifflin Co., 1963.

Fennema, E. "The Relative Effectiveness of a Symbolic and a Concrete Model in Learning a Selected Mathmatical Principle." *Journal for Research in Mathematics Education* 3 (1972): 233-238.

Graeber, A. O.; Rim, E. D.; and Unks, N. *A Survey of Classroom Practices in Mathematics: Reports of First, Third, Fifth and Seventh Grade Teachers in Delaware, New Jersey, and Pennsylvania.* Philadelphia: Research for Better Schools, Inc., 1977.

Jewell, W. F., Jr. "Hand Calculators in Secondary Education: Evaluation, Analysis and Direction." Ph.D. dissertation, State University of New York at Buffalo, 1979.

Judd, W. P. "A New Case for the Calculator." *Learning* 3 (1975): 41-48.

Kelley, J. L., and Lansing, I. G. *Some Implications of the Use of Hand Calculators in Mathematics Instruction.* Bloomington: Indiana University Mathematics Education Development Center, 1975.

Moser, J. M. *The Effect of Calculator Supplemented Instruction Upon the Arithmetic Achievement of Second and Third Graders.* (Technical Report No. 502). Madison: Wisconsin Research and Development Center for Individualized Schooling, 1979.

National Advisory Committee on Mathematical Education. *Overview and Analysis of School Mathematics, Grades K-12*. Washington, D.C.: Conference Board of the Mathematical Sciences, 1975.

National Council of Teachers of Mathematics. *An Agenda for Action: Recommendations for School Mathematics of the 1980s*. Reston, Va.: National Council of Teachers of Mathematics, 1980.

National Institute of Education and National Science Foundation. *Report of the Conference on Needed Research and Development on Hand-Held Calculators in School Mathematics*. Washington, D.C.: The Institute and the Foundation, n.d.

Roberts, D. M. "The Impact of Electronic Calculators on Educational Performance." *Review of Educational Research* 50 (1980): 71-98.

Shumway, R. J. "Hand Calculators: Where Do You Stand?" *Arithmetic Teacher* 23 (1976): 569-572.

Shumway, R. J.; White, A. L.; Wheatley, G. H.; Reys, R. E.; Coburn, T. G.; and Schoen, H. L. "Initial Effect of Calculators in Elementary School Mathematics." *Journal for Research in Mathematics Education* 12 (1981): 119-141.

Suydam, M. N. *Introduction to Research on Hand-Held Calculators, K-12 (Bulletin No. 9)*. Columbus, Ohio: Calculator Information Center, Ohio State University, 1977.

Suydam, M. N. *State-of-the-Art Review on Calculators: Their Use in Education*. Columbus, Ohio: Calculator Information Center, Ohio State University, 1978.

Suydam, M. N. *The Use of Calculators in Pre-College Education: A State-of-the-Art Review*. Columbus, Ohio: Calculator Information Center, Ohio State University, 1979a.

Suydam, M. N. *Calculators: A Categorized Compilation of References*. Columbus, Ohio: ERIC Clearninghouse, 1979b.

Suydam, M. N., ed. *Investigations with Calculators: Abstracts and Critical Analyses of Research*. Columbus, Ohio: Calculator Information Center, Ohio State University, 1979c.

Suydam, M. N., ed. *Investigations with Calculators: Abstracts and Critical Analyses of Research. Supplement*. Columbus, Ohio: Calculator Information Center, Ohio State University, 1979d.

Suydam, M. N., and Weaver, J. F. *Planning for Research in Schools* (Bulletin 11 in the series, *Using Research: A Key to Elementary School Mathematics*, rev.). Columbus, Ohio: ERIC Clearinghouse, 1975.

Weaver, J. F. "Some Suggestions for Needed Research on the Role of the Hand-Held Electronic Calculator in Relation to School Mathematics Curricula." In *Electronic Hand Calculators: The Implications for Pre-College Education*. Edited by M. N. Suydam. Columbus, Ohio: Ohio State University, 1976.

Wheatly, G. H.; Shumway, R. J.; Coburn, T. G.; Reys, R. E.; Schoen, H. L.; Wheatley, C. L.; and White, A. L. "Calculators in Elementary Schools." *Arithmetic Teacher* 27 (1979): 18-21.

Wittrock, M. C. "A Generative Model of Mathematics Learning." *Journal for Research in Mathematics Education* 5 (1974): 181-196.

Response

Richard J. Shumway

In order to facilitate the reader's interpretation of my remarks, I will attempt to summarize Weaver's major points as I see them:

1. The research on calculators and school mathematics is very *diverse* in scope and content.

2. We *need not fear* debilitating effects from student use of calculators.

3. We need to begin research efforts which explicitly study and make use of the *special capabilities* of calculators.

4. Could there be a *substantial impact* on school mathematics programs for student calculator use?

5. Can the calculator greatly *facilitate exploratory approaches* to mathematics?

6. Calculator research efforts should be quickly and effectively *communicated*.

7. Calculator research findings should be *tested by classroom* teachers in their own classrooms.

8. *All* students should have access to calculators.

9. "We are in a position to reformulate school mathematics programs in a manner that will free them from the shackles of the attainment of pencil-and-paper computational algorithms and skills as the basis upon which instruction is initiated, organized, and sequenced."

In spite of Weaver's warnings that he would not provide an extensive review of the research on calculators, I find his major points follow with scholarly care from current work and his own thoughtful experiences and deliberations. My own experiences and knowledge of the literature causes me to resonate fully with these points and add my wholehearted endorsement.

However, as a co-conspirator in the calculator revolution, I find Weaver strangely mute on several important points and school strategies for implementation of calculator use for mathematics instruction.

Attitude

One of the most powerful and consistently reported effects of student use of calculators is the high enthusiasm and valuing students have for calculator-aided mathematics activities. Any device which causes so much pleasure to be associated with mathematics and increases the probability appropriate mathematics strategies will be chosen for problem solving deserves special note. I also believe parental openness to calculator use can be influenced significantly by the enthusiastic response children exhibit for calculator-aided mathematics. So in the pragmatic real-world problems of making calculators available to all children, such attitude results are most important.

Computation

It would be well to note, obvious as it is, that calculators are the quickest, most accurate computational algorithms available to children today. In fact, the primary function of a calculator is to compute, and in the hands of children, the calculator serves the computational function better than any other technique or device in existence.

The effect of calculator use on measurement of mathematics achievement is important for two primary reasons. First, students' ability to perform computations is significantly improved through the use of a calculator. Secondly, few teachers see the logic of training students *with* calculators and then testing students and evaluating teachers *without* student use of calculators. For these reasons, I believe it is of the highest priority that all testing, classroom and standardized, should be done with student use of calculators immediately. Few teachers are going to make significant use of calculators, nor are tests and textbooks likely to be designed for calculator use until calculators are actually

used for testing. Consequently, the first needed step is to use calculators for all testing. The inconveniences of such a plan are minor compared to the significant delays failure to take such action will cause.

In my view, mathematics curriculum is most properly influenced by three factors: mathematical structure, learning theories, and societal needs. Weaver asserts mathematical structure can be supported by calculator use, calls for more research on learning and calculators, but is unfortunately mute on societal needs. It would seem societal needs might provide significant support for the use of calculators by all children. A recent survey [1] of 100 "random" occupations reports that fully 98 percent of those persons involved used a calculator. How many company presidents would support a personnel manager's recommendation that all new employees spend 120 hours learning skills by hand which are currently done by machines in the plant? Is our current education program guilty of such an error today? The argument seems strong that, based on societal needs, calculators should be used for mathematics instruction.

Communication

Weaver's points about the quick communication of results and the

---

[1]Saunders, H. "When Are We Ever Gonna Have to Use This?" *Mathematics Teacher* 73 (1980): 7-16.

testing of the ideas with teachers' own students are excellent. No research ideas should be introduced without careful, local evaluation. Such evaluation can do wonders for parent, teacher, and administrator receptivity to the use of calculators.

I was most excited by the quote near the end that Weaver cites admonishing one to "do something as a result of research." Sword in hand, I look for suggestions for what to do and all I find is "at the outset of the 1980's" I can embark on calculator use "with freedom from fear." In conclusion, however, Weaver supports the recommendations for action made by NCTM which suggest basically that all students should have access to calculators, they should be supported by budget, and integrated into the core mathematics curriculums. Weaver's citing of the NCTM Agenda for Action Recommendations is most appropriate and provides the best national endorsement for calculator use currently available. But here is what I wish he had said:

1. Calculators should be available for use at all times by all students.

2. All standardized testing should be done with calculators available. (Right now, do not wait for tests to be rewritten. Start with your school.)

3. Paper-and-pencil algorithms should only be taught to enhance and enrich mathematical thinking and not used or practiced as a computational tool.

4. Teachers should ask for, develop, and use supplementary materials that support students' ability to learn mathematics and solve problems. Dramatic changes ought to occur in mathematics curriculums. Material, trivially done with calculators, should be thrown out. The focus should be on activities that teach children mathematical thinking while using a calculator.

5. Programs should be evaluated carefully.

6. Expectations that children enjoy mathematics should be present.

Are these recommendations unsupported by research? Maybe; but as Weaver himself states, there is more research on calculators than on any other topic. Can researchers, teachers, and administrators not "do something" now? Assuming careful testing, are we not better prepared than ever before to take such a strong stand regarding the use of calculators? When I think of my own children, I want to say: "Don't waste their time doing trivia! Give them a calculator and get on to teaching them the mathematics they cannot now do." I was hoping Weaver would close with strong action items. Perhaps his quiet, reasoned approach is best, but I'm for a little dramatic action with careful testing.

About the Authors

GRACE M. BURTON is Associate Professor, Department of Curricular Studies, University of North Carolina at Wilmington.

THOMAS P. CARPENTER is Professor, Department of Curriculum and Instruction, University of Wisconsin-Madison.

THOMAS J. COONEY is Professor, Mathematics Education, University of Georgia, Athens.

MARY K. CORBITT is Assistant Professor, University of Kansas, Lawrence.

ROBERT B. DAVIS is Associate Director, Computer-Based Education Research Laboratory, and Professor, College of Education, University of Illinois at Urbana-Champaign.

DONALD J. DESSART is Professor of Mathematics and Mathematics Education, University of Tennessee, Knoxville.

M. VERE DEVAULT is Professor, Department of Curriculum and Instruction, University of Wisconsin-Madison.

ELIZABETH FENNEMA is Professor, Department of Curriculum and Instruction, University of Wisconsin-Madison.

KAREN C. FUSON is Associate Professor of Learning Development and Instruction, University of Kentucky, Lexington.

THOMAS L. GOOD is Professor of Education, University of Missouri, Columbia.

DOUGLAS A. GROUWS is Professor of Education, University of Missouri, Columbia.

172

JAMES HIEBERT is Assistant Professor of Mathematics Education, University of Kentucky, Lexington.

MARY GRACE KANTOWSKI is Associate Professor, Mathematics Education, University of Florida, Gainesville.

HENRY S. KEPNER is Professor, Department of Curriculum and Instruction, University of Wisconsin-Milwaukee.

MARY MONTGOMERY LINDQUIST is Chair, Mathematics Department, National College of Education, Evanston, Illinois.

ROBERT E. REYS is Professor, University of Missouri, Columbia.

RICHARD J. SHUMWAY is Professor, Mathematics Education, The Ohio State University, Columbus.

LARRY K. SOWDER is Associate Professor of Mathematical Sciences, Northern Illinois University, DeKalb.

DIANA WEARNE is Assistant Professor, Department of Curriculum and Instruction, University of Kentucky, Lexington.

J. FRED WEAVER is Professor, Department of Curriculum and Instruction, University of Wisconsin-Madison.

ASCD Publications, Fall 1981

Yearbooks

A New Look at Progressive Education
(610-17812) $8.00
Considered Action for Curriculum Improvement
(610-80186) $9.75
Education for an Open Society
(610-74012) $8.00
Evaluation as Feedback and Guide
(610-17700) $6.50
Feeling, Valuing, and the Art of Growing:
Insights into the Affective
(610-77104) $9.75
Life Skills in School and Society
(610-17786) $5.50
Lifelong Learning—A Human Agenda
(610-79160) $9.75
Perceiving, Behaving, Becoming: A New Focus
for Education (610-17278) $5.00
Perspectives on Curriculum Development
1776-1976 (610-76078) $9.50
Schools in Search of Meaning
(610-75044) $8.50
Staff Development/Organization Development
(610-81232) $9.75

Books and Booklets

About Learning Materials (611-78134) $4.50
Action Learning: Student Community Service
Projects (611-74018) $2.50
Adventuring, Mastering, Associating: New
Strategies for Teaching Children
(611-76080) $5.00
Approaches to Individualized Education
(611-80204) $4.75
Bilingual Education for Latinos
(611-78142) $6.75
Classroom-Relevant Research in the Language
Arts (611-78140) $7.50
Clinical Supervision—A State of the Art Review
(611-80194) $3.75
Curricular Concerns in a Revolutionary Era
(611-17852) $6.00
Curriculum Leaders: Improving Their Influence
(611-76084) $4.00
Curriculum Materials 1980 (611-80198) $3.00
Curriculum Theory (611-77112) $7.00
Degrading the Grading Myths: A Primer of
Alternatives to Grades and Marks
(611-76082) $6.00
Developmental Supervision: Alternative
Practices for Helping Teachers Improve
Instruction (611-81234) $5.00
Educating English-Speaking Hispanics
(611-80202) $6.50
Effective Instruction (611-80212) $6.50
Elementary School Mathematics: A Guide to
Current Research (611-75056) $5.00
Eliminating Ethnic Bias in Instructional
Materials: Comment and Bibliography
(611-74020) $3.25
Global Studies: Problems and Promises for
Elementary Teachers (611-76086) $4.50
Handbook of Basic Citizenship Competencies
(611-80196) $4.75
Humanistic Education: Objectives and
Assessment (611-78136) $4.75
Learning More About Learning
(611-17310) $2.00
Mathematics Education Research
(611-81238) $6.75

Measuring and Attaining the Goals of Education
(611-80210) $6.50
Middle School in the Making
(611-74024) $5.00
The Middle School We Need
(611-75060) $2.50
Moving Toward Self-Directed Learning
(611-79166) $4.75
Multicultural Education: Commitments, Issues,
and Applications (611-77108) $7.00
Needs Assessment: A Focus for Curriculum
Development (611-75048) $4.00
Observational Methods in the Classroom
(611-17948) $3.50
Open Education: Critique and Assessment
(611-75054) $4.75
Partners: Parents and Schools
(611-79168) $4.75
Professional Supervision for Professional
Teachers (611-75046) $4.50
Reschooling Society: A Conceptual Model
(611-17950) $2.00
The School of the Future—NOW
(611-17920) $3.75
Schools Become Accountable: A PACT
Approach (611-74016) $3.50
The School's Role as Moral Authority
(611-77110) $4.50
Selecting Learning Experiences: Linking
Theory and Practice (611-78138) $4.75
Social Studies for the Evolving Individual
(611-17952) $3.00
Staff Development: Staff Liberation
(611-77106) $6.50
Supervision: Emerging Profession
(611-17796) $5.00
Supervision in a New Key (611-17926) $2.50
Urban Education: The City as a Living
Curriculum (611-80206) $6.50
What Are the Sources of the Curriculum?
(611-17522) $1.50
Vitalizing the High School (611-74026) $3.50
Developmental Characteristics of Children and
Youth (wall chart) (611-75058) $2.00

Discounts on quantity orders of same title to
single address: 10-49 copies, 10%; 50 or more
copies, 15%. Make checks or money orders
payable to ASCD. Orders totaling $20.00 or
less must be prepaid. Orders from institutions
and businesses must be on official purchase
order form. Shipping and handling charges will
be added to billed purchase orders. *Please be
sure to list the stock number of each publication, shown in parentheses.*

Subscription to *Educational Leadership*—$18.00
a year. ASCD Membership dues: Regular (subscription [$18] and yearbook)—$34.00 a year;
Comprehensive (includes subscription [$18]
and yearbook plus other books and booklets
distributed during period of membership)—
$44.00 a year.

Order from:

**Association for Supervision and
Curriculum Development
225 North Washington Street
Alexandria, Virginia 22314**

THE COMMONWEALTH AND SUEZ

The Commonwealth and Suez

A DOCUMENTARY SURVEY

Selected, Edited, and with Commentaries by
JAMES EAYRS

LONDON
OXFORD UNIVERSITY PRESS
NEW YORK   TORONTO
1964

Oxford University Press, Amen House, London E.C.4

GLASGOW NEW YORK TORONTO MELBOURNE WELLINGTON
BOMBAY CALCUTTA MADRAS KARACHI LAHORE DACCA
CAPE TOWN SALISBURY NAIROBI IBADAN ACCRA
KUALA LUMPUR HONG KONG

© Oxford University Press 1964

*Printed in Great Britain
by the Camelot Press Ltd.
Southampton*

To Betty and Hartley Lofft

CONTENTS

PART III: COMMONWEALTH REACTIONS TO THE
INVASION OF SUEZ, OCTOBER–NOVEMBER 1956

PART IV: THE COMMONWEALTH AND THE UNITED NATIONS EMERGENCY FORCE

III. AUSTRALIA

IV. NEW ZEALAND

PART V: COMMONWEALTH RECKONINGS

ACKNOWLEDGEMENTS

I AM grateful to the following persons and publishers for permission to reproduce material: Cassell & Company Ltd., for the quotations from *Full Circle: The Memoirs of the Rt. Hon. Sir Anthony Eden*, and *Speech Is Of Time: Selected Speeches and Writings by the Right Honourable Robert Gordon Menzies*; the Rt. Hon. Lester B. Pearson, and S. J. Reginald Saunders and Company Limited, for the quotation from *Diplomacy in the Nuclear Age*; the Executor of the Estate of the late Lionel Shapiro, and *Maclean's Magazine*, for the quotations from 'Where Canada Stands in the World Crisis'; Lord Avon, for the quotations from his broadcasts of 16 August 1956 and 8 November 1956; Mrs. Hugh Gaitskell, for the quotation from the late Hugh Gaitskell's broadcast of 8 November 1956; the Rt. Hon. James Griffiths, for the quotation from his broadcast of 15 November 1956; the Rt. Hon. Selwyn Lloyd, for the quotations from his broadcasts of 16 August 1956 and 15 November 1956.

Research took me into many libraries in many cities, most often and most usefully into the Library of the Canadian Institute of International Affairs, Toronto, to whose unfailingly helpful staff it is once again a pleasure to express my gratitude.

To the Oxford University Press go my thanks for expert and friendly editorial assistance without which my rather complicated manuscript could never, across some 3,000 miles, have become a book.

Toronto
August, 1963   J.E.

ABBREVIATIONS USED IN FOOTNOTE REFERENCES

COMMONWEALTH PARLIAMENTARY DEBATES

Abbreviations used for the Parliamentary Debates of Commonwealth Member countries are as follows:

Australia, H.R. [House of Representatives] Deb. [Debates].
Canada, H.C. [House of Commons] Deb.
Ceylon, H.R. [House of Representatives] Deb.
New Zealand, H.R. [House of Representatives] Deb.
Pakistan, N.A. [National Assembly] Deb.
South Africa, H.A. [House of Assembly] Deb.
United Kingdom, H.C. [House of Commons] Deb. (All are those of the Fifth Series.)

Debates in the Indian House of the People are referred to as Lok Sabha Deb.

GOVERNMENT DOCUMENTS

Cmd. [Command Paper, London, H.M.S.O.] 9853. *The Suez Canal Conference: London, August 2-24, 1956. Selected Documents.*
Cmd. 47. *Damage and Casualties in Port Said. Report by Sir Edwin Herbert on his investigation into the effects of the Military Action in October and November, 1956.*
C.N.I.A. Current Notes on International Affairs. Department of External Affairs, Canberra.
C.N.L. Ceylon News Letter. Colombo. Government of Ceylon.
D.S.A.A. Digest of South African Affairs. Office of the High Commissioner of South Africa in the United Kingdom.
D.S.P. [Department of State Publication] 6392. *The Suez Canal Problem, July 26—September 22, 1956* (Washington, D.C.)
E.A. External Affairs. Department of External Affairs, Ottawa.
E.A.R. External Affairs Review. Department of External Affairs, Wellington.
F.A.R. Foreign Affairs Record. Ministry of External Affairs, New Delhi.
G.A.O.R. General Assembly Official Records. United Nations.
S. and S. Statements and Speeches. Department of External Affairs, Ottawa.
S.C.O.R. Security Council Official Records. United Nations.

PART I

COMMONWEALTH REACTIONS TO THE NATIONALIZATION OF THE SUEZ CANAL COMPANY

On this side of the House, we deeply deplore this high-handed and totally unjustifiable step. . . .

Mr. Hugh Gaitskell, 27 July 1956

It is best to keep our heads out of the beehive.

Mr. Johannes Strijdom, 27 July 1956

In crises of this kind we deal not merely with the issue in dispute, but we witness the upsurge and conflict of mighty forces. . . . In Asia as a whole, with its colonial memories, great resentment has been aroused [by the] military and naval movements ordered by the United Kingdom and France. . . .

Shri Jawaharlal Nehru, 8 August 1956

COMMENTARY

I

BETWEEN 27 June and 6 July 1956 the leaders of eight governments assembled in London for what had become their more or less regular annual or biennial meeting. They represented a quarter of the world's people, they were drawn from nations in nearly every continent, they reflected several varieties of non-communist opinion. The purpose for which the Prime Ministers of members of the Commonwealth were prepared to make so great a sacrifice of time and energy, most of them travelling vast distances to attend, was to enlarge their understanding of international affairs, and so to improve their respective external policies, by pooling information and exchanging ideas. As was (and remains) their custom, they discussed together the significant trends in world events and the crucial areas of world tension; and in their discussions—this much at least their traditionally laconic communiqué revealed—the Middle East occupied a central and disquieting position.

No aspect of Middle Eastern affairs was then more urgent than the unrest provoked by the exuberant nationalism of Abdul Gamel Nasser. The President of Egypt's military government was the forerunner of a type in the new nations of Asia and Africa to which the West has since become more or less accustomed; but it appeared at that time puzzling and unfamiliar. Much might be said in favour of Colonel Nasser, especially by comparison with predecessors. Certainly his national vision saw farther than the Riviera; his most bitter enemies could not accuse him of the decadent self-indulgence of a Farouk. If his early zeal for land reform and social legislation gave way too soon to preoccupation with apparatus for a police-state, it was arguable that uplifting the *fellaheen* required sterner direction than that provided by the easy-going Naguib. And if at first he paid more attention to stagecraft than to statecraft, offering the masses circuses and not much bread, might this be no more than the occupational hazard to which every fledgling Afro-Asian politician is in some degree prone? Might not a patient and understanding diplomacy divert him to better things?

The Egyptian dictatorship, admittedly, was more formidably equipped than most to wreak havoc abroad as well as at home. Colonel Nasser had written a book, *The Philosophy of the Revolution*, more often quoted, it was true, than read, but justifying the spread of Arab power in a way that called to mind careers of other literary dictators before him. Messages of disaffection beamed daily by the transmitters of Radio Cairo disturbed centres of British influence from Cyprus to Aden; while the bazaars from Marrakesh to Zanzibar City thrilled to the whispered intrigues of his agents. In all of this the Soviet Union took a keenly benevolent interest, and sought by friendly persuasion (and by bribes) to press so potent an instrument of disruption into the service of subversion. But Colonel Nasser possessed a weapon even more powerful than propaganda. Through his poor domain there ran a great high road of international commerce, controlled by a company incorporated under the law of his land. The Suez Canal, by custom and by legislation, was to be open without restriction to the shipping of all nations. But Nasser's régime, like those before it, had successfully defied resolutions of the United Nations in refusing passage to ships bound for or registered in Israel. And that tiny nation, like so many of its present citizens in former years, was marked out for destruction in the pages of a totalitarian's tract.

From such a man, what might a realist expect? Was he friend or enemy of freedom? Did he threaten peace? Could he be kept faithful to a benevolent if positive neutrality? Few more pressing questions confronted the chancelleries of the West. By the spring of 1956 they had become excruciating. For Washington and London had been caught up in the game of competitive co-existence. Their opponents were the political philanthropists of Moscow. The stakes of the game, in Egypt, were astronomically high. The fate of the *fellaheen* was there linked, as much in the popular imagination as in the brief-cases of economists, not to ploughs or pumps or fish-nets or any other modest device of technical assistance, but to a stupendous scheme of hydraulic development, centring on a proposal to harness the Nile River. A dam at Aswan, 110 metres tall, 5,000 metres across, would dwarf the mightiest of the pyramids. But here was to be a pyramid with a purpose, to make the desert blossom, to turn the wheels of a new Egyptian industry. No government could ask for more with which to lift up the spirit of a downtrodden populace—or to deflect its attention from political failure and the daily diet of poverty, ignorance, and disease. 'When we build the High Dam', declared Colonel Nasser, 'we are also building the dam of dignity, freedom and grandeur.'[1] Its total cost was estimated to be $1⅓ billions.

By the end of 1955 the Governments of the United States and the United Kingdom had decided, not without misgiving, to offer financial support to Colonel Nasser's régime to begin construction. Their aid was to take the form of an initial grant of $70 millions, $56 millions to be furnished by the United States, $14 millions by the United Kingdom, with an additional $200 millions to be loaned by the World Bank. This was agreement only in principle. Its fulfilment was dependent upon the continuing confidence of the donors that an honest, determined, and single-minded participation in their plans would be forthcoming from the recipient. Of this confidence Colonel Nasser seemed more than a little heedless. No progress was recorded upon the requisite negotiations with neighbouring authorities affected by any diversion of the Nile waters; an alternative offer by the Soviet Government to finance the scheme by a large low-interest loan received much pointed publicity; and, more nettling than anything else, foreign exchange needed for the Dam was being squandered on (of all things) heavy purchases of armaments from the Soviet bloc. Nasser had gone too far. 'For Dulles', the biographer of the American Secretary of State remarks, 'a moment of cold-war climax had come. It was necessary to call Russia's hand in the game of economic competition.'[2]

On 19 July, four days after the Egyptian Ambassador arrived in Washington to convey his master's anxiety that Western aid for Aswan be made quickly available, the Department of State announced that 'the ability of Egypt to devote adequate resources to assure the project's success has become more uncertain than at the time the offer was made' and that, accordingly, American financial assistance was 'not feasible in present circumstances'.[3] The President of Egypt received this disconcerting intelligence while conferring at Brioni with the President of Yugoslavia and the

---

[1] Quoted in *Documents on International Affairs, 1956* (London, 1959), p. 111.
[2] John Robinson Beal, *John Foster Dulles: A Biography* (New York, 1957), p. 258.
[3] *The New York Times*, 20 July 1956.

Prime Minister of India. As Sir Anthony Eden observed, benefiting by hindsight, 'the news was wounding to his pride'.[1]

The wisdom of Dulles's decision must remain a matter of opinion. 'For a foreign minister to handle a dictator in this way', two British authorities have written, 'must mean either that he is contemptuous of him, or else that the foreign minister is stupid.'[2] (It could conceivably mean both.) Against this may be set the view (expressed, it is true, in a general context) of Mr. George Kennan: 'When suggestions are made to us that if aid of one sort or another is not forthcoming, people will—as the saying goes—"go Communist", surely there is only one answer—"Very well then, go. Our interests may suffer, but yours will suffer first".'[3] Be that as it might be, Egyptian interests did not on this occasion suffer first. Questioned in the House of Commons whether the United Kingdom (whose offer of aid, like that of the World Bank, was withdrawn following the United States withdrawal) had been adequately consulted, the Foreign Secretary, Mr. Selwyn Lloyd, replied: 'We were in close consultation with the United States Government about this matter.'[4] A very different assessment was offered four years later by Sir Anthony Eden: 'We were informed but not consulted and so had no prior opportunity for criticism or comment. . . . We were sorry that the matter was carried through so abruptly, because it gave our two countries no chance to concert either timing or methods, though these were quite as important as the substance.'[5] These two different and even contradictory versions of a policy decision as crucial as any in modern times are not easily reconciled. Nor are they the only discrepancy in statements of British ministers during the months that lay ahead.

It was common, following Colonel Nasser's seizure of the Suez Canal Company on 26 July, to rank that event with the Berlin Blockade and the outbreak of war in Korea as the most critical of the many international crises since 1945. Such an assessment offers further testimony alike to the inscrutability of the future and to the myopia of foreign offices. For the Egyptian expropriation, like the Soviet pressure on Berlin and North Korea's assault

[1] *Full Circle: The Memoirs of Sir Anthony Eden* (London, 1960), p. 422.
[2] Guy Wint and Peter Calvocoressi, *Middle East Crisis* (Penguin Books, 1957), p. 69.
[3] George Kennan, *Russia, the Atom and the West* (London, 1958), p. 74.
[4] United Kingdom, H.C. Deb., vol. 557, col. 412.
[5] *Full Circle*, p. 422.

upon the South, was unforeseen by those whose interests were most directly affected by it. Nor could they well plead in their defence the novelty of Nasser's thrust. Proposals to nationalize the Suez Canal Company run through Egyptian history since the early 1920's, when the Socialist-Communist Party published them as a prime objective of policy. They had been earnestly reconsidered in the wake of the Naguib revolution in 1952. And Nasser was known to have been studying the prospects for taking over since February 1956.

In view of the British experience with Mossadeq in 1951, Nasser's more highly developed brand of nationalism, the historic controversy over the position of the Canal and the Company under international law, and the British evacuation of the Canal Zone in June, 1956, the Western allies should have been better braced for the possible nationalization of the Company as a riposte to the provocative character of the Aswan decision.[1]

What, throughout it all, were the nations of the Commonwealth doing? Only a fortnight before the withdrawal of the Anglo-American offer of aid, their Prime Ministers were ostensibly fortifying one another with information concerning the pressing international problems of the day. At meetings expressly designed to improve their knowledge of world events, did no one inquire of the United Kingdom Government what it proposed to do about aid for Aswan? Or, knowing that the proposed assistance to Egypt was then under reconsideration by Washington and London, did no one point out that Colonel Nasser's obvious method of retaliation was to seize the Canal and build the Dam out of its revenues?

[1] James E. Dougherty, 'The Aswan Decision in Perspective', *Political Science Quarterly*, vol. LXXIV, no. 1, March 1959, pp. 41–42. The comment of another American scholar, reflecting on an earlier occasion on which his country was taken unawares, is worth quoting in this connexion:

'Surprise, when it happens to a government, is likely to be a complicated, diffuse, bureaucratic thing. It includes neglect of responsibility, but also responsibility so poorly defined or so ambiguously delegated that action gets lost. It includes gaps in intelligence, but also intelligence that, like a string of pearls too precious to wear, is too sensitive to give to those who need it. It includes the alarm that fails to work, but also the alarm that has gone off so often it has been disconnected. It includes the unalert watchman, but also the one who knows he'll be chewed out by his superior if he gets higher authority out of bed. It includes the contingencies that occur to no one, but also those that everybody assumes somebody else is taking care of. It includes straightforward procrastination, but also decisions protracted by internal disagreement. It includes, in addition, the inability of individual human beings to rise to the occasion until they are sure it *is* the occasion—which is usually too late. . . .' Thomas C. Schelling, Foreword to Roberta Wohlstetter, *Pearl Harbor: Warning and Decision* (Stanford, California, 1962), p. viii.

Discussions at Prime Ministers' meetings proceed in secret; but
by the time an octet (today, a baker's dozen) of premiers have
reported to their parliaments (waylaid on their way home by
gentlemen of the Press in whom they may confide with greater or
lesser reticence), many of their secrets are spilled. It is thus no
secret that neither the future of Anglo-American aid to Egypt nor
the likely consequences of its abrupt curtailment were discussed at
the Prime Ministers' Conference of 1956. On this we have the
word of the Prime Minister of New Zealand. 'The Suez Canal . . .
was not even a prospective problem while I was in London', Mr.
Sidney Holland told the House of Representatives at Wellington
on 7 August. 'Not one word was said about it at the Conference.
There was no thought of this crisis developing.'[1] The account
offered to his Parliament by the Prime Minister of Canada con-
firms Mr. Holland's recollection. 'Our discussions in regard to the
Middle East', declared Mr. L. S. St. Laurent on 9 July, 'naturally
centered on the danger of conflict between the Arab States and
Israel. On the whole I was somewhat reassured by what I heard.'[2]
When, therefore, Mr. St. Laurent concluded that 'the Commonwealth is functioning as we expect it to function', he betrayed a
wholly unwarranted complacency. Mr. Holland's admission was
intended to excuse Commonwealth inaction and disarray. But
could he have offered a more damaging indictment of Commonwealth failure?

II

'In the past we were kept waiting at their offices—the offices of
the High Commissioner and the British Ambassador. Today . . .
they take us into account'.[3] So, exultantly, cried the President of
Egypt during his lengthy speech announcing the nationalization of
the Suez Canal Company on 26 July. And that, certainly, however
belatedly, was what the British Government was doing. On
27 July the Prime Minister confronted a tense and angry House of
Commons at Westminster. The Leader of the Opposition spoke
for his party and (indeed as never later) for all members in deploring 'this high-handed and totally unjustifiable step'.[4] Sir Anthony

[1] See below, p. 60.
[2] Canada, H.C. Deb., 1956, vol. VI, p. 5764.
[3] Quoted in *Documents on International Affairs, 1956*, pp. 80–81.
[4] See below, p. 25.

Eden was fully justified in concluding from this initial confrontation that Parliament was prepared to endorse stern measures of retaliation. His own language was more cautious than that used by any of his interrogators, partly because he had not yet met with his Cabinet. A Cabinet meeting took place immediately the House adjourned, and it enabled the Prime Minister to report privately to the President of the United States that 'we are all agreed that we cannot afford to allow Nasser to seize control of the Canal in this way' and that 'we must be ready, in the last resort, to use force to bring Nasser to his senses'.[1] The Commonwealth Prime Ministers were also informed of this evaluation, though whether in the same blunt language Sir Anthony Eden did not later see fit to disclose.[2]

On 30 July, the Prime Minister informed the House of Commons that foreign exchange controls had already been imposed to the disadvantage of Egypt; and that 'the Governments of the Commonwealth countries were given early information of the situation as it affects this country and the Commonwealth, and close touch is being maintained with them'.[3] On 2 August, the House was told that there were being taken 'certain precautionary measures of a military nature'.[4] Once again the Leader of the Opposition, expressing the support of the Labour Party, spoke more bluntly than had the Prime Minister hitherto of the nature of Nasserism. 'It is all very familiar', said Mr. Hugh Gaitskell, invoking a parallel that Sir Anthony Eden was later to make all very familiar. 'It is exactly the same that we encountered from Mussolini and Hitler in those years before the war.' But while he concluded that there were 'circumstances in which we might be compelled to use force, in self-defence or as part of some collective defence measures', he was careful to 'remind the House that we are members of the United Nations, that we are signatories to the United Nations Charter. . . . We must not, therefore, allow ourselves to get into a position where we might be denounced in the Security Council as aggressors, or where the majority of the Assembly was against us.'[5] Seldom, if ever, can an Opposition leader have more constructively criticized his Government. But from another Opposition critic, and a former Foreign Secretary at that, came less squeamish counsel. Recalling his experience with

---

[1] Quoted in *Full Circle*, pp. 427–8. [2] Ibid., p. 428.
[3] See below, p. 28. [4] See below, p. 32.
[5] See below, pp. 33–37.

Dr. Mossadeq, Mr. Herbert Morrison hinted broadly that there were circumstances in which the unilateral use of force by the United Kingdom might well be justified. 'Let them', he declared with reference to the Government, 'not be too much afraid.'[1] This was preaching to the converted. On 8 August the Prime Minister offered his own diagnosis in a broadcast to his people. 'This is how fascist governments behave, and we all remember, only too well, what the cost can be in giving in to fascism.'[2]

But by this time some sober second thoughts had begun to occur to members of the Parliamentary Labour Party. (Or, as Sir Anthony Eden later put it, 'The retreat then began amid a clatter of excuses.'[3]) Even before his speech of 2 August, Mr. Gaitskell had warned the Prime Minister that the Labour Party might not support the Government if it decided to use force unilaterally against Egypt; and he repeated this warning in two letters to Sir Anthony Eden written between 2 August and mid-September.[4] On 8 August, twenty-four members of the Parliamentary Labour Party (including Jennie Lee, Fenner Brockway, Sidney Silverman, Ian Mikardo, J. P. W. Mallalieu, and Sir Frederick Messer; Aneurin Bevan was excluded because of his position as a member of the 'Shadow Cabinet') issued a statement 'for ourselves in the belief that our view would be shared by all sections of the Labour movement'. The statement affirmed, among other things, that 'to attempt to carry out . . . internationalization of the Suez Canal by force against the resistance of the Egyptian Government and people would be an act of aggression under the United Nations Charter'.[5] It was reprinted in its entirety in *The Hindu* of Madras on 10 August. A Labour Party parliamentary delegation waited upon the Prime Minister and his colleagues in mid-August in a vain attempt to persuade them to renounce the use of force outside the United Nations. Thus, even during the earliest stages of the crisis, it had become apparent that there would be no bipartisan support for the tough policy that Sir Anthony Eden was resolved to follow in the weeks ahead.

---

[1] See below, pp. 37–38.   [2] See below, pp. 41–43.

[3] *Full Circle*, p. 445.

[4] Leon D. Epstein, 'Partisan Foreign Policy: Britain in the Suez Crisis', *World Politics*, vol. XII, no. 2, January 1960, p. 205.

[5] See below, pp. 43–44.

III

The Middle East is not a part of the world in which the Canadian people and their governments have tended to take much interest. Only a few Canadian troops fought there in the first of the two World Wars; fewer still in the second. On the rare occasions when Canada was called upon during peace-time to intervene—in 1885 when Sir John A. Macdonald refused to 'get Gladstone & Co. out of the hole' and in 1922 when Mackenzie King refused to help hold the line at Chanak—her responses were negative and are recalled today chiefly for their constitutional significance, milestones along the road to Dominion status. Despite some quickening of interest in the area after 1945 (mainly as a result of the birth of the State of Israel, which Canadian diplomatists attended as midwife's helpers), few Canadians thought of it as a region in which specifically Canadian commitments ought to be made and for which a distinctively Canadian policy was in any degree appropriate. 'I do not suppose', remarked Mr. Lucien Cardin, Parliamentary Assistant to the Secretary of State for External Affairs, on 27 March 1956, 'that we have yet got quite used to . . . the idea that an event on the frontier between Israel and Egypt might very well involve us in the gravest consequences.'[1] But so it was soon to prove.

The news of Nasser's *coup* reached Ottawa on the night of 26 July. When the House of Commons met the following morning there were questions about flax, box-cars, and the quality of food at an Army camp; but none about Suez. The Canadian correspondent of *The Round Table* noted 'the absence of crisis atmosphere on Parliament Hill . . . in sharp contrast with the tension reported from London'.[2] A cautious statement was issued to the Press.[3] On 28 July the Secretary of State for External Affairs informed the House of Commons that the Government was 'in consultation with Governments probably more directly concerned'; more than that it would be premature to say. To a question from the Conservative Party's foreign policy critic, Mr. John Diefenbaker, as to whether Canada should not have 'something to say on a matter that particularly affects the members

[1] *Statements and Speeches* (Information Division, Department of External Affairs, Ottawa), no. 56/10.
[2] *The Round Table*, no. 185, December 1956, p. 92.
[3] See below, p. 46.

of the Commonwealth, Britain, Australia, and New Zealand particularly, to indicate a degree of unity', Mr. Lester Pearson replied that the Canadian High Commissioner in London had been to see members of the United Kingdom Government not only to ascertain their views 'but also to [do] . . . what we can to concert our attitude and policy with other Commonwealth countries'. In these conversations, Mr. Pearson added, the High Commissioner had been instructed to stress the importance of settling the dispute peaceably through the United Nations.[1] A few days later, the Secretary of State for External Affairs resisted the demand of Opposition members that Canada lodge a formal protest against the seizure of the Suez Canal Company. 'A telegram from our representative in one of the Asian countries of the Commonwealth', Mr. Pearson noted, 'indicated that the feeling in that country on this matter was quite different indeed from that which exists in Canada or in certain other parts of the Commonwealth.'[2] In the view of the Canadian Government, the Commonwealth was a bridge to Asia in a world where nearly all the bridges had been blown; and it was not disposed recklessly to destroy that too. But Mr. Diefenbaker remained unsatisfied. 'The Canadian policy cannot be a waffling one', he declared on 8 August. 'An uncertain policy on the part of Canada might well bring about another Munich.'[3]

IV

The vocabulary of politics is still not so extensive that the introduction of new words is always a disservice. The term 'geopsychology' (geographers may prefer 'psychogeography'), however inelegant, usefully emphasizes that how people think they are situated is at least as important as their situation; and that these may not be the same. The people of Australia and New Zealand had by 1956 not yet wholly thrown off (perhaps they have still to throw off) the notion that their national survival depended upon an open Suez route to Britain and Western Europe; ultimately, upon peace and tranquillity in the Middle East. What Canadians ignored or sought to remain aloof from, Australasians had fought and died for. If Mr. Churchill's sure judgement deserted him

---

[1] See below, pp. 46–47.   [2] See below, pp. 48–49.
[3] See below, p. 50.

throughout the Chanak crisis of 1922, it reasserted itself when in his appeal for help from the Dominions he referred to 'soil which is hallowed by immortal memories of the Anzacs'. Only slowly, and against much popular resistance, was it accepted that a region for which two generations of Australians and New Zealanders had sacrificed their lives was no longer of decisive importance; that the Suez Canal could no longer be described, in a manner acceptable alike to orators and strategists, as Australasia's 'lifeline' or, yet more dramatically, as its 'jugular vein'. The really vital area was the Far East (or, as seen from the Antipodes, the 'Near North'), not the Middle East. At least partial awareness that this was so might be found in the Australian Government's decision to contribute troops to the defence of Singapore and the pacification of Malaya; or, using a different technique, to open schools and universities, 'White Australia' notwithstanding, to the youth of South and South-East Asia as part of the Commonwealth's contribution under the Colombo Plan. Economically, the Canal route remained important; but, as research stimulated by the Suez crisis was to show, not as important as that. 'Suez', two Australian experts concluded in 1957, 'is not vital to the Australian economy.'[1] Had so cool an assessment been available to, and accepted by, Mr. Menzies and his colleagues a year earlier, it is just possible that Australia's reactions both to Egyptian and to British policy would have left her less exposed and isolated (not that Mr. Menzies shrank from exposure or from isolation). Whether it would have similarly affected the response of Mr. Holland and his colleagues is more doubtful. For if Australia's relationship to the Old Country allowed occasionally for disagreement stemming from diverging strategic conceptions and a conflict of generations, New Zealand's partook characteristically of the devotion of Ruth for Naomi: 'Intreat me not to leave thee, or to return from following after thee. . . .'

An occupational hazard attending Australasian premierships is to be out of one's capital at moments of high crisis. Both Mr. Menzies and Mr. Holland were many thousands of miles removed from Canberra and Wellington when Colonel Nasser spoke at Alexandria on 26 July. Mr. Menzies was in Washington, and in his absence it fell to Mr. R. G. Casey, the Minister of External

---

[1] W. Woodruff and L. McGregor, *The Suez Canal and the Australian Economy* (Melbourne, 1957), pp. 19–20.

Affairs—in the course of a long and varied diplomatic career he had represented the United Kingdom in Cairo—to shape Australia's initial reaction. This he did non-committally: 'The Australian Government would view with concern any unilateral or precipitate departure from the terms of solemn understandings of direct concern to maritime states, or any move which might restrict the free use of the Canal by world shipping.'[1] His next statement, framed after consultation with Mr. Menzies, though still moderate, had a stiffer quality. It referred to the dependence of the Australian economy upon 'the free and unfettered flow of cargoes and ships through the Canal', and to the 'abrupt and high-handed action of the Egyptian Government'.[2]

Mr. Menzies, meanwhile, thought the situation serious enough to warrant his return, characteristically, not to Australia[3] but to the United Kingdom. Henceforth (as an Australian authority observed), 'the Prime Minister, rather than Mr. Casey, became the Government's official spokesman'.[4] In the view of another Australian authority, Mr. Casey's eclipse had unfortunate consequences for policy, for (writes Professor Norman Harper) 'there was in Canberra a clearer appreciation than in London of Asian reactions to Nasser's nationalization of the Canal and of the impact of the use of force on American opinion'.[5] Even at this early stage it appeared to observers on the scene that a difference of opinion existed between Mr. Menzies and Mr. Casey on what role Australia should play in any crisis over Suez.[6] Mr. Menzies, at any rate, entertained no doubts. He took the unusual step of airing his opinions to the British public, first in a statement issued in London on 10 August, and then, three days later, in a broadcast over the B.B.C.[7] These initiatives did not command the support of the

---

[1] See below, p. 51.   [2] See below, pp. 51–53.

[3] As long ago as 1941 Mackenzie King had observed of Mr. Menzies: 'I sensed the feeling that he would rather be in the War Cabinet in London than Prime Minister of Australia.' Quoted in J. W. Pickersgill, *The Mackenzie King Record*, I, 1939–1944 (Toronto, 1960), p. 214.

[4] W. Macmahon Ball, 'The Australian Reaction to the Suez Crisis', *The Australian Journal of Politics and History*, vol. II, no. 2, May 1957, p. 131.

[5] Norman Harper, 'Australia and Suez', in Gordon Greenwood and Norman Harper (editors), *Australia in World Affairs, 1950–55* (Melbourne, 1957), p. 343.

[6] Ibid.

[7] See below, pp. 56–57. The British Prime Minister was instrumental in getting Mr. Menzies on the air, and later described his negotiations in the matter with the British Broadcasting Corporation as 'a strange interlude'. Sir Anthony Eden's account of it is as follows:

'In conversation with Menzies, a day or two after his arrival, I asked him if he would comment publicly on the situation, preferably in a television broadcast.'

Australian Labour Party, which, in a statement of its own issued two days after Mr. Menzies' broadcast, denounced the Prime Minister's remarks as 'dangerously provocative'.[1] Their reception in India was more hostile still; in a speech in Madras, Mr. C. Rajagopolachari, a former Governor-General whose views commanded attention and respect, described Mr. Menzies' address of 13 August as 'the true voice of British colonialism, speaking from the grave'.[2]

The Prime Minister of New Zealand was even further afield, for the news of Colonel Nasser's act reached Mr. Holland in Los Angeles. In his absence his Government did not venture beyond conventional expressions of concern.[3] What in Canberra had been denounced as 'abrupt and high-handed' seemed in Wellington merely to fail to provide 'convincing evidence that [the Egyptian Government] will necessarily display the wisdom called for'.[4] By 7 August, Mr. Holland was able to report directly to the New Zealand House of Representatives. He recounted how in London on his way home he had told 'Sir Anthony Eden and Mr. Selwyn Lloyd that Britain could count on New Zealand standing by her through thick and thin'. His assessment was simple and traditional: 'The Suez Canal is vital to Britain, and Britain is vital to New Zealand.' He bestowed his approval upon the troop movements and naval preparations that the British Government had put in hand, and hinted broadly that New Zealand stood ready to support the United Kingdom by armed force if necessary.[5]

With all this the official Opposition, unlike that in Australia,

He replied that he would do anything he could to help. I said I had no doubt that this would be very welcome and passed on the information to the B.B.C. A little while later I learnt, almost by accident, that Mr. Menzies had not been asked to speak. I made inquiry why this was so. I was told that as I had already appeared and, I think, some other speaker who shared my point of view, it was felt that if Mr. Menzies were to be asked to speak, someone who clearly disagreed, Mr. Emrys Hughes was the name mentioned, should be put on the air to balance the presentation of views. I thought that this attitude was insulting to a Commonwealth Prime Minister, whatever his politics, and Mr. Menzies was invited to speak on television, which he did admirably.' *Full Circle*, p. 448.

Sir Anthony Eden's part in this episode has been described by a widely-read reconstruction of British policy throughout the Suez crisis as 'unparalleled and entirely improper'. It adds: 'His conduct in this alone points to an almost desperate wish to propagandise Britain.' Erskine B. Childers, *The Road to Suez* (London, 1962), p. 213 n.

[1] See below, pp. 57–58.   [2] *The Hindu* (Madras), 20 August 1956.
[3] See below, pp. 58–59.   [4] See below, p. 59.
[5] See below, pp. 60–61.

were in complete agreement. Indeed, the mood of the New Zealand Labour Party was if anything more militant than that of the Government; it was at any rate a Labour Party spokesman, Mr. P. G. Connolly, who made headlines in the New Zealand Press by declaring: 'I think every Member of the House would support the Prime Minister and the Government. . . . It is good at times that the roar of the British lion should be heard.'[1] Thoughtful New Zealanders, however, could not be content to allow loyalty and the conventional patriotism to substitute for hard and if necessary independent analysis by those responsible for policy. In the words of an editorial in the *Auckland Star*:

It is one thing to say that our sympathies are with Britain and that we regard our interest in this matter as practically identical with hers; it is quite another thing to suggest that whatever the British Government thinks, or may think, should be done, will have New Zealand's assent. If that were New Zealand's 'policy' we should spare ourselves the expense of maintaining an External Affairs Department and ambassadors abroad; all we should need would be someone to decode messages from the Foreign Office.[2]

V

The Government of the remaining 'Old Dominion' greeted these events very differently. The Nationalist Party, led first by D. F. Malan and then by Johannes Strijdom, had ruled South Africa since 1948; by the summer of 1956 it enjoyed a comfortable parliamentary majority over the United Party Opposition. Its reaction to Colonel Nasser's *coup* of 26 July was the product of conflicting emotions. No love at all was lost between the apostles of *apartheid* and the advocate of Pan-Arabism (and Pan-Africanism). What a resurgence of black nationalism might mean for the future of the white man in Africa preoccupied Pretoria as it did few other capitals of the West. On the other hand, much might be said for accepting Nasser's stroke without comment or riposte. Intervention might augment the principle of international concern at the expense of the principle of domestic jurisdiction, of which there had been no more staunch defender than successive South African governments since Smuts made his mark upon the Covenant. Moreover, little political advantage was to be derived from rushing too hastily to Britain's side, whatever straits she

---

[1] New Zealand, H. R. Deb., vol. 309, pp. 935-6.
[2] *Auckland Star*, 8 August 1956.

might be in. Finally, what for the Australasian Dominions was a grave economic threat, for South Africa brought increased revenue, and, more importantly, increased bargaining power: a Canal closed to the shipping of the West would bring the Cape route back into its own, adding not only to the prosperity of the Union's ports but enabling South Africa to press to better effect its case for the creation of a N.A.T.O.-type security organization linking what remained of the white man's Africa to the anti-communist powers of the North Atlantic area. These factors prevailed.

The Prime Minister's earliest, and perhaps reflexive, comment was that 'it is best to keep our heads out of the beehive'.[1] A more considered statement four days later by the Minister of External Affairs described the nationalization as 'a domestic affair of Egypt's in which South Africa would not presume to intervene'.[2] But it was evidently insufficiently considered, for within twenty-four hours Mr. Eric Louw felt it necessary to follow with a further statement making clear that South Africa's Government, though determined not to become embroiled in the dispute, was neither unconcerned nor indifferent.[3] In discussion with the Egyptian Ambassador on 4 August, the Union's Foreign Minister pleaded with Mr. Hamdy to urge his Government to avoid any breach of the 1888 Convention, 'even under provocation'.[4]

The United Party Opposition took, as oppositions may, a contrary line. The Egyptian expropriation it construed, in its leader's words, as 'no local issue' but rather as 'an ominous warning that we in South Africa cannot afford to blind ourselves to the intolerant extremism of Egypt's new nationalistic regime'. Its diagnosis differed further from the Government's in holding that 'events in Egypt are connected with the expansionist policies of the communist states'.[5] Above all, as the party to which most English-speaking South Africans gave their allegiance, it was critical of the Government's lack of support for the United Kingdom. 'The Nationalists', commented the *Cape Times* on 21 August, 'are far more interested in making it clear that they don't want to fight Britain's wars than in throwing in their moral weight strongly on the side of the West.'[6]

[1] See below, p. 62.   [2] See below, p. 62.
[3] See below, pp. 63–64.   [4] See below, pp. 64–65.
[5] See below, pp. 65–66.
[6] Editorial, 'Neutralism', *Cape Times*, 21 August 1956. Further examples of opposition press opinion may be found in Jitendra Mohan, 'South Africa and the Suez Crisis', *International Journal*, vol. XVI, no. 4, Autumn 1961, pp. 335–8.

But the overwhelming majority of South Africans sided not with
Britain but with Egypt. 'We pledge our solidarity with the Egyptian people', affirmed a statement of the African National Congress,
'and are confident that the people of Africa will not allow themselves to be used against their fellow Africans in any predatory
way.'[1]

VI

Even before the emergence of non-white nations in Africa, the
majority of Commonwealth citizens dwelt in those realms and
republics which came into existence only after 1947. In 1956 these
consisted of India, Pakistan, and Ceylon, accounting collectively
for three times the number of the inhabitants of Great Britain and
the so-called 'White Dominions'. To those among their 400
millions having any knowledge of the outside world, Colonel
Nasser's seizure of the Suez Canal Company appeared unambigu-
ously as an act of retributive justice, as militant nationalism com-
mendably asserting itself against the remnants of imperialism in
Asia. Their governments, therefore, could hardly condemn what
Nasser had done; indeed, for the most part, they condoned it.
Certain significant differences in their respective approbations
may, however, be detected.

Nowhere did the reputation of Abdul Gamel Nasser stand
higher than in India. There his position as leader of Arab nationalism could be appreciated (as it could not in Pakistan) without any
sense of jealousy.

Most Indians [an Indian scholar has written] inclined to view Colonel
Nasser in relation to Arab nationalism as they viewed Mr. Nehru in
relation to Asian nationalism. Though Mr. Nehru had never professed
any admiration for Colonel Nasser's military regime as a form of
government, he was sympathetic to its relative progressiveness. The
Egyptian leader's persistent refusal to respond to the slogan of Pan-
Islamism bore a certain surface resemblance to India's secularism and,
against the background of their common opposition to military alliances, had served to bring the two countries nearer.[2]

It was symptomatic of their close relationship that Colonel
Nasser should have learned of the Anglo-American withdrawal of

---

[1] See below, p. 66.
[2] Jitendra Mohan, 'India, Pakistan, Suez and the Commonwealth', *International Journal*, vol. XV, no. 3, Summer 1960, p. 186.

aid for Aswan while in conference with Pandit Nehru at Brioni. It was later suspected on this account that the Prime Minister of India had been privy to Colonel Nasser's plan before it was put into effect; this was a misconception which Mr. Nehru took the earliest opportunity to correct. 'The recent decision of the Egyptian Government in regard to the Suez Canal', he declared on 31 July, 'first came to my knowledge from the reports in the press after my return to Delhi.'[1] That statement did not exclude the possibility that Colonel Nasser might have ruminated about seizing the Company in the presence of his fellow positive neutralists; but a more likely reconstruction is that he kept his thoughts on the matter wholly to himself, a supposition which, if correct, explains a certain note of irritation marking Mr. Nehru's statement of 8 August.[2] The Indian Prime Minister may have felt annoyed that Colonel Nasser had not taken him into his confidence; but the underlying reason for the hint of disapproval of the *coup* of 26 July was India's dependence upon free transit through the Canal—directly for its own trade, indirectly for the unfettered flow of capital and technical assistance from the West. 'India', observed Mr. Nehru on 8 August, 'is not a disinterested party. She is a principal user of this waterway, and her economic life and development is not unaffected by the dispute.'[3] For the rest, he contented himself with warning, on several occasions, against its settlement by force. Being on good terms with both parties, India, he declared, 'is passionately interested in averting a conflict'.[4]

No significant opposition to this point of view emerged at this stage. But the Indian Press was inclined to go further than the Prime Minister in criticizing Britain's response, particularly as word spread of its military preparations. The *Amritsar Bazar Patrika* hoped that Britain and France would 'make the best of a bad job' by recognizing Nasser's act as a *fait accompli*; the *Bombay Chronicle* gave warning that 'diplomacy by threats will not work' and recommended instead 'a determined effort to acknowledge facts gracefully'; while Mr. Nehru's own paper, the *National Herald*, criticized Britain's freezing of Egypt's sterling balances:'At least the Asian members [of the Commonwealth] should advise not only restraint and caution but justice

---

[1] See below, p. 67.   [2] See below, pp. 67–69.
[3] See below, p. 69.   [4] See below, p. 69.

in the attitude which Britain is to maintain towards Egypt.'[1]

In Pakistan, admiration for Colonel Nasser and his new Egypt was not, as it was in India, wholly uncritical. Enjoined by their Constitution to 'endeavour to strengthen the bonds of unity among the Muslim countries', successive Pakistani governments had spared no effort since 1952 to achieve close and friendly relations with Egypt, based upon the common faith of their peoples. In particular Pakistan had assumed a position of uncompromising hostility towards Israel, hoping in return for some expression of Egyptian support of her claim to Kashmir.[2] This support was not forthcoming. India under Mr. Nehru was thought by Colonel Nasser to be, on balance, more worthy of cultivation, and there was indeed little enough in common between the secular nationalism of Nasser's revolution and the Islamic Republic of Pakistan. When Pakistan, more from desperation than from a careful calculation of strategic necessity, entered the Baghdad Pact in January 1955 (having already entered the South-East Asia Treaty Organization some months previously), the Egyptian Government abandoned all pretence of friendship and mounted upon Karachi a sustained and bitter propaganda attack. For these reasons the Pakistani Government, attempting to shape an appropriate response to Colonel Nasser's seizure of the Suez Canal Company in July 1956, confronted an option of difficulties unknown to New Delhi. Its alliances with the West, acquired in a last-ditch effort to find support in a hostile world, precluded too obvious rejoicing at the predicament in which the loss of control of the Canal might place the 'imperialists' of Europe and North America. On the other hand the Pakistani people could not (and did not) fail to respond sympathetically to any resurgence of nationalism in the Muslim world. Here was a dilemma, and, as an Indian scholar has suggested, it 'was primarily responsible for the

---

[1] Quoted in Taya Zinkin, 'India's Position in the Middle', *Manchester Guardian Weekly*, 9 August 1956.
[2] For these aspects of Pakistani foreign policy, see Keith Callard, *Pakistan: A Political Study* (London, 1957), chap. X, 'Pakistan and the World'; Keith Callard, *Pakistan's Foreign Policy: An Interpretation* (Institute of Pacific Relations, 1957); G. W. Choudhury and Parvez Hasan, *Pakistan's External Relations* (Karachi, 1958); K. Sarwar Hasan, *Pakistan and the United Nations* (New York, 1960); B. C. Rastogi, 'Alignment and Non-Alignment in Pakistan's Foreign Policy, 1947–1960', *International Studies* (New Delhi), vol. III, no. 2, October 1961, pp. 159–80; M.S. Rajan, 'India and Pakistan as Factors in Each Other's Foreign Policy and Relations', *International Studies*, vol. III, no. 4, April 1962, pp. 349–94.

cleavage that manifested itself at the outset of the crisis and was to be widened as it progressed'.[1]

On 2 August, while in Calcutta, the Pakistani Foreign Minister remarked that President Nasser had every right to nationalize the Suez Canal Company, if he so desired.[2] This statement, stronger than any emanating from Indian authorities at that stage of the crisis, did not go so far as did other prominent Pakistanis not responsible for policy. 'Colonel Nasser has done no more than pay America and England in their own coin', the President of the East Pakistan Muslim League declared at Dacca on 2 August. 'The curt and imperious manner in which these two great Western powers announced their refusal to finance the Aswan Dam project could not but wound the self-respect of resurgent Egypt, and small nations naturally feel gratified that Egypt has not taken it lying down.'[3] Still stronger was the statement of the leader of the East Pakistan Awami League on the same day: 'With great pride I assure President Nasser and the heroic people of Egypt that we, their brothers in Pakistan, will stand by them.'[4] The Pakistan Socialist Party in a statement of 31 July condemned the attitude of Britain and France, especially the 'aggressive utterances of Mr. Hugh Gaitskell in the House of Commons'.[5] There is little doubt that these utterances, more forceful and extreme than those of the Government, expressed the general will of the Pakistani people at this juncture of events, as later on.

Had Ceylon's United National Party (U.N.P.) remained in office for three months longer than it did, it is just possible that Nasser's nationalization would have found its only Asian critic. The U.N.P., first under D. S. Senanayake, then under his son, Mr. Dudley Senanayake, and most emphatically under Sir John Kotelawala, had steered for Ceylon a course unique among newly independent nations. The guiding principle of Ceylon's foreign policy between 1947 and 1956 had been hostility towards the communist bloc, manifested in such untypically Asian acts as entering into a defence pact with the United Kingdom by which her former rulers were allowed to retain military bases in Colombo and Trincomalee, or permitting United States warships to use her port facilities *en route* to fight the communists in Korea. It is true that

---

[1] Jitendra Mohan, 'India, Pakistan, Suez and the Commonwealth', op. cit., p. 186.
[2] *Dawn* (Karachi), 3 August 1956.
[3] See below, p. 70.   [4] See below, p. 70.   [5] *Dawn*, 1 August 1956.

this attitude did not preclude a sympathetic regard for the aspirations of colonial peoples. 'Despite the receding tide of colonialism in Asia,' commented the *Times of Ceylon* on 28 December 1954, 'the political, economic and cultural enslavement of millions of Asians and Africans remains one of the hardest and most stubborn facts of our times.' The hospitality offered to the American flotilla in 1950 had thus been denied to the Dutch warships bound for Indonesia in 1948. But in its unashamedly pro-Western orientation the Kotelawala régime was distinct in Asia, with its deliberate rejection of non-alignment and its conscious opposition (most strikingly evident at the Bandung Conference[1]) to the Indian outlook; and had it been responsible for the nation's external affairs when Nasser made his move of July 1956, it might well have confronted a conflict of loyalties more excruciating than for any other member of the Commonwealth.[2]

As it happened, Sir John Kotelawala was spared this decision by the results of the General Election in April 1956. He and his U.N.P. went down to overwhelming defeat, and there was installed in their place the People's United Front (Mahajana Eksath Peramuna, hence M.E.P.) led by S. W. R. D. Bandaranaike. The M.E.P., a Ceylonese authority has written,

reflected the political awakening and a conscious drive for power of a section of society which was frustrated because of the privileged position of the foreign educated group. . . . This socio-cultural renaissance was closely linked with nationalist fervour and had its political impact to the extent that it condemned everything which was foreign. Though it was not exactly anti-West, it did manifest the desire of the Ceylonese people to oppose any policy which subordinated its own interests to Western interests or which tied them to the West in such a way that

[1] Where Sir John Kotelawala delivered his justly famous speech attacking Soviet colonialism which, as he records, ruffled Jawaharlal Nehru even more than it did Chou En-lai. After the speech, Sir John has recalled, 'Nehru came up to me and asked me in some heat, "Why did you do that. . .? Why did you not show me your speech before you made it?" I have no doubt the remark was well meant, but the only obvious reply was, "Why should I? Do you show me yours before you make them?".' Sir John Kotelawala, *An Asian Prime Minister's Story* (London, 1956), p. 187. Neither Prime Minister seems to have thought of his obligation to consult another head of the government of a fellow Commonwealth country.

[2] On Ceylon's foreign policy, see W. Howard Wriggins, *Ceylon: Dilemmas of a New Nation* (Princeton, N.J., 1960), Part Three, 'The Search for a Foreign Policy'; Urmila Phadnis, 'Non-alignment as a Factor in Ceylon's Foreign Policy', *International Studies*, vol. III, no. 4, April 1962, pp. 425–42.

limited their freedom of action. As such it was more thoroughly Asian in its outlook than the policies of the U.N.P.[1]

It was not surprising, therefore, that one of the new Prime Minister's first acts in external affairs was to enter into negotiations with the United Kingdom with the object of repudiating the defence agreement concluded by his predecessor.

While these negotiations, which were both difficult and delicate, were in progress, Colonel Nasser moved against the Suez Canal Company. Mr. Bandaranaike, anxious not to say or to do anything which might prejudice a new bases agreement with Britain, was more cautious in his initial reactions to the nationalization than might have been expected, or than was expected by his followers. He did not attempt to conceal his concern at what had happened, describing it as 'probably . . . the most serious one single thing . . . since the end of the last war',[2] but he refused to be drawn into any expression of endorsement of Colonel Nasser's action, still less of the British Government's reaction. What made discretion appear all the more the better part of valour was, no doubt, the extent to which Ceylon's economic well-being depended upon the Canal route remaining open: some 60 to 70 per cent. of the nation's exports passed through Suez, as did some 40 to 50 per cent. of its imports and, unlike the case of the Pacific Dominions, the Cape route provided no practicable alternative. Valid as these considerations may have been, the Prime Minister's behaviour was too discreet for at least one sector of the opposition, the vocal Lanka Sama Samaja Party (L.S.S.P.), a splinter-group professing Trotskyite principles and loyalty to the 'Fourth International'.[3] Its official organ, *Sama Samajist*, declared editorially on 2 August:

Egypt's Nasser has achieved world fame overnight with what is probably the boldest step since World War II. Here is a proud assertion of the right to independence of a tiny country despite all the intimidation of the world's biggest and most powerful imperialist powers. The most important thing for us in Ceylon is that Egypt has set an example—in fact a challenge—to the Ceylon Government. Nasser did not go 'half way' as our P.M. did on the subject of bases. Nasser means business. Mr. Bandaranaike's hope that both sides will prove more accommodating is perhaps misplaced. Our Premier talked loftily in Parliament about 'bearing in mind our own position, principles and

---

[1] Urmila Phadnis, op. cit., pp. 430–1. [2] See below, p. 74.
[3] Wriggins, op. cit., pp. 119–43.

our interests'. We have heard that 'our principles' include nationalisation. Nasser has shown how to do it. 'Our honour' requires that we follow his example . . . in relation to our war bases and foreign capital investments. It is left to be seen whether our Premier and our 'People's Government' with 'socialists' like the Hon. Messrs. Philip Gunawardena, P. H. W. de Silva and T. B. Subasinghe[1] will have the guts of Col. Nasser. If all the M.E.P. Government's action is contained within the walls of conference rooms, then the Egyptian crisis will only have served to X-ray and expose the pretensions of our 'People's Government'.[2]

And:

The matter concerns Ceylon closely; and more than as a matter of trade. If the Imperialists have a right to prevent Egypt nationalising the Suez Canal Co., then they can also claim the right to prevent the nationalisation of foreign company properties in Ceylon. If they can claim a right to the Suez Canal in the name of their international military and commercial interests, then they can similarly claim a right over Trincomalee and Katunayake. The task of the Ceylon Government in this situation is clear and straightforward. As a member of the British Commonwealth, Ceylon should immediately inform the British Government of our opposition to their anti-Egyptian enterprise. . . .[3]

DOCUMENTS

I. UNITED KINGDOM

1. Answers by the Prime Minister, Sir Anthony Eden, to questions in the House of Commons, 27 July 1956.[4]

Mr. Gaitskell (*by Private Notice*) asked the Prime Minister whether he had any statement to make on the reported action of the Egyptian Government in regard to the Suez Canal.

The Prime Minister (*Sir Anthony Eden*): Yes, Sir. The unilateral decision of the Egyptian Government to expropriate the Suez

---

[1] These Ministers had broken from the United Front with the Communists and received their portfolios as the result of their efforts on behalf of S. W. R. D. Bandaranaike in the General Election of 1956.
[2] 'Nasser's Action X-Rays Banda', *Sama Samajist*, vol. XX, no. 54, 2 August 1956.
[3] 'M.E.P. Foreign Policy Under Test', ibid.
[4] United Kingdom, H.C. Deb., vol. 557, coll. 777–80.

Canal Company, without notice and in breach of the Concession Agreements, affects the rights and interests of many nations. Her Majesty's Government are consulting other Governments immediately concerned, with regard to the serious situation thus created. The consultations will cover both the effect of this arbitrary action upon the operation of the Suez Canal and also the wider questions which it raises.

Mr. Gaitskell: On this side of the House, we deeply deplore this high-handed and totally unjustifiable step by the Egyptian Government. Has the Prime Minister in mind to refer this matter to the Security Council? Has he yet come to any decision on that point? In view of the seizure of the property of the Suez Canal Company and the vague statement about future compensation, will he bear in mind the desirability of blocking the sterling balances of the Egyptian Government?

The Prime Minister: I am quite sure that the right hon. Gentleman will understand if I say that I would not wish to make a direct answer to his last question beyond saying that what he has mentioned has been already in our minds. As regards the Security Council, I would rather not say what action the countries concerned—we are in consultation, naturally, now with them—would wish to take, whether in the Security Council, or immediate diplomatic action, or whatever it may be.

Captain Waterhouse: Is my right hon. Friend aware that the statement that he has just made will be widely approved in all parts of the country? In his consultation, will he bear in mind that, under the Convention of October, 1888, Article III lays down that the High Contracting Powers 'undertake to respect the plant, establishments, buildings, and works of the Maritime Canal and of the Fresh-Water canal'; and, further, that Article VII of the same Convention gives each of the High Contracting Powers the right to put not more than two warships at the ports at either end of the Canal, that is to say, Port Said and Port Suez?

The Prime Minister: My right hon. and gallant Friend and the House can be assured that we have, not for the first time, examined international rights in this matter. There are a great many undertakings that have been given that are of later date and perhaps, in a sense, even more binding on the Egyptian Government than those which my right hon. and gallant Friend has quoted.

Mr. Clement Davies: May I be allowed to say, on behalf of myself

Dcs

and my colleagues, that we associate ourselves with what has been said by the Leader of the Opposition about this deplorable action? I assume that the Governments which will be consulted, and which are immediately concerned, will be the other eight who were parties to the Convention of 1888? Has the Prime Minister anyone else in mind besides those eight?

The Prime Minister: I am not sure about the immediate definition mentioned by the right hon. and learned Member for Montgomery. It is not only a question of the Convention. I do not want to go into detail, but, of course, there are later agreements than the Convention which are concerned here. All I can say to the House is that we got in touch, and are in touch, with the United States Government and the French Government, and we are in touch this morning with a number of Governments, I hope all, of the Commonwealth. I would ask the House not to press me to say more than that at the moment, if right hon. and hon. Members would not mind. The situation must be handled with both firmness and care, I think. I would undertake to give the fullest information to the House at every stage.

Mr. J. Amery: Is my right hon. Friend aware that he will have the overwhelming support of public opinion in this country on whatever steps he decides to take, however grave, to repair this injury to our honour and interests?

Mr. Paget: Is the Prime Minister aware that there do not exist in the world today sufficient tankers to move the oil required by Europe without using the Suez Canal, and that this is a threat to strangle the whole industry of Europe? Is he further aware that this 'weekend technique' is precisely the technique which we got used to in Hitler's day? Is he also aware of the consequences of not answering force with force until it is too late?

The Prime Minister: I made my statement and I have answered questions with some care. I think it would be a wiser judgment of the House—of course, hon. Members may say anything that is in their minds—that I should not go beyond anything I have said.

Viscount Hinchingbrooke: Will the Government take steps to reinforce the Suez Canal Zone base with civilian employees and stores to the fullest extent of our rights under Annexe II of the 1954 Agreement?

Mr. Robens: My right hon. Friend the Leader of the Opposition has requested the Prime Minister to look at the question of the sterling balances. I do not ask the Prime Minister to reply now, but will he also consider whether we should continue to send more arms to Egypt?

The Prime Minister: That matter, also, has been in our minds. The House can be assured that these matters are in our thoughts. I think I ought frankly to tell the House that we are having a Cabinet meeting the moment the House adjourns to take decisions on certain of these matters. I think I should really rather not say any more on the Floor of the House.

Several Hon. Members rose—

Mr. Speaker: I think that the House should respond to what the Prime Minister has said.

2. Extracts from telegram from the Prime Minister, Sir Anthony Eden, to the President of the United States, Dwight D. Eisenhower, 27 July 1956[1]

This morning I have reviewed the whole position with my Cabinet colleagues and Chiefs of Staff. We are all agreed that we cannot afford to allow Nasser to seize control of the canal in this way, in defiance of international agreements. If we take a firm stand over this now we shall have the support of all the maritime powers. If we do not, our influence and yours throughout the Middle East will, we are convinced, be finally destroyed.

The immediate threat is to the oil supplies to Western Europe, a great part of which flows through the canal. . . . If the canal were closed we should have to ask you to help us by reducing the amount which you draw from the pipeline terminals in the eastern Mediterranean and possibly by sending us supplementary supplies for a time from your side of the world.

It is, however, the outlook for the longer term which is more threatening. The canal is an international asset and facility, which is vital to the free world. The maritime powers cannot afford to allow Egypt to expropriate it and to exploit it by using the revenues for her own internal purposes irrespective of the interests of the canal and of the canal users. . . .

We should not allow ourselves to become involved in legal quibbles about the rights of the Egyptian Government to nationalize what is technically an Egyptian company, or in financial arguments about their capacity to pay the compensation which they

[1] *Full Circle*, pp. 427–8.

have offered. I feel sure that we should take issue with Nasser on the broader international grounds.

As we see it we are unlikely to attain our objective by economic pressures alone. I gather that Egypt is not due to receive any further aid from you. No large payments from her sterling balances here are due before January. We ought in the first instance to bring the maximum political pressure to bear on Egypt. For this, apart from our own action, we should invoke the support of all the interested powers. My colleagues and I are convinced that we must be ready, in the last resort, to use force to bring Nasser to his senses. For our part we are prepared to do so. I have this morning instructed our Chiefs of Staff to prepare a military plan accordingly.

However, the first step must be for you and us and France to exchange views, align our policies and concert together how we can best bring the maximum pressure to bear on the Egyptian Government.

3. Statement by the Prime Minister, Sir Anthony Eden, in the House of Commons, 30 July 1956[1]

The Prime Minister (Sir Anthony Eden): With your permission, Mr. Speaker, and that of the House, I wish to make a statement on the Suez Canal.

As a first step, measures have been taken, with effect from last Friday, in relation to Egypt's sterling balances and the assets of the Canal Company. An order has been made under the Exchange Control Act which has the effect of putting Egypt out of the transferable account area and generally making all transactions on Egyptian controlled sterling accounts subject to permission.

Secondly, a direction has been made under Regulation 2(a) of the Defence (Finance) Regulations safeguarding the securities and gold of the Suez Canal Company.

The Governments of the Commonwealth countries were given early information of the situation as it affects this country and the Commonwealth, and close touch is being maintained with them.

Yesterday afternoon the French Foreign Minister M. Pineau and Mr. Murphy of the American State Department came to London for consultations with Her Majesty's Government. My right hon. and learned Friend the Foreign Secretary took part in

[1] United Kingdom, H.C. Deb., vol. 557, coll. 918–19.

discussions with them which lasted until a late hour last night. They are continuing today. I cannot, therefore, say more about them at this stage.

This much, however, I can say. No arrangements for the future of this great international waterway could be acceptable to Her Majesty's Government which would leave it in the unfettered control of a single Power which could, as recent events have shown, exploit it purely for purposes of national policy.

4. Extracts from statement by the Prime Minister, Sir Anthony Eden, in the House of Commons, 2 August 1956[1]

... I am sorry that I am still not in the position to make the full statement that I should have liked to make, but discussions between the United States of America, France and ourselves are still proceeding. . . .

I am sure that the House will feel with me that it is necessary and desirable that full time should be taken with our Allies for all the aspects of this international problem to be considered and agreed while the three Foreign Ministers are in London. Meanwhile, there are certain observations which I should make to the House about the situation.

First, I think it is true to say that the cause for the anger and alarm felt, not only here but among the Governments and peoples of the democratic world, at the action of the Egyptian Government, is due to the special character of the Canal. It is right, therefore, that the House should be reminded of some aspects of this.

As the world is today, and as it is likely to be for some time to come, the industrial life of Western Europe literally depends upon the continuing free navigation of the Canal as one of the great international waterways of the world. I need give the House only one example. Last year, nearly 70 million tons of oil passed through the Canal—representing about half the oil supplies of Western Europe. Traffic through the Canal moved at the rate of 40 ships a day and amounted to 154 million tons of shipping—prodigious figures. Nor does this traffic affect the West alone. Australia, India, Ceylon and a large part of South-East Asia transport the major portion of their trade, or a large proportion of their trade, through the Canal.

---

[1] United Kingdom, H.C. Deb., vol. 557, coll. 1602–8.

Therefore, it is with these reflections in mind that I must repeat the carefully considered sentence which I used in the House on Monday last, if I may quote it again:

No arrangements for the future of this great international waterway could be acceptable to Her Majesty's Government which would leave it in the unfettered control of a single Power which could, as recent events have shown, exploit it purely for purposes of national policy.

That is still our position, and it must remain so.

However, that is not all the argument. The Egyptian Government have certain obligations in respect of the Canal. They are laid down in two principal instruments which I must mention. First, there is the Concession, which consists of a series of Agreements over the years between the Egyptian Government and the Suez Canal Company. The Concession defines the rights and status of the Suez Canal Company and the obligations of the Egyptian Government towards it. It is interesting to note that these were endorsed by the Egyptian Government as recently as 10th June this year, when a formal financial Agreement was concluded between the Egyptian Government and the Suez Canal Company.

This Agreement—made only a few weeks ago—was to continue in force until the end of the Company's Concession. That was in accord with the broadcast which was made on 17th November, 1954, by Colonel Nasser himself. He then said, according to our reports, that there remained only fourteen years until the end of the Canal Company's Concession. He added that good relations existed between the Egyptian Government and the Suez Canal Company. The Egyptian Government, he told us, had full confidence in the attitude of the Company and was sure that the latter would do everything to help the Egyptian Government in its task.

These undertakings are now torn up, and one can have no confidence—no confidence—in the word of a man who does that.

The second instrument is the well-known Suez Canal Convention of 1888, which lays down the vital principle that the Canal should always be open in time of war as in time of peace to every vessel of commerce or of war without distinction of flag. The Convention was signed by nine Powers, including the Sultan of Turkey on behalf of the Khedive of Egypt.

At this point I would ask the House to note, what is not generally known, that there is a link between the Convention of 1888 and the concession of the Canal Company. The Canal Company is specifically mentioned in Articles 2 and 14 of the 1888 Convention. Article 2 deals with the engagement of Egypt towards the Suez Canal Company as regards the freshwater canal, well known to many hon. Members.

Article 3 goes on to state that the contracting parties undertake to respect the plant, establishments, buildings and works of the maritime Canal and of the freshwater canal. Moreover, our own Anglo-Egyptian Treaty of 1954 states that the two Governments recognise that the Canal is a waterway economically, commercially, and strategically of international importance and expresses the determination of both parties to uphold the Convention of 1888. Now, Sir, I ask: how can this be reconciled with the Egyptian Government's action against the Suez Canal Company?

It is the free navigation of the Canal which is the solemn principle contained in the 1888 Convention, as the House well knows. That is not all, because free navigation does not depend only on the absence of discrimination or the absence of physical interference with the traffic in the Canal. The efficient functioning of the Canal, and its ability to deal with existing and future traffic, is also of decisive importance to us all.

The Canal is at present in need, as some hon. Members will probably know, of much new capital expenditure to enable it to cope with the increasing demands made upon it. Traffic through the Canal at present is increasing at a rate of about 7 per cent. a year. . . .

Therefore, the Company has, over the years, been accumulating capital to enable it to carry out the work of increasing the capacity of the Canal. . . .

Now, Sir, the reserves of the Company and the revenues from shipping would have been earmarked to finance these needs. But Colonel Nasser has now announced his intention to divert these revenues from this vital international waterway to build a dam in Egypt. . . . Here is a situation which neither Her Majesty's Government nor, I believe, the other maritime Powers, can accept, if they are to continue to live by means of the sources of supply which come through the Canal.

And there are some other considerations which we cannot ignore. Is it possible for us to believe the word of the present

Egyptian Government to the extent of leaving it in their power alone to decide whether these supplies shall reach the Western World through the Canal? I truly think that we have done everything in our power during the years—sometimes under criticism—by our actions and by our Treaty, to show our good will. I think that we have. Our reward has been a broken faith, and broken promises. We have been subjected to a ceaseless barrage of propaganda. This has been accompanied by intrigue, and by attempts at subversion in British territories.

Colonel Nasser's arbitrary action in breach of Egypt's solemn undertakings, many of them recently given, without previous consultation or previous notice, reveals the nature of the régime with which we have to deal; and I think that the action of the Egyptian Government in compelling the Canal Company employees to remain at their posts under threat of imprisonment is certainly, to say the least, a violation of human rights.

Sir, in these circumstances, and in view of the uncertain situation created by the actions of the Egyptian Government, Her Majesty's Government have thought it necessary—and I wanted to take this first opportunity to tell the House—to take certain precautionary measures of a military nature. Their object is to strengthen our position in the Eastern Mediterranean and our ability to deal with any situation that may arise. These measures include the movement from this country of certain Navy, Army, and Air Force units, and the recall of a limited number of Section A and A.E.R. Category I reservists, and also a limited number of officers from the Regular Army Reserve of Officers.

In addition, we shall have to recall—and I regret this, but it is inevitable owing to the categories in which they fall—a strictly limited number of men who are specialists, and skilled in certain essential tasks. These men are in Section B of the Regular Army Reserve and A.E.R. Category II. . . .

The principle of free navigation in peace and war is laid down in solemn international instruments to which the Egyptian Government is a party. It has been observed, let me say, by the Western Allies, even in two world wars. For instance, in the last war, when the effective control of the Suez Canal rested in the hands of His Majesty's Government alone, a search of cargoes in the Canal ports was instituted only for the purpose of ensuring that no damage was caused to the actual waterway. That was the action we took then and no attempt was made to seize cargoes in the Canal or its ports of access, even when they consisted of contraband. How could we look to the Egyptian Government alone to maintain these principles so scrupulously?—and if they are not maintained the life and the commerce of the whole free world is constantly at risk.

For all these reasons, I suggest to the House that the freedom and security of transit through the Canal, without discrimination, and the efficiency of the operation can be effectively ensured only by an international authority. It is upon this we must insist. It is for this that we are working in negotiation at this moment with other Powers deeply concerned. Nothing less than this can be acceptable to us.

5. Extracts from speech by the Leader of the Opposition, Mr. Hugh Gaitskell, in the House of Commons, 2 August 1956[1]

While the House will, of course, be a little disappointed that the Prime Minister was unable to tell us about the outcome of the talks now taking place, I am sure that we all appreciate and accept the reasons why he could not do so. I think, too, that we are all grateful to him for the statement he has made about the attitude of Her Majesty's Government to Colonel Nasser's action [Document 4].

For a long time, the Opposition has been critical of the Government's policy in the Middle East. . . . We have criticised from time to time the attitude of the Government on the question of the balance of arms in the Middle East, and I think that many of us feel that in the matter of the Aswan Dam the vacillations that have taken place are a subject for criticism.

But I do not propose today to develop these criticisms of the Government. May be, in the future, in a calmer situation, a further examination, a post-mortem, should take place, but I am sure that today what the country wants to hear about is the implication of Colonel Nasser's action and what should be done about it. Moreover, while I have not hesitated to express my disagreement with the Government in their policy in the past, I must make it abundantly plain that anything that they have done or not done in no way excuses Colonel Nasser's action in seizing the Canal.

[1] United Kingdom, H.C. Deb., vol. 557, coll. 1609–17.

I think that it is worth spending a moment or two on the question of why we do take such strong exception to this action. I know that some hon. Members may say, 'It is quite simple; it is an arbitrary act, a sudden act, something that involves dangers for us', but we have to recognise that the Egyptians are putting their point of view and that this point of view is being listened to elsewhere in the world. It is extremely important that the exact reasons why we resent and object to this action should be made clear at the start.

The Egyptian argument is perfectly clear. It is that this is an Egyptian company and that as the Government of Egypt they are perfectly entitled to nationalise any company they wish to nationalise, provided that they pay compensation; as regards the right of transit through the Canal they have given assurances that they regard it as necessary to observe the 1888 Convention. I should like to give my answers, as I see it, to that case put forward by the Egyptian Government.

First, so far as my hon. and right hon. Friends are concerned at any rate, we certainly do not say that the act of nationalisation in itself is wrong. Nor would we say that the act of nationalising a foreign-owned company was necessarily wrong, provided that the compensation was reasonable and fair. . . .

The real objections, it seems to me, are three. In the first place, as the Prime Minister has rightly emphasised, this is not an ordinary Company, conducting ordinary activities. It is a Company controlling an international waterway of immense importance to the whole of the rest of the world. It is, therefore, bound to be a matter of international concern when it changes hands. . . .

. . . It may be said there is no need for anxiety because we have had these assurances about the 1888 Convention. I am bound to say that it seems to me the strongest reason for having doubts in our minds as to whether we can accept those assurances has been the behaviour of the Egyptian Government in stopping Israeli ships from going through, and equally important—indeed, even more important—the clear defiance of the Resolution of the United Nations condemning this action, passed in September 1951.

The second reason why I think we must take strong exception to this is that any confidence we might have had in an action of this kind was profoundly shaken by the manner in which it was carried out. It was done suddenly, without negotiation, without discussion, by force, and it was done on the excuse that this was the way to finance the Aswan Dam project . . . Colonel Nasser . . . is proposing to take the whole of the gross revenues of the Canal—almost all of them transit dues—and divert them for the purpose of the Aswan Dam. Yet he has promised compensation. How can he at one and the same time both keep the Canal going, spend the necessary money on the repairs, extensions and reconstruction, pay the compensation or service the compensation loan to the shareholders, and also find money for the Aswan Dam? . . .

My third reason for thinking that we must object to this is that we cannot ignore—and this is a matter that the Prime Minister did not touch upon, no doubt for good reasons—the political background and the repercussions of the whole of this episode in the Middle East. We cannot forget that Colonel Nasser has repeatedly boasted of his intention to create an Arab empire from the Atlantic to the Persian Gulf. The French Prime Minister, M. Mollet, the other day quoted a speech of Colonel Nasser's and rightly said that it could remind us of only one thing—of the speeches of Hitler before the war.

Colonel Nasser has certainly made a number of inflammatory speeches against us and his Government have continually attempted subversion in Jordan and other Arab States; he has persistently threatened the State of Israel and made it plain from time to time that it is his purpose and intention to destroy Israel if he possibly can. That, if there ever was one, is a clear enough notice of aggression to come.

The fact is that this episode must be recognised as part of the struggle for the mastery of the Middle East. That is something which I do not feel that we can ignore. One may ask, 'Why does it involve the rest of the Middle East?' It is because of the prestige issues which are involved here. . . . I have no doubt myself that the reason why Colonel Nasser acted in the way that he did, aggressively, brusquely, suddenly, was precisely because he wanted to raise his prestige in the rest of the Middle East. He wanted to show the rest of the Arab world—'See what I can do'. He wanted to challenge the West and to win. He wanted to assert his strength. He wanted to make a big impression. Quiet negotiation, discussion around a table about nationalising the Company would not produce this effect.

It is all very familiar. It is exactly the same that we encountered

from Mussolini and Hitler in those years before the war. . . .
    I turn now to consider what kind of action should be taken. . . .
    . . . The first step is to call a conference of the nations principally
concerned. . . . I should like any control commission which may be
set up as a result of this conference to be a United Nations agency.
I am myself sure that, from the point of view of world peace and
development, it is far better that it should be done under the
United Nations than in some independent way. . . .

    Last of all, I come to a matter which cannot be ignored at this
moment just before the Recess. There has been much talk in the
Press about the use of force in these circumstances. First, I would
say that we need to be very careful what we say on this subject.
It is unwise to discuss hypothetical situations in present conditions. Obviously, there are circumstances in which we might be
compelled to use force, in self-defence or as part of some collective
defence measures. I do not myself object to the precautionary
steps announced by the Prime Minister today; I think that any
Government would have to do that, as we had to do it during the
Persian crisis.

    I must, however, remind the House that we are members of the
United Nations, that we are signatories to the United Nations
Charter, and that for many years in British policy we have stead-
fastly avoided any international action which would be in breach
of international law or, indeed, contrary to the public opinion of
the world. We must not, therefore, allow ourselves to get into a
position where we might be denounced in the Security Council as
aggressors, or where the majority of the Assembly were against us.

    If Colonel Nasser has done things which are wrong in the legal
sense, then, of course, the right step is to take him to the International Court. Force is justified in certain events. Indeed, if
there were anything which he had done which would justify force
at the moment, it is, quite frankly, the one thing on which we have
never used force, namely, the stopping of the Israel ships. We have
not done that; and it would, I think, be difficult to find—I must
say this—in anything else he has done any legal justification for
the use of force. What he may do in the future is another matter.

    I come, therefore, to this conclusion. I believe that we were
right to react sharply to this move. If nothing at all were done, it
would have very serious consequences for all of us, especially for
the Western Powers. It is important that what we should do should

be done in the fullest possible co-operation with the other nations affected. We should try to settle this matter peacefully on the lines of an international commission, as has been hinted. While force cannot be excluded, we must be sure that the circumstances justify it and that it is, if used, consistent with our belief in, and our pledges to, the Charter of the United Nations and not in conflict with them.

6. Extracts from speech by the Member for South Lewisham, Mr. Herbert Morrison, in the House of Commons, 2 August 1956[1]

. . . I went through this kind of business on another occasion and in a slightly different part of the world. I heard a lot about it and a good deal has been said about it since. Let me tell the House right away that it is an experience that I do not look back on with any great pleasure. That was another case in which one Government—if it were a Government, which is doubtful; more likely it was one man—unilaterally decided that an agreement which had been freely entered into between the Government of his country years before and the Anglo-Iranian Oil Company should be repudiated. . . .

Let it be remembered that in both cases there was no consultation with us or with our friends. There was no consultation with the citizens of the countries concerned; neither in the case of Mossadeq, who just whipped his decree out of his tail pocket, nor in the case of Colonel Nasser, who has done the same thing. There was no debate, or any effective debate, in the Parliaments of the countries concerned. There was no effective discussion among the citizens of those countries, and, after all, both these acts are calculated to damage the interests of the people within the country doing the damaging act. It was so in Persia, and it is so here.

I am bound to say that of all the uncomfortable experiences in the process of negotiation, negotiating with a man who happened to be the Prime Minister of Persia and who was utterly indifferent to the welfare of his own people—I will only say that it is a somewhat unfair form of negotiation. One does expect the man on the other side of the table to take some account of the interests of his own country, even if he does not take account of the interests of the country to which the man on the other side of the table is

---

[1] United Kingdom, H.C. Deb., vol. 557, coll. 1654-61.

attached; but there it was. That was how it was in that case.

I can well imagine the discussions which this present Government are having about the present position in Egypt, because I am familiar with the other negotiations, and there are arguments both ways—indeed, several ways—as to the right thing to do. . . .

Now, Sir, the question which has to be considered, and I think that some hon. Members on both sides are a little in danger of by-passing it—is whether, in the light of these discussions which are going on, force should or should not be a possible element in settling the matter.

I am in favour of taking this business to the United Nations. I am all in favour of it as long as the United Nations will be expeditious and effective about it. . . . I say to the United Nations that if it wishes—as we would wish it—to become the great moral authority of the world and the great decisive instrument, it must stop dodging vital international issues. If our Government and France and, if possible, the United States should come to the conclusion that in the circumstances the use of force would be justified, then I think that it is up to each hon. Member of this House to tell the Government whether we would support them or whether we would not. For my own part, in principle, if, after an elaborate and proper consideration, the Government and our friends come to that conclusion, I think that in the circumstances of this particular case it might well be the duty of hon. Members, including myself, to say that we would give them support. . . .

I therefore say to the Government that I wish them luck in solving this problem, but I ask them to be careful and judicious by all means and to mind how they are going; but let them not be too much afraid. This is a world in which great Powers can be a nuisance; and, by the way, the Prime Minister's cheerfulness about the Soviet Union does not quite seem to have come off. They are backing Egypt and so is China. By all means consider all the pros and cons, but I ask the Government not to be too nervous, because if they are too nervous we shall begin to evolve a situation in which countries can set themselves up against international practice, international morals and international interests, and, in that case, we are not helping the peace of the world; we are helping anarchy, conflict and bad conduct among the nations.

7. Statement by the Governments of the United Kingdom, France, and the United States, 2 August 1956[1]

The Governments of France, the United Kingdom and the United States join in the following statement:

1. They have taken note of the recent action of the Government of Egypt whereby it attempts to nationalise and take over the assets and the responsibilities of the Universal Suez Canal Company. This Company was organised in Egypt in 1856 under a franchise to build the Suez Canal and operate it until 1968. The Universal Suez Canal Company has always had an international character in terms of its shareholders, directors and operating personnel and in terms of its responsibility to assure the efficient functioning as an international waterway of the Suez Canal. In 1888 all the great Powers then principally concerned with the international character of the Canal and its free, open and secure use without discrimination joined in the Treaty and Convention of Constantinople. This provided for the benefit of all the world that the international character of the Canal would be perpetuated for all time, irrespective of the expiration of the concession of the Universal Suez Canal Company. Egypt as recently as October, 1954, recognised that the Suez Canal is 'a waterway economically, commercially and strategically of international importance', and renewed its determination to uphold the Convention of 1888.

2. They do not question the right of Egypt to enjoy and exercise all the powers of a fully sovereign and independent nation, including the generally recognised right, under appropriate conditions, to nationalise assets, not impressed with an international interest, which are subject to its political authority. But the present action involves far more than a simple act of nationalisation. It involves the arbitrary and unilateral seizure by one nation of an international agency which has the responsibility to maintain and to operate the Suez Canal so that all the signatories to, and beneficiaries of, the Treaty of 1888 can effectively enjoy the use of an international waterway upon which the economy, commerce, and security of much of the world depends. This seizure is the more serious in its implications because it avowedly was made for the purpose of enabling the Government of Egypt to make the Canal serve the purely national purposes of the Egyptian Government,

[1] Cmd. 9853, pp. 3–4.

rather than the international purpose established by the Convention of 1888.

Furthermore, they deplore the fact that as an incident to its seizure the Egyptian Government has had recourse to what amounts to a denial of fundamental human rights by compelling employees of the Suez Canal Company to continue to work under threat of imprisonment.

3. They consider that the action taken by the Government of Egypt, having regard to all the attendant circumstances, threatens the freedom and security of the Canal as guaranteed by the Convention of 1888. This makes it necessary that steps be taken to assure that the parties to that Convention and all other nations entitled to enjoy its benefits shall, in fact, be assured of such benefits.

4. They consider that steps should be taken to establish operating arrangements under an international system designed to assure the continuity of operation of the Canal as guaranteed by the Convention of 29th October, 1888, consistently with legitimate Egyptian interests.

5. To this end they propose that a conference should promptly be held of parties to the Convention and other nations largely concerned with the use of the Canal. The invitations to such a conference, to be held in London, on August 16, 1956, will be extended by the Government of the United Kingdom to the Governments named in the Annex to this Statement. The Governments of France and the United States are ready to take part in the conference.

ANNEX

Parties to the Convention of 1888

Egypt	Spain
France	Turkey
Italy	United Kingdom
Netherlands	U.S.S.R.

Other nations largely concerned in the use of the Canal either through ownership of tonnage or pattern of trade

Australia	Iran
Ceylon	Japan
Denmark	New Zealand
Ethiopia	Norway
Federal Republic of Germany	Pakistan

Greece   Portugal
India   Sweden
Indonesia   United States

8. Extracts from broadcast by the Prime Minister, Sir Anthony Eden, 8 August 1956[1]

The Suez Canal is a name familiar to everyone. I have come to talk to you tonight about what has happened there in the last few days and what it means to us. You, or some members of your family perhaps, have served there, or maybe one of you or more have helped to defend the Canal in one or other of the two Great Wars. For Britain the Canal has always been the main artery to and from the Commonwealth, bringing us the supplies we and they need. To many other nations throughout the world it has become the bearer of a traffic in ever-growing volume. The world's commerce depends upon it. It carries goods of all kinds for Europe and America, for Australia and New Zealand, and for eastern countries like Pakistan and India and Ceylon. It is, in fact, the greatest international waterway in the world, and what Colonel Nasser has just done is to seize it for his own ends.

. . . It may be said: 'Why is it so terrible to nationalize a company? It was done here.' That is perfectly true, but it was done, as Mr. Morrison rightly pointed out in the House of Commons, to our own British industries. Colonel Nasser's action is entirely different. He has taken over an international company, without consultation and without consent. The rights of this company were secured by repeated and solemn agreements, entered into by the Egyptian Government. There are, in fact, a whole series of them, to fill a book. The last was concluded some two months ago. Some people say: 'Colonel Nasser has promised not to interfere with shipping passing through the Canal. Why, therefore, don't we trust him?' The answer is simple. Look at his record. Our quarrel is not with Egypt, still less with the Arab world; it is with Colonel Nasser. When he gained power in Egypt, we felt no hostility towards him. On the contrary, we made agreements with him. We hoped that he wanted to improve the conditions of life of his people and to be friends with this country. He told us that he wanted a new spirit in Anglo-Egyptian relations. We welcomed that, but instead of meeting us with friendship Colonel Nasser

[1] 'A Very Grave Situation', *The Listener*, 16 August 1956, pp. 221–2.

Ecs

conducted a vicious propaganda campaign against this country. He has shown that he is not a man who can be trusted to keep an agreement. And now he has torn up all his country's promises towards the Suez Canal Company and he has even gone back on his own statements, because not so long ago he was speaking in praise of the company. He told them how satisfied he was with them, and now, in a night, they have been taken over by force, and their assets seized. By Egyptian law the company's employees, French and British, are ordered to stay at work under threat of imprisonment.

The pattern is familiar to many of us, my friends. We all know this is how fascist governments behave and we all remember, only too well, what the cost can be in giving in to fascism.

. . . If Colonel Nasser's action were to succeed, each one of us would be at the mercy of one man for the supplies upon which we live. We could never accept that. With dictators you always have to pay a higher price later on, for their appetite grows with feeding. Just now Colonel Nasser is soft-pedalling; his threats are being modified. But how can we be sure that next time he has a quarrel with any country he will not interfere with that nation's shipping, and how can we be sure that next time he is short of money he will not raise the dues on all the ships that pass through the Canal? If he is given the chance of course he will.

. . . An international conference is now to be held here in London next week, to which we have invited all the countries most closely interested in the use of the Canal. Let me tell you what the purpose of the conference is. It is just this: the Canal must be run efficiently and kept open as it has always been in the past, as a free and secure international waterway for the ships of all nations. It must be run in the interests not of one country but of all. In our view, this can only be secured by an international body. That is our purpose. . . .

Meanwhile, we have too much at risk not to take precautions. We have done so. That is the meaning of the movements by land, sea, and air of which you have heard in the last few days. My friends, we do not seek a solution by force, but by the broadest possible international agreement. That is why we have called the conference. We shall do all we can to help its work, but this I must make plain. We cannot agree that an act of plunder which threatens the livelihood of many nations shall be allowed to succeed. And we must make sure that the life of the great trading nations of the world cannot, in the future, be strangled at any moment by some interruption to the free passage of the Canal. These are our intentions. I am sure they will command your support.

9. Statement by 24 Members of the Parliamentary Labour Party, 8 August 1956[1]

The nationalization of the Suez Canal Company concerns Egypt alone. The only issue which is of international importance is the continued uninterrupted passage of the Suez Canal in accordance with the 1888 Convention. It is understood that the inviting Powers will present to the international conference a plan for the internationalization of the Canal.

No one doubts that if such a plan were agreed and accepted by all the Powers concerned, that would be a satisfactory solution: but Great Britain and France have made it clear by action as well as by speech that, with or without the consent of all interested Powers, they are prepared to put their plan or some modification of it into operation by force.

Unless recalled, Parliament will not be sitting when the conference meets. There has been no meeting of the Parliamentary Labour Party or the national executive to consider this matter, which may be fraught with the gravest consequences for the country and for the peace of the world.

Without seeking to arrogate to ourselves any right to formulate party policy, we wish to declare quite firmly for ourselves, in the belief that our view would be shared by the overwhelming majority of all sections of the Labour movement, that:

(1) International control of all international waterways, including Suez, would be an important contribution to world peace.

(2) To attempt to carry out such internationalization of the Suez Canal by force, against the resistance of the Egyptian Government and people, would be an act of aggression under the United Nations Charter.

(3) Such action would inflame the Arab nations against us and have the gravest repercussions in Asia and Africa and might well be fatal to the peace of the world.

(4) In any circumstances to which the Charter of the United

[1] *The Times* (London), 9 August 1956.

Nations applies, we are not prepared to support this or any British Government in the use of force not sanctioned by the Security Council in accordance with the Charter.

10. Statement by the Labour Party Parliamentary Committee, 13 August 1956[1]

The Labour Party Parliamentary Committee met this afternoon to review the Suez question in the light of developments since the debate in the House of Commons on August 2.

The Leader and the Deputy Leader reported on interviews and correspondence with the Prime Minister on this matter since the last meeting of the committee.

The committee endorsed the speech of the Leader of the party in the debate on August 2 [Document 5], and, in particular, the following points:

(a) His condemnation of Colonel Nasser's action not because the act of nationalization in itself was wrong but because of the arbitrary character of that action and the grave anxieties thereby created regarding the utilization of a vitally important international waterway and the political repercussions in the Middle East.

(b) That armed force, nevertheless, could not be justified except in accordance with our obligations and pledges under the Charter of the United Nations.

(c) That apart from the continued stopping of Israeli ships in disregard of the U.N. Security Council Resolution of 1951, Colonel Nasser has not done anything so far which justifies the use of armed force against Egypt.

In view of public anxiety, the committee call upon the Government to make plain that the military measures taken in the last 10 days are purely precautionary, solely intended for defence against possible aggression and not preparations for armed intervention outside and inconsistent with our obligations under the Charter of the United Nations.

The committee consider that the prime purpose of the international conference now assembling in London should be to prepare a plan which would guarantee the efficient operation and development of the canal, a fair and reasonable financial return to Egypt, no interference with the right of free passage, no arbitrary raising of dues and no discrimination among those using the canal.

[1] *The Times*, 14 August 1956.

To this end any such plan should, as suggested by Mr. Gaitskell on August 2, be associated with the United Nations.

The committee urge that the conclusions of the conference should be submitted to a special meeting of the General Assembly of the United Nations.

11. Extracts from broadcast by the Secretary of State for Foreign Affairs, Mr. Selwyn Lloyd, 14 August 1956[1]

. . . We have to remember that the present ruler of Egypt is a military dictator. He played a leading part in overthrowing the Egyptian monarchy by a military coup. He removed his own leader, General Neguib, by similar methods. He now rules supreme. He can change his mind overnight. He can denounce an international agreement or imprison a British subject according to his mood of the moment. He maintains himself in power by methods so well known to us from what happened in certain countries in the inter-war years.

. . . Colonel Nasser is misusing Arab nationalism to further his own ambitions. I read the other day a little book which he wrote about his revolution. Like other dictators before him he reveals on paper the pattern of his thoughts and ambitions. The three stages are clear. First, control of the Arab states and the oil. Second, control of the whole of Africa. Finally, control of all the Moslems throughout the world. Running through the book is the theme 'Who is the hero who will achieve all this?' I think we can guess his answer. And that is the problem with which we are faced. These thoughts, these actions, are all too similar to what we saw happen between the wars with other dictators.

I believe that there have been three critical times for us in the past ten years. First, there was the threat to Berlin in 1948, which was defeated by the Berlin air lift. Secondly, there was the communist aggression in Korea in 1950. That was repelled by the force of arms. The third threat, and, in my view, the most serious for all of us in Britain, is this act of aggression against this great international waterway. . . .

It is a deliberate challenge. All our friends in the Middle East are watching to see how we meet it. Although, having made his coup, Colonel Nasser's language is now milder and he makes all sorts of promises about how well he is going to behave in the

[1] 'The London Conference on the Suez Canal', *The Listener*, 16 August 1956, pp. 219–20.

future, that again is according to the pattern. I believe we have to take counsel from our past experience. If he is not checked now, what will his next step be? We have to be resolute in this situation. We must not permit the right of free passage through the Canal to depend upon the transient impulses of a single military dictator. . . .

II. CANADA

1. Statement by the Secretary of State for External Affairs, Mr. L. B. Pearson, 27 July 1956[1]

While Canada has no share in the ownership of the Suez Canal Company, as a trading nation we have a very real interest in the efficient and nondiscriminatory operation of this waterway of great and historic importance in peace and war. We would regret and be concerned about any action which interfered with such operations.

2. Answers by the Secretary of State for External Affairs, Mr. L. B. Pearson, to questions in the House of Commons, 28 July 1956[2]

Mr. J. G. Diefenbaker (Prince Albert): I should like to ask the Secretary of State for External Affairs if he is prepared to make a statement on the situation in relation to the Suez Canal. Would he also advise whether, in view of the unprecedented and shocking conduct of the Nasser Government, Canada ought not to join with Britain in condemnation of what has taken place there in a perversion of international contracts, and also indicate to Britain and the other nations Canada's agreement with the stand which they are taking to meet this situation.

Mr. L. B. Pearson: . . . Mr. Speaker, the violation by the government of Egypt of an international convention governing the use of an international waterway so important as the Suez Canal is, of course, to be condemned. Possibly it should be recalled at this time that the convention in question attempted to safeguard the free use of the waterway in war and in peace. In that sense, the convention was already violated by the Egyptian Government when Israeli vessels were prevented from using the canal.

We are exchanging views with governments probably more directly concerned with this matter than the Canadian government

---

[1] *The Globe and Mail* (Toronto), 28 July 1956.
[2] Canada, H.C. Deb., 1956, vol. VII, pp. 6607–8.

because of their association with the Suez Canal Company. I think it would be premature for me to make any statement beyond that at this time, except possibly to repeat what I said yesterday in answer to queries from the press [Document 1]. . . .

Mr. Diefenbaker: Might I just add a supplementary question? Does not the Minister consider that Canada, as a member of the Commonwealth, should have something to say on a matter that particularly affects the members of the Commonwealth, Britain, Australia and New Zealand particularly, to indicate a degree of unity in regard to this matter?

Mr. Pearson: Mr. Speaker, we have already had something to say on the matter, as I have indicated. And I should add that our High Commissioner in London, since the announcement of Egyptian action, had discussions in London with the Secretary of State for Commonwealth Relations, with the Foreign Secretary and with the Prime Minister of the United Kingdom, with a view, not only to ascertaining what United Kingdom policy is in this matter—and they are more directly concerned than we are because of their share in the ownership of the Suez Canal Company—but also to doing what we can to concert our attitude and policy with other Commonwealth countries on this matter. . . .

Mr. Alastair Stewart (Winnipeg North): Amongst the views which are being exchanged with the United Kingdom, will the Minister, on behalf of the government, stress to the United Kingdom the importance of using the good offices of the United Nations in an attempt to solve this problem?

Mr. Pearson: Mr. Speaker, our High Commissioner in London has already stressed that aspect of the question in his talks with the United Kingdom government.

3. Extracts from speech of the Secretary of State for External Affairs, Mr. L. B. Pearson, in the House of Commons, 1 August 1956[1]

. . . In recent days a new situation—I was going to say 'a new crisis'—has developed . . . in connection with the Suez Canal. A sudden arbitrary move on the part of the Egyptian Government has aroused fears that the right to use this international waterway in peace and war without discrimination may be prejudiced, a

[1] Canada, H.C. Deb., 1956, vol. VII, pp. 6787, 6831.

right which, as hon. members know, is guaranteed by an international treaty. Far more than the nationalization, or, if you like, the expropriation, of the Suez Canal Company is at stake in this matter; it is the future use for all nations without arbitrary or unnecessary interference of an essential international artery of trade and of communications, a waterway which was constructed by international agreement and with international co-operation and is now maintained and operated internationally.

As hon. members know, steps are being taken at the moment in London, by three powers very directly and importantly affected by the Egyptian decision, to bring about a satisfactory solution to the problem, the problem created by this action of the Egyptian Government, by establishing some form of permanent international control for this international waterway, by which the legitimate rights of all countries can be protected. Until the results of this London meeting are available . . . I think I should say nothing more about this matter, except possibly to express the support of our government for the principle of such international control, with the countries having the greatest interest in the operation of the canal sharing in that control, preferably, if this turns out to be practicable, under the aegis of the United Nations. . . .

[*Later:*]

. . . The hon. member for Prince Albert [Mr. Diefenbaker] felt that we should have taken, or should still take, no doubt, a stronger stand in support of the three governments which are now meeting in London in connection with the Suez crisis [Document 4]. The hon. member for Eglinton [Mr. Donald Fleming] complained that we were adopting . . . a lofty attitude of distant spectators, and he wondered whether we were taking the crisis seriously enough. Indeed we are, because it is a serious matter. Yet I think we were well advised in this government not to intervene. . . . This government, however, has already expressed its concern over this matter and has indicated our support for the principle of international control of the Suez Canal, a principle which they are trying to work out in London at the present time. But I doubt, even from the Commonwealth point of view, whether it would be wise for us to go much further at this particular moment. . . . This afternoon I received a telegram from our representative in one of the Asian countries of the Commonwealth which indicated that the feeling in that country on this matter was quite different indeed from that which exists in Canada or in certain other parts of the Commonwealth. . . .

4. Extracts from speech by the Member for Prince Albert, Mr. J. G. Diefenbaker, in the House of Commons, 1 August 1956[1]

. . . For some reason [the Secretary of State for External Affairs] did not deal at length with . . . the dangerous and truculent attitude on the part of General [sic] Nasser. . . . The arbitrary action on his part, challenging as it does international law, freedom under law among nations, breaking as it does the pledged word, may I say that I feel that Canada as the leading member of the British Commonwealth at this time should give public encouragement to the stand taken by Great Britain, France and the United States. We should indicate that the reasonable attitude of mind and the calm consideration which these nations are giving warrants the moral support of Canada in these dark hours. . . .

. . . If the Canal is closed the danger to Britain's lifeline and to freedom's defence is fearful to contemplate. The closing of the Suez Canal would bottle up freedom's defence in the Mediterranean. It would make almost impossible the maintenance of British trade which is being challenged today as the result of the economic offensive of the U.S.S.R. At the moment Nasser claims he has no such intention in mind, but it is only a few months ago that he with sweet words and in symphony with the U.S.S.R. spoke so warmly of his intentions with regard to Britain and to other free nations. His word has been proved to be false and valueless. . . .

. . . I ask even now that this government give to the world approval and support of what is being done at the meeting which is today convened in London to the end that those three nations will realize that we in the free world are united. . . .

5. Extract from answer by the Secretary of State for External Affairs, Mr. L. B. Pearson, to a question in the House of Commons, 4 August 1956[2]

Mr. J. G. Diefenbaker (*Prince Albert*): . . . Will the Minister say whether or not, having regard to the fact that Canada is a

---

[1] Canada, H.C. Deb., 1956, vol. VII, pp. 6792–4.   [2] Ibid., pp. 7047–8.

trading nation, she made any application or suggested that she would be willing to attend this conference so that the weight of her views might be expressed there and a degree of support given to the United Kingdom in this hour of great stress and difficulty?

Mr. Pearson: . . . Mr. Speaker, we did indicate to the meeting in London that we would be willing, of course, to attend the proposed conference if we were invited. The criteria which were laid down by the three governments for an invitation to that conference were ones to which we did not take any exception, namely, the signature of the Constantinople treaty, the importance to a country of its shipping going through the Suez and the importance of the Canal to the trade of a country. As it happened, Canada did not fall within those criteria, and therefore we did not receive an invitation, nor, indeed, did any other western hemisphere country except the United States, nor did South Africa. We have no complaints about that at all; but we would, of course, have been quite willing to attend the conference if we had been invited, and we indicated that to the governments that were meeting in London at that time.

6. Extract from speech by the Member for Prince Albert, Mr. J. G. Diefenbaker, 8 August 1956[1]

As I see the problem, Britain and France want the United States to back a plan for the International Control Board to guarantee that the Suez Canal shall remain a national [*sic:* international?] waterway in war and peace, a plan that would assure transit and technical competence. Canada, as a trading nation, cannot afford to take a position at once both posthumous and uncertain. In my opinion Canada should not be a mere tail on the American kite but should, as a senior nation of the Commonwealth, give to the Government of the United Kingdom moral support and encouragement. It is my belief that Canada's relationship to the British at this time should be one of closest co-operation and, as a nation which depends to an overwhelming extent on foreign trade, the Government should have made it known to Britain of its desire to be represented at the Conference of 24 nations which is to be convened on the 16th of this month.

[1] *Texts of Addresses delivered at the 25th Annual Couchiching Conference, August 4th–11th, 1956* (Toronto, Canadian Institute of Public Affairs), p. 72.

The Canadian policy cannot be a waffling one. As an interpreter between Great Britain, the Commonwealth and the United States, Canada, with no territorial ambitions, would exert a necessary and worthwhile influence in bringing into agreement the divergent policies of Britain and the United States and would bring to the Conference a spirit of conciliation and goodwill trusted by friend and potential foe alike. An uncertain policy on the part of Canada might well bring about another Munich.

The consequences of the Suez Canal being closed by Nasser would be the bottling up of freedom's defence in the Mediterranean, it would render impossible the maintenance of British trade which is being challenged today as a result of the economic consequences of the U.S.S.R.

III. AUSTRALIA

1. Report of statement by the Minister of External Affairs, Mr. R. G. Casey, 27 July 1956[1]

The Minister for External Affairs, Mr. Casey, referred to-day to reports that Egypt had nationalized the Suez Canal Company and had taken over the Canal installations.

Mr. Casey said that there was no necessity for him to stress the very great importance of the Canal as an international waterway linking Australia and countries of South and South-East Asia with the United Kingdom and Europe. The Australian Government would view with concern any unilateral or precipitate departure from the terms of solemn understandings of direct concern to maritime states, or any move which might restrict the free use of the Canal by world shipping.

Mr. Casey added that the full implications of the reports were being studied, and that the Australian Government was consulting urgently with the United Kingdom Government.

2. Report of statement by the Minister of External Affairs, Mr. R. G. Casey, 29 July 1956[2]

Mr. Casey said that it was essential to the welfare of all countries east and west of Suez who take part in the trade of the world that the [Suez Canal] Convention should be faithfully observed. For some countries, including Australia, it is vital, because their

[1] *C.N.I.A.*, vol. 27, no. 7, July 1956, p. 445.   [2] Ibid., p. 446.

economic life depends on the free and unfettered flow of cargoes and ships through the Canal.

'Colonel Nasser has, we understand, declared that there will be no interference with shipping, but with or without such a declaration, one fact is clear: the Egyptian Government, by seeking to take over the Company, has arbitrarily and unilaterally assumed the function of allowing or denying the use of the Canal and the power to charge whatever it thinks fit upon a large part of the world's commerce.

'This power does not pass to Egypt as the result of discussion of Egypt's interests and negotiations about them. It passes by an abrupt termination—without consultation—of the concession agreements between Egypt and the Suez Canal Company, the expropriation of the Company, and the establishment of Egyptian armed guards on the Canal itself. The abrupt and high-handed action of the Egyptian Government raises most serious apprehensions about their attitude to the economic and other interests of other countries.

'The issues go beyond the protection of the economic, commercial and strategic interests of the nations of the world.

'The Egyptian Government announces this act immediately following the decision of two Governments not to lend or give Egypt assistance in a developmental project. This aspect of the situation, which amounts to punitive action against others possessing legitimate rights, adds to the record of similar instances in recent times and must be a cause of profound concern to all countries whose policies are based on respect for essential principles in international relations.

'The Australian Government is not a party to the agreements establishing the Suez Canal Company, nor does it have a financial interest in the Company. Australia's interest in the situation arises from our dependence upon the right to use the Canal for most of our trade in peace and for our protection in war, and from our concern as a member of the United Nations to ensure that international understandings are observed.'

Mr. Casey said that the Australian Government was in urgent communication with the Governments of other interested countries. The Egyptian Government had rejected protests made by the United Kingdom and French Governments. Wider international consultation will no doubt take place to ensure that the economic and other vital interests of all countries are protected. The whole matter was most disturbing and was engaging the urgent attention of the Australian Government.

3. Report of statement by the Minister of External Affairs, Mr. R. G. Casey, 2 August 1956[1]

The Minister for External Affairs, Mr. Casey, spoke to-day about the discussions about the Suez Canal which are taking place in London between Britain, America and France.

Mr. Casey said that it was appropriate that these three powers should be meeting together, and the Australian Government welcomed their initiative. The Australian Government would expect to be included in any wider discussions that took place, which should clearly include the principal nations using the Canal in addition to countries possessing international mercantile marine fleets.

Australia had a direct, substantial and vital interest in the Canal. During the past few years, for example, nearly two-thirds of Australia's total imports and exports passed through the Canal. This was equivalent to nearly £A1,000 million worth of Australian trade passing through the Canal each year.

Mr. Casey said that while Australia, in terms of goods carried, ranked first among the users of the Canal in this part of the world, other neighbouring countries in Asia had a substantial interest in the unrestricted use of it.

Mr. Casey said that the essential question was whether this international waterway linking two oceans was to be allowed to come under the unfettered control of a single power which might operate it for its own purposes. It would be generally agreed that this should not occur. The Suez Canal was to be regarded as an international public utility and it should be administered as such— as it had been in the past. Its benefits should be available to all countries, and it should not be used for the particular advantage of any one country.

Mr. Casey said it was already evident that Egypt saw matters quite differently. Colonel Nasser has described the Canal as an Egyptian asset, and he has said that profits from it will be used to finance the Aswan High Dam—an undertaking the cost of which is estimated at £600 million sterling. Nor is any assurance

[1] *C.N.I.A.*, vol. 27, no. 8, August 1956, pp. 487-8.

to be found in the Egyptian stoppage, in defiance of the Suez Canal Convention, of Israeli shipping and ships bound for Israel. This action has been condemned by the Security Council, but Egypt has persisted in banning the passage of Israeli trade through the Canal.

Mr. Casey said that the objective to be aimed at was the maintenance of an international régime which would guarantee free navigation of the Canal at all times to all nations, reasonable and non-discriminatory dues and competent administration. He said it would be intolerable if the trade of Australia and other countries and the freedom of our vital communications in an emergency were to become liable at any time to capricious or arbitrary action, or that we should be confronted by the risk that by the decision of a single country one of the world's major highways might be denied to us.

4. Statement by the Acting Prime Minister, Sir Arthur Fadden, 7 August 1956[1]

Ministers to-day discussed at length the Suez Canal issue and have cabled further views to the Prime Minister who is at present in Washington.

The Prime Minister has been in close touch with the United States Administration and the Government of the United Kingdom, and is returning to London on Thursday of this week and will take part in consultations with the United Kingdom Prime Minister and other Ministers. He will be joined in London by the Minister for External Affairs, Mr. Casey, who will arrive in London about 15th August.

The Government to-day condemned the abrupt and high-handed action taken by the Government of Egypt and emphasizes the seriousness of the situation which has thereby been created. It commends the prompt and firm reactions of the United Kingdom Government and the convening of the Conference which is to begin in London on 16th August.

5. Report of statement by the Minister of External Affairs, Mr. R. G. Casey, 12 August 1956[2]

The Minister for External Affairs, Mr. Casey, said before leaving Sydney to-night that he was leaving at the end of a busy two

[1] *C.N.I.A.*, vol. 27, no. 8, August 1956, p. 488.   [2] Ibid., p. 489.

weeks, during which the Australian Government had given
continuous consideration, in consultation with other countries, to
the situation arising out of the abrupt and unilateral decision by
the Egyptian Government to nationalize the Suez Canal.

Mr. Casey said that he was now going to a conference where
Australia would join with other countries in an attempt to define
a position to meet the situation and which would be acceptable to
world opinion. He said he did not want to speculate as to what
would be decided at the conference, or to discuss the policy
of the Australian or any other Government.

Mr. Casey said that he thought there was universal agreement
with the three principles which he had stated publicly a fortnight
ago as defining Australia's basic position: freedom of passage
through the Canal for all countries without discrimination;
reasonable charges; and efficient operation. These principles
point to the necessity for the continuation of an international
régime.

Australia had a clear understanding of the vital importance of
the Suez Canal question to Britain—not only its value to Britain
herself and her position in the world—but also the role of the
Canal in world trade originating in or carried on by Britain. This
is something of which, as a British people, Australians are particularly conscious.

The Australian Government had been working away quietly
and continuously on the matters involved. The Prime Minister, Mr.
Menzies, had had most useful discussions in Washington and in
London. Australian diplomatic missions in all the interested
capitals, and particularly in Asia, had been discussing the questions
involved. He himself was now leaving for London to join the
Prime Minister, strengthened with a very full discussion in the
Australian Cabinet this week and with the close personal attention
that had been given to the problem throughout by the Acting
Prime Minister, Sir Arthur Fadden.

Mr. Casey said he was impressed by the community of interest
in this matter between Australia and the countries of Asia. The
Suez Canal had originally been created to shorten the distance by
sea between Europe and the countries east of Suez. Conditions
satisfactory to the world had been created and maintained for a
hundred years, by an international arrangement. All the countries
east of Suez—including Australia—had a vital interest in seeing

that the Canal was kept open to world shipping at reasonable charges, as it has been for a hundred years. If the Canal becomes the plaything of an individual country, the charges of imports and exports of Asian countries—and of Australia—might well be raised, to the detriment of all.

In conclusion, Mr. Casey said that he would watch with interest what President Nasser would say and whether Egypt would attend the conference. If Egypt did not attend, it would still further aggravate the grave responsibility that Egypt already bore for its affront to generally accepted standards of international order.

6. Extracts from broadcast by the Prime Minister, Mr. R. G. Menzies, 13 August 1956[1]

Colonel Nasser's action in respect of the Suez Canal Company has created a crisis more grave than any since the Second World War ended. The leading trading nations of the world are all vitally concerned. You in Great Britain are concerned, for a threat to the Suez Canal will, if not resisted, encourage other acts of lawlessness and so reduce the economic strength of your country that the whole standard of living may be drastically reduced.

This comment excludes the effect upon Britain's prestige and authority in the world. It is apparently not fashionable to speak of prestige. Yet the fact remains that peace in the world and the efficacy of the United Nations Charter alike require that the British Commonwealth and in particular its greatest and most experienced member, the United Kingdom, should retain power, prestige and moral influence.

So far, there may be a considerable measure of agreement. But I have been interested to observe, in both the United States and in London, a disposition in some quarters to find legal virtues in what Nasser has done, and to accuse national leaders either of trying to deny to Egypt its so-called legal right to nationalise the Suez Canal or of prematurely dealing with a risk of stoppage which may never arise. . . .

Not that I believe that the problem is purely or even mainly a legal one. On the contrary, it concerns great questions of international policy on which our views must, at our peril, be sensible, robust and firm. . . .

[1] *Speech is of Time: Selected Speeches and Writings by the Right Honourable Robert Gordon Menzies* (London, 1958), pp. 81–86.

International Law is not a precise body of jurisprudence. It is always in the making. But if there is one thing clear, it is that national contracts with the governments or citizens of other nations must be carried out unless there is legal excuse for the non-performance. If this were not so, all talk of International Law would become meaningless on the very threshold.

Nasser has, therefore, begun by violating the first principle of International Law. For people to conceal or excuse this violation by talking about the general power of governments to expropriate property within their own boundaries is therefore both irrelevant and absurd.

If, at any time and for any reason of real or supposed self-interest, a nation could claim that its sovereign rights entitled it to set treaties aside or violate international contracts, all talk of or reliance upon International Law would be a sham. . . .

We are about to try to deal, by negotiation, with a matter which is vital to the trade and economics of a score of nations. To leave our vital interests to the whim of one man would be suicidal.

We in Australia applaud the statement made by France, the United Kingdom, and the United States [I, Document 7, *supra*]. We support the Conference. We cannot accept either the legality or the morality of what Nasser has done. . . .

7. Extracts from statement by the Executive of the Federal Parliamentary Labour Party, 15 August 1956[1]

. . . If no solution is reached in London tomorrow, United Nations' jurisdiction must be invoked without further avoidable delay. The Australian Labour Party reaffirms the principle that the canal dispute could and should have been referred to the United Nations.

Mr. Menzies's statement in London on August 10 that Egypt's seizure of the Suez Canal was 'wrong in every way' and must be resisted 'at all costs', is dangerously provocative. In the first place he fails to understand that the situation between Egypt and the Suez Canal Company (registered in Egypt) is purely local and contractual and in no way prohibits Egypt from nationalizing the shares of the company, subject to compensation. . . .

Another error into which Mr. Menzies has fallen is his slighting reference to the United Nations, the Charter of which requires all

---

[1] *Sydney Morning Herald*, 16 August 1956.

Members, in settling their disputes, to refrain from the use of threats of force. . . .

The gravest anxiety is felt in many quarters as to the much-publicized military, naval and air movements by the United Kingdom and France. . . . Mr. Menzies has no authority to commit Australian forces in connection with this dispute without prior and express authority of the Australian Parliament.

IV. NEW ZEALAND

1. Report of statement by the Acting Prime Minister, Mr. Keith Holyoake, 28 July 1956[1]

When asked to comment on the press reports of the Egyptian Government's action in regard to the Suez Canal, the Acting Prime Minister, Mr. Holyoake, said today that the government was greatly concerned at the reported developments, which could have the most serious consequences, not only for the United Kingdom but also for New Zealand. 'The government is maintaining the closest contact with the United Kingdom and Australia through the New Zealand High Commissioners in London and Canberra.' Mr. Holyoake said he was not in a position at this stage to say anything further. He would await the receipt of fuller information about the facts of the situation.

2. Report of statement by the Acting Prime Minister, Mr. Keith Holyoake, 31 July 1956[2]

Although no decision to lodge a formal protest had yet been made, Cabinet at its weekly meeting today had expressed great concern at the implications of the nationalization of the Suez Canal, the Acting Prime Minister, Mr. Holyoake, said tonight.

He added that there had been a full discussion of developments by Cabinet, and though no formal action by the Government had been decided on at today's meeting, there had been an examination of a number of possible courses in concert with other interested nations.

Mr. Holyoake said that Cabinet had before it a survey of developments in the Suez Canal situation by the Minister of External Affairs, Mr. Macdonald, as well as a series of cables from the New Zealand diplomatic service giving an up to date picture of world reactions to the Egyptian decision.

[1] *E.A.R.*, vol. VI, no. 7, July 1956, p. 2.   [2] Ibid.

The Acting Prime Minister said that the New Zealand High Commissioner in London, Sir Clifton Webb, was being kept informed by the British Government of developments as they arose and the Government of New Zealand was now considering in some detail the view it would ask him to express in further discussions in London.

3. Statement by the Acting Prime Minister, Mr. Keith Holyoake, 2 August 1956[1]

The New Zealand Government considers it to be of the highest importance that international agreements governing the use of the Suez Canal should be observed, and that no arbitrary measures should jeopardize the principle of free transit.

The decision of the Egyptian Government is one which could have serious consequences for the Commonwealth and for New Zealand. Any interference with shipping there cannot but affect our own trade. If the United Kingdom were cut off from its oil supplies, to name only one of the important commodities that pass through the Canal, we in New Zealand could, within a short time, feel the effects ourselves, for the tankers which now supply New Zealand might have to be diverted to meet the emergency situation. Any disruption of the economy of the United Kingdom and Western Europe would directly affect our markets.

This emphasizes the importance which the question holds for us. The Government is therefore following developments closely and is keeping in touch with other interested governments.

The world has a right to be assured, as it was before the nationalization, that the authorities controlling the Canal will maintain it at an efficient level, and will not use their power in an arbitrary manner. The manner in which the Egyptian government has taken its decision, as well as some of its actions in the past, are not convincing evidence that it will necessarily display the wisdom called for. The New Zealand Government has, on past occasions, expressed its concern at instances of Egyptian interference with the right of all nations to use the Canal.

I am, of course, unable to comment on the various measures which are under consideration at present. I wish to emphasize, however, the grave concern felt by the New Zealand Government.

[1] *E.A.R.*, vol. VI, no. 8, August 1956, pp. 26–27.

4. Extracts from speech by the Prime Minister, Mr. S. G. Holland, in the House of Representatives, 7 August 1956[1]

. . . I think the House would expect me to say something about the Suez Canal and the problems associated with its seizure by the Egyptian Government. That was not even a prospective problem while I was in London [for the Prime Ministers' Conference in June–July 1956]. Not one word was said about it at the Conference. There was no thought of this crisis developing. Today I think we are facing a grave crisis with grave possibilities. It is characteristic of the world political scene that a crisis arises almost overnight, without any warning, and a matter of hours alters the whole outlook of nations towards events and towards policy. This seizure came about, as far as I can gather from my reading—and there seem to be hundreds of cables—because the United States and Britain refused to lend money to Egypt to build a certain dam. They had offered to do so, but for reasons that seemed adequate they refused.

. . . The immediate reaction, of course, was that within a week Colonel Nasser, the Egyptian President, announced the seizure of the Suez Canal. I happened to be in Los Angeles when I heard this news, and Sir Anthony Eden and the Secretary of State for Foreign Affairs, Mr. Selwyn Lloyd, telegraphed me a great deal of information which I had to study.

I had the good fortune to have had there with me the head of our Department of External Affairs, Mr. McIntosh. I was asked for our immediate reaction, and I at once cabled the Right Hon. Mr. Holyoake and the Cabinet what I had been told and suggested the answer that we should send. I was able to tell Sir Anthony Eden and Mr. Selwyn Lloyd that Britain could count on New Zealand standing by her through thick and thin. I am sure the House will applaud that announcement, as I am sure that we will not allow these people to get away with this. It was a very great man who coined the sentence, 'Where Britain stands, we stand'. I have said many times that we on this side of the House adopt that. I believe that that is the mood of the people of New Zealand. Where Britain stands, we stand; where she goes, we go, in good times and bad. I was able to convey that expression to the United Kingdom Government, and I have since had the most encouraging reply. . . .

[1] New Zealand, H.R. Deb., vol. 309, pp. 889–90.

. . . The Suez Canal is vital to Britain, and Britain is vital to New Zealand. Where she is in difficulty, we are in difficulty, and the Western world is in difficulty if Britain is not to have a free flow of her ships through this vital international waterway. . . .

. . . I am confident that the United Kingdom Government's decision to call a conference to see if some ways and means can be found of settling this problem peacefully is a right one. I do not think it is unfair to anybody to say there are some nations in the world that appreciate a show of strength, and in the problems of Cyprus and these other places I am sure the British have taken the proper course in their assembling of certain forces. I am confident, too, that New Zealand and her people will support the decision to stand as partners with the United Kingdom in this problem. The natural question arises, what does our support involve? I am not prepared at this stage to go further than to say that that means exactly what it says. If anyone reaches certain conclusions on those words he probably would be right. . . .

5. Extracts from speech by the Leader of the Opposition, Mr. Walter Nash, in the House of Representatives, 7 August 1956[1]

. . . I thank the Prime Minister for what he has said, and I should like to see the whole matter methodically discussed in the House, so that we can give some interpretation of New Zealand's thoughts to the Old Country and to other countries. I think all Members have a copy of Sir Anthony Eden's speech [I, Document 4, *supra*]. . . . It is as clear and as straightforward a statement as I have ever seen. That does not mean that we agree with every word in the text of it, but it does set out the position with regard to the Canal in as good a way as I have seen it by somebody who knows the whole situation. . . .

6. Extract from speech by the Minister of External Affairs, Mr. T. L. Macdonald, in the House of Representatives, 7 August 1956[2]

. . . A query may be raised as to why Britain is taking precautionary measures in the Mediterranean and adding to her forces there. After giving the matter considerable thought, I feel that in such a critical situation as this, where things can happen so

---

[1] New Zealand, H.R. Deb., vol. 309, pp. 902–3.   [2] Ibid., p. 908.

unexpectedly, maybe disastrously, Britain has every justification for preparing to meet any eventuality. . . .

7. Extract from statement by the Minister of External Affairs, Mr. T. L. Macdonald, 11 August 1956[1]

. . . The Government has made it very clear that it views with the greatest concern Colonel Nasser's high-handed repudiation of the agreement with the Suez Canal Company. The manner in which this arbitrary action was taken, and the selfish reasons which prompted it augur ill for the future control of the Canal. The implications are extremely serious and the United Kingdom, which, with France and the United States, has proposed the holding of the present conference, has New Zealand's firm and emphatic support in seeking to ensure that the ships of all countries will continue to have uninterrupted passage through the vital international waterway.

V. SOUTH AFRICA

1. Extract from statement by the Prime Minister, Mr. Johannes Strijdom, 27 July 1956[2]

. . . We are on friendly terms with the various States in that part of the world [i.e., the Middle East] and cannot favour one at the expense of another.

It is best to keep our heads out of the beehive. But the Middle East, which has always been a dangerous spot, is of the utmost importance to South Africa as geographically it is the gateway to this continent.

2. Extracts from statement by the Minister of External Affairs, Mr. Eric Louw, 31 July 1956[3]

The Union Government has been kept fully informed by the Government of the United Kingdom regarding developments following upon the decision of the Egyptian Government to nationalize the Suez Canal Company, and has given careful consideration to the issues involved. . . .

Unlike the United Kingdom, France and some other countries, the Union holds no shares in this Company and is, therefore, not affected by its nationalization, which is a domestic affair of Egypt's in which South Africa would not presume to intervene. . . .

---

[1] *E.A.R.*, vol. VI, no. 8, August 1956, p. 27.   [2] *Cape Times*, 28 July 1956.
[3] *D.S.A.A.*, vol. 3, no. 16, 3 August, 1956, p. 9.

The Union Government are nevertheless concerned about the situation which has arisen in a continent of which South Africa is a part, and trust that the different parties involved will explore all possible avenues which may lead to a satisfactory solution of the present difficulties.

3. Statement by the Minister of External Affairs, Mr. Eric Louw, 1 August 1956[1]

It appears that unjustified interpretations are being placed on the statement issued by me yesterday on behalf of the Government [Document 2] and the impression has been created that the Government is not concerned about the situation which has arisen in regard to the nationalization by Egypt of the Suez Canal Company—I emphasize the word 'Company'. In the concluding paragraph the Union Government expresses its concern about the situation, and in the statement I tried to put the position in the right perspective.

In so far as the original concession is concerned, nationalization of a joint stock company registered in Egypt is a matter for domestic concern—as was the case when British and American oil companies were nationalized by Mexico and when the British-owned Abadan refinery was nationalized by Persia. As is well known, the Union has always taken a strong stand in regard to matters of domestic concern.

In so far as the Suez Canal Convention is concerned, the position is that even if the Union of South Africa is considered to be a party to the Convention, which was signed when the Transvaal and the Orange Free State were independent republics, the fact remains that since the nationalization of the Suez Canal Company by a recent decree there has been no breach of the Convention. The Union Government is not prepared to take action which is based on presumption or the possibility of certain events taking place, and which may not take place at all.

I notice that in a statement issued on July 29 by Mr. Casey, the Australian Minister of External Affairs [III, Document 2, *supra*], apart from strongly criticizing the action taken by President Nasser he gave no indication of what Australia proposed to do, and concluded by saying: 'Wider international consultations will, no doubt, take place to ensure that economic and other vital interests of all

[1] 'South African Attitude on Suez', *The Times* (London), 2 August 1956.

countries are protected. The whole matter is most disturbing and is engaging the urgent attention of the Australian Government.'

Broadly speaking, these sentiments appear to reflect the sentiments expressed in the concluding paragraph of the Union Government's statement.

4. Extracts from statement by the Minister of External Affairs, Mr. Eric Louw, 4 August 1956[1]

The exclusion of Egypt from the transferable sterling account area by the United Kingdom has the effect of restricting the use of sterling on Egyptian account for international settlements.

It is implicit in South Africa's membership of the sterling area that similar action should be taken in the Union, and I have, therefore, asked the Reserve Bank to advise all banks that payments between the Union and Egypt may no longer be made over transferable accounts. . . .

Three criteria for determining direct interest in the Canal issue and thus qualifying for an invitation [to the London Conference on the Suez Canal] were considered by the conveners of the Conference.

They were: The principal users of the Canal; those countries for whom the Canal is an essential life line; and signatories of the 1888 Convention.

Canada and South Africa do not qualify under any of these criteria. In regard to the third condition, Canada, which, unlike South Africa, was entirely under British rule in 1888, does not succeed in being recognized as a 'successor country' and thus a signatory country.

South Africa is deeply concerned in the maintenance of peace in any part of the Africa continent and in the approaches to the continent, and for that reason I sincerely hope that a way out of the difficulties may be found.

5. Statement by the Government of the Union of South Africa, 8 August 1956[2]

The Egyptian Minister to the Union of South Africa, Mr. Self Ahmed Hamdy, called on the Union's Minister of External Affairs, Mr. Eric Louw, on Saturday, August 4, and handed to

[1] *The Star* (Johannesburg), 4 August 1956; *Cape Times*, 6 August 1956.
[2] 'Egypt and South Africa exchange Views on Suez Canal Issue', mimeographed release of the Union of South Africa Government Information Office, New York City, 8 August 1956.

him statements setting out the attitude of his Government in regard to the Suez Canal issue.

It is learned that Mr. Louw discussed the matter fully and frankly. He pointed out to the Minister that although the Union Government regarded nationalization of the Canal Company as a matter of domestic concern and although South Africa did not qualify as a country directly interested in the issue according to three criteria laid down by the tripartite powers, the Union nevertheless had a concern in observance of international agreements. He strongly urged upon Mr. Hamdy that in the interests of the maintenance of peace in the Middle East and also in Egypt's own interests, care should be taken to avoid a breach of the 1888 Convention, not even under provocation.

With regard to the conference convened by the three Western Powers, Mr. Louw hoped that Egypt would be represented at the conference. He suggested that the Egyptian Government might reserve its position in regard to nationalization of the Canal Company which in the view of the Union Government was a separate issue. Mr. Louw said that South Africa and Egypt were both sovereign states in Africa and for that reason it was most desirable that existing friendly relations should not be disturbed. He, however, reminded the Egyptian envoy of the close relations existing between the Union and those European powers which had interests in Africa, and said that with primary regard to South Africa's own interests, those relations would influence the Union Government's attitude towards events in or near the continent of Africa. Mr. Louw expressed the hope that there would be no precipitate action on either side and that the Government of Egypt would continue to observe the terms of the agreement.

The Egyptian Minister assured Mr. Louw that his Government was determined strictly to observe the terms of the 1888 Convention. As regards the proposed conference, he pointed out that the United Kingdom and France, by insisting on the internationalization of the canal as a basis for discussion, were making it very difficult for Egypt to attend the conference. Mr. Louw hoped Egypt would nevertheless attend the conference.

6. Extract from statement by the Leader of the United Party, Mr. J. G. N. Strauss, 9 August 1956[1]

[The United Party] demands a declaration of solidarity with all

[1] *The Star* (Johannesberg), 9 August 1956.

those whose aims are to maintain the Suez Canal as an open international waterway, not subject to the arbitrary whims of a dictator reverting to the methods of Fascism. What is happening in Egypt is of the most direct concern to South Africa. This view has been subscribed to by the Government of Dr. Malan, who said in the House of Assembly that South Africa has an interest in keeping the Suez Canal open and that there should be a military base somewhere in the Middle East to safeguard this gateway to Africa.

The United Party sees no reason for departing from that policy, which it welcomed in Parliament.

This is no local issue. The call of the Egyptian State Radio to the people of Africa to throw off an imaginary 'yoke of imperialism' is an ominous warning that we in South Africa cannot afford to blind ourselves to the intolerant extremism of Egypt's new nationalistic regime. It should also be clear to South Africans that events in Egypt are connected with the expansionist policies of the Communist States. Russia and her satellites are fishing in the troubled waters of the Middle East, and weakness by the Western democracies may well enhance the prestige of Communist diplomacy.

If the Union wishes to play any significant part in Africa, it should shoulder its responsibilities as the leading State in Africa and the declared ally of the Western democracies.

7. Statement by the African National Congress, 27 September 1956[1]

The threats of war against Egypt, the mobilization of armies and the actual transportation of troops and dispatching of battleships to the Mediterranean by the British Government, are a clear indication of the determination of these governments to maintain their decaying colonial systems in Africa, the Middle East, Asia, by brutal force and through military terrorism.

We pledge our solidarity with the Egyptian people and are confident that the people of Africa will not allow themselves to be used against their fellow Africans in any predatory war.

VI. INDIA

1. Extract from statement by the Prime Minister, Shri Jawaharlal Nehru, in the Lok Sabha, 31 July 1956[2]

At Cairo, President Nasser and his Ministers and I had further

---

[1] *Cape Times*, 28 September 1956.
[2] Lok Sabha Deb., 1956, Part II, vol. VI, col. 1561.

opportunities of talks. . . . These discussions did not relate to the Suez Canal or to any aspect of Anglo-Egyptian relations. The recent decision of the Egyptian Government in regard to the Suez Canal first came to my knowledge from the reports in the Press after my return to Delhi. . . .

2. Extracts from speech by the Prime Minister, Shri Jawaharlal Nehru, in the Lok Sabha, 8 August 1956[1]

On the 26th of July, President Nasser announced in a speech at Alexandria that the nationalisation of the Suez Canal Company had been effected. . . .

The announcement has had world-wide repercussions. A grave crisis which, if not resolved peacefully, can lead to conflict, the extent and effects of which is not easy to assess, has developed. In this crisis, the foremost consideration must be to strive for a calmer atmosphere and a rational outlook. When passions dominate, the real issues recede into the background, or are viewed or presented so as to emphasise the difference between the disputants and to rouse or feed the passions already engendered.

It is not easy for anyone, much less for the disputants, to escape this tragic involvement, and even for others, total objectivity is not possible. In crises of this kind we deal not merely with the issue in dispute, but we witness the upsurge and conflict of mighty forces. . . .

The French and the United Kingdom Governments reacted to the Egyptian announcement quickly, sharply and with vehemence. Hon. Members of the House have seen Press reports of military and naval movements ordered by the United Kingdom and France, and some military measures in Egypt. These have received much publicity and have aggravated the situation. All this has influenced public opinion not only in Egypt but over the Arab world. In Asia as a whole, with its colonial memories, great resentment has been aroused.

I have no desire to add to the passions aroused, but I would fail in my duty to this House and the country and even to all the parties involved in this crisis, and not least of all to Britain and France, if I do not say that threats to settle this dispute, or to enforce their views in this matter by display or use of force, is the wrong way. It does not belong to this age and it is not dictated by reason. It fails

[1] Lok Sabha Deb., 1956, Part II, vol. VII, coll. 2536–44.

to take account of the world as it is today and the Asia of today. If this were all, we could perhaps possess ourselves in patience and reflect that the mood will pass. But it would be unrealistic and imprudent not to express our deep concern at these developments and point to their ominous implications. We deeply regret these reactions and the measures reported to be taken in consequence, and we express the hope that they will cease and the parties will enter into negotiations and seek peaceful settlements.

We also much regret that, in the steps that have led up to this crisis, there has been no exercise by one side or the other of their respective or common initiative to inform or consult one another. . . .

The Government of India received an invitation from the United Kingdom on the 3rd of August to a conference in London 'on the Suez Canal question'. Prior to this, the United Kingdom Government kept the Government of India informed of developments.

Aware as they are of the extreme gravity of the situation that has developed and of the circumstances that obtain, the Government have given anxious and careful consideration to all aspects of this question, including the reply to the invitation. The Government have also been in contact with interested countries, including Egypt.

It has always been quite clear to the Government that they could not participate in any conference which bound its participants beforehand to the conclusions to be reached. The Government would equally decline participation in any arrangements for war-preparations or sanctions or any steps which challenged the sovereign rights of Egypt. They have also been concerned at the exclusion from the list of invitees of various countries who should be included. . . . Without seeking to make invidious distinctions, I would like to say to the House that the exclusion of Burma is to us a particularly regrettable omission. Yugoslavia . . . should also have found a place among the invitees. The Government of India, therefore, do not subscribe to the appropriateness of the list of invitees.

They have sought clarifications from the United Kingdom Government and feel assured that their participation in the conference does not in any way imply that they are restricted or bound by the approach and the principles set out in the joint communiqué [I, Document 7, *supra*]. . . . The Government of India had to take a decision in the situation as it confronted them. India is not a disinterested party. She is a principal user of this waterway, and her economic life and development is not unaffected by the dispute, not to speak of a worse development, in regard to it.

Even more, India is passionately interested in averting a conflict. She is in friendly relations with Egypt, and associated with her in the acceptance of the Bandung Declarations and the 'Five Principles'. India has also good and close relations with the principal Western countries involved. Both these relations are held in great esteem by us, as this House and all the world know. The considerations and the criteria on which the Government had to base their decision, and not an easy one, is how best they could serve the cause of averting conflict and obtaining a peaceful settlement before it is too late. . . .

The Government therefore obtained the necessary assurances from the United Kingdom and made their own position quite clear. They have satisfied themselves that their participation in the London Conference will not injure the interests or the sovereign rights and dignity of Egypt. With the sense of grave responsibility that rests on them, the Government have decided to accept the invitation and to send representatives to the conference.

They have kept in close contact with Indonesia and Ceylon and with others who broadly have a similar approach and attitude to that of India on this question.

The Government are well aware that this conference can reach no final decisions; for that requires the agreement of Egypt.

Sir, the House, I am aware, shares the grave concern of the Government in this matter. In all humility, I ask it to share with them the hope that the participation of India will assist in the endeavours for a peaceful settlement.

3. Extract from speech by the Prime Minister, Shri Jawaharlal Nehru, 15 August 1956[1]

. . . I hope that if this question of the Suez Canal is not solved in London, then a second or third attempt will be made to solve it. One thing is quite clear, that under no circumstances should this issue be tackled through the use of armed force or threats. If any effort is made, even by mistake of any power, to settle the Suez issue by force or by threats, then the results will be disastrous. . . .

[1] *The Hindu* (Madras), 17 August 1956.

VII. Pakistan

1. Statement by the President of the East Pakistan Muslim League, Tamizuddin Khan, 2 August 1956[1]

The Western Powers have become desperate at Egypt's action in nationalizing the Suez Canal concern. But it is apparent that Colonel Nasser has done no more than pay America and England in their own coin. The curt and imperious manner in which these two great Western Powers announced their refusal to finance the Aswan Dam project could not but wound the self-respect of resurgent Egypt, and small nations naturally feel gratified that Egypt has not taken it lying down.

Moreover, Egypt's action is also otherwise fully justified because her right to nationalize the Canal Company is nothing but a corollary to her national independence.

Egypt is completely within her rights, and the statesmanlike assurance she has given that the nationalization will mean no interference with the use of the Canal by international shipping should allay international anxiety, if any.

It is hoped that the Western Powers will not fail to make a realistic assessment of the situation and try to come out of the mess with as good grace as is yet possible. Russia is already reaping a rich harvest and if the Western Powers go on blundering it will not only be to the former's advantage, but the responsibility of the dispute developing into a major international conflict will lie heavily on their shoulders.

2. Extract from statement by the Leader of the East Pakistan Awami League, Maulana Abdul Hamid Khan Bhashani, 2 August 1956[2]

With great pride I assure President Nasser and the heroic people of Egypt that we, their brothers in Pakistan, will stand by them. [President Nasser] inspiringly echoes the feelings of the age-old exploited Asian and African nations when he proclaims that force will be met with force. . . . The underdeveloped countries of Asia and Africa which are groaning under the exploitation of the imperialist powers and are now making all attempts to organize and build up their economy, will find in Egypt's action a great source of inspiration and confidence.

[1] *Dawn* (Karachi), 3 August 1956.   [2] Ibid.

VIII. CEYLON

1. Statement by the Prime Minister, Mr. S. W. R. D. Bandaranaike, 27 July 1956[1]

I have not had sufficient time to consider the announcement that has been made that the Egyptian Government has nationalized the Suez Canal Company.

It is obvious that the step taken has many serious implications. I do not wish to make any detailed comment at this stage except to express the hope that immediate discussion amongst Powers most closely connected with this issue appears to be necessary to secure even at this late hour some friendly understanding and so avoid a grave set-back to recent prospects of peace that seem to be manifesting themselves.

I shall consider whether any urgent discussions between our countries in Asia, namely the Colombo Powers or even at a wider level, the Bandung Powers, might be necessitated by the present situation.

2. Extracts from questions by members of the Opposition, and answer by the Prime Minister, Mr. S. W. R. D. Bandaranaike, in the House of Representatives, 30 July 1956[2]

Dr. N. M. Perera: . . . I would like to draw the attention of the Prime Minister to a very grave situation that has arisen in the international field. I refer to the question of the taking over of the Suez Canal and the Company managing the Suez Canal by General [*sic*] Nasser of Egypt. . . . We are naturally sympathetic to any attempt on the part of Egypt to exercise her sovereignty over her own territory. There can be no question about that. But we are also very closely involved in the position of the Suez Canal in so far as it affects our trade. We import most of our goods from Europe. They come through the Suez Canal and I think it is only fair that we should have some voice in any deliberations concerning the question of the canal.

I would like the Hon. Prime Minister to tell this House and the country what the present situation is. We are not too happy about the attempt made by the Western Powers, the Imperialist Powers in particular, to browbeat Egypt in this regard and I think nobody

[1] *C.N.L.*, 30 July 1956, pp. 5–6.
[2] Ceylon, H.R. Deb., vol. 25, coll. 1099–1106.

can envisage the prospect of another war on this basis without serious misgivings. We should therefore like to know from the Prime Minister what steps he has taken to bring our own point of view to be felt in the discussions that are now apparently taking place among the Western Powers on this subject and we would like the Hon. Prime Minister to make clear that we are in some way concerned about the position of Egypt and the Suez Canal. . . .

Mr. P. G. B. Kevneman: Before the Prime Minister replies may I ask him what communications he has received on the question from the High Commissioner in London. We read in the newspapers that our High Commissioner in the U.K. has been called into conference by the British Government personally to apprise us of its views on the matter. If it is permissible this House will be glad to be informed of what proposals, if any, have been made to our Government and what attitude our Government has taken in the matter.

I would like to ask the Prime Minister in his communication to the British Government or any of the other Governments to make two things clear on behalf of the people of Ceylon. The first is that we regard the question of the nationalisation of the Suez Canal as a matter that falls within the sovereign jurisdiction of Egypt, that the right to nationalise that Canal or not is a matter for Egypt to decide and any incidental matters like compensation and so on are matters which could be settled between those involved. We would be failing in our duty as a country that aspires to maintain independence and support independent struggles if we do not make it quite clear to the British Government and other Governments that all sections of this country are united in believing that the question of the Suez Canal is a matter for Egypt to decide.

The second matter which I would most respectfully ask the Hon. Prime Minister to convey to the other Governments on behalf not only of the Government but also of all sections of the country is that we most strongly deprecate the talk of blockades, invasions, threats of war, use of warships and resort to military force in order to solve this problem. . . .

Dr. Colin R. de Silva: I only want to remind the Hon. Prime Minister of two issues that are involved in the question which are of direct relevance to the immediate question of policy of the Government itself.

The Hon. Prime Minister will remember that he is at present

occupied in relation to the British Government with regard to the question of bases. He will also remember that he is occupied at present with certain questions of nationalisation from the point of view of the announced programme of the Government involving the question of the nationalisation of the property of foreign companies in this country.

Both these matters raise the questions of sovereign states and the right attaching to the sovereignty to decide those questions as matters which are entirely within the jurisdiction of the country. There is a dangerous argument brought out . . . that Powers which claim to be world Powers with world maritime communications are entitled to special rights, for instance, in the Suez Canal and also that when the Egyptian government choose to nationalise the Suez Canal Company it is in some special category of its own. I do not wish to say anything on those matters because it is our intention and desire at this stage not to embarrass the Hon. Prime Minister in any way in whatever steps he is taking. But I would beg of the Prime Minister to remember that since these two very important issues are involved in this question it will be necessary for this country and its Government to take a position consistent with the rights which we ourselves hold that we have as an independent country, namely, the right to decide what we shall do with the bases in our country and whether we shall nationalise or not nationalise any particular concern.

The Hon. S. W. R. D. Bandaranaike: Mr. Speaker, my hon. Friends opposite have raised a question which I had indeed anticipated would be raised on the Floor of the House and that is the position arising out of the nationalisation of the Suez Canal Company by President Nasser. The House knows that this took place a few days ago and it came with a certain measure of surprise to most countries as it was not expected. . . .

I do not think that it was even a question that was in any way discussed even at the conference between the Prime Minister of India, Col. Nasser and Marshal Tito at Brioni. It was obviously clear at the outset that the action contained certain potentialities of a serious nature. Of course, the House will not expect me here to try to enter into any discussion of the merits on one side or the other. I am naturally most concerned in trying to see as far as we can that these disputes are settled in a reasonable spirit of accommodation and any danger of the situation deteriorating even perhaps to the point of war should by all possible means be avoided.

. . . The news of this action came to us, as I believe it did to most other countries, as a surprise. I realised naturally the dangerous potentialities of the situation and I immediately got in touch with certain other countries, particularly England and India, as to their views on the situation. . . .

. . . I will not disguise the fact from the House that the situation is serious. Probably it is the most serious one single thing that has happened to jeopardise the rather uneasy peace which we are enjoying at the moment in the world, since the end of the last war. . . .

I have explored the possibility of at least the Colombo Powers discussing this matter very early, but I feel—it is a feeling that is shared by others whom I have had an opportunity of consulting at the moment—that in the light of wider conferences in which we may all participate taking place early, it is better just for the present to watch the situation. . . .

3. Statement by the Prime Minister, Mr. S. W. R. D. Bandaranaike, 6 August 1956[1]

I received an invitation to attend the proposed conference on the situation that has arisen over the Suez Canal. The communiqué issued in this connection by England, France and the United States, as well as newspaper reports, make it evident that some further clarification is needed before deciding whether the invitation should be accepted or not. It appears as though the convening Powers have already made up their minds on a particular course of action, i.e., international control of the operation of the Suez Canal by securing certain rights to Egypt.

Meanwhile Egypt has rejected the suggestion of any international control of the Canal. It would thus appear that if the convening Powers stand by their point of view and Egypt stands by hers, there would be no alternative but to provoke war. If this is the correct position it is not very clear whether our participation in the proposed Conference would serve any useful purpose. I have therefore sought clarification of the position and I am still awaiting a reply before finally deciding whether to participate in the Conference or not.

Egypt's right to nationalize the Canal Company can hardly be disputed nor on the other hand the need for satisfactory guarantees

[1] *Ceylon Daily News* (Colombo), 7 August 1956.

for the international use of this important waterway. It may be possible to reconcile these two needs without insistence upon international control of the Suez Canal. If there is such insistence, it may hardly be possible to arrive at any peaceful solution without resort to force.

It would seem that India as well as some other countries also feel that a satisfactory clarification of the scope and objects of the Conference is necessary before deciding to participate in the Conference or not.

4. Extracts from questions by members of the Opposition, and answer by the Prime Minister, Mr. S. W. R. D. Bandaranaike, in the House of Representatives, 9 August 1956[1]

Dr. Colin R. de Silva: . . . Now there is a conference called over this, summoned by the aggressors under a formula which really requests the endorsement of their aggression but which they now seek, by explanation and exegesis, to suggest was never intended to be of that nature at all, and we have it from the Hon. Prime Minister that he has now instructed a representative from Ceylon to attend that Conference. . . . I would request the Hon. Prime Minister to state in this House that it is part of his instructions, or will be, to our representative at this conference to take a clear and firm public stand on the issue in line with what is obviously the general public opinion in this country and in line with the fundamental interests of this country, namely, that the great powers of the world have no right whatsoever to interfere in Egypt, or in relation to Egypt, with a view to compelling Egypt to give up her national rights to any international body that they may contemplate. . . .

. . . So long as we are within the British Commonwealth, we have a right to utilise that position also not only to bring our friendly pressure but also our direct influence to bear upon the British Government in particular in respect of this matter. The Hon. Prime Minister has that right. I hope he is exercising that right to advise the British Government. . . .

Mr. P. G. B. Kevneman: . . . The House will be aware that there has been a growing disquiet in the public mind about the purpose and aims of this conference. The Prime Minister obviously shares that disquiet because he was constrained only a couple of days ago

[1] Ceylon, H.R. Deb., vol. 26, coll. 112–42.

to seek clarification from those who were convening this conference on a number of salient points. I presume the clarifications
have been satisfactory; otherwise I cannot see any reason, in terms
of the clarifications which were sought, for the Prime Minister
agreeing to this country attending that conference. . . .

It is perfectly obvious that the main conveners of this conference
make no bones about the fact that they are out to establish international control over the Suez Canal and deny Egypt's sovereign
right to determine the future of that Canal. Therefore, the question naturally arises: what do we hope to attain by attending such a
conference? I read a statement this morning made by the Prime
Minister of India [VI, Document 2, *supra*]. . . . He stated that he
was dissatisfied with the composition of the conference. He said
that there were a number of countries which, in his opinion, should
have been invited but which had not been invited to attend it, and
he mentioned Burma and Yugoslavia. I must say that I agree
with him. . . . There is very heavy weightage given to the imperialist countries; the weightage given to Asian countries is extremely
slight and practically none at all to the Middle East countries.

Secondly, the Prime Minister of India pointed out that no final
decision could be taken at such a conference because final decisions
could only be taken at a conference to which Egypt herself was a
party. I think that is a reasonable point of view. An international
conference on Egypt, to which Egypt is not a party, cannot obviously take binding decisions which will affect those who are
involved in the dispute. If that is so, what is it that we are hoping
to achieve by agreeing to attend this conference? As far as I can see
we are hoping not to achieve something positive but to prevent the
situation getting worse. Lying at the back of the mind of the Prime
Minister of India—and presumably of our own Prime Minister as
well—is the idea that if there is not present at that conference
somebody who does not belong to the charmed circle of imperialism, heaven alone knows what will be decided and what will be
carried out in the name of internationalism. That seems to me to
be the only good reason for attending this conference—to prevent
a situation being created where Egypt will be turned into a second
Korea, where the name of internationalism will be used to disguise
the naked imperialist interests of certain powers. If that is the case,
I do think that the Prime Minister should, before he sends representatives to that conference, make our position clear beyond all

doubt. I would ask him to make a statement . . . that as far as this Government and this country are concerned we are attending this conference merely because we do not wish to see a war or a large-size military action breaking out over this issue, that we are anxious to do whatever we can in the circumstances to contribute towards such an end, and that this and this alone is our purpose in attending such a conference. I would also ask him to make it quite clear that, as far as we are concerned, Egypt's sovereign rights to do what she wishes to do with a company registered on her own territory are not questioned, and that we are not prepared to consider the idea of internationalising the Suez Canal or any formula which denies or whittles down in any way the sovereign national rights of Egypt. It is only on such a basis that we will be going to such a conference in our true colours. . . .

Mr. S. W. R. D. Bandaranaike: . . . If I may summarise the position, it is simply this: As soon as the Suez Canal trouble arose, I drew attention on the Floor of the House to the desirability of having a fairly wide conference on the matter [Document 2]. It may be at least partly as a result of what was said that a wider conference than perhaps was intended at the start was called; and invitations were sent for that conference.

The moment I received the invitation, it struck me that there were certain points that had to be clarified and certain others on which assurances had to be obtained before I was in a position to accept that invitation on behalf of my own country. As a matter of fact it appeared—hon. Members may have gathered from the afternoon papers of last Tuesday, 7th August, that I made a statement to the press also [Document 3]—at least on the face of the invitation, that the three main powers which were concerned in calling this conference, namely, the United Kingdom, the United States and France, had made up their minds with regard to one particular angle in view; that is, that there should be international control of the operation of the Suez Canal while at the same time guaranteeing certain rights to Egypt. Well, that was of course an unsatisfactory position. I could not possibly go to a conference the scope of which was restricted to a discussion of the international control of the Suez Canal, for the very good reason that it meant a complete change . . . of the step already taken by Egypt. . . . Any attempt to impose that view was bound to lead to a state of war. Therefore, I sought immediate clarification as well as an assurance from the British Government . . . that the position was not as it appeared and that those who did not hold the view that it was either desirable or indeed in practice possible to merely consider international control of the Suez Canal in the circumstances would be at liberty to express their views and have an assurance that those views would receive serious consideration.

Well, I not only took that step but also immediately addressed certain other countries of Asia that in my view the invitation couched in those terms was unsatisfactory and that I was seeking further clarification and assurances. I was pleased to see—I do not suggest it was as a result of my saying so—that other countries were seeking similar clarification—I was later informed of that fact. Well, those, clarifications and assurances have been forthcoming to the effect that this [view favouring international control] happens to be the view of England and France; that other countries attending the conference which held different views were at perfect liberty to express those views; and that those views will be given due consideration. In the circumstances, I felt it was reasonably possible for us to participate in the conference, as indeed we should in order to prevent our case going by default, if nothing else. And I observe that India, too, has been sufficiently satisfied with the assurance and clarification given to participate in this conference.

Our own attitude . . . is this: We do not question the right of Egypt to nationalise the Suez Canal Company with its implications. . . . We are also very much concerned, naturally, with two matters. We desire to obtain a peaceful settlement of a dispute that is, if I may say so, rather endangering peace. . . . We are also concerned with certain uses that we make of the Suez Canal for our trade, incoming as well as outgoing. We feel it may be possible to recognise the position of Egypt in that way. While at the same time obtaining reasonable guarantees for the use of the Suez Canal, guarantees indeed which I have no reason to believe Colonel Nasser himself will be unwilling to give. . . .

I think hon. Members can reasonably rest assured that from the point of view of national expression of sovereignty, the safeguarding of that against what might be considered as invasion from an imperialist or colonial power will be adequately resisted by the Government and that every effort will be made as far as we are able in this country to come to some satisfactory settlement and avoid the risk of an outbreak of hostilities. . . .

PART II

THE LONDON CONFERENCES, UNITED NATIONS NEGOTIATIONS, AND ANGLO-FRENCH PREPARATIONS

Under no circumstances should this issue be tackled through the use of armed force or threats. If any effort is made, even by mistake . . ., to settle the Suez issue by force or by threats, then the results will be disastrous. . . .

Shri Jawaharlal Nehru, 15 August 1956

Through all these negotiations, peace has been our aim, but not peace at any price. And I will tell you why. Because in dealing with a dictatorship peace at any price means to increase, step by step, the dangers of universal war.

Sir Anthony Eden, 13 October 1956

COMMENTARY

I

THE Conference of Twenty-two Nations which met in London from 16 August to 23 August 1956, to consider the situation brought about by Colonel Nasser's nationalization of the Universal Suez Maritime Canal Company, had its origins in a telegram from the President of the United States to the Prime Minister of the United Kingdom. In this message, sent within hours of Nasser's act and received by Sir Anthony Eden on the morning of 28 July, President Eisenhower proposed immediate consultation among the largest possible number of maritime nations affected by the Canal's new status. The United Kingdom Government agreed that a conference should be convened, but disliked the idea of a meeting that would be open to almost all comers. 'We favoured restricting representation', Sir Anthony Eden recalled afterwards, 'to the six or ten powers who were the principal users of the Canal in terms of tonnage and trade.'[1] Talks began almost immediately among the representatives of the United Kingdom, the United States, and France—respectively, Mr. Selwyn Lloyd, the British Foreign Secretary; Mr. Robert Murphy of the U.S. State Department; and M. Christian Pineau, the French Foreign Minister—to work out the composition and timing of the proposed conference; they were joined on 1 August by John Foster Dulles, the American Secretary of State, who thought it critically important that the three powers concerned 'make a genuine effort to bring world opinion to favour the international operation of the Canal'.[2] He therefore wanted a conference with as many participants as possible, and to allow enough time for an elaborate publicity build-up. Eventually a compromise both on composition and timing was accepted. Invitations were to be sent to the eight surviving signatories of the Constantinople Convention, and to sixteen other nations ranking as principal users of the Canal by virtue of the tonnage of their shipping and the extent of their trade. The conference of the twenty-four nations thus invited was to be convened in London a fortnight later on 16 August. Throughout these discussions, Dulles made upon Sir Anthony Eden the

[1] *Full Circle*, p. 433.   [2] Quoted ibid., p. 437.

impression that the United States would countenance the use of force if Egypt proved recalcitrant. 'Nasser must be made, as Mr. Dulles put it to me, "to disgorge". These were forthright words. They rang in my ears for months.'[1]

Of the eight members of the Commonwealth of Nations, two were not invited to the London Conference. Neither Canada nor South Africa fulfilled the requirements for invitation. The Canadian Government was not unhappy at its exclusion. One of its members, the Minister of National Defence, remarked on 3 August that the Suez seizure was 'primarily a European matter . . . not a matter which particularly concerns Canada. We have no oil there. We don't use the Canal for shipping.'[2] On the same day the Prime Minister, Mr. St. Laurent, informed the House of Commons that he had received

a personal message from Sir Anthony Eden explaining the real difficulties they have had in settling the list of countries to which invitations should be sent in order to have a basis which would keep the conference within reasonable limits and expressing the assurance that in spite of the fact that within that criteria [sic] Canada was not included we would be kept in the closest touch possible during the conference.[3]

Mr. Pearson, the Secretary of State for External Affairs, also spoke of Canada's exclusion. Had Canada been invited to send a representative, he explained, the Government would have been agreeable to doing so; but no invitation had arrived, and the Government felt no misgiving at being left out. There was no objection voiced in Parliament to this policy. But outside the House of Commons the Conservative Party's foreign policy spokesman (soon to become its Leader and not long afterwards Prime Minister) declared that 'the Government should have made it known to Britain of her desire to be represented', and warned that 'an uncertain policy on the part of Canada might well bring about another Munich'.[4]

In South Africa, opposition to the Government's decision not to press for representation at the London Conference was much more pronounced. In the view of the United Party, South Africa should have been present at the table. Apart from Egypt (which, though

[1] Full Circle, p. 437.
[2] Quoted in Canada, H.C. Deb., 1956 (Special Session), p. 13.
[3] Ibid., 1956, vol. VII, p. 6920.
[4] See above, pp. 50–51.

invited, did not attend) and Ethiopia, no other African state
would be there to consider the consequences of events from an
African point of view. Whatever the justification for Nasser's
act, whatever his motives, whether nationalization was to be
construed as a manifestation of nationalism or of international
communism, it was clearly of vital concern to South Africa's
interests.

The attitude of the Nationalist Government on Suez [commented
the *Cape Times* as the conference reassembled in London following
Mr. Menzies' unproductive mission to Cairo in September] grows
more and more extraordinary. . . . A conference of 18 nations is discussing in London plans which may include the massive diversion of
ships to the Cape route. . . . Where is the South African Government
expert at those talks? . . . A string of countries from Abyssinia to the
aurora borealis is attending that Conference, yet South Africa, which
claims to be the leading country on the African continent, remains
aloof. It almost seems as though the Cape Route has nothing to do with
this country. . . .[1]

II

The first London Conference lasted a week. All Governments
invited sent representatives except those of Greece and Egypt.
Two proposals emerged. One, endorsed by eighteen of the twenty-
two participating delegations, called for an international operating
Board for the Suez Canal. This proposal, put forward by the three
sponsoring powers, found its principal Commonwealth support
in the Governments of the United Kingdom, Australia, and New
Zealand. The objectives of British policy, as recorded by Sir
Anthony Eden, were twofold:

First, that it should reach agreement by a large majority on the international control of the canal. Secondly, that it should decide upon the
steps to take to effect this. I wished to secure a declaration from the
conference that, if Egypt rejected its recommendations for international
control, the powers using the canal would refuse to pay their transit dues
to the Egyptian company. A firm sanction would then be at our command. . . .[2]

It was Sir Anthony Eden's conviction at the time that these proposals would commend themselves to what would now be called

---

[1] 'Where do we Stand?', *Cape Times*, 21 September 1956.
[2] *Full Circle*, p. 449.

Afro-Asian opinion. 'According to our information', he has written, 'the Governments of several Arab States were alarmed that Nasser might be allowed to get away with his pillage'; he cited the case of 'a Nigerian chief of the Muslim faith' who, asked in London to sign a document endorsing Nasser's act, tore it into shreds.[1] But this gesture was no more representative of non-Western reaction than was the untypical policy of the Government of Pakistan (to be discussed below). Not for the only time, Sir Anthony Eden misread the evidence at his disposal. The overwhelming majority of attentive citizens throughout the Afro-Asian world, initially sympathetic to Nasser's act, had become if anything less qualified in their support as the deployment of troops and ships provided an ominous counterpoint to Anglo-French-American diplomacy in London.

Sir Anthony Eden's principal support within the Commonwealth, at the first London Conference as later, came from the Prime Minister of Australia. Mr. Menzies had immediately cancelled his impending visit to the Far East on learning of the crisis (he had been, it will be recalled, in Washington on 26 July), and Sir Anthony Eden 'at once invited him to join us in London and take part in our councils'. He attended three meetings of the United Kingdom Cabinet, and, for Sir Anthony Eden, there could not have been 'a wiser or more forthright colleague'. Forthright he certainly was. 'I have no disposition whatever', Mr. Menzies informed his fellow representatives in London on 18 August, 'to submit my people to all the chances and uncertainties that would arise from having their immense trading interests through this Canal made subject to the whim of the moment, to the judgement of one country, or the judgement of one man.' The Anglo-French-American proposals for its internationalization were, he insisted, eminently fair. In fact, he declared, 'when I began to analyse these things for myself I almost came to the conclusion that Egypt was getting rather too much out of this deal. . . . But I have resolved my doubts. I believe that the proposals ought to stand before world opinion as reasonable proposals. . . .'[2]

New Zealand was represented at the first London Conference by her Minister of External Affairs, Mr. T. L. Macdonald. He was, if anything, even blunter than Mr. Menzies, choosing to dwell upon

---

[1] *Full Circle*, p. 443.   [2] See below, pp. 122–4.

the manner of Colonel Nasser's expropriation as well as upon the deed itself. He recalled

the flamboyance and bitterness with which the decision was announced, the lack of warning, the failure to consult even neighbour states in the Middle East, the application of military measures in the canal zone, and the threats of imprisonment to company employees. . . . And over and above all this was the unmistakable note of menace, the threat of manipulation for national purposes of an economic agency of vital concern to at least half the countries of the world. . . .[1]

Of the three Asian members of the Commonwealth, Pakistan alone ranged herself on the side of the eighteen powers supporting internationalization. Torn between an instinctive sympathy for Egypt and its Western allies, the Government opted for its allies, and thereby incurred the profound hostility of its own public. *Hartals* (meetings of protest) took place in Karachi, Lahore, and Dacca. Pakistan's pro-Western policy was the cause of much resentment in Egypt. The Cairo newspaper *Al-Gomhouria*, known as an organ for promulgating government views, began an acrimonious Press campaign deploring the perfidy of a fellow Muslim nation; one of its cartoons depicted a bearded old man in a Jinnah cap surmounted by a coin-shaped halo, counting his *tasbih* (prayer readings) made of dollars.[2] Word of the sharp deterioration of Egyptian-Pakistani relations soon reached Karachi, and its chargé d'affaires in Cairo was roundly reprimanded in the Press, *Dawn* going so far as to declare that 'the activity on the part of Bharati [Indian] and Russian envoys in Cairo . . . is a rueful reminder to the Pakistan Government of the need to entrust our country's representation in important foreign capitals to competent and active hands and not to people who may be "political rejects" or trouble-makers to be kept at bay'.[3] In these unpromising circumstances, the Pakistani Foreign Minister, Mr. Hamidul Huq Choudhury, attempted, by a series of amendments and resolutions, to make the Anglo-American-French proposals before the London Conference acceptable to Colonel Nasser. This mediating effort was bound to fail.

Although Pakistani policy at this stage of events was dangerously out of step with domestic public opinion, there seems to have been

[1] See below, pp. 136–8.   [2] *Dawn* (Karachi), 8 September 1956.
[3] Ibid., 12 August 1956.

little connexion between the efforts made in London by Mr. Choudhury on behalf of his Western allies and the fall of his Government soon afterwards. The Prime Minister, Chaudhri Mohamad Ali, resigned on 8 September, in ill-health and weary of the constant turmoil and instability of his country's politics. To succeed him, the President, Iskander Mirza, called upon the leader of the Awami League, Mr. H. S. Suhrawardy. Elements of the Awami League (it no more than the Muslim League was at this stage a monolithic body) had indicated during previous months that they looked with less favour than did the Government upon Pakistan's alliance with the West. Mr. Suhrawardy, however, had not committed himself on this crucial issue of foreign policy; his new Foreign Minister (Mr. Firoz Khan Noon replaced Hamidul Huq Choudhury on 12 September), an immensely experienced politician who had served as Indian High Commissioner in London and on the Viceroy's Executive Council in the years before independence and partition, was not likely to become a party to lessening the nation's pro-Western attitudes. Between them, and guided by the President, these two seasoned members of the Pakistani 'Establishment' kept the country pretty much upon the course set by their predecessors, notwithstanding its unpopularity, until events compelled its alteration.

The opposition to the majority plan for international operation of the Suez Canal consisted of Ceylon, India, Indonesia, and the Soviet Union. It was led throughout the first London Conference by the Indian delegate, Mr. V. K. Krishna Menon, the Minister without Portfolio, known to some as Mr. Nehru's trouble-shooter and to others as Mr. Nehru's trouble-maker. While Mr. Menon deplored the absence of Egypt from the Conference, and lectured those whose approach to such problems was by 'first desiring a change of government or constitution or personnel in another country', he did not forget that he was representing India and not Egypt: 'My Government would like it to be stated that there are, in the manner in which the nationalization was carried out, features which have led to the present aggravated situation. We would like to have seen that nationalization carried out in the normal way of international expropriation, where there is adequate notice, and the way of taking over is less dramatic. . . .'[1] Colonel Nasser would not have put it that way.

[1] See below, p. 142.

Nevertheless, in its insistence that whatever kind of international supervision be devised it was not to 'prejudice . . . Egyptian ownership and operation', the Indian plan, acceptable to Egypt, was unacceptable to the majority of the London Conference. 'We were sorry that India was out of step', Sir Anthony Eden recorded later. 'In an attempt to persuade the Indian delegation to come nearer to us, we asked them whether they could endorse the principles upon which eighteen of the powers had agreed, even though they reserved their position as to the methods to give effect to them. This they felt unable to do.'[1] Ceylon followed suit. 'We would have preferred', declared Sir Claude Corea, Ceylon's High Commissioner in the United Kingdom and her delegate at the London Conference, 'to preserve our silence. We would have preferred to go with the rest of you . . . but we just cannot do that as things are now.'[2]

III

So the Twenty-two Powers became Eighteen. These next decided to depute from their number a delegation of five to present their plan for international control directly to the President of Egypt in Cairo. They chose as their principal spokesman—the suggestion was New Zealand's—the Prime Minister of Australia. Mr. Menzies and his companions remained in the Egyptian capital from 3 September to 9 September. The atmosphere was not improved by the arrests of British nationals throughout their visit. There was in fact little to negotiate about. The Menzies mission proposed that Nasser turn over the Canal to an international authority; Nasser was ready to discuss terms only on the basis of Egyptian ownership and control. 'With frightful reiteration he kept coming back to the slogans', Mr. Menzies reported to Sir Anthony Eden, in a letter which cannot have done other than reinforce the British Prime Minister's already highly unfavourable impression of his antagonist. 'Our proposal was "collective colonialism" which we were seeking to enforce; he constantly came back to "sovereignty"; to our desire for the "domination" of

---

[1] *Full Circle*, p. 451.

[2] See below, p. 160. Ceylon's stand brought a telegram from President Nasser to Prime Minister Bandaranaike, expressing 'the Egyptian people's and my own appreciation of the wise and fair attitude of your delegation and its support of the right of Egypt to nationalise the Suez Canal Company and to safeguard its own independence and dignity'. Quoted in Ceylon, Parl. Deb., H.R., 1956–7, vol. 27, col. 1158.

the canal; to our proposed "seizure" of the canal.'[1] Tiresome as this experience may have been, Mr. Menzies would have done well to remember that his side had its slogans too, and that the slogans of the other touched a responsive chord in the hearts and minds of Asia's millions. Instead, he produced for Sir Anthony's edification the following portrait of the man of the hour:

I was told that Nasser was a man of great personal charm who might beguile me into believing something foreign to my own thought. This is not so. He is in some ways quite a likeable fellow but so far from being charming he is rather *gauche*, with some irritating mannerisms, such as rolling his eyes up to the ceiling when he is talking to you and producing a quick, quite evanescent grin when he can think of nothing else to do. I would say that he was a man of considerable but immature intelligence. . . .[2]

During the Menzies mission's week in Egypt, the United Kingdom Government was busy with decision and diplomacy. It was decided to seek the approval for the Eighteen Power proposals of the United Nations Security Council, in the expectation that Colonel Nasser would not accept them. The N.A.T.O. Council was briefed (by the Foreign Secretary, on 5 September) on the London Conference. Allies were sounded out. On 3 September,

[1] Quoted in *Full Circle*, p. 472.
[2] Quoted ibid., p. 471. Mr. Menzies' appraisal of the Egyptian leader betrays an Antipodean bias. 'The irritating ways of Egyptians', remarks the official New Zealand War History, 'have a lot to account for. To an occidental, the habits of the oriental as seen in Egypt were often amusing, but just as often infuriating.' Major-Gen. W. G. Stevens, *Problems of 2 NZEF* (Wellington, 1958), p. 217. This traditional attitude towards 'gyppos'—contempt tempered by condescension—did not serve the United Kingdom well. The British were consistently handicapped by misleading intelligence in high places concerning Nasser's personality and power. Colonel Nasser himself alluded to this aspect of the crisis several years later: 'Mr. Selwyn Lloyd used to think I had a lot of buttons on my desk and I could start a revolution in any Arab country by just pressing one of them. This was absolute nonsense, but one of the many reasons which led to a clash between our two countries.' 'President Nasser Reviews his Policy', *The Times*, 15 May 1961.
'This kind of man [an informed commentator has written of the Egyptian President] . . . was psychologically a world away from an Eden, Lloyd, Macmillan or a Dulles—all of whom kept thinking they were dealing solely with him. For this reason, the special insights of one or two Western envoys might have been priceless to the West—but were also lost to the West early in the contest. There were two who could get through to and understand the moody, volatile young officer from Beni Mor village. One was U.S. Ambassador Henry Byroade. . . . The other, for a brief time after Byroade was reposted, was the brilliant Canadian Ambassador [Herbert] Norman—an exhausted victim of McCarthyism who killed himself in Cairo in April, 1957.' Erskine B. Childers, *The Road to Suez* (London, 1962), p. 271. Herbert Norman lived long enough to render to his country and to the world service which helped to keep the peace. See below, p. 284.

the Canadian Secretary of State for External Affairs, Mr. Pearson, and the High Commissioner in the United Kingdom, Mr. Norman Robertson, met in London with the British Foreign Secretary, Mr. Lloyd, and Mr. Anthony Nutting, Minister of State for Foreign Affairs. (Three days earlier, Mr. Pearson had issued a statement endorsing the majority proposals of the London Conference and expressing the hope that President Nasser would accept them.)[1] According to Mr. Pearson's later recollection of these discussions, he had been at pains, as was Mr. Robertson on subsequent occasions before 28 October, to urge the United Kingdom not to resort to force except with the authorization of the United Nations.[2] Sir Anthony Eden, however, claims that Mr. Pearson, 'though . . . averse to military sanctions, . . . did not exclude them in the last resort'.[3]

Meanwhile, Britain's other North American ally was causing Sir Anthony Eden some concern.

IV

At the extreme eastern end of Lake Ontario, on the Canadian side, lies an isolated lighthouse station named Main Duck Island. In 1941 it became the property of John Foster Dulles, and here, without even 'a battery radio to keep up with news of the outside world', the American Secretary of State (so his biographer has recorded) was wont to retreat to do 'his heavy thinking'. While Mr. Menzies was jousting with President Nasser in Cairo, Mr. Dulles was at Duck Island, where he conceived the project soon afterwards to be known as the Suez Canal Users' Association.[4]

Having returned to Washington, Dulles informed the British Ambassador on 4 September of the fresh conclusions to which his week-end's reflections had led him. A new convention with Egypt, which he had himself proposed and which had been endorsed at his suggestion by the Eighteen Powers in their London Conference, he now felt to be undesirable; the Convention of 1888 would suffice. The principal users of the Canal should form a club for the purpose of operating the Canal to their own, and hence to the world's, advantage. Their association was to hire pilots and

---

[1] See below, p. 121.
[2] Canada, H.C. Deb., 1956 (Special Session), pp. 52–53.
[3] *Full Circle*, p. 458.
[4] John Robinson Beal, *John Foster Dulles: A Biography* (New York, 1957), p. 267.

collect dues. Just how the project was to be made palatable to Colonel Nasser was not clear.

The British Government believed this, according to Sir Anthony Eden's recollection, to be 'a promising plan', but on the assumption that the users would have at their disposal a club in both senses of the word. It looked to the proposed Users' Association to provide a means of exerting pressure upon Egypt's President. If he violated its rules a resort to force could follow. This Sir Anthony Eden made clear in his speech in the House of Commons on 12 September[1] and clearer still in a speech there next day. The latter was, by all accounts, a remarkable performance. The Chamber was packed; Mr. Menzies was in the Distinguished Strangers' Gallery. It heard (in the words of the Parliamentary Correspondent of the *Manchester Guardian*) 'a different Eden from any we have hitherto seen', 'cool, self-possessed, resolute, never on the defensive or apologetic. . . . This is an Eden who would not shrink from using force if he decides there is no alternative.' He carried his party to a man, and alienated the Opposition to a man; there was now 'a clean-cut cleavage on foreign affairs such as we have not seen since the days of appeasement'.[2]

Nothing was more essential at this crucial juncture than the steadfast support and co-operation of Sir Anthony Eden's trans-Atlantic partner. This was not forthcoming. On the day following the Prime Minister's exposition of the functions of the Users' Association to the House of Commons, the American Secretary of State declared at his Press conference: 'It is not our purpose to try to bring about a concerted boycotting of the canal. I think under those conditions [blocking of the Canal by Egypt] each country would have to decide for itself what it wanted its vessels to do. . . . We do not intend to shoot our way through.'[3]

Sir Anthony Eden's retrospective comment upon this Dullesian gloss to the agreed text of the Anglo-American-French statement announcing the formation of the Users' Association was that 'it would be hard to imagine a statement more likely to cause the maximum allied disunity and disarray'.[4] It had certainly made his own position intensely difficult. He had accepted the Users'

---

[1] See below, pp. 109–12.

[2] 'Premier expounds his Suez Policy', *Manchester Guardian Weekly*, 20 September 1956.

[3] *The New York Times*, 14 September 1956.   [4] *Full Circle*, p. 483

Association in place of the Eighteen Power proposals partly for its value as a sanction against Egypt but increasingly because acceptance appeared to be the only way of keeping closely in step with the United States. He had agreed to the plan to retain the confidence of its author, and it now appeared as if its author was disowning his plan. He might well feel as a man betrayed. The righteous indignation experienced by Sir Anthony Eden at this time is undiminished in his narrative years afterwards:

. . . we accepted the Users' Club. We tried to make it work in the weeks that followed, only to find that the vital assurances on which the offer rested did not exist in substance at all. Her Majesty's Government fell in with the idea in order to keep unity with the United States. It was to lose us unity in the House of Commons. The paradox was that the United States Government made this lack of unity at home an increasing reproach against us.[1]

The Users' Club was an American project to which we had conformed. We were all three in agreement, even to the actual words of the announcement. Yet here was the spokesman of the United States saying that each nation must decide for itself and expressing himself as unable to recall what the spokesman of a principal ally had said. Such cynicism towards allies destroys true partnership. It leaves only the choice of parting, or a master and vassal relationship in foreign policy.[2]

On 15 September, President Nasser spoke in Cairo. The Eighteen Powers, he declared, were guilty of 'international thuggery and imperialism'; the proposed Users' Association was an 'association for waging war'. On 19 September the Eighteen Powers assembled in London for a second conference. This, unlike its predecessor, was paralysed by the diverging conceptions of the two principal Western participants, and by the knowledge that whatever emerged from a gathering described in such unflattering terms by President Nasser was unlikely to be acceptable to him. A certain disaffection was already manifest in the ranks of the Eighteen. 'In so far as the Users' Association is concerned', declared the Pakistani Foreign Minister (then Mr. Firoz Khan Noon), 'we are in the first place not sure what exactly is contemplated and secondly we have satisfied ourselves that our attendance at the Conference does not, in any way, bind us to be a party to this Association. . . . We cannot, in any case, associate ourselves with the use of force or with any solution imposed on Egypt against her

[1] Ibid., p. 479.   [2] Ibid., p. 484.

will.'[1] So the Eighteen became Fifteen, Ethiopia and Japan joining Pakistan in declining membership in the Users' Club.

V

Meanwhile, the Anglo-French military build-up proceeded apace. An Allied Headquarters had been established in London under the command of General Sir Charles F. Keightley; various task force commanders also set up their headquarters there. The British contribution included a parachute brigade group, a commando brigade of the Royal Marines, an armoured division, and an infantry division; medium and light bombers, fighter and ground attack aircraft, reconnaissance, transport, and helicopter aircraft; an aircraft carrier task group, cruisers, destroyers, frigates, mine-sweepers, and an amphibious warfare squadron. These forces were deployed at widely separated sites—Salisbury Plain, Malta, Cyprus—so that, as their commander observed in his official Dispatch, 'a great deal of travelling was required by all Commanders'.[2] Even had it been thought desirable to conceal the extent of these preparations, concealment would have been difficult, if not impossible. By the middle of September, therefore, it had become apparent to the world that a force capable of invading Egypt had been brought into being.

By the time the second London Conference had disbanded, on 21 September, only two Commonwealth Governments actively supported United Kingdom policy. One was the Australian. Mr. Menzies, back in Canberra after his long and eventful absence, spoke to the House of Representatives on 25 September. In a speech elaborately modelled on Sir Anthony Eden's speeches of 12 and 13 September, the Australian Prime Minister 'discussed and endorsed the possible emergence of circumstances which might necessitate, and thus justify, the translation of firmness into force'.[3] Of this analysis the Labour Party Opposition was strongly

[1] See below, p. 156. Pakistan's apparent realignment was immediately hailed in Egypt. Colonel Anwar Sadat wrote in *Al-Gomhouria* (Cairo): 'My absent Pakistani brother has at last returned. . . . What the West applauded most at the close of the first London Conference was the fact that Pakistan, Iran, Turkey and Ethiopia approved the Dulles scheme. Egyptians were shocked by the previous absence of the Pakistan and Iranian brothers but thanks to God they have returned.' Quoted in *Dawn*, 22 September 1956.
[2] Supplement to the *London Gazette*, 12 September 1957, p. 5328.
[3] Jitendra Mohan, 'Parliamentary Opinions on the Suez Crisis in Australia and New Zealand', *International Studies* (New Delhi), vol. II, no. 1, July 1960, p. 66. See below, pp. 128–33.

disapproving. Its leader, Dr. Herbert Evatt, in a speech no less powerful and forthright than that of Mr. Menzies, accused the Prime Minister of 'gunboat diplomacy' calculated to impede a peaceful settlement. 'We cannot talk to nations like that, no matter how small or weak they are, with any effect.'[1] There were persistent rumours, too widespread and too authoritatively stated to be wholly without foundation, that on the issue of using force even as a last resort the Australian Foreign Minister, Mr. Casey, stood closer to Dr. Evatt than to his own Prime Minister; the Canberra correspondent of *The Times* reported that it was Mr. Casey's view that 'recourse to force against Egypt ought to be avoided, because it would severely strain Anglo-American relations and create grave tension within the Commonwealth with Asian members refusing to participate in drastic measures'.[2] But if this was Mr. Casey's conclusion (and all his training and experience were such as to have led him to it), he did not make it public at any time; nor did subsequent events cause him to dissociate himself in any way from his leader's statements, still less to resign from his Government.

The New Zealand Government's reaction to the possible use of force by the United Kingdom against Egypt was perceptibly different from that of the Australian. It was true that Mr. Holland had expressed the characteristic response of New Zealanders when Britons are in trouble: 'New Zealand goes and stands where the Motherland goes and stands.'[3] It was true, as well, that he had matched these words with deeds: the New Zealand cruiser *Royalist* (it was announced on 26 September) would not proceed as scheduled to visit the Asian members of the Commonwealth but would remain in the Mediterranean, where her presence, according to the Admiralty, 'would be of great assistance at the present time'. But the New Zealand Government's analysis of the events

[1] See below, pp. 133–4. An Australian newspaper, describing Dr. Evatt's speech as 'a fierce, emotionally charged, spontaneous statement', pointed out that the Labour Party leader may have been aroused more by the tone than by the content of what the Prime Minister had to say. 'Many are placing the responsibility for Dr. Evatt's angry outburst on the manner in which Mr. Menzies expressed to the House his opinion on the possible use of force. Those who heard Mr. Menzies' speech agree that in discussing the question of force and economic sanctions he appeared to have a far more inflexible approach than is actually revealed in his statement.' 'Bitterness in Parliament—but Still Much in Common', *Sydney Morning Herald*, 28 September 1956.
[2] *The Times*, 11 September 1956. Quoted in Mohan, op. cit., p. 66 n.
[3] *New Zealand Government News Bulletin*, 12 September 1956.

which might embroil its people in war differed significantly from that of either Sir Anthony Eden or Mr. Menzies. There was no passionate denunciation of Colonel Nasser, no invocation of a communist menace in the Middle East, no parallel drawn with the years before the Second World War (years when New Zealand's Government distinguished itself by standing in splendid isolation against appeasement). 'Not a single speaker in the New Zealand parliament', notes Mr. Mohan, 'tried, as had been done previously in the United Kingdom and was to be done subsequently there and in Australia, to build up in advance a case in favour of the use of force against Egypt.'[1]

The Indian Government at this stage of the crisis was working tirelessly if ineffectually to bring about a peaceful negotiation of the dispute, notwithstanding the failure of the Menzies mission or its own refusal to take part in the second London Conference. On 13 September, Pandit Nehru addressed the Lok Sabha. The Users' Club, he declared, was not an instrument of negotiation but an instrument of force. Whatever it might bring about could only be 'in the nature of an imposed decision' and would therefore 'render peaceful settlement more difficult. . . .' He had appealed to both the United Kingdom and the United States to reopen negotiations with President Nasser. A basis for negotiation, he earnestly believed, did exist. 'There is here no question of appeasement of one side or another as what is to be sought and can in our view be obtained is a settlement satisfactory and honourable to all concerned.' He had read Sir Anthony Eden's speech of 12 September[2] with 'surprise and regret', for it seemed to discourage rather than call for further negotiation. Above all, there should be no resort, in any circumstances, to force of arms.[3] The Indian Government's proposals for a peaceful settlement were circulated to the powers concerned on 25 September.[4] Of these Indian efforts to avoid war over Suez, Sir Anthony Eden was of course not unaware, but he has recorded that he did not think them helpful:

The Indian Government . . . were constantly urging a negotiated settlement upon us. As we repeatedly explained to them, this is what we were seeking. This had been our aim in Korea, in Indo-China, in Trieste and in Iran, and we had gained it in all cases, sometimes with

---

[1] Mohan, op. cit., pp. 66–67.
[2] See below, pp. 109–12.
[3] See below, pp. 146–8.
[4] See below, pp. 148–51.

Indian help. We had no need to be ashamed of our record. Our difficulty now was that there could be no negotiation with Egypt unless there were some basis on which to negotiate. . . . Meanwhile, Mr. Krishna Menon made a number of journeys, between Cairo, London, and eventually New York. Her Majesty's Government considered fully and at length all suggestions put to them by India, but Delhi did not then share our view of the importance of keeping international agreements in the interests of all nations, or of the need to restore them when broken.[1]

VI

The decision to place the crisis over the Canal before the Security Council of the United Nations had been taken by the British Cabinet as early as 28 August, although the reference was not to occur until Mr. Menzies had reported on whatever might be the result of his mission to Cairo. Following the failure of that mission, Sir Anthony Eden proposed to the United States Secretary of State that the time had come for going to the United Nations. Mr. Dulles not only refused to sponsor the Anglo-French draft resolution, he refused even to support it. In Sir Anthony Eden's words, 'he accused us of trying to enlist the aid of the Security Council to force a new treaty on Egypt which would bestow new rights on the users of the Canal'.[2] In London this was hard to take. After these recriminations, the United States and the United Kingdom (as the Foreign Secretary told the American Ambassador on 8 September) were farther apart than at any time since the crisis began.[3] Partly in response to the pressure of public opinion at home, partly in response to that exerted by anxious members of the Eighteen Powers, the United Kingdom Government dispatched on 23 September a letter to the President of the Security Council,[4] requesting that 'the situation created by the unilateral action of the Egyptian Government in bringing to an end the system of international operation of the Suez Canal . . .' be inscribed upon the Council's agenda. Once again the United States declined to be a co-sponsor of this request.

On 26 September, the British Prime Minister and his Foreign Secretary, rebuffed by their trans-Atlantic ally, crossed the Channel for conversations with their French colleagues. Sir Anthony Eden has provided this account of what transpired:

---

[1] *Full Circle*, p. 491.　　[2] Ibid., p. 475.
[3] Ibid., p. 476.　　[4] See below, pp. 112–13.

At our meeting in Paris the French and ourselves were agreed that we must not allow our case to be submerged or manœuvred into a backwater at the United Nations. I wanted to reassure our [French] allies that we would not abandon our main objective, to remove the canal from the control of a single Government or man, and to secure enforceable guarantees for efficient navigation and maintenance. The French were sceptical about the United Nations, and more sceptical still about the Users' Club, which had been even less well received by their public opinion than by ours. They felt that the American Administration was not fulfilling its promises, that some of the utterances of the President and Mr. Dulles would make the Russians believe that they could back Nasser with impunity, that the delay was allowing Nasser to build up an even stronger position, thanks to a continuous supply of Russian arms. Finally, there was a danger of the whole Middle East coming under the dominance not so much of Egypt as of Russia. For all these reasons the French favoured action at an early date.

I had much sympathy with their views, but I was sure that we must first have recourse to the United Nations and do our best there. The Foreign Secretary and I undertook, however, that if the Security Council showed itself incapable of maintaining international agreements, Britain would not stand aside and allow them to be flouted. If necessary we would be prepared to use whatever steps, including force, might be needed to re-establish respect for these obligations. The French Ministers agreed, though with some reluctance, to try out all the resources of the Security Council, on the strict understanding that there should be no abandonment of the original proposals approved by the eighteen powers. This was an attitude with which we were ourselves in full agreement. . . .[1]

On 5 October, the long-awaited Security Council debate on the Suez Canal crisis began. The British Foreign Secretary put the United Kingdom's case,[2] supported by the representative of Australia[3] (which was the only other Commonwealth country at that time a member of the Security Council). India, in the person of Krishna Menon, was intensely active behind the scenes, though these activities were not appreciated by the United Kingdom. The Government of India, Sir Anthony Eden wrote afterwards with some asperity,

looked to the West for repeated concessions and found no difficulty in urging this course, while refusing the slightest concession to Pakistan over Kashmir. The Indian Government were canvassing their scheme, which they now put in writing, for attacking an international advisory

---

[1] *Full Circle*, p. 496.   [2] See below, pp. 113–14.   [3] See below, pp. 134–6.

body, which would only have vague powers of supervision, to the Egyptian nationalized canal authority. Mr. Menon had found ears in Cairo ready to listen to such a proposal, naturally enough, for this meant that any effective international element was eliminated. It might be that the Indians had sincerely convinced themselves that Nasser would not accept the eighteen-power proposals. Certainly the Indian Government had not supported them, but this did not seem a sufficient reason why all eighteen powers should, in deference, abandon their position. We had already considered Mr. Menon's ideas in London and found no substance in them. Thanks to the staunchness of the principal users of the canal, he now failed to sway the deliberations of the Security Council, but his activities caused a continuing superficial flurry.[1]

On 13 October, the Anglo-French draft resolution, having survived displacement by various alternatives, was put to the vote of the Security Council. The resolution was in two parts.[2] Part One laid down half a dozen principles, general enough to command broad acceptance. But Part Two declared that the principles corresponded to the more precise proposals of the Eighteen Powers, and stated as well that the Users' Association should receive all dues payable by the ships of its members. Part One was passed; Part Two was not, for Mr. Shepilov duly cast his veto for the Soviet Union.

This marked, for Sir Anthony Eden, a crucial turning-point.

It was clear enough to me [he wrote later] where we were. The powers at the London Conference had worked out, with care and forethought, a scheme which would have made the Suez Canal part of an international system giving security for all. The United States had put its whole authority behind the scheme and her Secretary of State had introduced the proposals himself before the London Conference. Now all this was dead. It was of no use to fool ourselves on that account. We had been strung along over many months of negotiation from pretext to pretext, from device to device, and from contrivance to contrivance. At each stage in this weary pilgrimage we had seen our position weakened. Now we had gone to the United Nations itself. It was not at our wish that we had been so late to make an appeal there. Here was the result. Two communist powers, Yugoslavia and Soviet Russia, had voted against the only practical scheme in existence for the creation of an international system for the Suez Canal. . . . The truth was starkly clear to me. Plunder had paid off.[3]

---

[1] *Full Circle*, p. 502.   [2] See below, pp. 114–15.
[3] *Full Circle*, pp. 505–6.

The course was now set for what, in a celebrated euphemism, Sir Anthony Eden called 'the crunch'.[1]

VII

How long, and in what way, before the failure of their resolution at the United Nations the British Cabinet had resolved that military action, so far from being a last and desperate resort, was the preferable method of dealing with the Suez crisis is not now known and may never be known. Whether the Government of the United Kingdom entered into collusion with the Government of Israel in advance of the tripartite strike against Egypt and, if so, the nature and extent of that collusion, are no less mysterious. The evidence that Sir Anthony Eden, several days and perhaps several weeks before the Anglo-French ultimatum of 30 October, had resolved both to use force and to collude with Israel in its use is impressive, though not conclusive.[1] The existence of a so-called 'Pretexts Committee' of the Cabinet, charged with seeking out acceptable *casus belli* with Egypt, was widely rumoured in London during late September and early October. Colonel Robert Henriques has revealed that in mid-September he had conveyed to Mr. Ben-Gurion at the request of a member of the British Cabinet the information that 'if, when Britain went into Suez, Israel were to attack simultaneously, it would be very convenient for all concerned'.[2] Field-Marshal Viscount Montgomery has related that 'when the Suez operation was being "teed up" I was asked by the Prime Minister of the day to go and see him. I said to the Prime Minister: "Will you tell me what is your object? What are you trying to do?" The Prime Minister replied: "To knock Nasser off his perch." '[3] And in his own Memoirs, Sir Anthony Eden makes no reference to his statement to Parliament of 20 December 1956 in which he declared: 'There was not foreknowledge that Israel would attack Egypt—there was not.'[4]

On 16 October (to return to what is known, as opposed to what is speculation), the Prime Minister and his Foreign Secretary paid their second visit to Paris within a month to confer with MM. Mollet and Pineau. It was a crucial conference, held under conditions of the strictest secrecy, with none of the usual Civil

[1] It is impressively marshalled in Childers, op. cit., pp. 225–80.
[2] *The Spectator*, 6 November 1959.
[3] Quoted in *The Times*, 29 March 1962.
[4] United Kingdom, H.C. Deb., 1956–7, vol. 562, col. 1518.

Service advisers present. Sir Anthony Eden has recorded that the French Ministers were asked to convey to the Government of Israel (that country then being subjected to greater harassment than usual by hostile propaganda and *fedayeen* commando raids from both Egypt and Jordan) his urgent request not to riposte against Jordan, as Britain would in that event have reluctantly to come to Jordan's aid. He told the French Ministers to make clear to the Israelis that 'if there were to be a break-out it was better from our point of view that it should have been against Egypt'.[1]

These pregnant, and enigmatic, words may be interpreted as evidence of collusive preparation for war. Or they may be taken as indicating no more (if no less) than the making of prudent precautions for 'certain possible eventualities'.[2] Facts are scarce. It is known that immediately after the meeting in Paris on 16 October, the movement of British and French troops, ships, and aircraft quickened perceptibly; that several 'senior British civil servants suddenly stopped getting the usual classified documents';[3] and that on 18 October Sir Walter Monckton, widely thought to be unsympathetic to using force at Suez, resigned as Minister of Defence.

On 25 October, the United Kingdom Cabinet received reports that Israel was about to mobilize. It immediately, according to Sir Anthony Eden's account,

discussed the specific possibility of conflict between Israel and Egypt and decided in principle how it would react if this occurred. The Governments of France and the United Kingdom should, it considered, at once call on both parties to stop hostilities and withdraw their forces to a distance from either bank of the canal. If one or both failed to comply within a definite period, then British and French forces would intervene as a temporary measure to separate the combatants. To ensure this being effective, they would have to occupy key positions at Port Said, Ismailia and Suez. Our purpose was to safeguard free passage through the canal, if it were threatened with becoming a zone of warfare, and to arrest the spread of fighting in the Middle East.[4]

This statement has been justly described as 'one of the most crucial . . . of all . . . made about the Sinai-Suez War'.[5] It reveals

---

[1] *Full Circle*, p. 513.
[2] The phrase used by Franklin Roosevelt in May 1940 to describe the prospect of the United Kingdom's surrender to, or successful invasion by, Nazi Germany.
[3] Childers, op. cit., p. 237.   [4] *Full Circle*, p. 523.
[5] Childers, op. cit., p. 236.

that the terms of the Anglo-French ultimatum, and what would be done if that ultimatum were rejected, were determined by the British Cabinet (albeit determined 'in principle') four days before Israel attacked and five days before the ultimatum was issued. It contradicts the version of events given out by Sir Anthony Eden and his colleagues at the time of the crisis—namely, that the ultimatum and the action following its rejection were decided on 30 October, under conditions of surprise and disarray. And it renders wholly unacceptable the excuse of the British Government and its apologists then and later that consultation with its Commonwealth partners could not be carried out because there was no time in which to consult.

DOCUMENTS

I. UNITED KINGDOM

1. Extracts from statement by the Secretary of State for Foreign Affairs, Mr. Selwyn Lloyd, at the 22-Power London Conference, 18 August 1956[1]

. . . Our cordial and constructive discussions here must not obscure the fact that there is a situation of the utmost gravity created by Colonel Nasser's action in regard to the Suez Canal. The circumstances in which he made his original statement and the tone of hostility in it were such that Her Majesty's Government in the United Kingdom felt that they had to take certain precautionary military steps. . . . British Governments and the British people do not like the use of force; it is for us always the last resort. As a nation we are very slow to anger, but when it is clear that vital interests are threatened by acts of deliberate hostility we are in the habit of standing fast whatever the odds.

. . . In our opinion, as I think you know, . . . the Egyptian Government have been guilty of an illegal act. We think that a breach of contract is not the less so because it is a government which commits the breach, and in these legal matters it is not a bad thing to look at the substance as well as the form, and I think anyone round this table really knows in his heart of hearts that the

[1] D.S.P. 6392, pp. 153–9.

Canal Company was an international company and, whatever the Egyptian Government could do with the assets of that company in Egypt because it was technically incorporated in Egypt, the fact is that the matter was handled in such a way as to disregard the rule of law between nations. . . .

. . . It has been inferred . . . that any international participation in the control or operation of the Suez Canal would be an infringement of Egyptian sovereignty. . . . That attitude, I submit, is based upon a complete misconception of the nature of sovereignty under international law. Sovereignty does not mean the right to do exactly what you please within your own territory. Sovereign government is the right in a general way to organise and govern the national territory and economy according to the wishes of the government of that territory. But the doctrine of sovereignty gives no right to use the national territory or to do things within the national territory which are of an internationally harmful character. . . .

I do not think there is any real substance either in the idea that a state suffers an infringement of its sovereignty by allowing an international authority to perform certain functions in its territory. History affords numerous examples of this sort of thing. . . . Still less is there substance in the idea that a state suffers a derogation from its sovereignty if it accepts by treaty certain limitations of its complete freedom of action. . . .

2. The '18-Nations Proposals', 21 August 1956[1]

The Governments approving this Statement, being participants in the London Conference on the Suez Canal:—

Concerned by the grave situation regarding the Suez Canal;

Seeking a peaceful solution in conformity with the purposes and principles of the United Nations; and

Recognising that an adequate solution must, on the one hand, respect the sovereign rights of Egypt, including its rights to just and fair compensation for the use of the Canal, and, on the other hand, safeguard the Suez Canal as an international waterway in accordance with the Suez Canal Convention of 29th October, 1888;

Assuming for the purposes of this statement that just and fair compensation will be paid to the Universal Company of the Suez

[1] Cmd. 9853, pp. 11–12.

Maritime Canal, and that the necessary arrangements for such compensation, including a provision for arbitration in the event of disagreement, will be covered by the final settlement contemplated below

Join in this expression of their views:—

1. They affirm that, as stated in the Preamble of the Convention of 1888, there should be established 'a definite system destined to guarantee at all times, and for all the powers, the free use of the Suez Maritime Canal'.

2. Such a system which would be established with due regard to the sovereign rights of Egypt, should assure:—

(a) Efficient and dependable operation, maintenance and development of the Canal as a free, open and secure international waterway in accordance with the principles of the Convention of 1888.

(b) Insulation of the operation of the Canal from the influence of the politics of any nation.

(c) A return to Egypt for the use of the Suez Canal which will be fair and equitable and increasing with enlargements of its capacity and greater use.

(d) Canal tolls as low as is consistent with the foregoing requirements and, except for (c) above, no profit.

3. To achieve these results on a permanent and reliable basis there should be established by a Convention to be negotiated with Egypt:—

(a) Institutional arrangements for co-operation between Egypt and other interested nations in the operation, maintenance and development of the Canal and for harmonising and safeguarding their respective interests in the Canal. To this end, operating, maintaining and developing the Canal and enlarging it so as to increase the volume of traffic in the interest of the world trade and of Egypt, would be the responsibility of a Suez Canal Board. Egypt would grant this Board all rights and facilities appropriate to its functioning as here outlined. The status of the Board would be defined in the above-mentioned Convention.

The Members of the Board, in addition to Egypt, would be other States chosen in a manner to be agreed upon from among the States parties to the Convention with due regard to use, pattern of trade and geographical distribution; the composition of the Board to be such as to assure that its responsibilities would be discharged solely with a view to achieving the best possible operating results without political motivation in favour of, or in prejudice against, any use of the Canal.

The Board would make periodic reports to the United Nations.

(b) An Arbitral Commission to settle any disputes as to the equitable return to Egypt or other matters arising in the operation of the Canal.

(c) Effective sanctions for any violation of the Convention by any party to it, or any other nation, including provisions for treating any use or threat of force to interfere with the use of operation of the Canal as a threat to the peace and a violation of the purposes and principles of the United Nations Charter.

(d) Provisions for appropriate association with the United Nations and for review as may be necessary.

3. Extract from telegram from the Prime Minister, Sir Anthony Eden, to the President of the United States, Dwight D. Eisenhower. (No date available: probably 28 August 1956)[1]

This is a message to thank you for all the help Foster [John Foster Dulles, U.S. Secretary of State] has given. Though I could not be at the conference myself, I heard praise on all sides for the outstanding quality of his speeches and his constructive leadership. He will tell you how things have gone. It was, I think, a remarkable achievement to unite eighteen nations on an agreed statement of this clarity and force [Document 2].

Before he left, Foster spoke to me of the destructive efforts of the Russians at the conference. I have been giving some thought to this and I would like to give you my conclusions.

I have no doubt that the Bear is using Nasser, with or without his knowledge, to further his immediate aims. These are, I think, first to dislodge the West from the Middle East, and second to get a foothold in Africa so as to dominate that continent in turn. In this connection I have seen a reliable report from someone who was present at the lunch which Shepilov [Dmitri Shepilov,

---

[1] *Full Circle*, pp. 452–3.

U.S.S.R. Foreign Minister] gave for the Arab Ambassadors. There the Soviet claim was that they 'only wanted to see Arab unity in Asia and Africa and the abolition of all foreign bases and exploitation. An agreed unified Arab nation must take its rightful place in the world'.

This policy is clearly aimed at Wheelus Field and Habbaniya [respectively, American and British air bases in Morocco and Iraq] as well as our Middle East oil supplies. Meanwhile the communist bloc continue their economic and political blandishments towards the African countries which are already independent. Soon they will have a wider field for subversion as our colonies, particularly in the West, achieve self-government. All this makes me more than ever sure that Nasser must not be allowed to get away with it this time. We have many friends in the Middle East and in Africa and others who are shrewd enough to know where the plans of a Nasser or a Musaddiq would lead them. But they will not be strong enough to stand against the power of the mobs if Nasser wins again. The firmer the front we show together, the greater the chance that Nasser will give way without the need for any resort to force. That is why we were grateful for your policy and Foster's expression of it at the conference. It is also one of the reasons why we have to continue our military preparations in conjunction with our French allies.

We have been examining what other action could be taken if Nasser refuses to negotiate on the basis of the London Conference. There is the question of the dues. The Dutch and the Germans have already indicated that they will give support in this respect. The Dutch may even be taking action in the next few days. Then there is the question of currency and economic action. We are studying these with your people and the French in London and will be sending our comments soon. It looks as though we shall have a few days until Nasser gives Menzies [R. G. Menzies, Australian Prime Minister and Chairman of the Five Power Committee presenting the '18-Nations Proposals' to President Nasser] his final reply. After that we should be in a position to act swiftly. Selwyn Lloyd [U.K. Foreign Minister] is telegraphing to Foster about tactics, particularly in relation to United Nations. . . .

Meanwhile I thought I should set out some of our reflections on the dangerous situation which still confronts us. It is certainly the most hazardous that our country has known since 1940. . . .

4. Telegram from the Prime Minister, Sir Anthony Eden, to the President of the United States, Dwight D. Eisenhower, 6 September 1956[1]

Thank you for your message and writing thus frankly.

There is no doubt as to where we are agreed and have been agreed from the very beginning, namely that we should do everything we can to get a peaceful settlement. It is in this spirit that we favoured calling the twenty-two power conference and that we have worked in the closest co-operation with you about this business since. There has never been any question of our suddenly or without further provocation resorting to arms, while these processes were at work. In any event, as your own wide knowledge would confirm, we could not have done this without extensive preparations lasting several weeks.

This question of precautions has troubled me considerably and still does. I have not forgotten the riots and murders in Cairo in 1952, for I was in charge here at the time when Winston was on the high seas on his way back from the United States.

We are both agreed that we must give the Suez committee every chance to fulfil their mission. This is our firm resolve. If the committee and subsequent negotiations succeed in getting Nasser's agreement to the London proposals of the eighteen powers, there will be no call for force. But if the committee fails, we must have some immediate alternative which will show that Nasser is not going to get his way. In this connection we are attracted by Foster's suggestion, if I understand it rightly, for the running of the canal by the users in virtue of their rights under the 1888 Convention. We heard about this from our Embassy in Washington yesterday. I think that we could go along with this, provided that the intention was made clear by both of us immediately the Menzies mission finishes its work. But unless we can proceed with this, or something very like it, what should the next step be?

You suggest that this is where we diverge. If that is so I think that the divergence springs from a difference in our assessment of Nasser's plans and intentions. May I set out our view of the position?

In the nineteen-thirties Hitler established his position by a series of carefully planned movements. These began with the occupation

[1] *Full Circle*, pp. 464–7.

of the Rhineland and were followed by successive acts of aggression against Austria, Czechoslovakia, Poland, and the West. His actions were tolerated and excused by the majority of the population of Western Europe. It was argued either that Hitler had committed no act of aggression against anyone, or that he was entitled to do what he liked in his own territory, or that it was impossible to prove that he had any ulterior designs, or that the Covenant of the League of Nations did not entitle us to use force and that it would be wiser to wait until he did commit an act of aggression.

In more recent years Russia has attempted similar tactics. The blockade of Berlin was to have been the opening move in a campaign designed at least to deprive the Western powers of their whole position in Germany. On this occasion we fortunately reacted at once with the result that the Russian design was never unfolded. But I am sure that you would agree that it would be wrong to infer from this circumstance that no Russian design existed.

Similarly, the seizure of the Suez Canal is, we are convinced, the opening gambit in a planned campaign designed by Nasser to expel all Western influence and interests from Arab countries. He believes that if he can get away with this, and if he can successfully defy eighteen nations, his prestige in Arabia will be so great that he will be able to mount revolutions of young officers in Saudi Arabia, Jordan, Syria and Iraq. (We know that he is already preparing a revolution in Iraq, which is the most stable and progressive.) These new Governments will in effect be Egyptian satellites if not Russian ones. They will have to place their united oil resources under the control of a united Arabia led by Egypt and under Russian influence. When that moment comes Nasser can deny oil to Western Europe and we here shall all be at his mercy.

There are some who doubt whether Saudi Arabia, Iraq and Kuwait will be prepared even for a time to sacrifice their oil revenues for the sake of Nasser's ambitions. But if we place ourselves in their position I think the dangers are clear. If Nasser says to them, 'I have nationalized the Suez Canal. I have successfully defied eighteen powerful nations including the United States, I have defied the whole of the United Nations in the matter of the Israel blockade, I have expropriated all Western property. Trust

me and withhold oil from Western Europe. Within six months or a
year, the Continent of Europe will be on its knees before you.'
Will the Arabs not be prepared to follow this lead? Can we rely on
them to be more sensible than were the Germans? Even if the
Arabs eventually fall apart again as they did after the early Caliphs,
the damage will have been done meanwhile.

In short we are convinced that if Nasser is allowed to defy the
eighteen nations it will be a matter of months before revolution
breaks out in the oil-bearing countries and the West is wholly
deprived of Middle Eastern oil. In this belief we are fortified by
the advice of friendly leaders in the Middle East.

The Iraqis are the most insistent in their warnings; both Nuri
and the Crown Prince have spoken to us several times of the
consequences of Nasser succeeding in his grab. They would be
swept away. . . .

The difference which separates us today appears to be a difference of assessment of Nasser's plans and intentions and of
the consequences in the Middle East of military action against
him.

You may feel that even if we are right it would be better to wait
until Nasser has unmistakably unveiled his intentions. But this
was the argument which prevailed in 1936 and which we both
rejected in 1948. Admittedly there are risks in the use of force
against Egypt now. It is, however, clear that military intervention
designed to reverse Nasser's revolutions in the whole continent
would be a much more costly and difficult undertaking. I am very
troubled, as it is, that if we do not reach a conclusion either way
about the canal very soon one or other of these eastern lands
may be toppled at any moment by Nasser's revolutionary
movements.

I agree with you that prolonged military operations as well as the
denial of Middle East oil would place an immense strain on the
economy of Western Europe. I can assure you that we are conscious of the burdens and perils attending military intervention.
But if our assessment is correct, and if the only alternative is to
allow Nasser's plans quietly to develop until this country and all
Western Europe are held to ransom by Egypt acting at Russia's
behest it seems to us that our duty is plain. We have many times
led Europe in the fight for freedom. It would be an ignoble end
to our long history if we accepted to perish by degrees.

5. Letter to the President of the Security Council from the Permanent Representatives of the United Kingdom and France to the United Nations, 12 September 1956[1]

In accordance with instructions received from Her Majesty's Government in the United Kingdom and from the French Government, we have the honour to address you, in your capacity as President of the Security Council, with regard to the situation created by the action of the Egyptian Government in attempting unilaterally to bring to an end the system of international operation of the Suez Canal which was confirmed and completed by the Suez Canal Convention signed at Constantinople on 29 October 1888.

2. Since the action of the Egyptian Government created a situation which might endanger the free and open passage of shipping through the Canal without distinction of flag as laid down in the above-mentioned Convention, a conference was called in London on August 16, 1956. Of the twenty-two States attending that conference, eighteen, representing between them over ninety per cent. of the user interest in the Canal, put forward proposals to the Egyptian Government for the future operation of the Canal. The Egyptian Government have, however, refused to negotiate on the basis of the above-mentioned proposals which, in the opinion of Her Majesty's Government and of the French Government, offer means for a just and equitable solution. Her Majesty's Government and the French Government consider that this refusal is an aggravation of the situation which, if allowed to continue, would constitute a manifest danger to peace and security.

3. We have the honour to request that the contents of this letter be brought to the notice of the members of the Security Council.

> (Signed) Pierson Dixon
> Permanent Representative of Great Britain and Northern Ireland to the United Nations
> (Signed) Louis de Guiringaud
> Acting Permanent Representative of France to the United Nations

[1] S.C.O.R., Eleventh Year, Supplement for July, August and September 1956, Doc. S/3645.

6. Extracts from speech by the Prime Minister, Sir Anthony Eden, in the House of Commons, 12 September 1956[1]

I think that the House will agree that the 22-Power Conference in London and the mission of Mr. Menzies' committee represented a very considerable effort to reach a solution by agreement. That offer has failed. In consequence, we have carefully considered, in consultation with our French and American allies, what our next step should be. We have decided, in agreement with them, that an organisation shall be set up without delay to enable the users of the Canal to exercise their rights.

This users' association will be provisional in character and we hope that it will help to prepare the way for a permanent system which can be established with the full agreement of all interested parties. Although discussions are still proceeding between the three Governments—the United States, France and ourselves—about the details of this plan, I can now give the House the broad outline, by accord with the other countries.

It will be as follows: the members of the users' association will include the three Governments I have already mentioned—the United States, France, and ourselves—and the other principal users of the Canal will be invited to join. We hope that the pattern of membership will be as representative as possible. The users' association will employ pilots, will undertake responsibility for co-ordination of traffic through the Canal, and, in general, will act as a voluntary association for the exercise of the rights of Suez Canal users.

The Egyptian authorities will be requested to co-operate in maintaining the maximum flow of traffic through the Canal. It is contemplated that Egypt shall receive appropriate payment from the association in respect of the facilities provided by her. But the transit dues will be paid to the users' association and not to the Egyptian authority. Through this organisation it should be possible to establish a system of transit of the Canal for a substantial volume of shipping. Of course, we recognise that a provisional organisation of this kind, designed to meet an emergency, cannot be in a position to provide for the major developments which are becoming urgently necessary if the Canal is to continue adequately to serve the interests of its users and we also recognise that the

[1] United Kingdom, H.C. Deb., vol. 558, coll. 10–15.

attitude of the Egyptian Government will have an important bearing on the capacity of the association to fulfill its functions [*Laughter*]. Yes. But I must make it clear that if the Egyptian Government should seek to interfere—

Mr. *Harold Davies* (Leek): Deliberate provocation.

The Prime Minister:—with the operations of the association, or refuse to extend to it the essential minimum of co-operation, then that Government will once more be in breach of the Convention of 1888. [*Hon. Members:* 'Resign.'] I must remind the House that what I am saying—[*An Hon. Member:* 'What a peacemaker.']—is the result of exchanges of views between three Governments. In that event Her Majesty's Government and others concerned will be free to take such further steps——

Mr. *S. O. Davies:* What do you mean by that?

The Prime Minister:—as seem to be required—

Mr. *Davies:* You are talking about war.

The Prime Minister:—either through the United Nations or by other means, for the assertion of their rights. [*Hon. Members:* 'Oh.'] I think that Hon. Members might let me develop this. There is more to come.

I shall no doubt be asked what are the intentions of Her Majesty's Government about the reference of this dispute to the United Nations. As I stated just now—if it was heard clearly—I certainly do not, and Her Majesty's Government do not, exclude that. Quite the contrary, it might well be necessary. Meanwhile, we have considered it our duty, jointly with the French Government, to address a letter to the President of the Security Council informing him of the situation which has arisen [Document 5].

That letter does not ask for any action at this stage, but it puts us in a position to ask for urgent action if that becomes necessary. At the same time, there are certain considerations about a reference to the Security Council which we should face—which the House should face—frankly.

Let me take the Abadan precedent as an example. . . . The Government took that issue to the Security Council. They tabled a resolution which was opposed by Russia and Yugoslavia. In an attempt to meet this, our resolution was whittled down. Even so, the Soviet Government would not have it and, consequently, the resolution was made dependent on the final findings of the International Court. I make no criticism of this; it is quite inevitable in view of the existence of the veto. When it reported many months later, the International Court said it had no jurisdiction, and, therefore, our resolution before the Security Council lapsed and we never went there again.

I say that because we have to face the realities of this situation. I also recall that the Labour Government of the day, in their wisdom, uttered a warning that the Security Council's failure to act effectively might create a most serious precedent for the future. I think that they were right. . . .

There has been much public discussion about the question of the use of force in relation to these events in the Canal. I must point out that in this instance it was Egypt who used force. The operation of the Canal has been taken over in complete disregard of Egypt's international obligations. The assets of the Company and its offices were seized by armed agents of the Egyptian Government. . . . The Company's employees were compelled to continue at their work under threat of imprisonment. To condone such actions is to invite their repetition and, I think, to bring international law into contempt.

In recent weeks certain military preparations have been made in the Mediterranean. They are limited in scope. On account of them we have been charged in some quarters with sabre rattling. How ludicrous that is. It might be regarded as provocative if I were to retail all the circumstances which could arise in Egypt. I will, therefore, mention only one, for hon. Members will recall it. I have not forgotten the appalling massacre of foreigners which took place in Cairo in 1952. . . . It is quite true that on that occasion the Egyptian Army intervened to stop further bloodshed, but hon. Members must judge for themselves how much that action was due to the knowledge that we had a plan to intervene by force in the last resort. If military precautions were justified a month ago, they are justified today, and I must make it plain that the Government have no intention of relaxing them. . . .

I should like to finish, if I may, on this personal note. In these last weeks I have had constantly in mind the closeness of the parallel of these events with those of the years before the war. Once again we are faced with what is, in fact, an act of force which, if it is not resisted, if it is not checked, will lead to others. There is no doubt about that. If Egypt continues to reject every effort to secure a peaceful solution, a situation of the utmost gravity will arise. . . .

Of course, there are those who say that we should not be justified and are not justified in reacting vigorously unless Colonel Nasser commits some further act of aggression. That was the argument used in the 1930s to justify every concession that was made to the dictators. It has not been my experience that dictators are deflected from their purpose because others affect to ignore it. This reluctance to face reality led to the subjugation of Europe and to the Second World War. We must not help to reproduce, step by step, the history of the 'thirties. We have to prove ourselves wiser this time, and to check aggression by the pressure of international opinion, if possible; but, if not, by other means before it has grown to monstrous proportions.

As has been rightly said by my right hon. and learned Friend the Foreign Secretary, for this country military action is always the last resort and we shall go on working for a peaceful solution so long as there is any prospect of achieving one; but the Government are not prepared to embark on a policy of abject appeasement, nor, I think, would the House—or most of the House—ask them to, because the consequences of such a policy are known to us. A stimulus is given to fresh acts of lawlessness. With the loss of resources the capacity to resist becomes steadily less, friends drop away and the will to live becomes enfeebled.

We will continue to make every effort, in concert with our Allies, to secure our rights by negotiation, but should those efforts fail the Government must be free to take whatever steps are open to them to restore the situation. That is the policy of the Government which I ask this House to approve.[1]

7. Letter to the President of the Security Council from the Permanent Representatives of the United Kingdom and France to the United Nations, 23 September 1956[2]

In accordance with instructions received from Her Majesty's Government in the United Kingdom of Great Britain and Northern Ireland and from the French Government, we have the honour to request you, in your capacity as President of the Security Council for this month, to call a meeting of the Council for Wednesday, 26 September 1956, in order to consider the following

[1] In the division on 13 September 1956, the Government's policy was approved by 321 votes to 251.
[2] S.C.O.R., Eleventh Year, Supplement for July, August and September, 1956, Doc. S/3654.

item: 'Situation created by the unilateral action of the Egyptian Government in bringing to an end the system of international operation of the Suez Canal, which was confirmed and completed by the Suez Canal Convention of 1888.'

The general nature of the situation referred to above was set out in the letter which was addressed to you on 12 September 1956 [Document 5] in accordance with instructions from our two Governments. Her Majesty's Government and the French Government consider that the time has now come when they must ask for a discussion of this situation by the Council.

(Signed) Pierson Dixon, Permanent Representative of the United Kingdom of Great Britain and Northern Ireland to the United Nations

(Signed) Bernard Cornut-Gentille, Permanent Representative of France to the United Nations

8. Extracts from statement by the Secretary of State for Foreign Affairs, Mr. Selwyn Lloyd, to the United Nations Security Council, 5 October 1956[1]

. . . That is the juridical aspect of the matter. Juridical aspects of matters are sometimes somewhat dull and dry to explain, but I have done my best to put as clearly as I could a situation which in law, I think, is clear beyond peradventure. But our apprehensions as to the future have been greatly increased by the way in which the Egyptian Government has behaved in practice. . . .

We have no quarrel with the right to nationalize as such. That is not an issue in this matter. Most countries, for better or for worse, have nationalized undertakings in their territories. But this was not an ordinary undertaking. Although technically registered in Egypt, it was in substance as in name an international company enjoying concessions built into an international treaty. The principle of the general right to nationalize is not in question. . . .

I pass over the point that this act of 'de-internationalization', as it has been called, is contrary to the trend of the times; I pass over the point that the cancelling of a concession, even if lawful, is hardly likely to encourage confidence; I pass over the point that this was a discriminatory action directed solely against a foreign entity. . . . The real issue in this matter is the sanctity of treaties and respect for international obligations.

[1] S.C.O.R., Eleventh Year, 735th meeting, 5 October 1956, pp. 7–14. Doc. S/PV 735.

It was not merely the disregard for a treaty, not merely the repudiation of an agreement made six weeks earlier, that worried us. The reasons given by the President of Egypt were most disturbing to any future confidence or reliance upon the undertakings of his Government. In an angry speech, he stated that the action was taken because the United States Government had refused aid to Egypt to build the Aswan Dam. He made it clear that this action against the Company was a form of retaliation. He described it as a triumph and said that his intention was to secure triumph after triumph, and he indicated that the revenues of the Canal would be used to build the Aswan Dam. . . .

. . . I say quite frankly that the United Kingdom Government does not feel, after what has happened, that we can take any chances in the future. The guarantees for the users must be clear and specific. . . .

. . . The situation which we are asking the Council to consider is indeed a grave one. It is one that threatens the very life and strength of countless nations. It is not just a quarrel between Egypt and a group of countries who own the ships that ply through the Suez Canal. It is a matter of vital importance to peoples in every part of the globe, the peoples of Asia and Africa no less than the peoples of Europe. No nation can or should stand aside indifferent, when the greatest international waterway in the world is subjected to the unrestricted control of one Government, when that is done in contravention of a treaty of long standing. . . .

9. Draft Resolution submitted to the United Nations Security Council by the United Kingdom and France, 13 October 1956[1]

The Security Council,

Noting the declarations made before it and the accounts of the development of the exploratory conversations on the Suez question given by the Secretary-General of the United Nations and the Foreign Ministers of Egypt, France and the United Kingdom,

1. *Agrees* that any settlement of the Suez question should meet the following requirements:

(*a*) There should be free and open transit through the Canal

---

[1] S.C.O.R., Eleventh Year, Supplement for October, November and December, 1956, Doc. S/3671.

without discrimination, overt or covert—this covers both political and technical aspects;

(*b*) The sovereignty of Egypt should be respected;

(*c*) The operation of the Canal should be insulated from the politics of any country;

(*d*) The manner of fixing tolls and charges should be decided by agreement between Egypt and the users;

(*e*) A fair proportion of the dues should be allotted to development;

(*f*) In case of disputes, unresolved affairs between the Universal Suez Maritime Canal Company and the Egyptian Government should be settled by arbitration, with suitable terms of reference and suitable provisions for the payment of sums found to be due;

2. *Considers* that the proposals of the eighteen Powers [Document 2] correspond to the requirements set out above and are suitably designed to bring about a settlement of the Suez Canal question by peaceful means, in conformity with justice;

3. *Notes* that the Egyptian Government, while declaring its readiness in the exploratory conversations to accept the principle of organized collaboration between an Egyptian authority and the users, has not yet formulated sufficiently precise proposals to meet the requirements set out above;

4. *Invites* the Governments of Egypt, France and the United Kingdom to continue their interchanges and in this connexion invites the Egyptian Government to make known promptly its proposals for a system meeting the requirements set out above and providing guarantees to the users not less effective than those sought by the proposals of the eighteen Powers;

5. *Considers* that pending the conclusion of an agreement for the definitive settlement of the regime of the Suez Canal on the basis of the requirements set out above, the Suez Canal Users' Association, which has been qualified to receive the dues payable by ships belonging to its members, and the competent Egyptian authorities, should co-operate to ensure the satisfactory operation of the Canal and free and open transit through the Canal in accordance with the Convention, signed at Constantinople on 29 October 1888, destined to guarantee the free use of the Suez Maritime Canal.

10. Extracts from speech by the Prime Minister, Sir Anthony Eden, 13 October 1956[1]

... A little progress has been made [at the United Nations] but there are still wide differences of opinion. If such progress has been made, it is due to the firmness and resolution we and those like us have been showing throughout this crisis. ...

Could any men have shown more patience than those five [members of the Menzies Mission] did? Was that wrong? Was that against the United Nations Charter? Or was that colonialism? ...

Through all these negotiations, peace has been our aim, but not peace at any price. And I will tell you why. Because in dealing with a dictatorship peace at any price means to increase step by step the dangers of universal war.

What is at stake is not just the Canal, important though that is. It is the sanctity of international engagements. This is the supreme lesson of the period between the wars. Let us never forget it.

... We thought it our duty to take certain military steps in the Eastern Mediterranean. The Government have no intention of modifying or withdrawing from this decision. Those precautions had to be taken to enable us to take action in defence of British life and interests or in any other emergency that might arise in the area concerned.

All that we have done since is completely in accord with what I said at the time. The movement of every unit has been either to British territory or to territories where we have treaty rights, not one of which we have infringed in the slightest degree.

In other words, we have shown the most absolute restraint. The presence of our forces in the Eastern Mediterranean, far from inflaming the situation, has greatly decreased the danger of further incidents and induced a measure of caution in some minds.

What would anyone in this country have said if we had not taken those precautions and if there had been a repetition of the massacre in Cairo in 1952? Better to be safe than sorry.

Some have protested that the precautions are on too large a scale. My reply is that if you are determined to defend national interests, nothing is more foolish than to do so with inadequate forces.

Others ask, do we need to keep it up? To relax now before a settlement is reached would be fatal.

[1] *The Times* (London), 14 October 1956.

There are some who argue that we should have acted more promptly by striking back the moment Colonel Nasser seized the Canal. I do not agree. By going through every stage which the United Nations Charter lays down we have given an example of restraint and respect for international undertakings.

But that does not alter the responsibility which rests upon us and our allies to ensure that justice is done and that international obligations are fulfilled. After all, the United Nations Charter was set up to discourage breaches of international engagements and not to allow them to pass with impunity.

I have seen it suggested that this dispute about the Canal has something to do with colonialism. No comment could be more misleading.

Colonialism has nothing to do with the matter one way or the other. We have never disputed Egyptian sovereignty. What is at stake in this dispute is whether the sanctity of contracts has to be respected or not. . . .

Let me sum up where we stand. From the very beginning we have never changed our position. It remains: no arrangement for the future of this great waterway could be acceptable to the British Government which would leave it in the unfettered control of a single Power which could, as recent events have shown, exploit it purely for purposes of national policy.

That is our position and it has not changed. It remains our intention to seek its acceptance by negotiation if we possibly can.

I cannot close my comments on this situation without expressing the satisfaction we all feel at the unity between ourselves and our French allies throughout these anxious weeks. One happy consequence of all this has been to increase the sense of partnership between the nations of Western Europe. The truth is, we all have a common interest in this, as in so many other matters.

For us, in Britain, the Commonwealth must always come first. Increasingly, our neighbours in Western Europe understand that this is true. And it would be greatly to the advantage of this country and the Commonwealth if, in commerce and other matters, we in Western Europe could draw closer together and always practice joint policies as well as observe joint treaties. . . .

11. Statement by the Secretary of State for Foreign Affairs, Mr. Selwyn Lloyd, and extracts from answers to questions, in the House of Commons, 23 October 1956[1]

With your permission, Mr. Speaker, and that of the House, I wish to make a statement on the Suez Canal question.

Since the House last met on 13th September, there have been, as hon. Members know, some important developments.

The Foreign Ministers of the 18 Powers which supported the proposals endorsed at the first London Conference met at a second Conference in London to consider the situation caused by the Egyptian Government's refusal to negotiate on the basis of those proposals. They, that is to say, the 18 countries, decided unanimously that certain observations made by Colonel Nasser after Mr. Menzies' departure were too imprecise to afford a useful basis for discussion. This Conference further decided to establish a Suez Canal Users' Association, to function in accordance with the principles laid down in the Declaration agreed on 21st September.

Fifteen of the 18 nations have now become members of the Association, which was officially inaugurated on 1st October. The remaining three have reserved their decision, but have continued to be represented at meetings of the Association as observers. The Users' Association has held a number of meetings, in which the problems of organisation, finance and operation have been thoroughly examined. Some progress has been made, and an Administrator has been appointed. It is our hope that the work of building up the Association will proceed rapidly.

These developments, including the discussions with our friends at the second London Conference, satisfied us that the moment had come to place this question before the Security Council. We had, a fortnight earlier, foreshadowed this move by our preliminary letter to the Security Council [Document 5]. On 23rd September the representatives of the United Kingdom and France in New York addressed a joint letter to the President of the Security Council [Document 7] and requested him to summon a meeting of the Council to consider the situation caused by the unilateral action of the Egyptian Government.

The Security Council met between 5th October and 13th October. I will not pretend to the House that Her Majesty's Government are fully satisfied with the results of these discussions.

[1] United Kingdom, H.C. Deb., vol. 558, coll. 491–6.

Once again, attempts to achieve a just settlement of a problem have been frustrated by a veto—the seventy-eighth—cast by the Soviet Union.

Nevertheless, certain solid advantages were gained. The most significant feature was that the case which the French Foreign Minister and I laid before the Security Council received overwhelming endorsement. Nine of the 11 members of the Council specifically approved the proposals of the 18 Powers and called upon Egypt either to accept them, or to put forward promptly an alternative system with no less effective guarantees.

Similarly, the non-Communist Powers also endorsed the Suez Canal Users' Association, recognised its competence to receive dues payable by ships of its members, and expressed the opinion that the competent Egyptian authorities should co-operate with the Association to ensure the satisfactory operation of the Canal, and free and open transit through it, pending a definitive settlement.

The position, then, at the end of the Security Council meeting was, and still is, that the Egyptian Government have been called on to put forward promptly their proposals. They have not done so. Part of the resolution at the Security Council was unanimously approved. That part set out six principles which must govern any solution of Suez Canal problems [Document 9].

We are told that the Government of Egypt accept those principles. Their representative at the Security Council said so. But there seems to be a gap between Egypt's acceptance of the principles and definition of her part of the means to apply them. What has to be done is to construct a system to provide the users of the Canal with adequate guarantees of efficiency and non-discrimination to replace the system which has been destroyed by the Egyptian Government. For this we need proposals, not the mere acceptance of principles. We still await these proposals.

Throughout the negotiations Her Majesty's Government have kept in close touch with the other members of the 18-Power group who represent over 90 per cent. of the user interest in the Canal from all five continents. It is our determination to continue to work for a solution under which there will be guarantees to the users not less effective than those sought by the proposals of the 18 Powers.

Mr. Robens: As the six principles advocated by the users have been accepted by the Egyptian Government, is it not right that the next step should be the opening of direct negotiations by the representatives of the Users' Association with the Egyptian Government? Does not the fact that the right hon. and learned Gentleman has indicated that the proposals made by Colonel Nasser after Mr. Menzies' departure were too imprecise seem to argue that it would be as well for there now to be direct talks rather than that both sides should be sitting and waiting for each other to act?

Mr. Lloyd: I spent four days in New York with the French Foreign Minister and the Secretary-General of the United Nations, and with the Egyptian Foreign Minister, having precisely those direct talks to see whether it was possible to achieve a basis for negotiation. Definite proposals must be put forward which can constitute a basis for negotiation. . . .

Mr. Gaitskell: May I ask two questions of the Foreign Secretary? First, is it the view of Her Majesty's Government that the necessary guarantees to users can be obtained only if there is an international board of management, day-to-day management, or can Her Majesty's Government conceive of the possibility of satisfactory guarantees by a different system? Secondly, has the right hon. and learned Gentleman received from the representative of the Indian Government the new proposals [VI, Document 5, *infra*], which were reported in the Press two days ago, and which seem to many of us to offer a very reasonable half-way house solution to the whole problem?

Mr. Lloyd: I still think that international co-operation is the best method of achieving the necessary guarantees. I think it is the best and, in fact, the simplest method, but we have never said that it is the only method. What the Security Council's resolution does is to cast upon Egypt the obligation to put forward proposals for guarantees. None the less, we should want to examine those proposals carefully to see whether they did give the same guarantees as international day-to-day operation.

It is quite true that I have received a copy of the Indian proposals. I think that the first question about them to be decided is whether they constitute the proposals of the Egyptian Government. As the right hon. Gentlemen will know, there are certain matters on which they need clarification and precision.

Mr. Gaitskell: While welcoming, so far as it goes, the answer to my first question, may I ask the right hon. and learned Gentleman whether he thinks—in view of the reasonably favourable reply about the Indian Government's proposals—it really would be desirable if a move were made to bring the parties together with a view to agreement on the basis of those principles? Is it the case that the Secretary-General of the United Nations has offered his services in this connection, and has the Foreign Secretary any other idea about clarifying the Egyptian view of the Indian proposals?

Mr. Lloyd: I really think that if right hon. and hon. Members want progress in this matter, the best way to get it is for the Egyptian Government, as quickly as possible, to put forward their proposals for consideration.

Mr. Robens: Since the right hon. and learned Gentleman told us that he spent four days in direct negotiation with the Egyptian Government—as, obviously, the principles were not a matter for argument, having already been agreed, and as the imprecise nature of Colonel Nasser's words must have at that stage been made more precise—would he tell the House exactly what is the sticking point between Egypt and ourselves?

Mr. Lloyd: The sticking point is that we have no proposals to implement the principles which emerged at the end of the four days' discussion. The suggestion that those principles were agreed to before we began is untrue; those principles emerged as the result of the four days' discussion. But it was made quite clear that there must be means to implement them. We had some discussion on the means to implement them, but at the moment the position is that we have had no precise proposals, and the Security Council recognised that. It received a report in private session from the Secretary-General, and it recognised where, in its view, the obligation now lies. I think that we shall make more progress by stating firmly that we await those Egyptian proposals.

II. CANADA

1. Statement by the Secretary of State for External Affairs, Mr. L. B. Pearson, 30 August 1956[1]

It is devoutly to be hoped that President Nasser will accept this invitation to negotiate a peaceful and permanent solution of this serious problem along the lines of the London majority proposals. A failure to do so would involve a very heavy responsibility indeed.

[1] *The Globe and Mail* (Toronto), 31 August 1956.

III. AUSTRALIA

1. Extracts from speech by the Prime Minister, Mr. R. G. Menzies, at the 22-Power London Conference, 18 August 1956[1]

. . . In many publications that I have looked at in the last few days, I have noticed a disposition to talk about nations 'east of Suez' and nations 'west of Suez' as if they had conflicting interests. That, Sir, I venture to describe as deplorable. The truth is that we all have a common interest in having a free, open, competently managed, and ever-improving Suez Canal. I speak for Australia, which is one of the nations east of Suez. In common with our colleague countries who are here, we have a very large trade through the Suez Canal, and as all the countries east of Suez grow in their national development, and in their economic strength and trade, they will have an increasing share of the traffic through the Suez Canal. Indeed, looking into the future, it may very well be that the countries east of Suez will have a dominating share of the traffic through the Suez Canal. Therefore, to create artificial grounds of debate between countries east and countries west seems to me to be a very grave disservice to the settlement of this controversy. . . .

. . . The people who in the long run are affected by the advantages of an open Canal or the disadvantages of an uncertain one are the ordinary people of our countries, the people who in the long run pay for the goods that arrive and get whatever benefit there is from the goods that go. We are here dealing with the interests of the ordinary people of our own countries. I have no disposition whatever, on behalf of my own country, to submit my own people to all the chances and uncertainties that would arise from having their immense trading interests through this Canal made subject to the whim of the moment, to the judgment of one country, or the judgment of one man. . . .

. . . I detect occasionally an undercurrent of feeling that in some way these proposals are unfair to Egypt—that, in our own homely phrase, they amount to pushing something down Egypt's throat. Well, we have all had a few feelings in the last week or two; I do not want to express my own because they are now immaterial; but I am going to ask this Conference to say that these proposals by Mr. Dulles are in fact eminently fair to Egypt, and that Egypt will

[1] *Speech is of Time*, pp. 92–102.

get from them advantages which deserve to be understood and, in particular, deserve to be understood by Egypt. . . .

. . . In the first flush of enthusiasm, in the first flush of saying: 'Well, we have taken this Canal and it is ours', and a few other rhetorical remarks about it and about sovereignty, there may be a tendency to forget that this Canal is a business operation. It involves engineering skill, business organization, and traffic control of the highest possible order; and, if it is to be carried out effectively, it will obviously require, in the future, large capital amounts in order that it may growingly serve its purpose. Is Egypt alone, in the events that have happened, likely to be able to manage that financial problem? Does anybody seriously suppose it? Capital, after all, is a very shy thing—it runs away very rapidly if it feels a loss of confidence. The events of the last month in respect of Suez have made it almost a certainty that the future capital expansion of the Canal, to say nothing of the maintenance of its ordinary revenue by the physical maintenance of the Canal itself, will depend on how far there is genuine confidence in its future. I do not believe there is a man sitting at this table, or a woman—bowing in the direction of my friend Mrs. Pandit—who does not believe that the future of this Canal is pretty dismal unless there is a complete feeling of confidence in its independent future, in the impartiality of its administration, and in the fact that of a large number of nations each has an interest in maintaining it, keeping it out of politics, keeping it out of wars, keeping it out of disputes, letting it serve the interests of the commerce of mankind.

Sir, I have no doubt that only international management will produce that result. . . .

When I began to analyse these things for myself I almost came to the conclusion that Egypt was getting rather too much out of this deal. It seemed to me that the benefits to Egypt were so great. But I have resolved my doubts. I believe that the proposals ought to stand before world opinion as reasonable proposals, as proposals which protect the vital interests of all of us who are here, and at the same time are not tyrannical or unjust in relation to Egypt.

After all, Sir, we did not produce this crisis. We did not make this Conference necessary. We have come here because of what one country has done. We might have been pardoned if our speeches had contained a good deal of resentment. On the contrary, without exception, I believe that they have been moderate and restrained, that we have all addressed ourselves to the true task, and that if we accept the proposals that have been put forward we will have made a first-class contribution to the settlement of this whole problem and, by the settlement of this problem, to the whole future peace and intercourse of the world.

2. Extracts from report of statement by the Minister of External Affairs, Mr. R. G. Casey, 29 August 1956[1]

The Suez Canal Conference in London was very much more successful than might have been expected. The overwhelming proportion (18 out of 21) of the nations represented supported the policy of co-operative control in the running of the Suez Canal. The 18 countries which supported the American resolution represented nearly 95 per cent. of the tonnage passing through the Canal.

Even the four dissenting countries were not very far removed from the sense of the American resolution, in that they agreed on the 'user' countries participating in the control and having a say in the fixing of Canal charges, as well as there being some arbitral body to settle matters in dispute.

The atmosphere of the Conference was responsible and remarkably calm. Practically no resentment was expressed at what had happened, although the seriousness was universally admitted. Considering the dramatic nature of the subject, there was no fireworks. However, it was universally agreed that the present position could not be allowed to continue. . . .

The attitudes taken by India and Pakistan, in particular, were important. Pakistan took a positive and realistic attitude in support of the 18-nation resolution. India took the other view, although Mr. Krishna Menon, the Indian delegate, stated after the Conference that it would be a mistake to consider that the Indian minority proposals were still being advanced as an alternative to the majority proposals of the 18 nations, and that if they were acceptable to Egypt, India would not oppose them. He said that India agreed that the association of 'user' interests was essential. . . .

Another satisfactory result of the Conference was the fact that the 18 supporters of the American resolution comprised Eastern, Western and Middle Eastern countries of a wide variety of backgrounds and associations. There was no line up of countries by race, religion or interests. . . .

[1] *C.N.I.A.*, vol. 27, no. 8, August 1956, pp. 502–3.

The London Conference on the Suez Canal provided opportunity for contact between the Australian and many other delegations. Although the press was not present, the Conference itself was in effect a public one, in that each delegation was free to make public its attitudes and its statements. But the constant contact between delegations outside the Conference gave opportunity for private and confidential exchanges on all aspects of the Suez problem. Mr. Casey said that he availed himself freely of these opportunities for personal contact, particularly with the Foreign Ministers of countries of the Middle East and Asia whom he knew personally. . . .

3. Questions by the Leader of the Opposition, Dr. Herbert Evatt, and reply by the Acting Prime Minister, Sir Arthur Fadden, in the House of Representatives, 12 September 1956[1]

Dr. Evatt: I desire to ask the Acting Prime Minister a question of urgency and importance in relation to the Suez dispute, pointing out that there have been reports in Australia that yesterday the Australian Cabinet was acutely divided on the question of whether force might be used in connexion with the dispute and also pointing out that, on the wireless during the day, President Eisenhower has been reported as having said that he was certain that the question would be referred to the United Nations before force was used, or words to that effect. Having in view what the Acting Prime Minister has said and what the Minister for External Affairs said in that important statement that he made on his arrival in India, in effect advocating conciliation, and in view of the statement of President Eisenhower and the views expressed by the Labour movement, including the supreme governing body of the Labour movement, in favour of reference of the matter to the United Nations, I ask the following questions:—First, will the Acting Prime Minister make a statement in relation to the suggestion of a dispute in the Australian Cabinet on this vital question of principle? Secondly, would the reference of the dispute to the United Nations, as advocated by honourable members on this side and, I hope, on the other side of the House, and by President Eisenhower, be in accord with the view of the Government?

[1] Australia, H.R. Deb., 1956 (Second Period), p. 411.

Sir Arthur Fadden: I recommend the public to disregard completely unauthorised press reports which purport to be an account of Cabinet proceedings. Those are, of their very nature, speculative and usually mischievous in their effects. They are the more deplorable when the matters alleged to be under discussion by the Cabinet are of a delicate character and possess grave international implications. One such report, published yesterday, had actually gone into print before the Cabinet even met. Any report of discussion in Cabinet yesterday over the policy to be followed by the Australian Government in the Suez Canal dispute has no basis in fact. Yesterday's meeting of the full Ministry was for the purpose of giving Ministers the latest information that had reached the Government from our own Prime Minister and other official sources. No question arose at that stage of the future course to be followed by the Government, nor was any decision taken. We are maintaining the closest contact with the Prime Minister and are being kept fully informed on the views of the British Government. Any decisions required of us will be taken at the appropriate time and in the light of the best information and advice available to us. The Parliament and the public will be informed as fully as possible of developments as they occur.

Dr. Evatt: Parliament will be consulted?

Sir Arthur Fadden: Yes.

4. Questions by the Leader of the Opposition, Dr. Herbert Evatt, and reply by the Acting Prime Minister, Sir Arthur Fadden, in the House of Representatives, 13 September 1956[1]

Dr. Evatt: I desire to ask the Acting Prime Minister whether he, as acting leader of the Government, was informed or consulted by the representatives of the United Kingdom Government, or by the Prime Minister, who is in London, as to the course of action indicated by the Prime Minister of the United Kingdom in his speech in the debate in the House of Commons [I, Document 6, *supra*]. Has any consideration been given to these proposals by the Australian Government? In the circumstances, will arrangements be made for a debate on this vital matter in this House? I think that the time has now come for such a debate.

---

[1] Australia, H.R. Deb., 1956 (Second Period), pp. 477–8.

Sir Arthur Fadden: The Australian Government was not consulted with regard to the suggested committee to be set up, but the Government will give the closest consideration to the proposals. As to the problem of the Suez Canal being the subject of debate, I am sure that the Prime Minister will take the first opportunity after his arrival next week to bring that about. I have received an official communication on this subject this morning, and, with the permission of the House, I shall make a statement in connexion with it.

The Prime Minister of the United Kingdom announced in a speech to the House of Commons yesterday afternoon a plan to form a new organisation of users of the Suez Canal. Sir Anthony Eden explained that the operation of the canal by the Suez Canal Company was an essential part of the system to ensure the use of the canal by all powers as provided by the convention of 1888. But he added also that the rights of user countries had not derived only from the convention. They have been established by long and uninterrupted use, and include not only free passage but also the efficient operation, administration and maintenance of the canal without discrimination. Consequently, the United Kingdom Government, together with the Governments of the United States of America and France, has decided to set up a users' association to exercise these rights. The other principal users of the canal will be invited by those three governments to join the association. The association will make arrangements for the co-ordination of traffic through the canal, the provision of pilots, and the exercise of other rights, and will collect the transit dues payable by the nations of its member countries. Egypt will be called upon to co-operate, and will be entitled to an appropriate payment for the facilities provided by it. Sir Anthony Eden added that the United Kingdom Government believes that this system will enable a substantial volume of shipping to move through the Canal, and will limit the economic dislocation which might otherwise occur in both Europe and Asia. If the Egyptian Government seeks to interfere with the operations of the association it will be in breach of the convention of 1888. The United Kingdom will invite all the governments concerned to join the proposed association. The Australian Government will, of course, give full and immediate consideration to this proposal.

Dr. Evatt: I propose to say very little. The question of the desirability of the proposed arrangement is one thing. There will be room for debate on it, and I do not wish to anticipate it. Egypt will be asked to co-operate in this entirely novel arrangement. The proposal seems to me to be contrary to the international agreement, but this is a question of international law, and I say nothing further about it. The serious thing is the implication that if Egypt does not co-operate force will be used to compel it to do so. I do not know the facts——

Sir Arthur Fadden: I do not know whether that can be read into the proposal.

Dr. Evatt: I think that is its intended meaning. I hope it is not. It is the duty of the governments concerned to consult each other, and the Acting Prime Minister should have been consulted before such a proposal was made, particularly as the Prime Minister was on the scene. I ask the Acting Prime Minister to have the matter considered most carefully before irrevocable action is taken contrary to the wishes of the Government, the Parliament and the people of Australia.

5. Extracts from speech by the Prime Minister, Mr. R. G. Menzies, in the House of Representatives, 25 September 1956[1]

. . . From the point of view of the canal-using nations, there were great and urgent issues at once created by Colonel Nasser's act of repudiation and confiscation.

First, such a grave breach of international law, if overlooked or condoned, would encourage further acts of lawlessness, bringing immense damage to the whole economy of the free world.

Secondly, it would be folly to regard the canal seizure as a single act, to be dealt with in isolation. As an isolated act, it would in all truth be dramatic and crucial enough. But Colonel Nasser, acting in a similar fashion to other dictators before him, has made no secret of his particular ambition to be the acknowledged head of the Arab world, to encourage confiscations of outside investments and installations and to humiliate and drive out the foreigner. The canal seizure is, in plain English, the first shot in a campaign calculated, unless it is promptly and successfully resisted, to make the peoples and economies of Great Britain and Western Europe dependent literally from week to week on one

1 Australia, H.R. Deb., 1956 (Second Period), pp. 817–26.

man's whim. In the literal sense, the Suez Canal is, for millions, a question of survival.

Thirdly, as Colonel Nasser's 'acquisition' of the Suez Canal Company was achieved by the repudiation of a long-standing contract, it was clear that Egypt's credit would be so weakened that she could not obtain or spend the many scores of possibly hundreds of millions needed for the much-needed expansion of the canal. The canal would, therefore, become more and more inadequate to cope with rapidly increasing traffic, which would accordingly need to follow longer and less economic routes.

Fourthly, Colonel Nasser had, at his very first announcement, made it clear that the Suez Canal was in future to be the political instrument of Egypt, losing its specially impressed international and non-political quality. . . .

. . . These are but a few of the grave and critical implications of Colonel Nasser's action. It is small wonder that the reaction in the world was so sharp, and that 22 nations came to the London Conference with such anxiety about the future. The reaction of the United Kingdom Government was both prompt and vigorous. It denounced Colonel Nasser's action. It concerted measures with both the United States and France for the calling of a world conference, of interested nations, including Egypt herself. It put into train military measures of mobilisation and preparation.

In these steps, as I should like to remind the House, the United Kingdom Government secured, in Parliament at Westminster, the swift support of all parties. . . .

. . . Before I conclude, I want to speak quite frankly about three other matters. One is the question of force. That question calls for a cool and clear answer. There has been a great variety of vocal opinions, ranging between what I will call two extreme views. One view is that force should at once have been used to defeat a confiscation by force. That view is out of harmony with modern thinking, or, at any rate, this side of the iron curtain.

The other view is that force can never be employed, except presumably in self-defence, except by and pursuant to a decision of the United Nations Security Council. . . . This I would regard—and I speak quite bluntly—as a suicidal doctrine for, having regard to the existence of the veto, it would mean that no force could ever be exercised against any friend of the Soviet Union except with the approval of the Soviet Union, which is absurd. . . .

Each of these extreme views must, I believe, be rejected. The truth is that, in a world not based on academic principles, a world deeply affected by enlightened self-interest and the instinct of survival, but nevertheless a world struggling to make an organisation for peace effective, force, except for self-defence, is never to be the first resort, but the right to employ it cannot be completely abandoned or made subject to impossible conditions.

Let me say, quite plainly, that the whole lively and evolving history of the British Empire and the British Commonwealth of Nations was not the product of any theory. It has been, from first to last, a practical matter, an inductive process, like the slow creation of the common law and of all the great instruments of self-government. It would be a sad day if it allowed itself to be theorised out of existence. We need not get into a timid state of mind in which the very mention of the word 'force' becomes forbidden. There is no community of nations which can say, with a clearer conscience, that it has set a great twentieth century example of using force only when forced into it, and then not for conquest but for resistance to aggression.

But does this mean that we are to be helpless in the presence of an accomplished threat to our industrial and economic future? I believe not. Is our task to 'patch up' peace and no more? Surely our task is not merely to prevent hostilities but to build up a firm order to law and decency, in which 'smash and grab' tactics do not pay. We must avoid the use of force if we can. But we should not, by theoretical reasoning in advance of the facts and circumstances, contract ourselves out of its use whatever those facts and circumstances may be. We are to seek peace at all times, but we are not bound to carry that search so far that we stand helpless before unlawful actions which, if allowed to go unchecked, can finally dissipate our own strength and deprive the world of that power and authority, both moral and physical, which reside in the free nations, and are still vital to the free world and the human interests which the free world protects.

What, then, should be our programme of action in relation to the Suez Canal? . . .

First, negotiation for a peaceful settlement by means of honourable agreement. So far, we have tried this without success. The failure, let me repeat and emphasise, has not been due to any unfairness or illiberality on our side, but to dictatorial intransigeance on the other. . . .

Should we continue to negotiate on a watered-down basis, in the spirit which says that any agreement is better than none? I cannot imagine anything more calculated to strengthen Colonel Nasser's hand, or weaken our own.

Secondly, the putting on of pressure by co-operative effort on the part of the user-nations. Colonel Nasser must be brought to understand that his course of action is unprofitable to his country and his people, and that he is abandoning the substance for the shadow. This is one of the great merits of the users' association now established by the second London conference. The more canal revenue that is diverted from the Egyptian Government, the less will the Egyptian people believe that it pays to repudiate.

Thirdly, should the United Nations, by reason of the veto, prove unable to direct any active course of positive action, we may find ourselves confronted by a choice which we cannot avoid making. I state the choices . . . in stark terms—

(a) We can organise a full-blooded programme of economic sanctions against Egypt, or . . .

(b) We can use force to restore international control of the canal, or . . .

(c) We can have further negotiation, provided we do not abandon vital principles, or . . .

(d) We can 'call it a day', leave Egypt in command of the canal, and resign ourselves to the total collapse of our position and interests in the Middle East, with all the implications for the economic strength and industrial prosperity of nations whose well-being is vital to ours.

This is, I believe, a realistic analysis of the position. It does not matter whether people like it or not, it is true. . . .

The second matter concerns the attitude and activities of the Soviet Union. . . . My observations in London and since have convinced me that: First, the Soviet Union is not looking for a world war, but is willing to stir up and foment trouble in those regions where the strength of the Western democracies can be materially weakened. Secondly, it is anxious to increase its influence in Egypt, by the provision of arms and the development of economic ties. How to reconcile this with Egypt's sovereignty is a problem it will leave to Colonel Nasser. Thirdly, it has been in constant and persuasive touch with Colonel Nasser during the recent negotiations. Mr. Shepilov openly declared the argument for Egypt, in terms, in phrases, in slogans, which I was later to hear used, word for word, by Colonel Nasser himself, at Cairo.

The third matter concerns the impact of the Suez Canal confiscation on Australia and on the great new nations of South and South-East Asia, whose interests we respect and have done something to help. So far as Australia is concerned, I need hardly say that an open canal is essential to British prosperity, and that a closed canal could mean mass unemployment in Great Britain, a financial collapse here, a grievous blow at the central power of our Commonwealth, and the crippling of our greatest market and our greatest supplier.

We are not alone in this. The nations and peoples of South-East Asia, being much nearer to Suez than we are, are even more dependent on it than we are. Further even than this, Asia contains great populations which need the developmental assistance of foreign capital and friendly co-operation. Colonel Nasser's policy of repudiation in the name of sovereignty is not calculated to help the very countries whose admiration and support he is now claiming. Indeed, it is ironical that, in the guise of their leader, he is now taking steps to deprive some of the great Middle East powers of the natural and established markets for the product of their oil wells.

A final note of warning is necessary. In or out of the United Nations, there are great principles and vital interests at stake. A matter of this kind is not disposed of by being sent to the Security Council, or, under present procedures, to the General Assembly. Nothing could suit the Egyptian dictator better than for the free world to lose interest, or a sense of crisis or a sense of urgency. There must be both speed and realism. We must also look ahead, keep our sense of direction, and maintain our impetus.

Should the United Nations' machinery fail to produce an early settlement, are we then to wash our hands of the whole matter, saying, 'Well, it is too bad; but we can do nothing. Colonel Nasser must be left with his spoils; retreat in the Middle East must go on.' I decline to believe it. The principle of internationally assured non-political control of the Suez Canal is vital. It cannot be watered down without being washed away. To abandon it would be suicidal.

Therefore, if the United Nations, once more frustrated by Soviet action, proves ineffective; if it cannot impose economic sanctions or direct any other course of effective action, we, the user nations, must in the absence of willing and proper negotiation, be ready to impose sanctions ourselves. For the central and unforgettable fact in all this unhappy business is that unless Egypt's action is frustrated and the international status of the canal assured, a score of nations, great and small, will have put their fortunes into pawn. We are, indeed, at one of the cross-roads of modern history. We will take the wrong turning at our peril. . . .

6. Extracts from speech by the Leader of the Opposition, Dr. Herbert Evatt, in the House of Representatives, 25 September 1956[1]

I want to say immediately that I think it is appalling that on the very eve of the discussion of this dispute by the chosen representatives of the United Nations, the permanent and elected members of the Security Council, a speech of this kind [Document 5] should have been made in Australia. . . .

I think that I can best help the House and the country by dealing at once with the so-called choices which the Prime Minister indicated towards the end of his speech. . . .

I shall answer these questions asked by the Prime Minister, because each of them has been drafted in a cunning and misleading way. This is the stark choice. He said that if a veto prevented a decision in the Security Council, we had several choices, the first of which was to organise a full-blooded programme of economic sanctions against Egypt. We do not object to economic sanctions because they are provocative. We object to economic sanctions, unless authorised by the Charter of the United Nations, because they are a form of economic warfare—the form which, very often, is most cruel and most wicked. . . .

That first alternative is rejected. No one will accept it. Certainly the Labour movement in this country will not accept it. It will not be accepted in Great Britain. That has been made clear by a unanimous decision of the Trade Union Council of the British Labour Party. It will not be accepted by the Labour movement in New Zealand. Only a day or two [ago] Mr. Nash forwarded to me a communication to that effect. . . .

[1] Australia, H.R. Deb., 1956 (Second Period), pp. 826–35.

The second point in the Prime Minister's programme of action is a suggestion of the use of force. I say that is tantamount to a suggestion of war, and that the people of Australia will adopt the same attitude towards it as have the people of Great Britain, who have shown, in public opinion polls, overwhelming opposition to the use of force. . . .

The Prime Minister has said that we can negotiate, provided we do not abandon vital principles. That language is not the language of the twentieth century. It is the language of the years before there was a League of Nations or a United Nations.

Mr. Curtin: Gunboat diplomacy!

Dr. Evatt: Yes, gunboat diplomacy. We are often critical of other powers that use that sort of diplomacy. We cannot talk to nations like that, no matter how small or weak they are, with any effect. Even though a nation's cause is unpopular with other people, no one will stand for that method of approach. . . .

Now, Mr. Speaker, I want to say a few things about what ought to be done. The trouble with the speech of the Prime Minister is that it undoubtedly was calculated to impede the processes of the Security Council. . . . I suggest that what the Government ought to be doing is working out a scheme under which mediation would be commenced in a real sense, so that there could be true negotiations and meetings round the table, as the result of which agreement might be reached between Egypt and the Security Council. . . . We do not want what one English politician called 'a run through of the Security Council', just to get the hands going up to indicate the numbers necessary to apply the veto, and then act, as the Prime Minister indicates. . . .

7. Extracts from statement by the Permanent Representative to the United Nations, Dr. Ronald Walker, to the Security Council, 8 October 1956[1]

. . . Although the Suez Canal runs through Egypt, it has long been, in a most direct way, a part of the national life of many countries. The Suez Canal is no mere accident of geography. Its creation represents a great human achievement, bestowing a benefit on all mankind, so long as it is operated on the basis of mutual advantage.

[1] S.C.O.R., Eleventh Year, 737th meeting, 8 October 1956, pp. 15–19, Doc. S/PV 737.

We all know the great importance of the Canal to the trading nations of Europe, and Australians are naturally concerned over the effect of any interruption of the Canal's operations upon the United Kingdom. However, as one of the few representatives at this table from a country east of Suez, I am also particularly conscious of what Suez means to the people of those distant lands. At school, we learned of the Canal as one of the first great steps in modern man's conquest of distance. Until the development of air travel, the Canal provided the shortest and most natural route to Europe. Thousands of Australians still pass through it every year, and, to most of the million immigrants who have come to Australia since the war, the Canal has been the highway to the new life they were seeking. . . . The Suez Canal is therefore something so vividly real to us in Australia that the man in the street takes a deep personal interest in any proposal to detract from the Canal's international character.

What is true of Australia in this regard is also true, in varying degree and with varying emphasis, of country after country throughout the Far East. In fact, for many of these countries of Asia the loss of assured passage through the Canal would be a far more serious blow than for Australia, situated as we are so far to the south that the Cape of Good Hope offers an alternative route. There are countries in Asia and Africa whose domestic economy will be most seriously disturbed if the Canal is not freely available to their trade. There are many people in Asia, and for that matter in the Middle East, whose livelihood could be placed in jeopardy and who could go hungry if the conversion of the Suez Canal from an international waterway to an instrument of purely national policy in the hands of a single country became an accomplished fact. . . .

The Egyptian Government has seized the Suez Canal under the guise of nationalization. One of the most disturbing features of this action, and of the arguments offered in defence of it, is the hostility of President Nasser towards international operation of the Canal, in the name of national sovereignty. In discussing nationalization, we must be on our guard against the tyranny of ambiguous words. In my country, and in most others, the word 'nationalization' is applied to the governmental ownership and operation of enterprises—in contrast with their private ownership and operation. . . . But when President Nasser speaks about nationalization—as he has in recent weeks—he apparently means something quite different. He seems to mean the conversion to purely national control of an enterprise that was previously an international undertaking. . . .

The Australian delegation stands ready to participate in a constructive manner in such negotiations as may be possible here. We consider that negotiations should be based on the principles drawn up by the eighteen Powers. At the same time, as Mr. Menzies stated in Cairo, we do not believe that there must be between the users of the Canal and the Government of Egypt any irreconcilable difference of principle. We look forward to a peaceful settlement upon a basis of justice to both sides.

The Australian delegation believes that the draft resolution submitted by the United Kingdom and France [I, Document 9, *supra*] offers a sound foundation for the Council's work. The Australian delegation considers its proposals fair and reasonable, and will, therefore, vote in favour of it.

IV. NEW ZEALAND

1. Extracts from statement by the Minister of External Affairs, Mr. T. L. Macdonald, at the 22-Power London Conference, 17 August 1956[1]

I hope it will not seem out of place to speak of intangibles like confidence and trust, at a conference which centres upon something very tangible, that is, the Suez Canal. . . .

But these intangibles—confidence and trust—are at the heart of the issue before this conference, and we shall ignore them at our peril. International confidence has been seriously shaken. In great degree it depends upon those countries represented here whether confidence among nations is to be restored, or whether with possibly disastrous consequences it is to remain disturbed. In still greater degree it depends upon Egypt.

The international interest in the Suez Canal is generally admitted. Colonel Nasser himself has acknowledged it. This conference will be considering various proposals for giving practical expression to this international interest. Some proposals will doubtless provide for a fairly stringent international control and guarantee; others for a looser supervisory system. Obviously, the looser the system, the greater the need for confidence and trust.

---

[1] D.S.P. 6392, pp. 111–15.

Equally obviously, if trust is non-existent we can hardly be content with vague assurances and a loose system. In considering each proposal we shall not be able to avoid, nor should we avoid, considering the amount of trust in Egypt which it implies. . . .

The significance of Colonel Nasser's announcement concerning nationalisation of the Universal Suez Canal Company did not lie solely in the fact that it affected an enterprise of international character. . . . There were other distinctive features, the flamboyance and bitterness with which the decision was announced, the lack of warning, the failure to consult even neighbour states in the Middle East, the application of military measures in the canal zone, and the threats of imprisonment to company employees should they seek to leave the company service. And over and above all this was the unmistakable note of menace, the threat of manipulation for national purposes of an economic agency of vital concern to at least half the countries of the world. . . .

It is important, I suggest, that these points be kept in mind when we consider Egypt's recent offers of guarantees concerning freedom of navigation through the Suez Canal. It is essential that they be kept in mind when substantive proposals are placed before us. . . .

I wish to emphasise that New Zealand's support for a soundly based international system in the Suez Canal is not accorded without regard for the interests of the Egyptian people. New Zealand forces fought in and near Egypt in two wars, often close to the Suez Canal itself, and at that time many of them gained an understanding of Egypt's problems and were acquainted with its people. I do not wish to labour this point, but there are few of us who have seen the fellaheen at work in the fields of the Nile Delta who would not wish with all sincerity to see their skilled and unremitting industry fittingly rewarded. . . .

I can understand that a people, on achieving independence, might well have reason to annul agreements made in past colonial days which cramp their present capacity to prosper; but the Suez Canal was a constantly expanding asset to Egypt; its value derived from its international connections and its increasing contribution to the enrichment of Egypt grew out of the confidence of the nations of the world that they would have assured use of it. By attempting to destroy the international character of the operations of the canal and in thus undermining the international confidence on which the steady expansion of trade through the canal was taking place, Colonel Nasser has taken the risk of jeopardising his people's prosperity. . . .

2. Extracts from statement by the Minister of External Affairs, Mr. T. L. Macdonald, in the House of Representatives, 11 September 1956[1]

Sir, I thank the House for the courtesy it has extended to me in permitting me to make a statement concerning the conference on the Suez Canal which was held recently in London and which I attended on behalf of New Zealand. . . .

Early debate within the conference revealed that, initially at least, the countries concerned fell into three groups. In the first of these were those countries which felt that little reliance could be placed in Colonel Nasser's promises, and that confidence could be re-established only if the operation of the canal were divorced from national policies and made the responsibility of an international authority. New Zealand, which considered that the operation of the canal should assure free transit and be efficient, financially stable, and proof against arbitrary misuses, was a member of this group. In the second were those countries which felt that reliance could be placed on Colonel Nasser's promises of free passage for all shipping in the canal, and, therefore, favouring the institution of only a loose form of international association with the canal's operation. In the third group were countries which, while concerned to ensure future freedom of passage through the canal, were undecided in what degree it was necessary for the management of the canal to become an international rather than an exclusively Egyptian responsibility.

The outstanding development within the conference was that the countries of the third group, containing notably Pakistan, Ethiopia, Iran and Japan, ultimately joined the countries of the first, making eighteen in all, in supporting the creation of an international Suez Canal Board which would have the responsibility of operating, maintaining, and developing the canal and enlarging it so as to increase the volume of traffic in the interests of world trade and of Egypt. The proposal for the creation of such a board was criticised within the conference, particularly by the Soviet Union, as a Western attempt to exploit the East, to impose collective

[1] New Zealand, H.R. Deb., vol. 309, pp. 1667–9.

imperialism, to deprive Egypt of much profit, and to intrude upon Egypt's sovereign rights. Much of this criticism was destroyed at its foundation, however, by the fact that amongst the eighteen powers supporting the proposals for international management of the canal were the countries already mentioned, countries which were not of the West but of the East, or which, having themselves achieved hard won independence, could not conceivably be suspected of contemplating imperialist exploitation of Egypt. . . .

The most substantial criticism of the eighteen Power proposals was in fact that relating to Egypt's sovereign rights, and it was to this aspect that, as New Zealand's representative, I gave particular attention. I am convinced that the proposals were fully consistent with Egyptian sovereignty. In a statement to the conference I expressed the view that in an age of interdependence such as ours, inflexible insistence on considerations of national sovereignty is out of date. I said that it was an abuse, not an exercise of national sovereignty, to seek to use resources in one's national territory in a way which could be harmful to the international community. I pointed out that what the eighteen Powers asked of Egypt was only the practical expression of what, in her undertaking to abide by the Suez Canal Convention of 1888, she had already promised to guarantee. The fact that no infringement of Egyptian sovereignty was contemplated by the eighteen Powers is, however, most convincingly established by examination of the procedure by which it was agreed that the proposal for an international authority should be submitted to Egypt.

In accordance with a declaration submitted to the conference by New Zealand, the eighteen Powers decided to appoint five from amongst their number to approach Colonel Nasser to place their proposals before him, to explain their purposes and objectives, and to ascertain whether Egypt would agree to negotiate a convention on the basis thereof. Certainly Egypt was to be asked to surrender sovereignty in favour of the international authority, but she was to do so by signing a treaty—an action which in itself would involve, in the highest degree, the exercise of Egyptian sovereignty.

. . . It is now clear that the discussions have failed. . . . It was understood that if Egypt was not prepared to negotiate upon the basis proposed by the London Conference a new situation would be created which must be considered according to the circumstances then prevailing. That situation now exists, and consultation is taking place among interested states, including New Zealand. . . .

3. Extracts from report of statement by the Prime Minister, Mr. S. G. Holland, 17 September 1956[1]

The Prime Minister, Mr. Holland, stated today that Cabinet had spent a great deal of the morning discussing the Suez situation in the light of messages received from various parts of the world over the weekend.

Mr. Holland said that New Zealand had the fullest sympathy with the British Government in their efforts to find an equitable solution of this problem, and New Zealand had accepted an invitation to attend the conference in London on Wednesday composed of representatives of the eighteen Powers which supported the proposals which Mr. Menzies and his committee had put to the Government of Egypt. . . .

'I should like to emphasize that the Government continues to regard the proposals for international management, which were supported by the overwhelming majority of the first London conference, as eminently fair and reasonable. President Nasser's refusal of the proposals has made it essential for user countries to explore other ways and measures to keep their ships moving through the Canal. I continue to hope that President Nasser may yet recognize the essential justice of the original 18-power proposals.'

The Prime Minister said that the New Zealand Government fully supported Sir Anthony Eden's statement in the House of Commons [I, Document 6, *supra*] when he said that it was the United Kingdom's intention if circumstances allow, that is except in an emergency, to refer any act of hostility or interference by Egypt to the United Nations. 'I feel the qualification expressed by Sir Anthony is an essential one, especially in view of the likely need in an emergency to take action in defence of British lives and property in Egypt.'

[1] *E.A.R.*, vol. VI, no. 9, September 1956, pp. 5–6.

V. SOUTH AFRICA

1. Extracts from speech by the Prime Minister, Mr. Johannes Strijdom, 18 August 1956[1]

. . . Our actions must be such as not to make enemies of the 200 million non-whites on the continent of Africa. . . .

. . . Apart from its strategic position between West and East, this area [near the Suez Canal] is also regarded as the most important gateway from the East to Africa.

In view of these facts I want to express the hope very strongly that in the present and any future negotiations between various countries directly concerned with the Suez Canal because of their shipping, an acceptable agreement will be reached which will obviate friction in the future. . . .

2. Extracts from statement by the Minister of External Affairs, Mr. Eric Louw, Pretoria, 5 October 1956[2]

We are anxious about the situation both from the standpoint of seeing peace preserved and the 1888 Convention maintained, keeping in mind that Egypt is the gateway to Africa. . . .

We see nothing illegal in Egypt's nationalization of the joint stock company registered in Egypt, but we are concerned that the 1888 Convention be observed.

To the present time it has been observed except in the case of the interference with Israeli shipping. But no action was taken by the Western Powers in this connection for years before the nationalization.

VI. INDIA

1. Extracts from statement by the Minister without Portfolio, Mr. V. K. Krishna Menon, at the 22-Power London Conference, 20 August 1956[3]

. . . We have reservations with regard to the composition and the character of this Conference, and we deeply regret the absence of Egypt from our midst. We have recorded the view before the Conference and afterwards, that no final solutions, or even approaches to final solutions, are possible without the participation of the country most concerned. . . .

---

[1] *The Star* (Johannesburg), 20 August 1956.
[2] Ibid., 5 October 1956.   [3] D.S.P. 6392, pp. 159–78.

My Government is particularly concerned to point out that the failure by us to resolve this problem peacefully would have consequences which would go far beyond Egypt or far beyond any of the countries concerned and, in the present state of the world, where no problem can be isolated from the context of international events and international relations generally, those consequences deserve our very serious consideration. . . . So far as my delegation is concerned it is not our business at the present stage to enter into discussion of the rightness or wrongness of other things. . . . That is to say, we would like to confine this question to the sole problem of how the functioning of the Suez Canal can be ensured if there is a danger to it, and secondly to how the difficulties that have now arisen can be resolved. Therefore, we must avoid any temptation to go into other fields of connected problems, either involving the prestige of countries, or their ambitions, or their fears, or of the merits or otherwise of one man or one government. . . . When we have to deal with countries, it is the approach of my Government that we have to take their internal structures and their administration, and their Governments and their leaders as they are; it is not possible for us to approach problems by first desiring a change of government or constitution or personnel in another country. . . .

. . . The question of nationalization, in the view of my Government, was an act which was within the competence of the Egyptian Government. I think however that my Government would like it to be stated that there are, in the manner in which the nationalization was carried out, features which have led to the present aggravated situation. We would like to have seen that nationalization carried out in the normal way of international expropriation, where there is adequate notice, and the way of taking over is less dramatic and does not lead to these consequences. . . .

. . . My delegation have deliberately refrained from putting forward suggestions which are too specific. The reason is that we believe that it is possible by negotiation to come to arrangements that are mutually satisfactory. We believe equally that the achievement of that end will be considerably obstructed by any attempt *ex parte* to prescribe this remedy, and therefore the proposals we are making are in the way of opening the door to bringing about these arrangements. Having said that, Sir, I want to read out the draft that my delegation has made which we consider would be a basis for negotiating a peaceful settlement. . . :

'Realizing that it is imperative that a peaceful and speedy solution to the situation concerning the Suez Canal in accordance with the principles of the Charter of the United Nations must be found and the way for negotiations opened without delay on the basis of:

'1. The recognition of the sovereign rights of Egypt.
'2. The recognition of the Suez Canal as an integral part of Egypt and as a waterway of international importance.
'3. Free and uninterrupted navigation for all nations in accordance with the Convention of Constantinople of 1888.
'4. The tolls and charges being just and equitable and the facilities of the Canal being available to all nations without discrimination.
'5. The Canal being maintained at all times in proper condition and in accordance with modern technical requirements relating to navigation.
'6. The interests of the users of the Canal receiving due recognition.

'Recalling that the Convention of 1888 sets out as its purpose the establishment of a "definitive regime destined to guarantee at all times and for all powers the free use of the Canal".

'Taking note that Egypt has declared even as late as 31 July, 1956, that she is determined to honour all her international obligations and recalling the Convention of 1888 and the assurances concerning it given to the Anglo-Egyptian Agreement of 1954.

'Make the following proposals in the belief that they will provide the basis for negotiation for a peaceful settlement.

'*The Freedom of Navigation, Charges, Maintenance and Efficiency*

'I. That the Constantinople Convention of 1888 be reviewed to reaffirm its principles and to make such revisions as are necessary today, and more particularly incorporating the provisions in regard to just and equitable tolls and charges and the maintenance of the Canal as set out in 4 and 5 above.

'*The Conference*

'II. That all steps not excluding a conference of the representatives of the signatories of the 1888 Convention and all user nations of the Canal for the above (I) be considered.

'*International User Interests:*

'III. That consideration be given, without prejudice to Egyptian ownership and operation, to the association of international user interests with the 'Egyptian Corporation for the Suez Canal'.

'IV. A consultative body of user interests be formed on the basis of geographical representation and interests charged with advisory, consultative and liaison functions.

'*United Nations:*

'V. That the Government of Egypt transmit to the United Nations the annual report of "The Egyptian Corporation for the Suez Canal".'

. . . These are the submissions that my delegation makes, Sir, and we want with the utmost sincerity to request you and this conference not to embark on steps that would mean dictation. It is not possible, whether it be in the international authority, or whether it be in the maintenance of an uninterrupted service, to have a peaceful arrangement that does not bring Egypt into negotiations.

What is the alternative? The alternative is the imposition by Great Powers of this position, with all the consequences of the use of force that would necessarily be involved. We of the countries of Asia are alarmed at the situation—I do not say others are not—we are aware of the impact and of the reaction of nationalism to attempted imposition. We are aware of the expressions of nationalism, sometimes desirable, sometimes undesirable, in the countries of Asia and Africa, particularly in the Arab countries. Therefore we say that any proposals that emerge from this conference, that is by way of making the decision here, even if it is called a basis of negotiation, are likely to impede the path towards settlement. . . .

Once again I repeat, my Government and my country are not unconcerned in this matter. We would be the victims whether the canal went out of service by the action of Egypt or by the act of hostility or war or conflict. Whatever the cause, the results will flow to us from the fact that the service goes out of operation. But we are even more concerned, if I may say so, at the dreadful consequences which would in effect reverse the currents that have been set in motion in regard to the relations between the Western countries and peoples, including the peoples of Asia and Africa, during the last thirty or forty years. We would hate to see the reversal of these currents, we would like that the reactions of peoples should be taken into account and anything that is done here would have to take this factor into account. And we bring this to you, Sir, not as parochial citizens of one country or as parochial people in one part of the world, but with the full responsibility and realization of our obligations to the international community and our appreciation of the ways in which settlement can be reached. I plead with you to adopt the part of conciliation and not the part of dictation.

2. Extracts from statement by the Minister without Portfolio, Mr. V. K. Krishna Menon, at the 22-Power London Conference, 22 August 1956[1]

. . . My Government is firmly convinced that, given the right approach, the Governments concerned, the Government of Egypt and the others, would make a contribution in a calmer atmosphere in the weeks to come. It is possible to establish the kind of arrangement that, while everybody may not regard it as the most ideal one (neither Egypt nor the others), will enable the Suez Canal to function satisfactorily. . . . It is with that in view, Sir, that I, even at this late stage, ask this Conference not to endorse proposals that would draw a barrier across the possibility of co-operation.

The Indian Government does not ask you to accept our views as to how it should be done. I do not ask the distinguished representative of the United Kingdom or anyone else to accept our views on what may be the intentions or the propensities of the Government of Egypt. We are not here to stand guarantee for them; but we do say it is the considered view of the Government of India that this matter should be shifted from this struggle to force the matter one way or the other. . . .

And, Mr. Chairman, I want once again to remind you, as I have already reminded you I do not know how many times in the last three or four days, that there are other issues to be taken into account. There is a continent that is awake, and whether they are objectively in the right or not, it is very wrong, it is very dangerous, to disregard their susceptibilities. That does not mean we are to pander to every kind of mob opinion that turns up from everywhere. In dealing with countries formerly subject to empires, where the rights of nationalism and sovereignty have received an

[1] D.S.P. 6392, pp. 238–49.

exaggerated outlook in many cases, the way to deal with them would be to convince them of their interest in co-operation, and that can only be done by bringing them into the field of negotiation in the first instance. . . .

3. Extracts from statement by the Prime Minister, Shri Jawaharlal Nehru, in the Lok Sabha, 13 September 1956[1]

. . . I should like to say a few words in regard to the latest developments relating to the Suez Canal issue. . . .

The House knows of our earnest efforts to bring about a negotiated settlement. . . . It has been clear to us that any other approach or any attempt to impose a decision would not only not bring about the results aimed at but might lead to much graver consequences, the extent of which it is not possible to foresee. At the Conference held in London we pleaded with all the force at our command for steps to be taken to bring about negotiations, and certain broad proposals were set out by us [Document 1]. We were supported in these proposals by Ceylon, Indonesia and the Soviet Union. The majority of those present at the Conference, however, adopted, as is known, a different line. . . .

The action proposed to be taken by the [governments of the United Kingdom, the United States and France] which purports to be in the interests of the users of the Canal and to maintain the freedom of use of the Canal seems, to say the least, surprising, and the consequences that may flow from it may well be very grave. One thing is clear and that is that the action proposed is not the result of agreement, co-operation or consent, but is to be taken unilaterally, and thus is in the nature of an imposed decision.

The Government of India deeply regret this development, which is very unusual and which will render peaceful settlement more difficult of realisation. It is not calculated to secure to the users peaceful and secured use of the Canal which should be and is what is required by the users and the international community.

The Menzies Mission which recently visited Cairo asked the Egyptian Government to accept international control of operation and administration and the establishment of an international corporation displacing the Egyptian National Corporation. Egypt has declined to accept them as being contrary to her sovereign rights. . . .

[1] Lok Sabha Deb., 1956, Part II, vol. VIII, no. 45, coll. 6963–7

The reply of the Egyptian Government has opened a way to negotiations. In the view of the Government of India, such negotiations could have led to a settlement which would have met all requirements of the users and the international community without prejudice or derogation to the sovereignty of Egypt. . . .

I have in the last few days communicated to the Prime Minister of the United Kingdom and the President of the United States our view that the situation that emerged after the Menzies Mission and the statement made by the Egyptian Government accepting all international obligations and inviting negotiations opened a way to settlement.

We appeal to both the United Kingdom and the United States to consider all this and enable the development of negotiations which will lead to a settlement. We hope that despite all that has happened and the tensions that have been engendered the path of peace will be followed. There is here no question of appeasement of one side or another as what is to be sought and can in our view be obtained is a settlement satisfactory and honourable to all concerned.

The Government of India earnestly hope that the appeal we have made will not be in vain. The Government have right through the course of this development used their influence with all parties for restraint, negotiations and a peaceful settlement. To seek to impose a settlement by force or by threats of force is to disregard the rights of nations even as the failure to observe international treaties and obligations would be. The Government also regret to learn from Press reports that pilots of British, French, Italian and other nationalities are being withdrawn. This is an action not calculated to promote the use of the Canal and is not in the interest of user nations. The Government of India are desirous that no statement of theirs should come in the way of the efforts to lower tension and to open the way for negotiations. But, they cannot fail to point out that the steps announced to assume the operation of the Canal without the consent and co-operation of the Egyptian Government are calculated to render a peaceful approach extremely difficult and also carry with them the grave risk of conflict. I should like to say that I have read the report of Sir Anthony Eden's speech [I, Document 6, *supra*] with surprise and regret as it appears to close the door to further negotiations. The action envisaged in it is full of dangerous potentialities and far-reaching consequences. I earnestly trust that, even now, it is not too late to refrain from any such action and to think more in terms of a peaceful negotiated settlement which can only achieve the results aimed at in regard to the proper functioning of the Suez Canal for the good of all countries concerned as well as for the maintenance of friendly relations in the Middle Eastern region and the whole of Asia. . . .

4. Extract from statement by the King of Saudi Arabia, Ibn Saud, and the Prime Minister, Shri Jawaharlal Nehru, 27 September 1956[1]

. . . Both Saudi Arabia and India are deeply interested in a peaceful settlement of the dispute relating to the Suez Maritime Canal, which is a waterway of vital importance to their own economic well-being as also to that of many other countries in the world. There can be no settlement of the dispute by methods of conflict or by denial of the sovereign rights of Egypt over the Suez Canal. The right of all countries to free navigation through the Canal on payment of reasonable dues has been accepted. His Majesty the King and the Prime Minister are convinced that, in spite of the difficulties and tensions that have arisen over this question, it is possible to reach a settlement negotiated between the parties concerned without any derogation from Egyptian sovereignty and authority and maintaining the interests of other countries in the unrestricted use of the Canal as an open waterway. They share the hope that there will be no recourse to political and economic pressure in dealing with this matter, as such pressure would only retard a peaceful settlement, apart from having other undesirable and far-reaching consequences. . . .

5. Proposals of the Government of India, 25 October 1956[2]

Desirous that a peaceful and adequate solution of the situation, which has arisen in respect of the Suez Canal, in accordance with the Charter, the principles and purposes of the United Nations, and consistently with the sovereignty of Egypt, must be found and the way for negotiations opened on the basis of

---

[1] *Foreign Policy of India: Texts of Documents 1947–59* (New Delhi, 1959), p. 253.
[2] Ibid., pp. 257–9.

1. The recognition of the Suez Canal as an integral part of Egypt and as a waterway of international importance.
2. Free and uninterrupted navigation for all nations in accordance with the Convention of 1888.
3. The tolls and charges being just and equitable and the facilities of the Canal being available to all nations without discrimination.
4. The Canal being maintained at all times in proper condition and in accordance with modern technical requirements relating to navigation, and
5. Co-operation between the Canal Authority and the users of the Canal receiving due recognition.

Recalling that the Convention of 1888 sets out as its purpose the establishment of a definitive regime with a view to guaranteeing for all times and for all the Powers the free use of the maritime Suez Canal, we make the following proposals as the basis for a peaceful settlement:

1. The Convention of 1888 to be re-affirmed and also reviewed and revised to bring it up-to-date.
2. The review and revision of the Convention to provide for (*a*) maximum of tolls leviable by Egypt as under the last agreement between the Egyptian Government and the Suez Canal Company, (*b*) Egypt's responsibility for the maintenance and development of the Canal in accordance with modern requirements, more particularly the carrying out of the eighth and ninth programmes as the minimum and during the period as set out in the programmes, and (*c*) Egypt to transmit to the United Nations for information the annual report of 'The Suez Canal Authority'. (*a*) and (*b*) above will be in the schedules or annexures to the Convention.
3. The signatories to the Convention to affirm their respect for the Charter and the principles and purposes of the United Nations in the observance and execution of the Convention by each and all of them.
4. (*a*) Disputes or disagreements arising between the parties to the Convention and in respect of it shall be settled in accordance with the Charter. (*b*) Differences arising between the parties to the Convention in respect of the interpretation of its provisions, if not otherwise resolved, will be referred to

the International Court of Justice under Article 36 of its statute or, by agreement, to an appropriate organ of the United Nations.

5. The Convention as thus reviewed and revised to be registered with the United Nations.

Compensation and Claims: The question of compensation to be paid by Egypt and claims by Egypt against the parties arising from nationalization will, unless otherwise agreed to between the parties concerned, be referred to and settled by arbitration.

Co-operation:

1. The administration, operation and management of the Canal is vested by the Egyptian Government under Egyptian law in 'the Suez Canal Authority'.

2. (*a*) The Canal Authority, with the approval of the Egyptian Government, will recognize a Users' Association for the purpose of promoting co-operation between the Canal Authority and the users. (*b*) The functions of the Association will be consultation and liaison.

3. (*a*) Joint sittings will be held between the representatives of the Canal Authority and the representatives of the Users' Association periodically or at the request of either side; (*b*) the representatives of the Users' Association will include the principal Users and provide also for geographical representation and be constituted on the following basis: France, the U.K., the U.S.A., the U.S.S.R., Egypt, India, Japan, one representative from Australia, one from South-East Asia, one from the Middle East, one from Africa, one from Eastern Europe, one from Southern Europe, one from Northern Europe, one from Western Europe, and one from Latin America; (*c*) The purpose of holding joint sittings will be to promote and effect co-operation between the Canal Authority and the Users; (*d*) the Users' representatives at the joint sittings may discuss and make recommendations on all matters affecting or concerning User interests, more particularly: tools; condition of the Canal; observance of the Canal Code and Breaches thereof by either side, and complaints by either side; (*e*) The Canal Authority may refer to the Users' Representatives at the joint sittings any matter for discussion or advice; (*f*) The Users' Association or its representatives at

the joint sittings will not in any way interfere with the administration; (*g*) The constitution of the Canal Authority which is regulated under Egyptian law cannot be within the competence of the Users' Association or its representatives at the joint sittings.

Tolls: The Canal Authority in effecting any increase in tolls beyond an agreed limit, say, within any 12 months, will do so only by agreement at the joint sitting; in case of disagreement, the matter will be referred to arbitration.

Discrimination: Allegations or complaints of discrimination will be referred to the Canal Authority by the aggrieved party; if not resolved, the aggrieved party may take such allegations or complaints (*a*) either to the appropriate Court in Egypt, (*b*) or to Users' representatives at the joint sittings. In the event of [the] matter not being resolved at the joint sitting, either side (the Canal Authority or the representatives of the Users) may refer it to arbitration.

Canal Code: The regulations governing the Canal, including the details of its operation, and the obligations of the Authority and the users as well as the penalties for breaches thereof by either side, will be contained in the Canal Code which will be the law of the Canal.

The Egyptian Government has decided, in the exercise of its authority, to appoint high level experts through the United Nations to the three main departments of the Canal Authority for three years in the first instance.

VII. PAKISTAN

1. Extracts from statement by the Minister of Foreign Affairs and Commonwealth Relations, Mr. Hamidul Huq Choudhury, at the 22-Power London Conference, 18 August 1956[1]

... I should like to impress our Government's view that the act of nationalisation of the Suez Canal Company on the part of Egypt was an exercise of her sovereignty. It is the considered view of my Government that irrespective of other issues involved the sovereign right of Egypt in her dealing with a commercial concern within her own territory cannot be challenged or contested. I do not

---

[1] D.S.P. 6392, pp. 150–3.

propose to comment on the merits of the timing of the action, nor do I intend to dwell on the reactions of the various participants. . . .

My delegation . . . proposes for the consideration of this conference: That the nationalisation of the Universal Suez Canal Company by Egypt be accepted as *fait accompli*, whether we like it or not: financial settlement and questions of compensation can be considered separately between the parties and hereafter. (2) An effective machinery be set up in active collaboration with Egypt to ensure the efficient, unfettered and continuous freedom of navigation, without discrimination and within the capacities of the trade of all nations, while at the same time the legitimate interests of Egypt should be fully protected. (3) A Committee be set up to negotiate with Egypt by this conference on the basis of our second proposal, and to report back to this conference the result of their negotiations.

In conclusion, Mr. Chairman, I would like to emphasise that Pakistan is bound by close ties of common faith, religion and culture to the countries of the Middle East, and Egypt particularly, and as such it cannot remain indifferent to a situation which may adversely affect the welfare and progress of these countries or jeopardise their legitimate interests and aspirations.

2. Extracts from statement and proposals by the Minister of Foreign Affairs and Commonwealth Relations, Mr. Hamidul Huq Choudhury, at the 22-Power London Conference, 21 August 1956[1]

I have got a few proposals to make on behalf of my delegation and the delegations of Ethiopia, Iran and Turkey. . . .

The countries of Asia have just emerged from a long period of foreign domination, and it must be said to the praise of contemporary statesmanship of the Western world that they have shown great wisdom and a profound appreciation of the realities of our times in so far as they have recognised, and given effect to their recognition, of the need for the political emancipation of subject nations. As is only natural, there is a tremendous upsurge of nationalism in the countries of the Middle East and Asia. We should be failing in our duty if this fact was not recognised, and I would appeal to the statesmen of the Western countries to recognise this fact, and help the cause of peace and welfare of mankind

[1] D.S.P. 6392, pp. 182–7.

by giving due weight and consideration to the emotion and sentiment of peoples who have recently achieved their independence.

Much has been achieved over the last decade to allay fear and suspicion on the part of these peoples, but much more remains to be done to build confidence among these nations and establish a sense of security, and of their sovereign dignity. I would therefore request the distinguished delegates to bear this background in mind while considering the proposals which have been placed, and which we are also placing, for consideration.

We must start with the unqualified acceptance of negotiation as the basis of settlement with Egypt. There should be no question whatsoever of any imposition of the will of this conference on the Egyptian Government. My Government joined this conference on that assurance and on this basic understanding. I have made it clear that we will only participate if a negotiated peaceful solution is the aim of the conference. In reaching a solution the rights of all parties will have to be fully ensured.

I am happy to see that Mr. Dulles has emphasised the need for the rejection of any idea of imposing any settlement on Egypt. . . .

We have considered very carefully the various proposals that have been placed before this conference, and have come to the conclusion that there is considerable unanimity in the matter of essentials. The main conflict seems to centre round the composition and authority of the body which shall be responsible for the working of the Canal as an international highway, and, if the interests of the maritime nations are to be associated with that body, the question of how this shall be done arises. It is agreed by all that user interests should be represented in the body set up in the management of the Canal. I quote the proposal of the Indian delegate [V, Document 1]: 'That consideration be given, without prejudice to Egyptian ownership and operation, to the association of international user interests with the Egyptian Corporation for the Suez Canal.' I would also like to quote from the proposal of Mr. Dulles: 'There should be established by convention, institutional arrangements for co-operation between Egypt and other interested nations in the operation, maintenance and development of the Canal.'

I would submit for the consideration of this conference that there is no substantial difference between the proposals of the

Mcs

delegates of the United States, India, and the proposals that we are making. . . .

Mr. Krishna Menon's proposals have visualised the setting up of a consultative body in addition to the association of international user interests for the Egyptian Corporation for the Suez Canal. It is the opinion of my delegation that the association of these interests in the body set up for the management would be the best solution as it would ensure the restoration of confidence among the maritime users of the Canal and would have none of the fundamental disadvantages of the system of dual management. The setting up of dual bodies would inevitably complicate matters by setting up discord and interrupting the smooth functioning of the machinery.

Acceptance by Egypt would be the cornerstone of any settlement, and I would request the conference to consider the proposals of Mr. Dulles as a basis of negotiation with Egypt, for the purpose of securing the co-operation of Egypt, by securing its sovereign rights and interests. We can only achieve our object after obtaining Egypt's acceptance of the proposal. My delegation proposes, together with the delegations of Ethiopia, Iran and Turkey, the following amendments to the proposals of Mr. Dulles which, if accepted, would make the proposals acceptable to all of us. I will refer now to the changes we propose to be introduced. . . .

. . . Members will kindly refer to Mr. Dulles' proposal 2, subparagraph (e). This reads: 'Payment to the Universal Suez Canal Co. of such sums as may be found its due by way of fair compensation'. We want to take this phrase out of this paragraph, which we submit should be entirely devoted to future arrangements, and introduce this subject in the preamble. . . .

And then we take up paragraph 2, wherein it is stated: 'Such a system . . . should assure.' We wish to suggest that . . . after the words 'Such a system' the following words should be introduced: 'which would be established with due regard to the sovereign rights of Egypt' . . .

In conclusion I would like to emphasise that this conference should do everything possible to achieve unanimity. We would be doing a great disservice to the cause of promoting international peace and the lessening of tension if we did not give due recognition to the necessity of arriving at a substantial agreement on the principles that are to govern the future negotiations with the Government of Egypt. . . . The problem which confronts this conference is one which affects every nation equally. If we fail to solve it peacefully we will all suffer equally. On the other hand if we succeed we will lay a foundation of the future settlement of other problems in other spheres on the basis of equity, justice and peace. . . .

3. Statement by Mian Mumtaz Mohammad Khan Daultana, ex-Chief Minister, West Pakistan, 15 September 1956[1]

The crisis over Suez has reached a stage when no one in Pakistan can afford to ignore it. The world is faced with an immediate peril of war, and in an issue where Pakistan should all along have played, and can play today, a decisive role. Not only as a Moslem country, but also as one pledged to an international order based on justice and liberated from all taint of domination and colonialism, Pakistan must support Egypt.

This support means acceptance of Egypt's invitation to the Conference called by her regardless of who else attends or does not attend. It means using all our influence and taking an initiative, which we have so often left to others to take, to convince and persuade friendly Powers, particularly the U.S.A., that war over the Suez would be a terrible human disaster.

Happily there are indications that give rise to the hope that Mr. H. S. Suhrawardy, with his political maturity, his training as a politician, and his commitment to the manner of thought [sic], will give a lead in harmony with the sentiments of his people and the interests of world peace, and will not hesitate to retrace false steps that may have been taken hitherto.

4. Statement by the Minister of Foreign Affairs and Commonwealth Relations, Mr. Malik Firoz Khan Noon, 16 September 1956[2]

There are three different subjects to be discussed at this conference: First, the Menzies Report; secondly, Colonel Nasser's proposals; and thirdly the proposed Users' Association.

We are a party to the five power proposals, which were agreed upon at the last Conference and on the basis of which Mr. Menzies

---

[1] *Dawn*, 16 September 1956.   [2] *Dawn*, 17 September 1956.

led his mission to Cairo. We have no doubt that Colonel Nasser's counter-proposals will be given careful thought.

In so far as the Users' Association is concerned, we are in the first place not sure what exactly is contemplated and secondly we have satisfied ourselves that our attendance at the Conference does not, in any way, bind us to be a party to this Association.

In this context, we have noted with particular satisfaction the British Prime Minister's statement of September 12 [I, Document 6, *supra*] affirming that the Association cannot operate in practice without Egyptian acquiescence and that if any untoward incident arises, the three sponsoring Powers will take the matter to the Security Council.

We cannot, in any case, associate ourselves with the use of force or with any solution imposed on Egypt against her will.

We feel that in view of the above considerations and assurances given by the sponsoring Powers, we may be able to play a useful role at the Conference.

Pakistan has throughout stood for a negotiated settlement satisfactory to the parties concerned. It is in this spirit that we have already agreed to attend the Conference to be called by Egypt provided a majority of the countries similarly addressed agree to do so.

Meanwhile, in London, my delegation and I shall continue to strive to achieve a peaceful solution of this problem.

I take this opportunity to re-affirm once again our determination that Pakistan will not associate itself in any way with the use of force.

5. Statement by the Minister of Foreign Affairs and Commonwealth Relations, Mr. Malik Firoz Khan Noon, 18 September 1956[1]

We are out to have a peaceful understanding of the dispute. If any compromise can be arrived at with the willing co-operation of Egypt and the Canal users, no one will be more pleased than I.

But if the Canal Users' Association have any intention of enforcing their wishes on Egypt, that, in our view, is against the United Nations Charter.

We are members of the United Nations, and have pledged

[1] *Dawn*, 19 September 1956.

ourselves to resolve peaceful disputes peacefully. My Government and people would be against the imposition of any plan by force. The Pakistan Government and people wish to co-operate in any scheme which leads to a peaceful settlement of the Suez problem.

We are frightfully concerned with the free passage through the Suez Canal, but it is necessary that whatever solution is arrived at must be arrived at peacefully.

6. Extracts from speech by the Minister of Foreign Affairs and Commonwealth Relations, Mr. Malik Firoz Khan Noon, at the Second London Conference on the Suez Canal, 19 September 1956[1]

My delegation would like to commend to the consideration of the Conference that the users should adhere to their resolve to arrive at a peaceful settlement of this dispute through negotiations with Egypt. The scope of negotiations with Egypt might be somewhat enlarged. Since any settlement that is to be arrived at in this behalf [sic] has to be between the users and Egypt, of which the Canal is an integral part, ways and means should be found to bring about such a conference. Should President Nasser see fit to reject this proposal also, then the sponsoring Powers should immediately take the matter to the Security Council, which is already seized of the situation on Egyptian request.

. . . Being located in the Middle East region and bound to it by indissoluble ties of faith and culture, my country and people are vitally concerned with the maintenance of peace in the region. An emergency or untoward incident can conceivably arise in the present situation charged with risk as it is. Even if that were not to come to pass, the people of Pakistan, as indeed people throughout Asia and the Middle East, rightly or wrongly believe that it might. They also believe that the present proposal [for setting up the Suez Canal Users' Association] means an imposed settlement to which we have declared our opposition all along. For this reason I regret to inform the Conference that my country cannot associate itself with the proposal.

. . . My country irrevocably stands for a peaceful settlement and would in no circumstances associate itself with any proposals which imply the use of force. . . .

[1] *Dawn*, 20 September 1956.

7. Extracts from statement by the Minister of Foreign Affairs and Commonwealth Relations, Mr. Malik Firoz Khan Noon, 3 October 1956[1]

My attention has been drawn to a statement made by the Bharati Home Minister, Pandit Govind Vallabh Pant, on September 28, in Srinagar. . . .

. . . Pandit Pant says that the Muslim world loves Bharat because of their stand on Suez. I can say without fear of contradiction that Bharat has followed a double-faced policy of hunting with the hound and running with the hare. Why has Bharat propagated that the matter of tolls cannot be left in the hands of Egypt? I am not discussing the merit of this statement. But after this how can Bharat claim to be a friend of Egypt? . . .

If it is true, as has been suspected, that Bharat advised President Nasser to raise the Canal issue, it was a double-edged sword: if, in this controversy, Egypt's freedom was destroyed (which, thank God, has been saved by Pakistan's intervention), Bharat would have said: 'Thank God! One more Muslim country wiped off the face of the earth.' But if succeeded [sic], Bharat would claim the credit for it.

Egyptian Muslims are our brothers and let not Pandit Pant forget that blood is thicker than water. Our soul is one with the Muslims of the world. . . .

The part played by Pakistan in the Suez Canal dispute is well-known to the Egyptian Muslims and the world. . . .

VIII. CEYLON

1. Extracts from statement by the High Commissioner in the United Kingdom, Sir Claude Corea, at the 22-Power London Conference, 18 August 1956[2]

. . . I want to stress at the very beginning our great satisfaction that this conference has been called because it demonstrates once again, and very clearly, that the world today is tending towards the complete acceptance of the principle . . . that when these international disputes arise they must be settled by the method of negotiation. . . . We are proud as a very young member of the Community of Nations to be associated with this conference. . . .

. . . I favour an approach which will lead to a kind of compromise, which will recognise the factual position of Egyptian

[1] *Dawn*, 4 October 1956.   [2] D.S.P. 6392, pp. 136–44.

sovereignty and at the same time provide machinery which will enable the canal to be operated in a satisfactory way. For this purpose we have before us the United States plan. . . . The position I want to put to this conference is that there should be such modification of this proposal that we should proceed first from the basis of the sovereignty of Egypt. . . .

Now, fellow delegates, when you speak of sovereign rights of countries, there is an element of difference between what it means to old-established sovereign countries and newly-established sovereign countries. I quite concede that these newly sovereign countries are perhaps a little too jealous and are too apt to stick to the idea of their rights of sovereignty, but this is inescapable. This idea of the sovereign right of a country to a new country is something far more important; . . . far more importance is attached by those countries than in the case of the older sovereign countries. . . . There is an element not only of sentiment but of the idea of patriotism, which may be a wrong idea, let me concede for the purpose of this argument, but which is nevertheless still there and operating very strongly in the minds and hearts of the people. That is a factor which must be taken into account when you think of a situation of this kind and think, after all, what does it matter if the canal can be operated in a practical satisfactory manner, if it ensures to all of us, to me and my country, which requires the canal as much as anybody else, in a way that will give efficient service—is not that sufficient and why bother about questions of sovereignty? . . . We can brush these things aside, we can try to enforce or impose a settlement of our own thinking. We may succeed for a time in doing that, but the embers of resentment which will be kindled as a result of that action will not only not die but will go on being kindled and kindled until they result in a conflagration. . . .

2. Extracts from statement and proposal by the High Commissioner in the United Kingdom, Sir Claude Corea, at the 22-Power London Conference, 21 August 1956[1]

Nothing would give my delegation greater pleasure than to be able to join with all our fellow delegates to make a unanimous recommendation as to the lines on which negotiations should be undertaken. We have listened very carefully and given very

[1] D.S.P. 6392, pp. 218–25.

grave consideration to the various statements made by the different delegations. We are still not without difficulty in reconciling the two clear points of view which have been presented to us. . . .

The amendment proposed by the Hon. Foreign Minister of Pakistan [VI, Document 2, *supra*], supported by some of his colleagues, to our mind does not make any material alteration in substance of the proposal of the Secretary of State for the United States. As has been said, it might have improved the draft; it might have given emphasis to the spirit of the United States declaration, but it has not materially changed it. We were hoping that there would be some material change which would have permitted us to accept it; but I am sorry, so far as the amendment goes, we still have the main original United States statement.

There is one point I would like to refer to at the very beginning in the now amended proposals. I refer to Article 2 of the amendment submitted by the Pakistan delegation. It reads like this: 'Such a system, which would be established with due regard to the sovereign rights of Egypt, should assure. . . .' What is meant by the sovereign rights of Egypt? . . . Clearly, among others, territorial rights; and territorial rights would include right to the canal which, as has been admitted, is an integral part of Egypt. If it concedes this territorial right as one of the sovereign rights which are to be paid due regard to, then it means we must conserve the right of Egypt to the control of that part of her territory which is the Suez Canal. Is that meant or intended in this statement? If it is so, it affects the whole question. . . .

The Indian proposals [V, Document 1, *supra*] leave the position open . . . but this [United States] proposal [as amended by Pakistan] definitely takes the position of international control. . . . That immediately takes away the right of the Government of Egypt to any action in regard to the operation of the Canal. Does not that contravene and contradict the statement here, that any system established will be established with due regard to the sovereign rights of Egypt? How can we reconcile the two positions? . . .

. . . We are very sorry to mention this. We would have preferred to preserve our silence. We would have preferred to go with the rest of you in subscribing to these proposals, but we just cannot do that as things are now. I would suggest therefore that something of this kind be considered even at this late stage . . . :

'The Governments represented at the London Conference on the Suez Canal: Seeking in conformity with the purposes and principles of the United Nations a peaceful solution of the situation that has arisen in connection with the Suez Canal: Suggest for negotiation with the Government of Egypt the following proposals:

1. They recognise the Sovereignty of Egypt over the Suez Canal, which is an integral part of Egypt, and its legal right to nationalise the Universal Suez Canal Company.
2. They also recognise that the Canal is an international waterway and is subject to the Convention of 1888 which guarantees at all times and for all the Powers its free use.
3. They further recognise that it is in the interests not only of Egypt but of all user countries that:

   (a) The Canal should be operated and maintained efficiently and improved to meet the needs of all users now and in the future.
   (b) That the dues collected for use of the Canal shall be reasonable and fair, and sufficient for the proper maintenance and improvement of the Canal and consistent with the right of Egypt to obtain a reasonable income from the international use of the Canal and to apply it for the benefit of Egypt.

4. They therefore recognise that it is to the advantage of Egypt and all users that in the operation of the Canal there should be co-operation between them with a view to obtaining the fullest support of technical skill and financial aid in the operation of the Canal.
5. To this end they recommend that the Government of Egypt should consider the establishment by it of a Board or Corporation consisting of representatives of Egypt and of all user Countries, in which will be vested all rights and facilities to operate the Canal efficiently. The Board or Corporation shall be under the general direction of the Government of Egypt.'

3. Statement by the Prime Minister, Mr. S. W. R. D. Bandaranaike, 23 August 1956[1]

The position that has arisen at the London Conference is not surprising, nor is it surprising that all the members of the Conference could not come to a unanimous decision. That position was expected by me.

It is clear that even opinion in England has strengthened against the hasty use of force.

The main issue regarding the settlement of the dispute still remains; an issue on which opinion is sharply divided, i.e., insistence by the Western powers on international control, while certain other countries such as ourselves feel that an insistence on this will make a negotiated settlement very difficult if not impossible. We feel that if nationalization of the Suez Canal is recognized even as a *fait accompli* what follows from that position, i.e., the control of operation by Egypt, must also be recognized. On the other hand we realize that international interests particularly in the use of the Canal must also be safeguarded.

That is why Sir Claude Corea, on my instructions, suggested at the Conference that the nationalization of the Canal by Egypt together with implicit control of the Canal by Egypt should be recognized, and that another international convention be entered into between Egypt and the other countries concerned, which would reiterate Article 1 of the Convention of Constantinople in 1888, i.e., guaranteeing free use by all ships through the Canal, and also under the proposed convention to set up a consultative or advisory international body which Egypt could consult on matters of importance relating to the operation of the Canal. If any dispute arose between Egypt and the consultative body provision could be made for suitable arbitration by the United Nations.

I think that on these lines a negotiated settlement would be possible. But if there is insistence on international control of the Canal, a serious deadlock is likely to arise.

I hope that when the representatives to the London Conference go to Egypt to place their proposals before the Egyptian Government, it will be done in a manner to keep the door open for further negotiation after the reactions of Egypt are obtained and assessed.

[1] *Ceylon Daily News* (Colombo), 24 August 1956.

4. Statement by the Prime Minister, Mr. S. W. R. D. Bandaranaike, 13 September 1956[1]

Britain and France have taken a giant stride towards war, which has given rise to a situation from which it is difficult to see how war can be avoided.

The setting up of the 'Canal Users' Association' means in fact that Britain, France and America are going to operate the Canal, and surely the President of Egypt and the Egyptian Government cannot help but resist it with force of arms. The fat is in the fire.

5. Statement by the Leader of the Opposition, Dr. N. M. Perera, 13 September 1956[2]

It is clear from the statements of Sir Anthony Eden that both France and the United States are working jointly with regard to the new policies set up by him. The least that could be said is that the intended action, far from being a solution to the Suez Canal problem, is more likely to aggravate the situation and create explosive possibilities.

Instead of actually declaring war, Sir Anthony proposes to adopt a line of action that would give the appearance of Egypt as an aggressor and thus provide the excuse for those imperialistic powers to intervene and deprive Egypt of her legitimate sovereign rights over her own territory. The proposed new 'Canal Users' Association' is nothing more than a parallel organization to run and control the Canal. Sir Anthony must know that this is provocative and an incitement to Egypt to take counter measures.

If great powers are to be allowed to take the law into their own hands to safeguard what they consider to be their own interests, then we might as well scrap the United Nations organization. England has no excuse for not referring the issue to the U.N. for a just solution in the question.

I can only hope that world public opinion would be mobilized to prevent England and France from following this disastrous course of action which they intend to take. In this setting of a possible war in the Middle East, it becomes more urgent than ever for Ceylon to take complete control of her bases so as to prevent Trincomalee and Katunayake being used by England or any other power and thus drag us into the vortex of war.

[1] *Ceylon Daily News*, 14 September 1956.   [2] Ibid.

6. Questions by Mr. Leslie Goonewardena and answers by the Prime Minister, Mr. S. W. R. D. Bandaranaike, in the House of Representatives, 21 September 1956[1]

Mr. Goonewardena: . . . We all know—and it is a matter of great concern to us—that with the decision of the Western Powers to form a Users' Association in connection with the Suez Canal the possibility of war has unfortunately come closer; and in so far as we are concerned the matter that affects us most is the existence of foreign bases in Katunayake and Trincomalee. The newspapers reported a few days ago that the Hon. Prime Minister was making efforts to obtain an assurance from the United Kingdom in regard to this matter. We do not, of course, know how much such an assurance will be worth when it comes to a question of actual war but he was at least, in the present situation, trying to obtain an assurance that in the event of the outbreak of hostilities the bases here would not be used for war operations. I would like to know if he could tell us whether there have been any further developments in the situation; whether he has received a reply or is expecting one in the near future in regard to this matter.

There is also another matter which, although not of such great importance, is worrying the minds of not a few in this country. At an interview the Hon. Prime Minister gave the press—I think it was one of his weekly 'Kiribath Interviews'; I believe it was at his last interview—he stated that our attitude towards the Suez Canal issue was one of neutrality. But in the minds of certain people, and I must confess in my own mind too, I begin to wonder whether that attitude of the Government has changed. I have a very clear recollection that on this question the Government took the position at the very commencement that the attitude of the Egyptian Government to nationalise the Suez Canal was its sovereign right to do so [sic]. It is not a question of neutrality with regard to hostilities but a question of neutrality in relation to the rights and wrongs which is another matter. Since this matter, as I know, created a certain amount of, shall I say, speculation, I shall be glad if the Hon. Prime Minister will make use of this opportunity to say that there is no change in the position that he had taken up and that he will continue to adhere to the position that the Egyptian Government, as a sovereign nation, has the right to nationalise the Suez Canal. . . .

[1] Ceylon, H.R. Deb., vol. 27, coll. 422–6.

The Hon. S. W. R. D. Bandaranaike: . . . With regard to the question of our bases, the House knows that it was decided that we should take them over. The British Government agreed to that. I have received today from them particulars regarding certain facilities that we decided to discuss and which will be discussed very early in London. We will send a deputation from here to discuss those matters.

In the meanwhile, with regard to the uses of the bases, I had sought for and received an assurance from the British Government at the outset of the Suez Canal dispute, that our bases at Trincomalee and Katunayake would not be used by them for any purpose connected with any military action in the event of the outbreak of hostilities. I have sought reaffirmation of that assurance and the Acting High Commissioner came this morning and gave that assurance to me on behalf of the Prime Minister of Britain, Sir Anthony Eden. So that I do not think that on that score we need worry. I do not think that an assurance given in that way will be materially broken; in the meantime, I do not believe that hostilities will break out and I think the position is somewhat more favourable now, if I may say so, by the cooling down of feelings in this matter, at least to the extent of not indulging in the very dangerous and far-reaching method of the use of force in the Suez Canal dispute. I do not think that issue will arise. If it does, there will be no utilisation of our bases.

With regard to the second question asked me regarding our neutral attitude, of course, the attitude was that if hostilities broke out I was not prepared to plunge our country into war and that we would remain neutral in that event. It did not imply—I am sorry if anything has been reported that has created a false impression— in any way that we had gone back on the attitude which we had adopted, which our representative put forward at the London Conference, which I still believe is the best way of solving this problem and which we will press on any and every occasion that we get in order to obtain some honourable and peaceful solution of this dispute. It was only a statement regarding the neutral position of this country as far as war was concerned in the unfortunate event—let us hope it will not materialise—that war does break out. I am glad to have this opportunity, if there was such a misunderstanding, of clearing it here.

PART III

COMMONWEALTH REACTIONS TO THE INVASION OF SUEZ, OCTOBER–NOVEMBER 1956

. . . Our ties [with the United Kingdom] are never closer than in times of stress and danger.

Sir Leslie Munroe, 1 November 1956

Britain's action, I personally say—and I will say it if I am the only one left to say it—was brave and correct.

Mr. R. G. Menzies, 12 November 1956

I have no hesitation in applauding the United Kingdom Government for its timely recognition of the claims for freedom and self-determination in India, Pakistan, Burma and Ceylon. This act earned for them the admiration and respect of the whole world. I would appeal to the United Kingdom even at this late stage not to jeopardize this well-earned good will in the pursuit of an objective that is fraught with disaster.

Mr. R. S. S. Gunawardene, 1 November 1956

. . . Events have happened which I would have thought could not possibly occur in this modern age.

Shri Jawaharlal Nehru, 16 November 1956

COMMENTARY

I

ON the morning of 30 October, the British Cabinet had before it reports that Israeli troops had entered Egyptian territory during the night and were striking deeply at Ismailia on one front and at Suez on another. The Cabinet proceeded to draft the terms of the ultimatum in accordance with the decision taken five days earlier.[1] After approval by the French Prime Minister and Foreign Minister, who had set out immediately for consultations in London, a message in identical terms was dispatched to the Governments of Egypt and of Israel in the name of the Governments of France and the United Kingdom.[2] The belligerents were ordered to cease their fire, and to withdraw their forces to a distance of ten miles from the Suez Canal. Failure to comply with these instructions within twelve hours, the ultimatum declared, would result in the occupation of Port Said, Ismailia, and Suez by Anglo-French forces. The message was delivered to the Egyptian and Israeli Governments at 4.15 p.m. (G.M.T.). At the same time the Secretary of State for Commonwealth Relations informed the Commonwealth High Commissioners in London of the terms of the ultimatum, while Sir Anthony Eden conveyed them to the leaders of the two opposition political parties. At 4.30 the Prime Minister announced them in the House of Commons.[3] The Prime Minister had also dispatched two messages to the President of the United States,[4] from whom he received a reply expressing deep concern at the prospect of such drastic action. 'I was not', wrote Sir Anthony Eden afterwards, 'surprised at this sentiment.'[5]

It was hardly to be expected that the Government of Egypt would comply with the demand of two unfriendly foreign powers (B. + Fr.) that it evacuate its own territory as a consequence of being attacked. It did not do so. The deadline came and went. The British Cabinet thereupon ordered its Commander, General Sir Charles Keightley, to put his battle plan into effect. This he did forthwith:

[1] *Full Circle*, p. 527. See above, pp. 99–100. [2] See below, p. 204.
[3] See below, pp. 206–8. [4] See below, pp. 205–6.
[5] *Full Circle*, p. 528.

At 0430 hours 31st October I was informed that the Israeli Government had agreed the requirements [sic] and that Egypt had refused.

My object was defined as follows:

(i) To bring about a cessation of hostilities between Israel and Egyptian forces.

(ii) To interpose my forces between those of Israel and Egypt.

(iii) To occupy Port Said, Ismailia and Suez.

The agreement to our requirements by the Israelis and the refusal by the Egyptians meant that we were now involved in operations against the Egyptians but with limited objectives.

My instructions were that air operations against the Egyptians would start on 31st October. . . .

At 1615 hours GMT on 31st October, 1956, Valiant and Canberra bombers under the command of Air Marshal Barnett began their attacks on Egyptian airfields at Almaza and Inchas near Cairo and at Abu Suier and Kabrit in the Canal Zone. . . . By the end of the day the Egyptian Air Force had been severely treated. . . .[1]

Before the British and French forces began their air strikes and naval actions (an Egyptian frigate was sunk in the Red Sea on the night of 30 October by H.M.S. *Newfoundland*), the United Nations Security Council had been summoned into emergency session on the insistence of the United States. The United Kingdom representative had been instructed to join with the American in seeking this meeting, but the United States (as Sir Anthony Eden noted reproachfully afterwards) 'refused to amend the letter summoning the Security Council so that the French and ourselves could also sign it. To denounce and neither to offer nor to accept any constructive suggestions was the core of American policy'.[2] When the Security Council met, the British delegate, Sir Pierson Dixon,[3] assisted by the Australian, Dr. Ronald Walker,[4] asked its members

---

[1] Supplement to the *London Gazette*, 12 September 1957, pp. 5329–30.

[2] *Full Circle*, p. 529.

[3] Sir Anthony Eden has observed: '. . . I might have been unhappy at leaving an official [rather than a Minister] in such a position, but I was quite sure that if this had to be, no better choice [than that of Sir Pierson Dixon] was possible. . . . It was said of him when the period of the Suez crisis was over, "He went through the whole business without ever missing a trick and without making an enemy".' *Full Circle*, p. 531. It was said of Mr. Menzies, on the other hand, that he preferred to allow Australia to be represented at the United Nations by career officials, as he could not rely upon his Foreign Minister, Mr. Casey, to argue his case with the necessary conviction. See Norman Harper, 'Australia and Suez', in Gordon Greenwood and Norman Harper (editors), *Australia in World Affairs, 1950–55* (Melbourne, 1957), p. 351.

[4] See below, pp. 225–9.

not to take action until the Anglo-French strike had had an opportunity to achieve its objectives. But the United States representative would not consent to delay. As Sir Anthony Eden put it,

Mr. Cabot Lodge ... pressed his resolution to a vote with all speed and included in it phrases explicitly directed against Anglo-French action. His only reply to the arguments of the British representative, in public and in private, was to ask the Council to take the vote at once. As a result, Britain used her veto for the first time in her membership of the United Nations. The American resolution secured seven votes, we and the French voted against it, Australia and Belgium abstained.

The Russians then moved a resolution substantially the same as the American draft, but without its most offensive paragraph directed at the French and ourselves. We were willing to abstain on this vote, in the hope of taking some of the heat out of the debate and inducing reflection on the wider issues at stake. It would certainly have been helpful if this had been possible, but here we were up against the difficulties of rapid consultation between the capitals and New York, and the insistence on immediate decisions by the United States delegate. The French delegation were under instructions to use their veto and no time was allowed for further discussion between the Governments. For the sake of solidarity, therefore, we acted together. Tension mounted at the United Nations and our friends among the maritime Powers were appalled at the rift which was revealed between us and the United States.[1]

The next move at the United Nations was to invoke the so-called 'Uniting for Peace' procedure, permitting the General Assembly to take up in emergency session any dispute or situation authorized by the affirmative votes of any seven members of the Security Council. No one, least of all the American originators of the device —which had been adopted in November 1950 as a means of escaping the paralysing consequences of successive Soviet vetoes— had expected that its first use would be directed against two leading members of the Atlantic Alliance. The vote on the proposal to go to the Assembly ranged Britain and France against the remainder (Australia and Belgium abstaining). The crisis was now before the eighty-odd members of the General Assembly. (How they dealt with it there is the subject of Part IV.)

On 31 October, the House of Commons at Westminster faced the fact of Anglo-French military action against Egypt, by some

[1] *Full Circle*, p. 530.

for so long desired, by others for so long feared. Sir Anthony Eden sought to defend what had been done.[1] Mr. Gaitskell, for the Labour Party Opposition, described the Government's decision as 'an act of disastrous folly whose tragic consequences we shall regret for years', and announced that his party would do everything constitutionally within its power to oppose it and if possible reverse it.[2] On 1 November, the Minister of Defence (of a fortnight's standing) gave to the House of Commons details of the results of the first raids against the Egyptian Air Force. 'There was much interruption and disorder', Sir Anthony Eden has recalled, 'and the Speaker was obliged to suspend the sitting for half an hour.'[3]

By 2 November, the Egyptian Air Force was thought by the Allied Commander to have been pretty well destroyed, and he now directed his air power against other military installations. 'Lack of suitable targets in areas away from the civilian population', he records in his official Dispatch, 'whose safety was from the outset one of our primary concerns, naturally restricted . . . activities.'[4] Air strikes were mounted against armoured concentrations and military movements on the roads, principally round about the Canal and Port Said. This was the second stage of the plan. The third called for invasion, by airborne troops reinforced, if necessary, by forces put ashore from sea.

On Sunday, 4 November, the British Cabinet met to decide whether to put stage three into effect, in the face of General Assembly resolutions calling for an immediate cease-fire, massive opposition to the Anglo-French operation throughout the Commonwealth and in the United States, and an evenly and deeply divided public opinion at home. Three courses were open to it: to call off all further military action; to postpone the impending airdrop; or to proceed. The choices resolved themselves into alternatives: to postpone or to proceed.

It was clear to me [Sir Anthony Eden writes retrospectively] that a postponement could not be accepted. I was confident that the Commander-in-Chief, if consulted, would urge overwhelming arguments against it, which he properly and promptly did. The political objections were also obvious, once action was halted it could hardly be resumed.

---

[1] See below, pp. 209–10.   [2] See below, pp. 208–9.
[3] *Full Circle*, p. 535.
[4] Supplement to the *London Gazette*, loc. cit., p. 5331.

Postponement would, in fact, have meant calling off the operation. If we postponed now we should be doing so before we had gained our main physical objective, to insert an impartial force between Egypt and Israel along the Canal. There was no United Nations force yet in being to fulfil our aims for us. Unless we persisted, it might never come into existence. The Cabinet was in favour of going forward and shouldering the political risks. It was determined by the news that reached us towards the end of our meeting. Though Israel had accepted a cease-fire in principle, she had not yet agreed to the United Nations force, wishing no doubt for information as to its location, duties and composition. All was still uncertain, there was every reason for proceeding with our action.[1]

Early on 5 November, the planned airborne assault by British and French paratroopers was accordingly carried out. After about seven hours of fighting, the Egyptian commander in Port Said issued orders for a cease-fire, but these were countermanded from Cairo, and fighting continued. 'On the resumption of operations', General Keightley recorded, 'the garrison and population were encouraged to resist by loudspeaker vans which toured the town announcing that Russian help was on the way, that London and Paris had been bombed and that the third World War had started.'[2] It was now all too clear that the airborne force alone could not accomplish its objectives, and that the invading force standing by offshore (having steamed six days from Malta) would have to attempt a landing on the beaches. The assault force made its landing on the following morning (6 November) under covering fire from destroyers; about 20,000 men, one-third of them French, were put ashore. Fighting in and about Port Said was heavy throughout the day, but towards evening resistance seemed to be overcome, and the Commander was considering the next stage of the operation, which was to turn south, and advancing along the Canal, seize Ismailia and Suez. His troops had in fact started upon an advance to the south which had carried them twenty-three miles when 'at 1700 hours GMT orders were received from London that a United Nations Force would take over from us and that a cease-fire was to take effect at 2359 hours GMT, and that no further move of forces would take place after that hour'.[3]

[1] *Full Circle*, pp. 551–2.
[2] Supplement to the *London Gazette*, loc. cit., pp. 5332–3.
[3] Ibid., p. 5334.

Sir Anthony Eden has provided the following account of the factors which led him and his Cabinet colleagues to call off the attack. Its abandonment was not due, he has insisted, either to the threat of Soviet intervention or to defection on the part of his supporters in the House of Commons. Of the former he writes: 'We considered that the threats in Marshal Bulganin's note need not be taken literally.'[1] Of the latter: 'There were reports at this time of a dissident minority in the Conservative Party in the House of Commons. I was told that if a cease-fire were not announced that day, some of them would not vote with us. I was not influenced by these reports. . . . The overwhelming majority was firmly loyal. There are always weak sisters in any crisis. . . .'[2] 'Weak sisters' neither justly nor accurately describes Mr. Anthony Nutting and Sir Edward Boyle, the two junior but important members of the Eden Government (the former, Minister of State for Foreign Affairs, the latter Economic Secretary to the Treasury) who resigned from it on 31 October and 5 November respectively. (It was widely rumoured that had the fighting continued, other and more damaging resignations would have followed, notably those of Mr. R. A. Butler, Lord Privy Seal, Mr. Heathcoat-Amory, Minister of Agriculture, and Sir Walter Monckton, Paymaster-General.[3]) The crucial considerations, as Sir Anthony Eden has since stated them, were, first, the run on the pound 'at a speed which threatened disaster to our whole economic position';[4] and, second, the fact that fighting between Egypt and Israel had by this time ceased. This second factor he claims to have been decisive:

We had intervened to divide and, above all, to contain the conflict. The occasion for our intervention was over, the fire was out. Once the fighting had ceased, justification for further intervention ceased with it. I have no doubt that it was on this account more than any other that no

---

[1] *Full Circle*, p. 555.

[2] Ibid, p. 557. The number of potential Conservative parliamentary defectors has been variously estimated. One study declares that 'the Suez Critics consisted of 24 Members who seem to have been hostile to, or to have had misgivings about, the British intervention'. S. E. Finer, H. B. Berrington, and D. J. Bartholomew, *Backbench Opinion in the House of Commons, 1955–59* (London, 1961), p. 92. Professor Barraclough records that 'By 6 November it was believed that some thirty Conservatives had made up their mind to vote against the Government unless there was an immediate cease-fire.' Geoffrey Barraclough, *Survey of International Affairs, 1956–1958* (London, 1962), p. 63.

[3] Barraclough, op. cit., p. 63.   [4] *Full Circle*, p. 556.

suggestion was made by any of my colleagues, either then or in the hours which elapsed before my announcement in the House that evening, that we and the French should continue our intervention.[1]

What had been bought, and what had been the cost? A balance sheet which ignored political consequences (and no self-respecting historian-auditor could be satisfied by such a balance sheet) might reckon it as follows. The fighting between Israelis and Egyptians had ended, and a United Nations 'presence' in an as yet undetermined form was about to be injected into the battle area. Against this achievement (if such it might be called) there had to be set three deleterious facts.

First, and most grievously, people had been killed and wounded. The British forces suffered casualties amounting to 16 killed and 96 wounded; the French, 10 killed and 33 wounded.[2] The extent of Egyptian casualties was both harder to ascertain and heatedly disputed: early British estimates placed them at 100 killed and 500 wounded, Egyptian estimates as high as 12,000 killed and wounded.[3] A British Government inquiry, after a careful and (there seems little reason to doubt) a fair examination on the spot, later estimated Egyptian casualties in the Port Said area as 650 killed, 900 wounded, 1,200 slightly wounded. An additional 100 Egyptians were said to have been killed in the Port Fouad fighting (where the French forces were engaged). It was thought improbable by the investigator that 'the true civilian casualties exceeded 25 per cent. of the whole'.[4]

A second consequence was that the Suez Canal, through which shipping had passed without hindrance between 26 July and 29 October (save for Israeli vessels which had been denied passage since 1948), was blocked by the wrecks of vessels sunk in the channel and in the harbour by the Egyptians following the Anglo-French attack.

A third consequence was that pipelines in the neighbouring states of Iraq and Syria, carrying oil to the West from the great fields owned by the Anglo-Iranian Oil Company and Iraqi

[1] *Full Circle*, p. 557.
[2] Supplement to the *London Gazette*, loc. cit., p. 5335.
[3] Guy Wint and Peter Calvocoressi, *Middle East Crisis*, p. 81.
[4] *Damage and Casualties in Port Said*. Report by Sir Edwin Herbert on his investigation into the effects of the Military Action in October and November 1956 (London, 1956, Cmd. 47), p. 28.

Petroleum, had been sabotaged by agents friendly to President Nasser.

Though the dead would not be raised, the wounded might be healed, the Canal cleared, and the pipelines restored. The political consequences, however, omitted from the reckoning, were at once less easy to ascertain and more difficult to overcome. What these consequences were, as they were manifest throughout the Commonwealth, the following sections attempt to assess, beginning with the Canadian and ending with those in the United Kingdom itself.

II

It is not often in Canada that one hears foreign policy being talked about on street corners and in the lobbies of hotels.[1] So deeply was the nation stirred by the news of the Anglo-French ultimatum and of the military action which followed. Churchmen preached sermons; university students debated and passed resolutions; trade union councils were roused to petitioning pitch; public libraries reported a run on *The Philosophy of the Resolution*; speakers on Middle Eastern affairs were in great demand and found audiences of unprecedented size and wakefulness. An avalanche of mail fell upon editors' desks and into the offices of Members of Parliament; while the East Block reported the most heavy volume of correspondence since Mr. Pearson had become Secretary of State for External Affairs.

Two aspects of British policy provoked Canadian criticism. The critical public tended mainly to deplore the decision to resort to force; official opinion was more distressed by the failure of the United Kingdom to let its allies and its Commonwealth partners know in advance what it was up to. Mr. Pearson declared in the House of Commons: 'There was no consultation with other members of the Commonwealth . . . and no advance information that this very important action, for better or for worse, was about to be taken.'[2] In Mr. St. Laurent's confidential dispatch to Sir Anthony Eden of 5 November, setting forth the reaction of the Canadian Government to the events of the past week, it is likely

---

[1] I have drawn throughout this Section upon 'Suez, Britain, and the Canadian Conscience', the part of my book *Canada in World Affairs, 1955–1957* (Toronto, 1959) in which Canadian reactions to the Anglo-French intervention are discussed.

[2] Canada, H.C. Deb., 1956 (Special Session), p. 53.

that a complaint at not having been consulted by the United Kingdom figured prominently.[1]

In the public response of the Government, however, it figured not at all. At the outset the Government was extremely reluctant to make any public response: alone among Commonwealth Governments, it did not issue a prepared statement for more than forty-eight hours following the news of the Anglo-French ultimatum, and the reticence of Mr. St. Laurent and Mr. Pearson was in sharp contrast to the alacrity with which the Australian Prime Minister moved to support the United Kingdom, and the Indian Prime Minister to condemn it. It is not hard to understand the reason for their reticence. They were now presented with an issue threatening general war, on the merits of which the United Kingdom and France, on the one hand, and the United States, on the other, were sharply divided—though no more sharply than their own people. It was a kind of Canadian Foreign Minister's nightmare, one wherein hinges stuck, linch-pins snapped, and bridges fell into the sea. By emulating Francis Bacon's interrogators ('They will so beset a man with questions, and draw him on, and pick it out of him, that, without an absurd silence, he must show an inclination one way'[2]), reporters were able to extract from Mr. Pearson, on 30 October, an admission of 'regret' that Britain and France 'found it necessary to take this action while the Security Council was discussing the matter'.[3] With Mr. St. Laurent, who was plainly disturbed to the point of distraction by the international situation, they were even less successful. Asked on 31 October, following a Cabinet meeting, whether he also 'regretted' the Anglo-French action, Mr. St. Laurent turned in some anger upon the inquiring reporters to reply (as one of them recorded) 'that he was having nothing further to say about that. When there was any decision by the Canadian Government, he would let it be known. In the meantime, he was not going to inform his questioners about what might be running through his own mind'.[4] On

---

[1] This dispatch was drafted at a meeting of the Cabinet, which was rumoured to have been sharply divided on what to say and how to say it. The contents were widely thought to have been strongly worded; Mr. St. Laurent described the tone of his message as 'friendly but frank', and resisted the efforts of Opposition Members of Parliament to have it made public.

[2] *Of Simulation and Dissimulation.*

[3] 'Canada opposes Plan of Britain and France to Intervene in Egypt', *The Globe and Mail* (Toronto), 31 October 1956.

[4] George Bain, 'Minding Your Business', *The Globe and Mail*, 1 November 1956.

1 November, waylaid once again by reporters, this time *en route* to a Cabinet meeting, the Prime Minister's attention 'was brought to the Australian statement, with its support of the British position, and he was asked for comment. "That is their affair", he barked. "We have no criticism or commendation of what other Governments choose to do".'[1]

It was not until 4 November that the Prime Minister of Canada put before the Canadian people his Government's reaction to the events of the past few days. Most of his address on this occasion, which was broadcast over the national radio and television network, was devoted to a recitation of the events concerning Canada's participation in the proceedings of the United Nations General Assembly Emergency Session.[2] On the issue of the Anglo-French action, Mr. St. Laurent restricted himself to an expression of 'regret . . . that, at a time when the United Nations Security Council was seized of the matter, the United Kingdom and France felt it necessary to intervene with force on their own responsibility'.[3] A few days later, during the debate in the House of Commons, Mr. St. Laurent was goaded into stating what may have been his real feeling when he confessed that he had been 'scandalized' by the action of the leaders of the Great Powers, the 'supermen of Europe'.[4] It was to be a long time before the Prime Minister was to hear the last of these possibly ill-chosen words; and by then he was no longer Prime Minister. In the same debate Mr. Pearson prefaced his defence of Canadian policy by observing that Canada was no longer 'a colonial chore-boy running around shouting "ready, aye ready" '.[5] As part of that defence, however,

---

[1] *The Globe and Mail*, 2 November 1956.   [2] See below, pp. 222–5.
[3] Ibid.   [4] See below, p. 417.
[5] See below, p. 421. 'Ready, aye ready' was the celebrated cry with which the then Conservative Leader of the Opposition, Arthur Meighen, had urged Canadian support for the United Kingdom during the Chanak crisis of 1922. During the Suez crisis, Arthur Meighen briefly re-entered public life to deliver once again, unrevised and unrepented, his famous message of a quarter-century ago.
   'Canada [he was reported to have stated] should have sought without delay alignment unmistakably and strongly with Britain and Commonwealth councils in the current crisis.
   'Prime Minister Eden, for whom I have the highest regard and respect, merits the support of the Commonwealth in his endeavour to maintain Britain's honour and her place in world affairs. We should surely have as much interest in that as does Britain herself.
   'The conduct of the Australian Government has been most exemplary, and a bulwark for the Eden Administration, which seems to have been having more trouble than it has any right to expect. The Australian Government added a

he pointed out that 'instead of indulging then or since in gratuitous condemnation we expressed our regret and we began to pursue a policy . . . which would bring us together again. . . .'[1] Two days later he added: 'I do not for one minute criticise the motives of the Governments of the United Kingdom and France in intervening in Egypt at this time. I may have thought their intervention was not wise, but I do not criticise their purposes.'[2] Not all his country-men were disposed to be as magnanimous.

There is no way of telling with precision how the Canadian people were divided in their reaction to the Anglo-French inter-vention and to their own Government's response to it.[3] The fol-lowing impressions, based on parliamentary debate, editorials and letters to newspapers, and what one learns from observation and hearsay at the time, suggest some rough indication of regional opinion and of the kinds of argument which were to be heard in what may without exaggeration be described as a nation in debate.

In the House of Commons, the Government and its supporters dissociated themselves from the policies of the United Kingdom and France; they were joined by the members of Canada's party of democratic socialism, the Co-operative Commonwealth Federa-tion (C.C.F.). The Progressive Conservative and Social Credit Parties declared their approval of British policy.[4] The Conserva-tive Opposition moved an amendment to the Address which proposed to add to it a fourfold expression of regret, the first part deploring the fact 'that Your Excellency's advisers (1) have

great deal of strength and effectiveness to Mr. [*sic*] Eden's course which already had deserved the widest support on its merits. . . .

'And, to my view, it is highly unfortunate that here in Canada . . . we have been providing ammunition for . . . critics of the Eden Government and its policies. This simply helps to impair the effectiveness of the course it has been pursuing . . .' 'Advice from Meighen: Stand by British', *The Globe and Mail*, 5 November 1956.

[1] See below, p. 423.   [2] Canada, H.C. Deb., 1956 (Special Session), p. 168.

[3] The Gallup Poll conducted a survey revealing that of those Canadians asked whether they approved of British and French policy in the Middle East, 43 per cent. said 'Yes', 40 per cent. 'No', and only 17 per cent. 'No opinion'. This finding could indicate not only that Sir Anthony Eden was receiving proportion-ately greater support among Canadians than among his own countrymen, but also that the Canadian Government was not proceeding according to the wishes of the majority of its electorate. It developed, however, that the sample on which these findings were based was obtained by means of a 'quick spot check', and that taken in Toronto.

[4] The party standing in the House of Commons was at this time as follows: Liberals, 170; Progressive Conservatives, 53; C.C.F., 23; Social Credit, 15; Independents, 4.

followed a course of gratuitous condemnation of the action of the
United Kingdom and France which was designed to prevent a
major war in the Suez area. . . .'[1] This amendment was energetically supported by Mr. Earl Rowe and Mr. Howard Green, two
prominent Conservatives who were, however, not candidates for
their party's leadership;[2] of the three who were (Mr. John Diefenbaker, Mr. Donald Fleming, and Mr. Davie Fulton), only Mr.
Fleming displayed an advocacy of comparable vigour. None of
the amendment's strongly worded criticism found its way into the
Conservative Party platform assembled only a fortnight later.
The leader of the Social Credit group thought it 'a pity that the
Government did not find it possible to provide Britain and France
with political and moral backing', but its members did not support
the Conservative amendment, which was defeated by a vote of
171 to 36.

No markedly regional grouping of opinion is evident in editorial
comment, with the exception of the French-language Press in
Quebec. Pro-British sentiment was strong in the Maritime
Provinces, particularly in New Brunswick where United Empire
loyalism roused itself for the occasion. But if the Fredericton
Gleaner denounced the assailants of British policy and the St. John
Telegraph-Journal deplored Ottawa's failure to see that 'Britain
and France acted in good faith and as responsible nations should',
the Halifax *Chronicle Herald* considered the policy of the Canadian
Government to be 'above reproach'. It was in Southern Ontario,
the traditional rallying ground for 'Tory imperialism', that the
most perfervid supporters of British policy were to be found, their
views most pungently expressed in a series of forthright editorials
in *The Globe and Mail* of Toronto.[3] But it was a Toronto newspaper,

---

[1] Canada, H.C. Deb., 1956 (Special Session), p. 18.

[2] Mr. Howard Green became, in June 1959, Secretary of State for External
Affairs.

[3] Which denounced, with equal fervour, the political behaviour of John Foster
Dulles and Jawaharlal Nehru. The latter's failure to condemn, with what *The
Globe and Mail* believed to be the requisite degree of indignation, Soviet intervention in Hungary, caused that newspaper to remark editorially on 6 November
'that under Nehru's influence this section of Asia has aligned itself with Moscow.
It no longer stands for freedom—or for neutrality. . . . But the free world has
this consolation: it knows now, at long last, just where it stands with the Nehrus,
the Bandaranaikes, and the rest of the shifty crew. Knowing that, seeing where
their sympathies really lie, it no longer needs to endure their pharasaic cant, or
be influenced by their mealy-mouthed hypocrisy. It can treat them as what they
have shown themselves to be—whited sepulchres, "which indeed appear beautiful outward, but are within full of dead men's bones".'

the *Star*, which denounced in intemperate language the 'incredibly reckless and arrogant conduct of the Eden Government'.

The French-language Press in Quebec unitedly opposed Anglo-French policy. Moreover, Quebec editors almost unanimously supported the reaction of the Canadian Government, although there was little disposition among them to reprove Mr. St. Laurent for not having more forcibly dissociated Canada from the Suez adventure. They were also notably restrained in their criticism of the British and French Governments, partly because of a reluctance to chide France, partly because of a belief that the operation had been primarily directed against the expansion of Soviet influence in the Middle East rather than against the kind of genuinely nationalist régime for which French-speaking Canadians traditionally have had nothing but sympathy.

Those Canadians—and there were many of them—who wrote to newspapers to express their views were less restrained both in criticizing and in defending British policy. Some of the critics did not hesitate to say that the Canadian Government ought to have censured Britain far more forthrightly than it appeared to have done. The tone of most of the correspondents disposed to disapprove was, however, one of pain and sorrow rather than of anger —'almost tearful', as *The Economist* put it, 'like finding a beloved uncle arrested for rape'.[1]

The defenders of British policy employed a variety of arguments, couched in terms of passionate conviction less evident among the critics. That the 'ready, aye ready' school was not yet dead was shown by the number of those who argued that the Dominion ought to stand by the Mother Country right or wrong. More often it was argued, however, that the Mother Country was not wrong. How could Sir Anthony Eden, at the pinnacle of so long and so honourable a career, commit so serious a blunder in an area of which he was an acknowledged master? 'Let us be wary of making common cause with Britain's critics', warned the Ottawa *Journal*. 'Downing Street is not without sense, experience and courage. The decision that has so surprised her friends was taken open-eyed; she must have known the consequences.'[2] Many Canadians

---

[1] 'Shock and Distress in Canada', *The Economist*, 10 November 1956.
[2] Ottawa *Journal*, 5 November 1956. This newspaper, which traditionally supports the Conservative Party, broke with that tradition 'after an appraisal troubling and agonizing' to support the Canadian Government's stand on Suez.

echoed Eden's own diagnosis. Nasser was a bully, a thug, a megalomaniac dictator. History had shown nothing if not that such men are to be stopped in their tracks, by force if necessary, before they threaten wider destruction. 'What difference is there', asked Mr. John B. Hamilton, the Conservative Member for the Toronto constituency of York West, 'between a Nasser in 1956 at the Suez and a Hitler in the Rhineland or a Mussolini in Ethiopia?'[1] This appraisal was widespread, there being little if any tendency to argue that the parallel was ill-chosen and that a more apt comparison might be made with the early Mustapha Kemal or the younger Nehru.

Following the cease-fire, pro-British opinion strengthened noticeably. The quantity and kinds of Russian arms uncovered in Egypt appeared to many Canadians, as to Sir Anthony Eden's supporters in the United Kingdom, to vindicate the theory that the intervention had been designed to forestall an impending Soviet
→ coup. But the critics were able to point to the absence of any such motive in the United Kingdom Government's original justifications to Parliament and public, and to Mr. Pearson's statement in the House of Commons that as recently as a week before the Israeli attack the Canadian Government had had no 'knowledge or intimation about anything which could be called a Russian plot to seize Egypt and take over the Middle East'.[2] Many Canadians, whether or not approving of British policy, vented their frustration upon the United States; making uncomplimentary remarks about the American Secretary of State became at this time a form of tensional outlet on almost a national scale.

III

The news of 30 October startled and divided the publics of the Pacific Dominions hardly less than it did the people of Canada. Despite Mr. Menzies' close and even intimate partnership with

Because my *Canada in World Affairs, 1955–1957*, while citing its remark about Downing Street, did not take note of this break, the *Journal* charged me editorially with 'serious misrepresentation' and of distorting the record by quoting 'a single fragment wrenched from one editorial' (Ottawa *Journal*, 7 January 1960). One can only write history by quoting people out of context. But I am pleased to have this opportunity to set down for posterity the *Journal's* position over Suez, and to reveal its subsequent indignation as an illustration of how intensely Canadians came to feel about the issue.

[1] Canada, H.C. Deb., 1956 (Special Session), p. 82.
[2] Ibid., p. 53.

Sir Anthony Eden during the preparatory phases of the crisis, he, like Mr. St. Laurent, received no advance warning either of the Anglo-French ultimatum or of the military action which was sure to follow its inevitable rejection by Egypt. Unlike his Canadian colleague, however, Mr. Menzies took no umbrage at not being consulted. There was no time, he explained, echoing Sir Anthony Eden's explanation (before Sir Anthony Eden fell upon Mr. Menzies' words to excuse his own conduct in the matter), for consultation:

> . . . is the United Kingdom at fault in not having engaged in a pre-consultation with the other British Commonwealth countries? . . . she was not at fault at all. . . . Effective consultation—and I say 'effective' because a mere 'form of consultation' would have been quite useless—would plainly have occupied considerable time and the urgent position might have fallen into irretrievable disaster. . . . We are not living in an academic world. . . .[1]

As to the propriety of the Anglo-French action, Mr. Menzies declared, with an absence of qualification unique among government leaders throughout the Commonwealth, that 'the action was proper. . . . It seems to us to be quite realistic and to pay due regard to the moving and inexorable facts of life.'[2] So uncritical an assessment was hardly surprising in view of the Australian Prime Minister's continuous association with Sir Anthony Eden's policy throughout its formative stage, more especially in view of Mr. Menzies' unprompted defence on 25 September of a unilateral resort to force. On 6 November, in the heat of the Port Said battle, the Egyptian Government found time to sever diplomatic relations with Australia. Mr. Menzies continued to defend the 'military activities' of Britain and France. 'It is', he insisted, 'because they took strong action that the United Nations itself has been galvanised into action.'[3]

At the United Nations, Australia's moves did not quite so firmly register its Prime Minister's attitude of unqualified support for the United Kingdom, perhaps because of the difficulty of maintaining quick communication between Canberra and New York City ('An odd thing it was', Mr. Menzies later recalled, 'to learn in my own country about what the motion was or the amendment was only after it was all over. . . . They do not always appear to realize in

---

[1] See below, p. 232.   [2] See below, p. 232.   [3] See below, p. 237.

New York that the clock is different [in Australia] . . . .'[1]), perhaps, also (there is still no firm evidence on the point), because of hesitation and division within the Cabinet.[2] Australia, the only Commonwealth state on the Security Council apart from the United Kingdom, abstained from voting on the United States resolution of 30 October calling upon all members to refrain from using or threatening force, but supported the resolution, curiously sponsored by the Soviet Union, which, while calling for a cease-fire, placed less restriction upon unilateral action should no one cease firing. On the Yugoslav motion to place the dispute before the General Assembly the Australian representative again abstained from voting, although he declared that he (not his Government) could see no good coming out of the Assembly's deliberations.[3] (Australia's first abstention was duly noted and commented upon by Mr. Gaitskell at Westminster: 'It is a remarkable and most distressing fact that Australia was unable to support us in the United Nations Security Council.'[4]) However, the events of the next few days found Australia ranged less equivocally at the side of Britain and France, and Israel against a hostile General Assembly. On 2 November, it voted against the cease-fire resolution of the United States in the company of only the three belligerents and New Zealand; sixty-four other states voted for it, and six states abstained. Australia abstained as well from voting on the Canadian resolutions of 3 and 4 November concerning the formation of a United Nations police force. 'For ten days in November', the authors of *Australia and the United Nations* state in a manner which barely conceals their reproach, 'Australia and New Zealand were isolated in company with the three "aggressors". . . .'[5]

In the eyes of the Australian Labour Party forming the official Opposition, this isolation appeared anything but splendid. On 31 October, in the House of Representatives, Dr. Evatt, with great heat, challenged the Prime Minister's uncritical approval of Sir

---

[1] See below, p. 445.

[2] 'The Israeli invasion of Egypt and the Anglo-French ultimatum to Egypt on 30 October took the Australian Government completely by surprise in spite of their knowledge that Israel was mobilizing. This was reflected in the initial uncertainty as to Australia's policy in the Security Council, an uncertainty arising out of sharp divisions within the Australian cabinet.' Norman Harper and David Sissons, *Australia and the United Nations* (New York, 1959), p. 131.

[3] See below, p. 227.   [4] See below, p. 209.

[5] Harper and Sissons, op. cit., p. 134.

Anthony Eden's action. '. . . Innocent people are being killed because of the action of the British and French Governments. It is disgraceful. . . .'[1] On 8 November, he spoke, now more moderately than earlier, not changing his mind but denying that he had previously implicated the British people in his condemnation of British policy. 'The people of Great Britain are our kinsmen and I believe that in this matter they take the same view as that expressed by the Labour Party here.'[2] In the opinion of one Australian observer at least, Dr. Evatt had 'overreached himself' in some particulars of his attack; his apparent readiness to identify British action at Suez with the Soviet repression of Hungary's rebellion was to many Australians 'a monstrous affront'. At any rate a by-election at Wentworth (New South Wales) on 8 December resulted in overwhelming defeat for the candidate dubiously favoured by Dr. Evatt's support.[3]

Opinions of the Press and (it may be presumed) of the general public were as deeply divided on the propriety of British action and of their Government's response to it as were the two key figures of the parliamentary debate. (Unlike the Canadian debate in the House of Commons, into which leading members of the Government and all Opposition groups plunged with fervour, the debate in the House of Representatives at Canberra was restricted in participation to Mr. Menzies and Dr. Evatt; I do not know why this should have been so, but it was.) The immediate reaction of most Australian newspapers was to express support for and sympathy with the United Kingdom. British action, stated the Melbourne *Age*, 'may cause deep anxiety in Australia but it will be understood as politically inevitable'. The United Kingdom had been on the point of having to choose 'between economic oblivion or becoming a mendicant of the American oil companies'; for her 'swift and bold' attempt to escape this dilemma she was to be congratulated.[4] But this mood of understanding quickly passed after the news of the attack on Port Said and of Egyptian casualties. 'The Egyptian war confuses Australians with the barbarity of its beginnings', commented the Brisbane *Courier-Mail* on 2

---

[1] See below, pp. 233–4.   [2] See below, p. 241.

[3] 'Australia: The Suez Canal Crisis', *The Round Table*, vol. 47, 1956–7, p. 186.

[4] Quoted in Norman Harper, 'Australia and Suez', in Gordon Greenwood and Norman Harper (editors), *Australia in World Affairs, 1950–1955* (Melbourne, 1957), p. 348.

November, 'and the inadequacy of the reasons given for starting it. . . . The best service Australia can now give to Britain is to get her people to get out of an unjust war into which the Eden Government has thrust them.' The Melbourne *Argus* condemned the attack as lacking 'moral or legal authority'. Newspaper support for the Eden-Menzies position was firm but scarce. 'The Commonwealth of Nations should be delighted with the way Britain has asserted herself', declared the Sydney *Sun* on 1 November. 'This is more like the old Britannia.' So it was, but many Australians seemed to be more chagrined than delighted by her reappearance. 'Among all political parties', affirmed the Brisbane *Courier-Mail* on 5 November, 'a powerful public opinion wants to stop it.' When it was stopped, twenty-four hours later, the sigh of relief was almost audible from Perth to Sydney.

New Zealand, too, was something less than ecstatic over the unexpected turn taken by events after 30 October—more unexpected by her Government than by the Australian. Although caught by surprise, the New Zealand Ministry showed greater resourcefulness than the Australian in overcoming the problem of time and space between itself and the United Nations in New York. Officials of the Department of External Affairs at Wellington listened to the debates of the General Assembly broadcast on short-wave radio, frequently telephoning to the homes of the Prime Minister and the Minister of External Affairs to inform them of the course of events and to ascertain their instructions, which were then relayed to the New Zealand representative at the United Nations by radio-telephone.[1] When, therefore, Sir Leslie Munroe voted in support of the United Kingdom in the General Assembly on 2 November against the cease-fire resolution, he was doing what he had been told to do by his Government, unlike the Australian representative, Dr. Walker, who had done what he thought he might have been told had there been time to tell him. New Zealand did not, however, vote against the Canadian resolutions of 3 and 4 November dealing with the proposed United Nations force. Unlike Mr. Menzies, whose attitude towards the force was one of thinly veiled contempt, Mr. Holland and his colleagues saw it as the only means of achieving an honourable settlement; the Prime Minister of New Zealand continued to hope

---

[1] *The Dominion* (Wellington), 5 November 1956.

to be allowed, as the Prime Minister of Australia never allowed himself to hope, to contribute troops to the United Nations Emergency Force once this had been set in place.[1]

New Zealand's support of the United Kingdom at the United Nations concealed from all but the closest students of her affairs the mood of profound disquiet back home which assailed not only the Opposition party and its supporters but members of the Government as well. 'Behind the scenes', a New Zealand commentator wrote afterwards, 'the events of November still gave cause for anxious, if little publicized, thinking.'[2] One does not even have to read between the lines to detect this anxiety in the Prime Minister's public statements on the crisis. Mr. Holland acknowledged, in his statement of 1 November, that 'several features of the present situation are disturbing'; it was, he went on to say, 'a matter of grave concern' that the United Kingdom and the United States stood so far apart in their Middle East approaches. New Zealand, he declared (without rancour but equally without enthusiasm), had not been consulted about what was taking place. He expressed confidence not (as Mr. Menzies had done) in Britain's methods but in her 'intentions'.[3] This 'element of reserve' (as it appeared to the New Zealand commentator quoted above) may be discerned throughout his subsequent statements. On 6 November, Mr. Holland declared that he had 'never been in the slightest doubt that both the U.K. and the U.S. have as their main objective the preservation of peace'. It was only to be expected, he added, 'that strong-willed, independent nations would vary in the method by which they seek to obtain this end'.[4] Some weeks later he affirmed that New Zealand 'had never wavered in its belief in the sincerity of Britain's motives'.[5] That way of putting it allowed the implication that sincerity in these matters might not be quite enough.

All these feelings were fully shared by the Labour Party Opposition. From the outset of the critical stage of the Suez drama, members of that Opposition had been taken into the Government's confidence. There did not prevail between Mr. Holland and Mr.

---

[1] See below, pp. 354–7.
[2] 'New Zealand: Echoes from the Middle East', *The Round Table*, vol. 47, 1956–7, p. 185.
[3] See below, p. 242.
[4] *The Dominion*, 6 November 1956.
[5] See below, p. 357.

Nash the personal animosity marring the relations of Mr. Menzies and Dr. Evatt in Australia. Whereas the latter, like Mr. Gaitskell in the United Kingdom, had sought to oppose and if possible to reverse the policy of the Government, the leader of the New Zealand Parliamentary Labour Party adopted, and was encouraged to adopt, a bipartisan approach. The Government shared its information not only with members of the Shadow Cabinet, but with the External Affairs Committee of Parliament, which met in secret session on 5 November. On that same day the entire Cabinet had assembled to listen to a short-wave radio broadcast of Mr. Gaitskell's address to the British people over the B.B.C. [I, Document 10, *infra*]. Mr. Menzies' Cabinet, it is fair to suppose, would have preferred some other form of diversion. That evening, the Auckland *Star* expressed what must have been the unspoken thoughts of more than one New Zealand Minister as British soldiers were landing at Port Said:

New Zealand is a Pacific country—off the coast of Asia it might almost be said. Its future is irrevocably bound up with that of Asia, and for its long-term security it must largely depend on the existence of a strong United Nations capable of effective action. . . . While New Zealanders retain the strongest ties of kinship with Britain and find themselves largely in sympathy with her in her present hazardous role it must be realized that Britain's interests are not necessarily and wholly identical with our own.

IV

The minority of South Africans allowed by *apartheid* to put forward its views on events such as those on and following upon 30 October was, as nowhere else in the Commonwealth, deeply in sympathy with Israel's pre-emptive invasion of Egypt. A population of 110,000 South African Jews contributed much to this feeling. But there was also something in common between the outlook of the Afrikaner and the Zionist which predisposed many more South Africans than these to favour Israel in her warfare with her Arab neighbours. The involvement of the United Kingdom in that struggle produced something of a conflict of loyalties for the Nationalist Government and its supporters. On the one hand, they could without difficulty identify their cause with Israel's; for were they not (in their own view) like the Jews of

Palestine, an embattled minority struggling for self-expression and survival in a corner of a hostile continent? But on the other, it was not in the Nationalist tradition to rally to Britain's side. Prudence quickly reinforced habit. The bees were now swarming round the hive (to employ Johannes Strijdom's simile of the previous July), and it appeared to the Nationalist Government to be more sensible than ever to keep one's distance so as to avoid being stung.

At the United Nations, South Africa's representative pleaded lack of time and absence of information to excuse himself from voting upon the cease-fire resolution of 2 November.[1] At Pretoria, the Cabinet delayed its consideration of the crisis until 5 November. After its meeting on that date, the Minister of External Affairs, Mr. Louw, issued a statement expressing the Union Government's belief that it was 'not involved' in the hostilities, although it was 'deeply concerned about the maintenance of peace in the Middle East'. It hoped a solution would be found—without indicating what form a solution might take—'or that in any case the hostilities will remain limited and localized'.[2] The Afrikaner Press took at first the same line. An editorial in *Die Transvaler* for 2 November remarked: 'Just as the Union has nothing to do with the Suez dispute so it also stands outside the feud between Israel and Egypt and outside the "punitive" measures that England and France are at present enforcing on Egypt with bombs from the air.'[3] When, however, it became evident that these 'punitive' measures might not succeed, *Die Transvaler* became perceptibly more alarmed and cast aside its pose of neutrality. 'Failure now', it declared on 5 November, 'would without a doubt bring catastrophe to the West.'[4]

Before the Nationalist Government had fully defined their stand, the United Party Opposition had defined theirs. Mr. Strauss' statement of 2 November dissociated his party from 'the harsh condemnation directed at Britain and France'. These powers had been subjected to extreme provocation; one could not blame them for their action, one could only support and applaud them. 'The aggressor is not always he who fires the first shot.'[5] The

---

[1] See below, p. 246.   [2] See below, pp. 247–8.
[3] Quoted in Jitendra Mohan, 'South Africa and the Suez Crisis', *International Journal*, vol. XVI, no. 4, Autumn 1961, p. 343.
[4] Quoted ibid., p. 344.   [5] See below, pp. 246–7.

English-language Press took generally the same view of Anglo-French policy, although the *Cape Times* proved to be an exception. In an editorial of 1 November it criticized Britain and France for having by their unilateral resort to force weakened the United Nations and the force of the Free World's criticism of what Soviet Russia was doing in Hungary.[1]

Perhaps the most interesting aspect of South Africa's reaction from the standpoint of Commonwealth relations was its concern with the significance of consultation—or rather of the lack of consultation. In his first statement on this stage of the crisis, Mr. Louw had stressed the complete lack of forewarning from London. He had learned, he stated, of the ultimatum from a news broadcast over the South African radio; and when he had telephoned the Prime Minister, he had discovered that Mr. Strijdom knew no more than he did.[2] Three days later he issued a further statement in which he remarked that the absence of consultation 'would seem to indicate a major change of policy' on the part of the United Kingdom Government. He added that its failure to consult with the Union Government over the crisis absolved South Africa from any responsibility in the event of a wider conflict.[3] This statement evidently upset the High Commissioner for the United Kingdom in South Africa, Sir Percivale Liesching, who, as a former Permanent Under-Secretary of State in the Commonwealth Relations Office, remained a true believer in its gospel of consultation at all costs. He accordingly sought out the Minister of External Affairs on 4 November to proffer the explanation of extenuating circumstances for the lapse in consultation over Suez and to assure the Union Government that the United Kingdom had no intention of departing from this crucial convention of the modern Commonwealth. Mr. Louw replied to Sir Percivale Liesching in a letter of 5 November, of which only the final paragraph was made public. It read as follows:

In this connection, I would remind you that while the Union Government could reasonably have expected to be taken into the confidence of your Government in regard to a proposal which involved the risk of a Middle East conflagration, and even a Third World War, the South African Premier, Mr. Strijdom, made it clear to you that it was, of course, for the United Kingdom Government to decide whether or not

---

[1] Quoted in Mohan, op. cit., p. 345.   [2] See below, p. 245.
[3] See below, p. 246.

the circumstances required a departure from the procedure of consultation, or of giving prior information. The Union Government's attitude in this matter was governed primarily by the fact that Sir Anthony Eden had stated in public that consultation had taken place and, as you are aware, that statement was not corrected by him.[1]

Sir Percivale Liesching, still not content to allow the matter to rest, countered with a public statement of his own, reiterating that 'the United Kingdom's object has always been and will remain to consult all Commonwealth countries about issues of importance affecting them'.[2]

The High Commissioner's intervention inevitably attracted attention and comment in the South African Press. *Die Transvaler* took the view that the absence of consultation, so far from being deplorable, was only to have been expected: the Commonwealth could function solely on the basis of the strictest observance of the principle of non-intervention in each other's affairs by those belonging to it. The Johannesburg *Star*, however, regretted the absence of consultation, hoped it would prove to be only an isolated lapse, and lectured the Minister of External Affairs for his indifference: 'If Mr. Louw believes that the practice may fall into disuse as far as South Africa is concerned, his duty is not to welcome it but to do everything in his power to oppose such a tendency.'[3]

V

When newly independent nations enter the international arena for the first time, particularly if they enter by the way of revolution, they characteristically believe themselves to be above the struggle. Not for them the sordid power politics of older nations. Conflict is seen as a disease from which the ancient and the degenerating may suffer but to which the revolutionary young, if they keep their distance and their principles, may with luck remain immune. To the newly independent nations of Afro-Asia, the contagion of conflict was endemic in the West. Power politics are thought to be 'European or Europeanized politics, introduced into Asia, perforce copied by Asians in moments of temptation,

[1] *Cape Times*, 10 November 1956; *Digest of South African Affairs*, vol. 4, no. 24, 23 November 1956, p. 7.

[2] See below, pp. 219–20.

[3] Quoted in Mohan, op. cit., pp. 347–8.

but essentially alien to them'.[1] The West is an arena of conflict, the East an arena of concord. Where there is conflict in Asia—between India and Pakistan, for example—it is the fault of the West; it was the United States which had lured Pakistan from the paths of righteousness and the principles of *Panch Sheel* into the clutches of an imperialist alliance. 'Are we going to continue to be dragged into Europe's troubles, Europe's hatreds and Europe's conflicts?' Such was the question put by the Prime Minister of India to the Afro-Asian representatives convened at Bandung in 1955; it was a rhetorical question, and Pandit Nehru supplied his own answer: 'I hope not', he said. 'Europe has got us into the habit of thinking . . . that their quarrels are the world's quarrels. . . . Are we copies of Europeans or Americans or Russians? What are we? We are Asians or Africans. We are none else.'[2] His sister has even more explicitly sought to exempt Asia from the recurring crises of the West.

There has been no history of prolonged and painful conflict between the Asian countries [writes Madame Pandit] as there has been in the Western world. While conflict has been the general rule throughout European history, and peace has been confined to periods between wars, in Asia the reverse has been true. Because of this background the Asian nations, now independent, can renew their old cultural contacts with ease, regarding each other as peaceful neighbours rather than as potential enemies. There has been no inherited legacy of conflict to mar their present outlook.[3]

The Western student will register surprise that the modern history of South Asia, including as it does the forcible seizure of Hyderabad, the struggle between India and China and, above all, the bitter and protracted conflict over Kashmir that has been the despair of friends of India and Pakistan alike, may be so readily placed in the tradition of non-violence. Further in the Indian past he will encounter, certainly, the doctrine of the renunciation of force as an instrument of policy that runs in a direct line from the Buddha to the Mahatma; but he will encounter as well, in any fair-minded foray, the classical treaties of Indian political theory which for cool, unprincipled *realpolitik* match anything in the

[1] Alan de Rusett, 'On Understanding Indian Foreign Policy', *International Relations*, April 1959, p. 553.

[2] Quoted in George McT. Kahin, *The Asian-African Conference* (Ithaca, N.Y., 1956), pp. 64 ff.

[3] Madame V. L. Pandit, 'India's Foreign Policy', *Foreign Affairs*, vol. 34, no. 3, April 1956, p. 434.

writings of Machiavelli.[1] How nations see themselves, however, is often more important than how they are seen by others. To the governments and peoples of the Asian members of the Commonwealth, the assault upon Egypt by Britain and France appeared not as a manifestation of the code of behaviour to which all nations are subject but rather as a reversion to the kind of Western imperialism of which the new nations believed themselves incapable. The Anglo-French ultimatum of 30 October and the resort to force during the following week were seen in India, Pakistan, and Ceylon as a grim and terrible vindication of their way of looking at the world.

Of the three governments, that which was the most obviously distressed was the Government of India. It had every right to feel distress, for no government had worked more energetically during preceding months to prevent the kind of calamity which now had burst upon it. The hard news was brought to the Prime Minister's home by the United Kingdom Deputy High Commissioner, Mr. W. A. W. Clark, at six in the morning of 31 October. Mr. Nehru immediately summoned the members of the foreign affairs committee of his Cabinet: he, Pandit Pant, and Maulana Azad conferred for about an hour (Mr. Krishna Menon, the fourth member of the committee, was absent in Madras). He then requested the British High Commissioner to come to see him; that afternoon Mr. Malcolm MacDonald was told, doubtless in no uncertain terms, of the Indian Government's 'surprise and indignation at developments in the Suez area, and its intention to stand by Egypt'.[2] The Ministry of External Affairs, in the course of a day of unprecedented diplomatic activity (it received representatives of twenty-five governments, including those of the United

[1] So rarely is this point conceded in Indian scholarship that it is worth quoting from a notable exception:
'It would be highly partial to imagine [writes Mr. Zakir Husain] that this dominance of the interest of the State over the canons of established morality was a characteristically modern or an exclusively European phenomenon. While we remember Machiavelli we may not forget our own famous Kautilya of the Arthashastra. . . . So far as inter-statal relations were concerned, the State in India too "roamed", in the words of Mühar, "in an ethical jungle". . . . In foreign policy, expediency is recognized as the golden rule. He who is losing strength shall make peace; he who is growing strong shall make war; he who thinks that neither the enemy can beat him nor he the enemy shall observe neutrality; he who has an excess of advantage shall march; he who is wanting in strength shall seek protection; he who undertakes work, requiring assistance, shall adopt a dual policy. . . .' Zakir Husain, 'Ethics and the State', Malavankar Memorial Lecture, 1960, Harold Laski Institute of Political Science (Ahmedabad, 1960), pp. 8–9.
[2] The Times (London), 2 November 1956.

Kingdom, Pakistan, and Canada), issued a statement containing
the words 'aggression' and 'flagrant violation of the United
Nations Charter'.[1] A telegram, if anything stronger than this first
statement, was dispatched by the Prime Minister to the Secretary-
General of the United Nations.[2]

Mr. Nehru's indignation and anger mounted rather than abated
during the next few days. On 1 November, in a speech in Hyderabad, he declared that 'after fairly considerable experience in foreign
affairs, I cannot think of a grosser case of naked aggression than
what England and France are attempting to do'. His 'sorrow and
distress', he added, were all the greater because of the 'many
liberal gestures' of the British Government during previous years
when it had been, in his view 'a force for peace'. 'In the middle of
the 20th century we are going back to the predatory method of the
18th and 19th centuries. But there is a difference now. There are
self-respecting independent nations in Asia and Africa which are
not going to tolerate this kind of incursion by the colonial powers.'[3]
These were not idle words for, alone among Commonwealth
nations, the Government of India dispatched a formal protest to
the Government of the United Kingdom against the Anglo-French
bombing of Egypt 'as being against all considerations of humanity'.[4]
A former Governor-General, Mr. C. Rajagopalachari, publicly
recommended that India leave the Commonwealth in protest, and
urged that Pakistan and Ceylon follow suit.[5] Whether so drastic a
step was at any time seriously considered by the Prime Minister
and his Cabinet is not known; the Canadian Government evidently
believed it was, for, as Mr. Pearson later told the House of Commons at Ottawa, a prime reason for its sense of urgency in the
crisis was its conviction that the Commonwealth had been brought
to 'the verge of dissolution'.[6] The force of the Indian Govern-

---

[1] See below, p. 248. A sympathetic student of Indian foreign policy describes
this reaction as 'harsh and quick'. Michael Brecher, 'Neutralism: An Analysis',
International Journal, vol. XVII, no. 3, Summer 1962, p. 233.

[2] See below, p. 249.   [3] See below, pp. 249–51.

[4] M. S. Rajan, 'Stresses and Strains in Indo-British Relations, 1954–6',
International Studies, vol. II, no. 2, October 1960, p. 164.

[5] See below, p. 256.

[6] See below, p. 424. A correspondent of The Round Table alleged not long afterwards that Mr. Pearson's revelation of strain among Commonwealth members
'was widely understood both in Canada and in Britain and elsewhere' to refer
'to the possibility of Canada's seceding'. 'Commonwealth and the Crisis',
The Round Table, vol. XLVII, 1956–7, p. 117. This assertion evidently attached
altogether too much significance to the charges of the Leader of the Social
Credit Party in the General Election Campaign of 1957. See below, pp. 384–5.

ment's protest to the United Kingdom was blunted there (as elsewhere) by the feeling that Delhi had turned a blind eye towards far more outrageous events in Eastern Europe. On this Sir Anthony Eden has written:

> The Indian reaction was remarkable. Mr. Nehru declared in a speech that whereas in Egypt 'every single thing that had happened was as clear as daylight', he could not follow 'the very confusing situation' in Hungary. He then proceeded to read out the excuses which Marshal Bulganin had sent him for the Russian intervention. These Mr. Nehru described as 'facts'.[1]

In Canada, too, the Indian Prime Minister's continued refusal to express an unqualified condemnation of the Soviet action puzzled his friends and moved his critics to astonishing outbursts of *schadenfreude*.[2]

Ceylon's reactions to the events of the black week of 30 October to 6 November were in all essential respects similar to those of India. Mr. Bandaranaike was less quick off the mark than Mr. Nehru. 'I waited before making a statement', he explained at a Press conference on 1 November, at which reporters had asked him why Ceylon had not protested as soon and as severely as India had done, 'until the position became clear. It has now become painfully clear.' He was immediately addressing messages to the President of the United States, the Prime Ministers of Britain and France, and the Secretary-General of the United Nations, containing 'an urgent appeal that British and French troops should, even at this late stage, be withdrawn from Egyptian territory before the situation worsens'.[3] The Opposition party— the most radical in the Commonwealth—demanded, a few days later, that United Kingdom forces be expelled from their bases in Ceylon as a protest against British policy. Mr. Bandaranaike refused to consider such a step, noting that he had received assurances from the British Government that the bases would not be used for any purpose in any way connected with the action in the Middle East.[4]

Pakistan's reactions were more complex. Throughout the earlier stages of the crisis, public indignation had not been fully reflected

---

[1] *Full Circle*, p. 545.   [2] See above, p. 180, n. 3.
[3] *The Times* (London), 2 November 1956.   [4] Ibid., 6 November 1956.

in the policies of the Government, which attempted to reconcile its position as the spokesman of a Muslim power in the Afro-Asian world with its position as a member of a Western military alliance. Their reconciliation required the skill of a funambulist, and the Government had on more than one occasion been in danger of losing its balance. The events which now assailed it threatened to sweep Mr. Suhrawardy off his tightrope altogether.

The Prime Minister's earliest pronouncement showed him to be fully aware of the need to propitiate the public's anger. 'So far as Pakistan is concerned', he stated at Karachi on 3 November, 'the fact that a Muslim country towards which Pakistan has always entertained fraternal feelings should be the victim of aggression has further exercised public opinion. . . . The Government of Pakistan unreservedly condemn this aggressive action.' There can be little doubt, however, that on this day Mr. Suhrawardy was as much concerned about the action of extremists in Pakistan as he was about extremists at Westminster. The British Information Service offices in Dacca had been set on fire, police guarding the High Commissioner's offices in Karachi had been stoned by a hostile mob and had used tear-gas in defence, while Karachi students had burned an effigy of Sir Anthony Eden. The people were in an ugly mood, and the Prime Minister had accordingly to temper his denunciation of British policy by appeals to their reason. He urged them 'to remain calm and refrain from violence', and to respect diplomatic immunities 'whatever the provocation'.[1] To an audience of students he pledged Pakistan's opposition to aggression, but warned them that peace would not be won 'in a day. . . . I ask you to wait, and I assure you that you will be pleased when you find that you now have a Government in Pakistan who can face a situation like this fearlessly and take a bold stand on an issue of this nature. I ask you to leave it to us.'[2]

Having it left to him, at least for the moment, the Prime Minister set off on the following day (4 November) with his Foreign Minister for Teheran, where they conferred with their colleagues from the three other Muslim members of the Baghdad Pact. A message was drafted and dispatched to their errant ally requesting an immediate end to hostilities. The cease-fire came twenty-four hours later. The British Ambassador in Teheran assured the Four

[1] See below, p. 262.   [2] See below, p. 263.

Powers that their intervention had been decisive; the Foreign Office, in a formal statement issued on 8 November, declared that their representations had 'weighed heavily in the decision to bring an end to military action in Egypt'.[1] Their efforts are not, however, listed by Sir Anthony Eden among the factors which, he states, caused him to give the order to cease fire.

Pakistan's involvement in the diplomacy of the Baghdad Pact did not commend itself to the two other Asian members of the Commonwealth. One consequence of Mr. Suhrawardy's presence in Teheran was his absence from New Delhi when the Colombo Powers met there at about the same time that the Four Powers were conferring, an absence duly noted and unfavourably commented upon by the Colombo Powers' communiqué issued on 14 November.[2]

Nor did it commend itself to Pakistani public opinion. The people of a country, in their concern for foreign affairs, may be attentive or inattentive.[3] In Pakistan, as everywhere else, the latter greatly outnumber the former. Their characteristics are universal: protracted periods of apathy and indifference punctuated by bouts of frenetic concern in times of crisis. The news that the United Kingdom—Pakistan's creator, its ally, its Commonwealth partner —had joined forces with France and with the hated Israelis in an attack upon a Muslim people roused the entire nation to a frenzy. The tone of public discussion was set at the outset by the editorial in *Dawn* for 1 November, entitled 'Hitler Reborn':

. . . In the middle of the twentieth century enlightened countries like Britain and France have suddenly turned the clock back hundreds of years, unwritten much of what has since been written in the book of human civilization, and decided to act as self-chartered libertines with the gun and the bomb, killing and conquering the weak like cowards. And so this is the second 'Elizabethan age' of which so much was talked when Britain's young Queen ascended the throne!

[1] See below, p. 403.

[2] See below, p. 462. While Mr. Suhrawardy gave as his reason for declining the invitation to attend the New Delhi talks his prior commitment in Teheran, he did in fact return from the Iranian capital to his own on 10 November, two days before the Colombo Powers began their conversations. An Indian scholar goes so far as to describe the Pakistani policy as one which 'in effect supported the Anglo-French stand as against the Indian stand'. M. S. Rajan, 'India and Pakistan as Factors in Each Other's Foreign Policy and Relations', *International Studies*, vol. III, no. 4, April 1962, p. 383.

[3] See, for justification and elaboration of this distinction, Gabriel A. Almond, *The American People and Foreign Policy* (New York, 1950).

It is not our custom to print an editorial on the front page; but when Britain produces a Hitler who throws his own country's honour and all cherished moral, human and international values which it has itself fostered in the past out of the windows of the Houses of Westminster and into the Thames and proceeds to shed innocent Muslim blood to dye red the Nile, opening a new and unbelievable chapter of perfidy and violence in the history of the human race—a little editorial custom is a small thing to disregard for the sake of proclaiming without losing a single day what the people of Pakistan feel. . . .

Such inflammatory prose expressed the mood of the moment much more accurately than Mr. Suhrawardy's pleas for tolerance and restraint. There had always been in Pakistan a conflict, more or less latent, between the Establishment, on the one hand—speaking always in an 'air of ministerial dignity and caution'—and, on the other, the 'angry young men', found mainly but not exclusively in East Pakistan, who

view themselves as the victims of history, victims of slavery under the British, victims of exploitation by the Hindu middle-class of Calcutta, victims of unjust discrimination by West Pakistan. They use language, in political speeches, not to prove a point, but to create the desired emotion. Their oratory reaches its peak in mass gatherings but they attempt to use a similar style in the legislature. . . .[1]

The Suez crisis brought this conflict into the open. The malcontents found their spokesman in Maulana Bhashani, the President of the Awami League; he, and other dissidents, claimed that Pakistan's policy of alignment with the West had served it ill during the crisis, and urged the Government to sever its commitments as a member of the Baghdad Pact and of the South East Asia Treaty Organization. Typical of the kind of criticism directed at the Government by this group is the following extract from a statement issued by Chaudhri Mohammed Hussain Chattha, an Awami League member of the National Assembly:

One is not surprised at Malik Firoz Khan Noon. He has served the British too long, too faithfully, and too docilely. . . . But the recent indications given by Mr. Suhrawardy himself, the sentimentality with which he has spoken of the British connection, the fantastic fruits he has attributed to the Baghdad Pact, and the general docility with which he seems to have succumbed to the iron hand behind the curtain, makes us

[1] Keith B. Callard, *Pakistan's Foreign Policy: An Interpretation* (Institute of Pacific Relations, New York, 1957), p. 24.

fear a more disastrous betrayal. The people of Pakistan are determined to renounce the British connection, and all pacts which tie it to colonialism and the soulless service of Britain and France. If the highest in the land are living in illusions let them wake before they are kicked awake, for since they served and fawned under their British patrons, too much water has flowed under the bridges both of the Indus and the Nile. . . .[1]

If Mr. Suhrawardy's moderation was not appreciated at home, it gave offence as well in Egypt. It was true that on 4 November, just before setting out for Teheran, the Prime Minister was paid an official visit by the Egyptian Ambassador (the first such visit since 26 July), who thanked him on behalf of President Nasser for Pakistan's support during the crisis.[2] But, following the Pakistani participation in the deliberations of the Baghdad Pact powers, the attitude of the Egyptian Government became distinctly hostile. Pakistan's offer of men and material for the United Nations Emergency Force was not accepted by Egypt; and when Mr. Suhrawardy requested a meeting with the Egyptian leader, President Nasser refused to see him. It is difficult to overestimate the storm of anger and indignation aroused among the Pakistani people by this rebuff. The Government itself was shocked and taken aback. The Minister of Commerce and Industry, Mr. Abdul Mansur Ahmad, on learning of the news, declared in Karachi on 20 November:

President Nasser's reported refusal to receive the Prime Minister of Pakistan, if true, is shocking for more reasons than one. . . . It is particularly painful because Prime Minister Suhrawardy's only intention in visiting Egypt was to effect unity amongst the Muslim countries. This unity is essential for the benefit of the Muslim countries themselves in general and Egypt in particular. If the report is correct, the refusal must have been inspired by countries not well disposed towards Pakistan and for the matter of that towards any idea of unity amongst the Muslim countries. . . . The least I can say for the present moment is that President Nasser has been ill advised to yield to such inspirations. By this action he has injured himself and his country the most. I only pray that President Nasser will reverse his decision.[3]

But he did not. Spurned by Nasser, Mr. Suhrawardy returned again to his Baghdad Pact partners, rejoining the premiers of its Muslim members at Baghdad from 19 November to 23 November

---

[1] *Dawn*, 13 November 1956. [2] Ibid., 6 November 1956.
[3] Ibid., 21 November 1956.

this time in the company of the President, Mr. Iskander Mirza. In their communiqué, the Four Powers 'reaffirmed their conviction that recommendations made by them in their Teheran communiqué remained the only basis on which a just, honourable, peaceful and lasting settlement of Middle East problems could be achieved'[1]. As for the Egyptians, the President remarked glumly on 11 January, 'despite all our endeavours we were not successful in convincing them of our friendship and it looks as if it is quite useless trying to convince one who will not be convinced'.[2] This explanation did not satisfy the angry young men of Pakistan.[3]

VI

And so back, full circle, to Britain where, the distinction between publics attentive and inattentive being less easily drawn than elsewhere in the Commonwealth (in which category does Bagehot's bald-headed gentleman in the back of the omnibus belong?), it is even harder to state with any precision how people responded to events.

Sir Anthony Eden has cited, not an omnibus passenger, to be sure, but at any rate an omnibus driver in support of his contention that, following its initial unfavourable reaction, public opinion, by the day before the cease-fire, had swung in favour of his policy. The Prime Minister (wrote the bus-driver to the Prime Minister's wife)

[had] done the only thing possible—my opinion and also that of a great number of fellow bus employees—only a small proportion of London's multitudes—but if a bus driver agrees—he must be right—I personally thank God we've got one man who's not afraid to do the right thing. As regards the rioting of the other evening—as I was on a bus (driving) right in the middle of it—I saw possibly more than most people— Eighty per cent. of the crowd were of foreign extraction so that was no true census of opinion and can be ignored.[4]

This eye-witness report—which derives its significance solely from the fact that Sir Anthony Eden believes it to have been significant—was in one important respect technically inaccurate. The 'foreigners' to which it disparagingly alludes were not

---

[1] *New York Times*, 24 November 1956.
[2] Quoted in 'Pakistan: Rebuffs from Egypt', *The Round Table*, vol. 47, 1956–7, p. 174.
[3] See below, pp. 458–60.   [4] Quoted in *Full Circle*, p. 546.

foreigners at all, but rather members of the sizeable settlements of Indians, Pakistanis, and other Commonwealth citizens who had gathered in the capital of their former rulers to try to acquire the benefits of its higher education and its higher wages. Moreover, whatever the bus driver might report, Trafalgar Square had its share of Englishmen (and Englishwomen) that Sunday afternoon of 4 November. Some had been led thither, no doubt, by that craving for excitement which impels a proportion of any crowd whatever the occasion; some by the inarticulate need to protest against injustice, real or imagined, experienced by Jean Rice in John Osborne's play:

JEAN. I went to the Rally in Trafalgar Square last Sunday.
BILLY. You did what?
JEAN. I went to the Rally in Trafalgar Square.
BILLY. What for, for God's sake?
JEAN. Because, Grandad, somehow—with a whole lot of other people, strange as it may seem—I managed to get myself steamed up about the way things were going.[1]

And some—who can say they were more, or less?—by a more profound and rigorous conviction that Britain had betrayed her trust: the same conviction that had moved Coleridge over a century before them:

> We have offended, Oh my countrymen,
> We have offended very grievously,
> And been most tyrannous. From East to West
> A groan of accusation pierces Heaven!
> The wretched plead against us; multitudes
> Countless and vehement, the sons of God,
> Our brethren. Like a cloud that travels on,
> Steamed up from Cairo's swamps. . . .

All this will not satisfy behavioural scientists. These will be more interested in the findings of pollsters and in the contents of the Prime Minister's mail-bag. Both, as it happens, offer evidence favourable to the Government. The poll published by the *Daily Express* (not, to be sure, an impartial guide, and one thankfully

[1] John Osborne, *The Entertainer* (London, 1957), pp. 27–28. Billy Rice speaks poignantly for his own generation: 'What d'you make of all this business out in the Middle East? People seem to be able to do what they like to us. Just what they like. I don't understand it. I really don't . . .' (p. 17).

Pcs

seized upon by Sir Anthony Eden in his Memoirs, but which differs little if at all from rivals) shows the proportion of citizens interrogated approving of the Anglo-French action as having risen from 48·5 per cent. on 30 October to 51·5 per cent. on 5 November; those disapproving fell from 39 per cent. to 30 per cent. during the same period.[1] These figures may demonstrate, as Sir Anthony Eden understandably would have them demonstrate, a dramatic vindication of his policy on the part of the people he was elected to serve. Or they may demonstrate, as Professor Max Beloff has suggested, 'that the broad masses of popular opinion whose leanings are registered by the pollsters were hardly moved from their accustomed political allegiances'.[2] On the contents of the Prime Minister's mail-bag, the only authority is the then Prime Minister. 'In the first few days after our decision', Sir Anthony Eden has recorded, 'the letters were heavily adverse, at the outset something like eight to one against. With the passage of time, this majority weakened and finally disappeared until, in the later stages and on the day before the cease-fire, the majority was heavily in favour of the action we had taken to the extent of about four to one.'[3]

It has been asserted that the 'crisis of conscience' brought upon the British public by its Government's unilateral resort to force was 'very much an affair of the intellectuals'.[4] If that is so, it makes the task of identifying significant trends of opinion less difficult. Not all intellectuals are Oxford dons (as not all dons are intellectuals); but how the dons reacted may indicate something of the thoughts and feelings of the intellectual community as a whole.

The Oxford dons, one of their number has reported, by noon on 3 November had been sufficiently stirred by the events of the four days since the Anglo-French ultimatum for 325 of them to sign a resolution 'deploring the government's action on the grounds that it was morally wrong, endangered the solidarity of the Commonwealth, put a grave strain upon the Atlantic alliance, and was a flagrant violation of the U.N. Charter'. The number of signatories rose later to 350. 'The counter-offensive took longer to mount and the signatories to a rival resolution were often more prepared to

---

[1] *Full Circle*, p. 546.

[2] Max Beloff, 'Suez and the British Conscience', *Commentary*, vol. 23, no. 4, April 1957, p. 310.

[3] *Full Circle*, p. 546.   [4] Beloff, op. cit., p. 310.

question the right of the government's opponents to be so categorical, and the propriety of criticizing policy once British troops
were actually engaged, than to affirm the wisdom as such of the
government's decision. A resolution drafted to allow for this line
of thought obtained 136 signatures.' Most dons, Professor Beloff
notes, signed neither resolution. He presumes 'that the abstention-
ists did not include any large number of persons who felt morally
outraged by what had been done'.[1] Either that, or many were
guilty of *trahison des clercs*.

Among the names of those supporting the Government, none
was more distinguished than that of Dr. Gilbert Murray. That so
revered a champion of collective security should now support the
cause of unilateral force made something of a sensation in both
camps—elation among the pro-Government ranks, consternation
among the critics. He explained his position in these words:

> The real danger was that, if the Nasser movement had been allowed
> to progress unchecked, we should have been faced by a coalition of all
> Arab, Muslim, Asiatic, and anti-Western States, led nominally by
> Egypt but really by Russia; that is, a division of the world in which the
> enemies of civilization are stronger than its supporters. Such a danger,
> the Prime Minister saw, must be stopped instantly, and, since the U.N.
> has no instrument, it must be stopped, however irregularly, by those
> nations who can act at once.[2]

They are gratefully quoted by Sir Anthony Eden in his Memoirs;
and they represent (as Professor Beloff has pointed out)

> almost the precise opposite of . . . the main emotional force behind the
> opposition to the Government: the extraordinary strength of the 'anti-
> imperialism' or 'anti-colonialism' which has come to possess 'en-
> lightened' intellectual circles. . . . While an earlier generation, even of
> liberal-minded men like Dr. Murray, were on the whole proud of what
> Europeans had done to bring civilization to other parts of the globe,
> younger men are conscience-stricken about the evils which accompanied
> the process, sceptical about its alleged motives, and proud, if at all, only
> of the fact that Britain has more rapidly than other colonial powers
> accepted the necessity of a severance of all imperial ties. No argument
> was heard more often than the one that the Suez action had interrupted
> the smooth progress of substituting relations of equality for those of
> 'white' domination in the whole Asian and African world.[3]

---

[1] Ibid., p. 312.   [2] *Time and Tide*, 10 November 1956.
[3] Beloff, op. cit., p. 313.

For some Englishmen, mostly younger Englishmen, the 'white man' could do only wrong; for others, as for Dr. Murray, 'that the slave-holders of Saudi-Arabia should sit in judgment upon Britain or France or Israel was quite unacceptable'.[1] In Britain, as in no other Commonwealth country, the conflict over Suez was a conflict between generations.

DOCUMENTS

I. United Kingdom

1. Communication addressed by the Governments of the United Kingdom and France to the Government of Egypt, 30 October 1956[2]

The Governments of the United Kingdom and France have taken note of the outbreak of hostilities between Israel and Egypt. This event threatens to disrupt the freedom of navigation through the Suez Canal, on which the economic life of many nations depends. The Governments of the United Kingdom and France are resolved to do all in their power to safeguard the free passage of the Canal. They accordingly request the Government of Egypt:

(a) To stop all warlike action on land, sea and air forthwith;

(b) to withdraw all Egyptian military forces to a distance of ten miles from the Canal; and

(c) in order to guarantee freedom of transit through the Canal by the ships of all nations and in order to separate the belligerents, to accept the temporary occupation by Anglo-French forces of key positions at Port Said, Ismailia and Suez.

The United Kingdom and French Governments request an answer to this communication within twelve hours. If at the expiration of that time one or both Governments have not undertaken to comply with the above requirements, United Kingdom and French forces will intervene in whatever strength may be necessary to secure compliance.

A similar communication has been sent to the Government of Israel.

[1] Beloff, op. cit.   [2] S.C.O.R., Eleventh Year, 749th Meeting, para. 5.

2. Telegram from the Prime Minister, Sir Anthony Eden, to the President of the United States, Dwight D. Eisenhower, 30 October 1956[1]

We have never made any secret of our belief that justice entitled us to defend our vital interests against Nasser's designs. But we acted with you in summoning the London Conference, in despatching the abortive Menzies mission and in seeking to establish S.C.U.A. As you know, the Russians regarded the Security Council proceedings as a victory for themselves and Egypt. Nevertheless we continued through the Secretary-General of the United Nations to seek a basis for the continuation of negotiations.

Egypt has to a large extent brought this attack on herself by insisting that the state of war persists, by defying the Security Council and by declaring her intention to marshal the Arab states for the destruction of Israel. The latest example of Egyptian intentions is the announcement of a joint command between Egypt, Jordan and Syria.

We have earnestly deliberated what we should do in this serious situation. We cannot afford to see the canal closed or to lose the shipping which is daily on passage through it. We have a responsibility for the people in these ships. We feel that decisive action should be taken at once to stop hostilities. We have agreed with you to go to the Security Council and instructions are being sent this moment. Experience however shows that its procedure is unlikely to be either rapid or effective.

3. Extract from telegram from the Prime Minister, Sir Anthony Eden, to the President of the United States, Dwight D. Eisenhower, 30 October 1956[2]

My first instinct would have been to ask you to associate yourself and your country with the declaration [Document 1]. But I know the constitutional and other difficulties in which you are placed. I think there is a chance that both sides will accept. In any case it would help this result very much if you found it possible to support what we have done at least in general terms. We are well aware that no real settlement of Middle Eastern problems is possible except through the closest co-operation between our two countries. Our two Governments have tried with the best will in the world all sorts of public and private negotiations through the

[1] *Full Circle*, p. 525.   [2] Ibid., pp. 525–6.

last two or three years and they have all failed. This seems an opportunity for a fresh start.

. . . Nothing could have prevented this volcano from erupting somewhere, but when the dust settles there may well be a chance for our doing a really constructive piece of work together and thereby strengthening the weakest point in the line against communism.

4. Statement by the Prime Minister, Sir Anthony Eden, in the House of Commons, 30 October 1956[1]

With your permission, Mr. Speaker, and that of the House, I will make a statement.

As the House will know, for some time past the tension on the frontiers of Israel has been increasing. The growing military strength of Egypt has given rise to renewed apprehension, which the statements and actions of the Egyptian Government have further aggravated. The establishment of a Joint Military Command between Egypt, Jordan and Syria, the renewed raids by guerillas, culminating in the incursion of Egyptian commandos on Sunday night, had all produced a very dangerous situation.

Five days ago news was received that the Israel Government were taking certain measures of mobilisation. Her Majesty's Government at once instructed Her Majesty's Ambassador at Tel Aviv to make inquiries of the Israel Minister for Foreign Affairs and to urge restraint.

Meanwhile, President Eisenhower called for an immediate tripartite discussion between representatives of the United Kingdom, France and the United States. A meeting was held on 28th October, in Washington, and a second meeting took place on 29th October.

While these discussions were proceeding, news was received last night that Israel forces had crossed the frontier and had penetrated deep into Egyptian territory. Later, further reports were received indicating that paratroops had been dropped. It appears that the Israel spearhead was not far from the banks of the Suez Canal. From recent reports it also appeared that air forces are in action in the neighbourhood of the Canal.

During the last few weeks Her Majesty's Government have thought it their duty, having regard to their obligations under the

Anglo-Jordan Treaty, to give assurances, both public and private, of their intention to honour these obligations. Her Majesty's Ambassador in Tel Aviv late last night received an assurance that Israel would not attack Jordan.

My right hon. and learned Friend the Foreign Secretary discussed the situation with the United States Ambassador early this morning. The French Prime Minister and Foreign Minister have come over to London at short notice at the invitation of Her Majesty's Government to deliberate with us on these events.

I must tell the House that very grave issues are at stake, and that unless hostilities can quickly be stopped free passage through the Canal will be jeopardised. Moreover, any fighting on the banks of the Canal would endanger the ships actually on passage. The number of crews and passengers involved totals many hundreds, and the value of the ships which are likely to be on passage is about £50 millions, excluding the value of the cargoes.

Her Majesty's Government and the French Government have accordingly agreed that everything possible should be done to bring hostilities to an end as soon as possible. Their representatives in New York have, therefore, been instructed to join the United States representative in seeking an immediate meeting of the Security Council. This began at 4 p.m.

In the meantime, as a result of the consultations held in London today, the United Kingdom and French Governments have now addressed urgent communications to the Governments of Egypt and Israel. In these we have called upon both sides to stop all warlike action by land, sea and air forthwith and to withdraw their military forces to a distance of 10 miles from the Canal. Further, in order to separate the belligerents, and to guarantee freedom of transit through the Canal by the ships of all nations, we have asked the Egyptian Government to agree that Anglo-French forces should move temporarily—I repeat, temporarily—into key positions at Port Said, Ismailia and Suez.

The Governments of Egypt and Israel have been asked to answer this communication within 12 hours. It has been made clear to them that, if at the expiration of that time one or both have not undertaken to comply with these requirements, British and French forces will intervene in whatever strength may be necessary to secure compliance.

I will continue to keep the House informed of the situation. . . .

[Later:]

Mr. Healey: . . . I hope that this Government take seriously the views of their friends who are also members of the Commonwealth. I would therefore ask the Prime Minister whether the Commonwealth Governments have been consulted about this decision and whether they have approved of it. . . .

The Prime Minister: . . . We have . . . kept in close consultation with the Commonwealth Governments, but the responsibility for the decision was that of the French and British Governments owing to the information reaching us of the situation in the neighbourhood of the Canal. I do not believe that any other course would have been open to any Government.

5. Extracts from speech by the Leader of the Opposition, Mr. Hugh Gaitskell, in the House of Commons, 31 October 1956[1]

. . . In taking this decision the Government, in the view of Her Majesty's Opposition, have committed an act of disastrous folly whose tragic consequences we shall regret for years. [*Hon. Members:* 'Oh'.] Yes, all of us will regret it, because it will have done irreparable harm to the prestige and reputation of our country.

Sir, this action involves not only the abandonment but a positive assault upon the three principles which have governed British foreign policy for, at any rate, the last ten years—solidarity with the Commonwealth, the Anglo-American Alliance and adherence to the Charter of the United Nations. . . .

The Prime Minister said yesterday that he had been in close consultation with the Commonwealth [Document 4]. What were the results of this close consultation? I do not think that there was ever much doubt about what the attitude of the Government of India was likely to be, and we now know. There has now been a special announcement, and in case hon. Members have not seen it, I will read it, stating that the Government of India considers Israel's aggression and the ultimatum of Britain and France a flagrant violation of the United Nations Charter and opposed to all the principles of the Bandoeng Conference [VI, Document 1, *infra*]. . . .

I do not think that there is much doubt that substantially the same attitude is likely to be adopted by Pakistan and Ceylon. But it is not only the Asian members of the Commonwealth who are

1 United Kingdom, H.C. Deb., vol. 558, coll. 1454–5, 1462.

concerned. There are the older Dominions. It is a remarkable and most distressing fact that Australia was unable to support us in the United Nations Security Council. On one resolution Australia abstained, on the other resolutions she voted against us. The Australian Government have said that they are still not in sufficient command of the facts to be able to make a full statement. So it does not seem as though the close consultation has been so very close after all.

The Canadian Government, through the mouth of their Foreign Secretary, have expressed in the coldest possible language their regret at the situation which has arisen [II, Document 1, *infra*]. They have also made it plain, through Mr. Pearson, that they were not consulted in advance before this ultimatum was sent. [*Hon. Members:* 'Shame'.] The New Zealand Prime Minister has said, in substantially the same words as the Canadian Foreign Minister, that he regrets the situation which has arisen, and that he was unable to say whether he supported the United Kingdom or not [IV, Document 2, *infra*].

This is a tragic situation, and I cannot but feel . . . that hon. Members, some of whom I know to be sincerely concerned with the maintenance of this unique institution the British Commonwealth, must too, in their hearts, feel the deepest anxiety at what has happened. . . .

We, as Her Majesty's Opposition, have had to consider what attitude we should adopt to the war on which the Government have so recklessly embarked; we understand, let me say, the gravity of the decision we have to take. We were not, I repeat—and I make no complaint, I merely state it—consulted by the Government in this matter. They did not seek our consent and they indicated last night that we were completely free to make our own decisions.

I must now tell the Government and the country that we cannot support the action they have taken and that we shall feel bound by every constitutional means at our disposal to oppose it. . . .

6. Extracts from speech by the Prime Minister, Sir Anthony Eden, in the House of Commons, 1 November 1956[1]

Much has been said in the course of this debate . . . about the effect of our actions on the unity of the Commonwealth, and, particularly yesterday, Australia was many times mentioned. I

---

[1] United Kingdom, H.C. Deb., vol. 558, coll. 1649–50.

therefore propose to quote to the House a passage from a speech delivered by the Prime Minister of Australia—[*Interruption*]— who certainly did more than anyone else to try to bring success to the 18-Power proposals and is for that deserving of the support and approval, I should have thought, of every hon. Member of the House. . . .

On the question of whether the United Kingdom was at fault in not having consulted other British countries in advance, he said that the answer to this question seemed to him to be that she was not at fault at all. The circumstances were those of great emergency. Hostile armed forces were approaching each other and extensive combat was imminent, and in that combat vital interests in passage of the Canal were quite likely to suffer most serious damage. The Canal was an international waterway with guaranteed freedom of passage for ships of all nations, but that guarantee would cease to have much value if the Canal itself became part of a theatre of active war.

He said that there was literally no time to be lost if any action was to be taken to keep the combatants out of the Canal area and afford it proper protection. In the Australian Government's opinion, Great Britain, whose interests were so vast, was correct in proceeding upon her own judgment and accepting her own responsibility. We were not living in an academic world. The Government of Australia believed that the action taken by the United Kingdom and France was proper. It was quite clear that normal processes of the Security Council were such that even assuming that some resolution could be carried, the Canal would have been involved in war long before any United Nations action could have been effective. Finally, the Australian Government saw nothing sinister in all this. On the contrary, it seemed to them realistic and to pay due regard to the moving facts of life.

The Prime Minister of New Zealand has also spoken in a similar sense of support. . . .

7. Statement by the Prime Minister, Sir Anthony Eden, and extract from speech by the Leader of the Opposition, Mr. Hugh Gaitskell, in the House of Commons, 3 November 1956[1]

The Prime Minister: We move the Adjournment of the House in

---

[1] United Kingdom, H.C. Deb., vol. 558, coll. 1857–62.

order that I may make the following statement in accordance with the undertaking that I gave yesterday to give the House as soon as I possibly could an indication of the reply we propose to send to the Resolution of the Assembly of the United Nations.

I should first recall a statement which I made in the House in the course of my speech on 1st November [Document 6] when I said this:

The first and urgent task is to separate these combatants and to stabilise the position. That is our purpose. If the United Nations were then willing to take over the physical task of maintaining peace in that area, no one would be better pleased than we. But police action there must be to separate the belligerents and to prevent a resumption of hostilities.

Since that statement was made, I have had consultations in London with the French Foreign Minister. As a result, Her Majesty's Government and the French Government are sending the following reply to the Resolution of the United Nations General Assembly:

The British and French Governments have given careful consideration to the Resolution passed by the General Assembly on 2nd November. They maintain their view that police action must be carried through urgently to stop the hostilities which are now threatening the Suez Canal, to prevent a resumption of these hostilities and to pave the way for a definitive settlement of the Arab-Israel war which threatens the legitimate interests of so many countries.

They would most willingly stop military action as soon as the following conditions could be satisfied:

(i) Both the Egyptian and the Israeli Governments agree to accept a United Nations force to keep the peace;

(ii) The United Nations decides to constitute and maintain such a force until an Arab-Israel peace settlement is reached and until satisfactory arrangements have been agreed in regard to the Suez Canal, both agreements to be guaranteed by the United Nations;

(iii) In the meantime, until the United Nations Force is constituted, both combatants agree to accept forthwith limited detachments of Anglo-French troops to be stationed between the combatants.

We have been in consultation with the Governments of Australia and New Zealand. [*Hon. Members:* 'And Canada?'] I am coming to that. The House will understand the difficulties of timing in these consultations, but I have good reason to believe that those Governments will welcome my statement. We have also communicated the substance of the statement at once to the Governments of Canada and the United States, and to the Secretary General of the United Nations.

Mr. Gaitskell: The first paragraph of the Resolution carried by 64 votes to 5 in the General Assembly of the United Nations calls upon all parties now involved in hostilities to agree to an immediate cease-fire and to halt the movement of military forces and arms into the area.

It is unfortunately perfectly clear, both from the reports of the continuing and, indeed, intensification of bombing by British planes and from the Prime Minister's statement this morning, that the British Government are not carrying out the recommendation of the Assembly. We are, therefore, faced with the position that our Government are defying a Resolution of the United Nations Assembly, carried by a majority which is larger, I believe, than that on any other Resolution previously carried by the Assembly. We can only say that, for our part, we regard this as utterly deplorable.

As regards the conditions laid down by the Government, it is no part of the business of Her Majesty's Government to lay down conditions in this matter. It is their duty, as loyal members of the United Nations—if they were loyal members—to accept that majority decision. [*Hon. Members:* 'And sell Britain?']

I must ask the Prime Minister a number of questions on his statement. First of all, is he aware that the Egyptian Government have already announced that they are prepared to agree to an immediate cease-fire if all the other parties do so as well? Therefore, one of the combatants at any rate has already agreed to this.

Secondly, is the Prime Minister aware, as he should be, because the Minister of Transport and Civil Aviation has announced it, that the Suez Canal is now blocked and that the consequence of the intervention by Her Majesty's Government, far from facilitating the passage of ships through the Canal, has had precisely the opposite effect?

Is the Prime Minister further aware that the Israeli Government have announced that the fighting in the Sinai Desert area is virtually at an end, and that, therefore, the original situation, from that point of view, has substantially changed?

The Canal is blocked, there has been no rescue operation for British ships, no British lives have been saved, and all that has happened is that the intervention of Her Majesty's Government on behalf—or, rather, against Egypt—has no doubt prematurely brought the operations in the Sinai Desert to a close. [*Hon. Members:* 'Warmonger.'] . . . What Her Majesty's Government have undoubtedly done, of course, is to intervene against Egypt, which was clearly attacked by Israel. I do not know whether they regard that as a matter of which they should be proud. I do not know whether they regard that as separating the combatants. I do not know whether they regard that as settling hostilities. What they have done is to bomb a number of civilians as well as military installations in Egypt. What they have done is to destroy all faith in collective security. What they have done now, by refusing to accept the United Nations Resolution, is virtually to destroy that institution, which the Prime Minister once described as the hope of mankind. . . .

All this is unquestionably in defiance of the Resolution of the General Assembly. One cannot get away from that. For our part, we regard the Government's reply today as the most tragic statement that has been made in this House since 1939. . . .

Up to this moment, I for my part had hoped for a change in Government policy. I had hoped originally that the Government would have accepted our first proposal to defer action. They refused. I hoped then that the pressure of world opinion upon them would have made them change their mind, and I hoped finally that the passing of this Resolution by such a vast majority in the United Nations Assembly would have brought them to their senses.

Alas, that is not so, and we can draw only one conclusion. That is that if this country is to be rescued from the predicament into which the Government have brought it, there is only one way out, and that is a change in the leadership of the Government. Only that now can save our reputation and re-open the possibility of maintaining the United Nations as a force for peace. We must have a new Government and a new Prime Minister. . . .

8. Extracts from broadcast by the Prime Minister, Sir Anthony Eden, 3 November 1956[1]

I know that you would wish me, as Prime Minister, to talk to you tonight on the problem which is in everybody's mind; and to

[1] *The Listener*, 8 November 1956, pp. 735–6.

tell you what has happened, what the Government has done, and why it has done it. . . .

As a Government we have had to wrestle with the problem of what action we should take. So have our French friends. The burden of that decision was tremendous but inescapable. In the depths of our conviction we decided that here was the beginning of a forest fire, of immense danger to peace. We decided that we must act, and act quickly.

What should we do? We put the matter to the Security Council. Should we have left it to them? Should we have been content to wait to see whether they would act? How long would this have taken? And where would the forest fire have spread in the meantime? Would words have been enough? What we did was to take police action at once: action to end the fighting and to separate the armies. We acted swiftly and reported to the Security Council, and I believe that before long it will become apparent to everybody that we acted rightly and wisely.

Our friends inside the Commonwealth, and outside, could not in the very nature of things be consulted in time. You just cannot have immediate action and extensive consultation as well. But our friends are coming—as Australia and New Zealand have already done and I believe that Canada and the United States will soon come—to see that we acted with courage and speed, to deal with a situation which just could not wait.

There are two things I would ask you never to forget. We cannot allow—we could not allow—a conflict in the Middle East to spread; our survival as a nation depends on oil and nearly three-quarters of our oil comes from that part of the world. . . .

The other reflection is this. It is a personal one. All my life I have been a man of peace, working for peace, striving for peace, negotiating for peace. I have been a League of Nations man and a United Nations man, and I am still the same man, with the same convictions, the same devotion to peace. I could not be other, even if I wished, but I am utterly convinced that the action we have taken is right.

Over the years I have seen, as many of you have, the mood of peace at any price: many of you will remember that mood in our own country and how we paid for it. Between the wars we saw things happening which we felt were adding to the danger of a great world war. Should we have acted swiftly to deal with them—

even though it meant the use of force? Or should we have hoped for the best, and gone on hoping and talking—as in fact we did?

There are times for courage, times for action—and this is one of them—in the interests of peace. I do hope we have learned our lesson. Our passionate love of peace, our intense loathing of war, have often held us back from using force even at times when we knew in our heads, if not in our hearts, that its use was in the interest of peace. And I believe with all my heart and head—for both are needed—that this is a time for action, effective and swift. Yes, even by the use of some force in order to prevent the forest fire from spreading—to prevent the horror and devastation of a larger war.

The Government knew, and they regretted it, that this action would shock and hurt some people: the bombing of military targets, and military targets only; it is better to destroy machines on the ground than let them destroy people from the air. We had to think of our troops and of the inhabitants of the towns and villages. After all, it was our duty to act and act swiftly, for only by such action could we secure peace. . . .

So finally, my friends, what are we seeking to do? First and foremost, to stop the fighting, to separate the armies, and to make sure that there is no more fighting. We have stepped in because the United Nations could not do so in time. If the United Nations will take over the police action we shall welcome it. Indeed, we proposed that course to them. And police action means not only to end the fighting now but also to bring a lasting peace to an area which for ten years has lived, or tried to live, under the constant threat of war. . . .

9. Letter from the Member for Woodford, Sir Winston Churchill, to the Chairman of the Woodford Conservative Party Association, 3 November 1956[1]

The British connexion with the Middle East is a long and honourable one. Many of the states there owe their origin and independence to us.

In peace we have assisted them in many ways, financially, technically, and with our advisers in every sphere. In war we have defended them at great cost. Above all, we have endeavoured to confer on them the benefits of justice and freedom from internecine

[1] *The Times* (London), 5 November 1956.

wars. In the last few years the United States, France and we ourselves have been principally concerned with keeping the peace between Israel and her neighbours.

In spite of all our endeavours, the frontiers of Israel have flickered with murder and armed raids. Egypt, the principal instigator of these incidents, had openly rejected and derided the Tripartite Declaration by which we, the French and the Americans sought to impose restraint. The last few days have brought events to a head. Israel, under the gravest provocation, erupted against Egypt. In this country we had the choice of taking decisive action or admitting once and for all our inability to put an end to strife.

Unfortunately, recent months have shown us that at present it is not possible to hope in this area for American co-operation on the scale and with the promptness necessary to control events. Her Majesty's Government and the Government of France have reacted with speed. I regret profoundly that the Egyptian reaction has forced the present course on us. But I do not doubt that we can shortly lead our course to a just and victorious conclusion.

We intend to restore peace and order to the Middle East, and I am convinced that we shall achieve our aim. The American alliance remains the keystone of our policy. I am confident that our American friends will come to realize that, not for the first time, we have acted independently for the common good. World peace, the Middle East, and our national interest will surely benefit in the long run from the Government's resolute action. They deserve our support.

10. Extracts from broadcast by the Leader of the Opposition, Mr. Hugh Gaitskell, 4 November 1956[1]

It has been a tragic, terrible week: indeed, a tragic and terrible day, with the news coming in about Hungary. It has been, I think, by far the worst week, for the world and for our country, since 1939. . . .

Make no mistake about it: this is war—the bombing, the softening up, the attacks on radio stations, telephone exchanges, railway stations, to be followed, very, very soon now, by the landings and the fighting between ground forces.

We are doing all this alone, except for France: opposed by the

---

[1] *The Listener*, 8 November 1956, pp. 737–8.

world, in defiance of the world. It is not a police action; there is
no law behind it. We have taken the law into our own hands. That
is the tragic situation in which we British people find ourselves
tonight. We would all have thought it inconceivable a week ago.

Why was it done? The Prime Minister justifies it on these
grounds: first of all, he says, to protect British lives and property.
But there has been no rescue operation. Instead, to tell the truth,
thousands of British civilians now living in Egypt have been put
in grave danger because of what we have done.

The Prime Minister says it was to safeguard the Canal and the
free passage through it. What has happened to the Canal? It is
blocked because of what we have done. Was the Canal indeed ever
really menaced before we began bombing? I very much doubt it.
There is no evidence to show that it was. I am afraid the real reason
for going to war with Egypt was different. I have seen the text of
the first broadcast of the Allied Command to the Egyptians.
This is what it said—in Arabic, of course: 'Oh Egyptians, why has
this befallen you? First, because Abdul Nasser went mad and
seized the Canal.' The broadcast was right—it was this which
really induced the Prime Minister to decide on intervention.

The Prime Minister has said we were going in to separate the
two sides, but you do not separate two armies by bombing airfields
and landing troops a hundred miles behind one side only. No. This
is a second onslaught on a country which was already the victim of
an attack.

Now a new idea has been put forward: the idea that we are
going in to make way for a United Nations force. But nothing was
said about this in the ultimatum to Egypt. Nothing was said about
this at the Security Council. If this was the Government's plan,
why on earth did they not put it forward before? Why did they not
propose it right at the beginning, accepting the rest of the Security
Council's resolution. . .? I will tell you why they did not do this.
If the Prime Minister had agreed to this Britain would not have
been able to occupy the Canal; for the idea of the United Nations
police force, proposed by Canada yesterday, is quite different. It
would not give us control of the Canal: it has another aim—the
aim of keeping the Israeli and Arab forces within their own
frontiers, of patrolling the borders of Israel and the Arab States;
and these are one hundred miles from the Canal. . . .

What are the consequences? We have violated the Charter of

the United Nations. . . . A deep, deep division in the Commonwealth—only Australia and New Zealand support us. Canada and
South Africa have abstained. India, Pakistan and Ceylon are all
against us. This is a very grave consequence. For I believe, as do
millions of others, that this Commonwealth of ours was—and
could have been—the greatest force for peace and unity in the
world: above all, a bridge between East and West, of incalculable
value. That bridge is now almost destroyed. . . .

Here at home the Government policy of war with Egypt has
produced terrible heart-searchings. The Archbishop of Canterbury
has led a deputation, of all denominations of the Churches, to the
Government. The all-party United Nations Association has
denounced the policy in strong terms. . . . Mr. Nutting, the Minister of State, whose job was especially concerned with United
Nations affairs, has resigned from the Government because he
thinks the policy is indefensible. . . .

. . . I do not think there is any doubt as to what the policy should
be now. We should, surely, without qualification, argument or
conditions, accept the resolution of the Assembly of the United
Nations calling for an immediate cease-fire. . . . We should also
give full support to the new resolution on which we abstained
to-day, for a United Nations force to police the Arab-Israel
borders until a proper peace settlement has been reached.

But—make no mistake—this means abandoning the idea which has
been at the root of this policy: the idea of trying to solve the Suez
Canal problem by force. It means going back to negotiating. . . .

I do not believe the present Prime Minister can carry out this
policy. I bear him no ill-will. We have been personally quite
friendly. But his policy this last week has been disastrous; and he
is utterly, utterly discredited in the world. Only one thing now can
save the reputation and honour of our country. Parliament must
repudiate the Government's policy. The Prime Minister must
resign. . . .

11. Extract from message of the Prime Minister, Sir Anthony Eden, to the Chairman of the Council of Ministers of the U.S.S.R., N. A. Bulganin, 6 November 1956[1]

I have received with deep regret your message of yesterday.
The language which you used in it made me think at first that I

---

[1] *Commonwealth Survey*, 13 November 1956, pp. 966–7.

could only instruct Her Majesty's Ambassador to return it as entirely unacceptable. But the moment is so grave that I feel I must try to answer you with those counsels of reason with which you and I have in the past been able to discuss issues vital for the whole world.

. . . You accuse us of waging war against the national independence of the countries of the Near and Middle East. We have already proved the absurdity of this charge by declaring our willingness that the United Nations should take over the physical task of maintaining peace in the area. You accuse us of barbaric bombardment of Egyptian towns and villages. Our attacks on airfields and other military targets have been conducted with the most scrupulous care in order to cause the least possible loss of life. Some casualties there must have been. We deeply regret them. When all fighting has ceased, it will be possible to establish the true figure. We believe that they will prove to be small. They will in any event be in no way comparable with the casualties which have been, and are still being, inflicted by the Soviet forces in Hungary. . . .

Her Majesty's Government have repeatedly said that the essential aim of the action taken by the British and French Governments was to stop the fighting between Israel and Egypt and to separate the combatants. That aim has now been virtually achieved. As regards the future, you know that the Canadian Government have proposed the establishment of an emergency international United Nations force in the area. The General Assembly has taken the first steps to organize such a force. Her Majesty's Government fully approve the principle of an international United Nations force. Indeed, we suggested this ourselves. . . .

12. Statement by the High Commissioner in South Africa, Sir Percivale Liesching, 6 November 1956[1]

It is true that there was no prior consultation with the Union [of South Africa] or with any other Commonwealth Government. Hostilities had already broken out. There was a grave danger to the Suez Canal, and there was an immediate need to prevent the hostilities engulfing the whole Middle East. This called for the swiftest possible action and allowed no time to observe the established practice of prior consultation.

[1] *Cape Times*, 7 November 1956.

But it would be a great mistake to infer from this case, caused as it was by circumstances beyond our control, that there has been any change of policy over consultation with other Commonwealth Governments or that any change is contemplated. Nothing could be further from the facts. The United Kingdom's object has always been and will remain to consult all Commonwealth countries about issues of importance affecting them. The only exceptions (and those much against their will) must be those cases, such as the present situation, in which the speed of events puts it beyond their power to do so.

I should add that Commonwealth consultation rests on a constant flow of information on every matter of international concern and that this flow has continued throughout, and still continues.

13. Extracts from broadcast by the Secretary of State for Foreign Affairs, Mr. Selwyn Lloyd, 7 November 1956[1]

. . . Law and order cannot be maintained in any country without policemen. The burglar is not deterred because a society of property-owners passes a resolution condemning house-breaking. Unlawful wounding is not stopped because the victims may all condemn violence. So it is in international society. . . .

What, then, should we have done last week? A war was taking place. It was a local war, certainly. But it was in an area of the most vital concern, both to ourselves and to the world. The Suez Canal, the main artery to and from the Commonwealth between Europe and Asia, was in imminent danger. The local war could only too easily have spread into a wider war, and then into a world conflict. With so much at stake, should we have sat back and done nothing? Should we have contented ourselves with resolutions? Should we have talked endlessly and to no avail? Or should we have acted? I believe that we did the right thing and the courageous thing. We acted in the cause of ultimate peace. We had the forces ready. We gave notice of what we were intending to do. We did it promptly. We did it with a minimum of force and casualties. And we did it successfully—in seven days. . . .

---

[1] *The Listener*, 15 November 1956, pp. 800–1.

14. Extract from broadcast by the Deputy Leader of the Parliamentary Labour Party, Mr. James Griffiths, 8 November 1956[1]

. . . Last night Mr. Selwyn Lloyd spoke of the seven fateful days [Document 13]. He seemed to speak of them with pride. I regret that I cannot, and [know] that many of you, my fellow citizens, do not share his pride. We think of those days, yes, and we shall remember them as the seven days of humiliation for our country; the days in which we lived through the bitter experience of seeing a British Government, Mr. Selwyn Lloyd's Government, waging war in violation of our solemn pledges under the Charter of the United Nations; defying the resolutions of the General Assembly; using the veto—the first British Government ever to use the veto —in the Security Council; tearing up the Tripartite Pact which bound us, and France, to act in unison with the United States in the Middle East; and bringing the Commonwealth to the brink of dissolution.

During the past half-century we have been slowly but surely achieving an ideal unique in human history. We have been transforming an Empire into a Commonwealth of free and equal nations, and this has become the greatest interracial body ever known to mankind. It is a force for peace, for it shows by example how people of different colour, creeds and tongues can live together in friendship and equality. Whenever the Commonwealth has been united, it has had an influence on world affairs vastly greater than any influence that Britain could have had alone.

Now let us look at what the action of the British Government has done this last week to the Commonwealth. The Canadian Prime Minister has declared that the present crisis has stretched the bonds of the Commonwealth more than any event since the Second World War. The Government of Pakistan has unreservedly condemned our aggressive action, and strong demands have been made in Pakistan that they should leave the Commonwealth. One of our oldest friends in India, a past Governor-General, has called for India's withdrawal from the Commonwealth [VI, Document 6, *infra*]; and one of our newest Dominions, Ceylon, has felt compelled to vote against us in the United Nations. Australia and New Zealand are, like us, deeply divided. The whole concept of the Commonwealth, and its contribution to world peace, are in danger. . . .

[1] *The Listener*, 15 November 1956, pp. 801–2.

II. CANADA

1. Extracts from broadcast by the Prime Minister, Mr. L. S. St. Laurent, 4 November 1956[1]

I think it my duty to speak to you tonight about the very grave events of the last two weeks. I should like first to talk about the Middle East crisis. I would like to explain to you the Government's recent actions in the context of our general policy in the Middle East. For the last few years peace has been precarious in this area, especially around the borders of Israel, whose creation as a state was recommended by the United Nations General Assembly with Canada's support in November 1947.

While the tensions arising out of the situation in the Middle East have continued, Canada has steadily encouraged efforts to secure a fair settlement based on the principle that Israel should live and prosper—but not the principle that it should expand at the expense of its Arab neighbours.

A recent communist intervention in the Middle East has contributed directly to the present crisis. By supplying offensive weapons in large quantities to Egypt, the communist world threatened to upset the balance of power between Israel and its Arab neighbours. In order to help redress the potential imbalance Canada agreed a few weeks ago to authorize the export of 24 F-86 jet fighter planes to Israel over a six-month period. We realized however that a permanent settlement between Israel and its neighbours arranged by the United Nations was the only way in which peace could be preserved in the long run.

Egypt's nationalization of the Suez Canal Company increased the dangers inherent in the Middle East situation. The Egyptian action introduced a threat to the trade on which the economic life of many countries depends. It placed the control of shipping in the Canal in the hands of a government which for some years has been denying access to the Canal for Israeli ships in defiance of a Security Council resolution.

In the crisis which resulted from the nationalization of the Canal Company, the Canadian Government has followed a definite and consistent policy in public statements and in private discussions with the nations concerned. We have advocated that a settlement of the issues relating to the Canal which directly affect so

[1] *E.A.*, November 1956, pp. 322–5.

many countries should be achieved under the auspices of the United Nations and that there should be no resort to force. The Canadian Government welcomed the 18-Power proposals agreed to at the London Conference in August as a sound basis for settlement. We have stated our belief that this settlement should respect the legitimate sovereign rights of Egypt. It should also safeguard the right of ships of all nations to pass through the Canal. At the same time it should protect the international waterway from arbitrary and unjustified intervention by any country, including Egypt. We have stated our belief that this settlement should be embodied in co-operative arrangements with which the United Nations should be associated in an appropriate manner.

Because we believe that a permanent settlement of Israel's relations with its neighbour and of the future of the Suez Canal should be reached by peaceful negotiations under the aegis of the United Nations, the Canadian Government regrets that Israel proceeded last week to use force against Egypt, although we recognize that Israel has been subject to grave threats and provocations during the last few years. Though we recognize the vital importance of the Canal to the economic life and international responsibilities of the United Kingdom and France, we could not but regret also that, at a time when the United Nations Security Council was seized of the matter, the United Kingdom and France felt it necessary to intervene with force on their own responsibility.

Your Government has acted promptly in this crisis. We have taken immediate steps to further the safety of Canadian civilians in the Middle East. We have suspended the shipment of jet interceptor aircraft to Israel. The Canadian Government voted for consideration of the Israeli attack at the Special Session of the United Nations General Assembly on November 1 which was called after Security Council action was made impossible by the negative votes of two of its permanent members.

A United States resolution was introduced which called for an immediate cease-fire, the prompt withdrawal of forces and the end of military shipments to the area. On Friday morning this resolution was carried by 64 votes in favour to 5 against, including the United Kingdom and France. Canada and five other nations abstained in the vote on this resolution.

In explaining the reasons for this abstention, I should like to quote part of what Mr. Pearson said in the General Assembly [Part IV, II, Document 1, *infra*]:

I regret the use of military force in the circumstances which we have been discussing but I regret also that there was not more time, before a vote had to be taken, for consideration of the best way to bring about that kind of cease-fire which would have enduring and beneficial results.

He later added:

I therefore would have liked to see a provision in this resolution . . . authorizing the Secretary-General to begin to make arrangements with member governments for a United Nations force large enough to keep these borders at peace while a political settlement is being worked out.

We have swiftly followed up this suggestion. At another special session of the United Nations General Assembly in New York last night, Mr. Pearson introduced a resolution on behalf of Canada which requests the Secretary-General to submit within 48 hours a plan for a United Nations force to secure and to supervise the cease-fire arrangements which were referred to in the United States resolution. Mr. Pearson explained that no members of the United Nations are to be asked to provide forces without their previous consent. The Canadian Government is ready to recommend Canadian participation in such a United Nations force if it is to be established and if it is thought that Canada could play a useful role.

The Canadian resolution was passed by the General Assembly early this morning without a single dissenting vote although there were a number of abstentions. At the same time the General Assembly passed a resolution sponsored by 19 nations; it reaffirmed the United States resolution about cease-fire arrangements and authorized the Secretary-General to arrange with the nations concerned the implementation of this resolution and asked him to report on their compliance.

The establishment of the United Nations force will be to ensure an effective cease-fire in the affected area. The Governments of the United Kingdom and France have signified their willingness, under certain conditions, to suspend their military intervention if a United Nations truce force is given responsibility. According to present information, Israel and Egypt have stated their willingness to accept cease-fire arrangements provided other parties also co-operate.

We have strong reason to believe that a United Nations command will be established within the 48 hours set in the Canadian resolution. This is only the first step toward a permanent settlement of Middle East problems. In the General Assembly last night the United States introduced two new resolutions which seek to establish United Nations committees to consider the future of Israel's relations with its neighbours and the future of the Suez Canal. We believe these resolutions represent a constructive approach to these problems. We will actively participate in efforts to make progress on the lines which the Assembly has approved.

We have spent anxious days of late and I am sure you all share our anxiety. The present crisis has strained both the Western Alliance and the bonds of the Commonwealth more than any other event since the Second World War. If we can use it as the opportunity to dissipate the black cloud which has hung over the Middle East these many years, the present danger and strains may prove to have been a price worth paying. . . .

. . . In conclusion, I wish to assure my listeners that all the members of their Government have been in full agreement at all times as to what should be done and what could be said and when it should be done and when it could be said. And I am sure that, if and when any action of ours requires, according to our practices, the approval by Parliament, that approval will be given in no uncertain terms.

Let us all hope that this approach to unanimity of men of good will of so many nations may help to realize that part of our daily prayer to a Power greater than any here below: 'Thy will be done on earth as it is in heaven.'

III. AUSTRALIA

1. Extract from statement by the Permanent Representative to the United Nations, Dr. Ronald Walker, to the Security Council, 30 October 1956[1]

The Australian delegation welcomes the action of the United States delegation in bringing this matter urgently before the Security Council. We believe that the Security Council must assume its responsibilities in relation to the increasingly difficult and dangerous situation in the Middle East.

The military operations by Israel that have been reported over

[1] S.C.O.R., 748th Meeting, pp. 6–7.

the last day give rise to the very greatest concern throughout the whole world, a concern that is fully shared by the Australian Government. The information and details at our disposal are of course far from complete, and the rapidity with which events are moving and the number of various reports make it not too easy for Governments to determine rapidly just what is the wisest course of action in the present circumstances. I say 'what is the wisest course of action' because undoubtedly the objective that we all have in our hearts and minds is to arrest the fighting which has begun and to endeavour to restore peace to this troubled area.

Although the details are far from complete, it is quite evident that the operations undertaken by Israeli military forces have been clearly in contravention of the Armistice Agreements. There is apparently no contesting of these basic facts, and it seems to us that clearly puts the Israel Government in the wrong in this particular matter.

We have always taken the view that the problem of Israel must be seen in the broad context of the menaces and threats and actions taken against Israel. But we have all along felt that the violence of the reprisals on various occasions has not been justified by the particular events that have led up to them.

In this particular case, in contrast with the situation that the Council has been discussing along the Jordanian frontier, we have not had reports of any recent actions on the side of Egypt that would afford justification or provocation for an action of this kind. This seems to be quite apparent. It is true that Israel has suffered at the hands of Egypt in various ways and that Egypt may be regarded as in the wrong in certain very definite respects arising from the Armistice Agreements. But that, in our view, does not justify what has been taking place in recent hours. . . .

2. Statement by the Permanent Representative to the United Nations, Dr. Ronald Walker, to the Security Council, 30 October 1956[1]

The statement which I made in the Council this morning [Document 1] was based upon the latest instructions that I had received from my Government. I made it quite clear that our objective in this matter and the objective of the Security Council should be to arrest the fighting as soon as possible. At the same

[1] S.C.O.R., 749th Meeting, pp. 21–22.

time, I said that my Government considered that the first task of the Council was to obtain an objective and official account of the facts as a preliminary to considering what practical steps in the Council were most likely to achieve the objectives on which we are all agreed.

In the course of the debate, we have had a great deal of additional information, and we have heard important statements from the representatives of Egypt and Israel. But it is also a fact, I am sorry to say, that my Government has not had an opportunity to consider and to advise me of its views on any of these most recent developments and the facts brought before us in the last few hours. Even the text of the United States draft resolution came into my hands only this afternoon. And although it was dispatched immediately to Canberra, it would, at the earliest, only now be available there for ministers to consider.

Meanwhile, we have had the news of the course of action envisaged by the United Kingdom and France, and the representative of the United Kingdom has argued with some force that such action is more likely, in his view, to arrest the fighting—and this is our objective—than the adoption of the United States draft resolution. Our relations with the United States are so close and our respect for its efforts to maintain peace is so great that it is extremely unpleasant for me to have to raise any objection against Mr. Lodge's insistence upon an immediate vote. It is not, however, in my view, unreasonable for my delegation to urge that no vote be taken this afternoon, and that the decision be deferred until later, thereby making it possible for me to receive instructions based on my distant Government's consideration of the facts we have learned today. If we are required to vote at this moment, I would feel obliged to abstain.

I should also like to remark that a pause giving time for reflection and conversations might help to avoid a tragic divergence of views between our friends, who together bear so much of the burden of maintaining freedom and building a peaceful world.

3. Extracts from statement by the Permanent Representative to the United Nations, Dr. Ronald Walker, to the Security Council, 31 October 1956[1]

. . . In connexion with the proposal that an emergency special session of the General Assembly should be called, I should like to

---

[1] S.C.O.R., 751st Meeting, pp. 20–21.

say, first of all, that we do recognize the value of such procedures in ordinary circumstances, where effective action by the United Nations is being blocked by the use of the veto. But I should like to remark, further, that the present situation is by no means an ordinary one. If the United Nations had developed along the lines we envisaged when it was founded, there would have been no question of such an action as that taken and announced by the United Kingdom and France, which has led to today's discussion. . . .

We are meeting here today because of the tragic situation which has existed in the Middle East—a situation the gravity of which has been increasing but which, as I say, has existed for a very considerable time. Heaven knows that the Security Council has done its best to maintain peace in the Middle East. It has investigated, it has endeavoured to conciliate, it has recommended, it has issued calls to the parties. It has been defied by Israel and defied by the Arab neighbours of Israel again and again. The Security Council sent the Secretary-General to the area, and he had achieved, we believed, a considerable degree of success in holding the position. . . . But there has still been deterioration in this tragic situation. . . .

Yesterday we met to discuss the latest phase in the struggle that had developed between Israel and its neighbours. And we have heard, yesterday and today, statements on behalf of the United Kingdom and French Governments regarding the action which they are taking and which is under consideration at this time. The representatives of the United Kingdom and France have made it clear, on behalf of their Governments, that the measures they are taking are temporary measures, that they are measures not directed against the sovereignty or territorial integrity of Egypt, that they are measures which will be terminated as soon as peace is restored, and that their objective is to restore the peace. . . .

. . . The record of the United Kingdom and France in the work of the United Nations, and their contributions in the past to our attempts to solve these problems, have been such as to merit dispassionate and even sympathetic consideration for the statements that they have made in this Council, and I would hope that the Council is able to accept the assurances which the representatives of these great States have given us.

I ask myself whether an emergency special session of the General Assembly, called as is now proposed, will contribute to the solution of this conflict—a conflict, as we were considering it yesterday, between Israel and Egypt, for since the Council has felt it appropriate to admit this proposal for an emergency special session, and has rather rejected the view that the question dealt with yesterday and the question dealt with today are entirely different, it seems to me that, if there is an emergency session, it must be concerned not only with the action taken by the Governments of the United Kingdom and France today but with the continuing conflict between Israel and its neighbours, for the whole situation is just one phase of that unresolved conflict which has plagued the world and the Security Council for so long. I ask myself, then, whether a special session will contribute to a solution of these problems. I very much doubt it.

4. Questions by the Leader of the Opposition, Dr. Herbert Evatt, and answers by the Prime Minister, Mr. R. G. Menzies, in the House of Representatives, 31 October 1956[1]

Dr. Evatt: The question I wish to ask the Prime Minister concerns a very urgent matter. Can the right honourable gentleman say whether the United Kingdom Government communicated to him or to the Australian Government its intention prior to its taking certain action at the meeting called by the United Nations Security Council to discuss the Israel-Egypt situation? Secondly, what instructions were given to the Australian representative at that meeting, who, having expressed the opinion that Israel was guilty of aggression against Egypt [Document 1], did not record a vote on the matter? Were any instructions given to him? Have any consultations taken place, either with Australia's representative at the United Nations or with the British and French Governments? I ask the Prime Minister, more broadly, whether he will take the opportunity of making a general statement of Australia's views in relation to this critical situation, whether Australia has any obligation in the matter, express or implied. Will the right honourable gentleman ensure that the House is kept informed and in session, if there is any possibility of Australia becoming involved in the obligations referred to in some of the messages?

Mr. Menzies: This problem in the Middle East has, as honourable members know, arisen somewhat suddenly, and the reports

[1] Australia, H.R. Deb., 1956 (Second Period), pp. 1941–2.

are even now somewhat conflicting. We are not yet in a position, on the basis of official reports, to say anything definitive to the House. The Prime Minister of the United Kingdom made a statement in the House of Commons within the last few hours [I, Document 4, *supra*], the text of which I am prepared to read to the House, which exhibits the approach of the United Kingdom Government. As to the other aspect of the matter referred to by the Leader of the Opposition, I may say that we have been in constant communication, not only with the United Kingdom Government, but with our own posts and representatives, and have so far exhibited a very keen desire to know what the facts are. . . . As soon as we have the facts reasonably clearly available on the matter, I shall naturally take the first opportunity of making a statement to the House about it. . . .

Dr. Evatt: I desire to ask the Prime Minister a supplementary question arising out of his answer to my previous question. Was the Australian Government consulted in any way before this ultimatum containing an intimation of the use of force was issued by the French and British Governments? Further, did the Australian Government give instructions to Australia's representative on the United Nations Security Council? Since our representative abstained from voting, will the right honourable gentleman tell the House what instructions were given to him?

Mr. Menzies: The United Kingdom Government has been in communication with us, and we with it, since the matter arose, and we have exchanged ideas. We have been in communication with our posts, and our posts have been in communication with us. We have been in communication also with the Australian representative at the United Nations. I think that it would be completely unwise, in respect of a matter which is at present at a very critical stage, to take any one of these matters out of context. That is why I propose to make a comprehensive statement at the earliest possible moment. But I do not propose to make it except on a clear official basis of fact.

5. Extracts from statement by the Prime Minister, Mr. R. G. Menzies, in the House of Representatives, 1 November 1956[1]

. . . When it first became known to us that the Security Council was meeting urgently, our instructions to our representative on the

---

[1] Australia, H.R. Deb., 1956 (Second Period), pp. 2057–9.

Council, Dr. Walker, were that before any resolution was passed the Council should satisfy itself about the facts which, at that time, were, in Canberra, completely obscure; we pointed out to him that judgment by the Security Council should not be too hasty and should follow a quick ascertainment of the facts rather than precede it. The Council had placed before it by the representative of the United States a resolution....[1] Great Britain and France voted against this resolution, being plainly of the opinion that it was aimed at imposing disabilities upon Israel, and Israel only. The Australian representative abstained from voting, for, by the time the terms of the resolution reached us, it was too late to add to the instructions given and, in any event, the investigation of the facts asked for by Dr. Walker had not occurred. His abstention was, therefore, the sensible and proper course. . . .

. . . We were early yesterday morning advised from our Acting High Commissioner in London that the United Kingdom had in mind calling upon both Israel and Egypt to cease fighting and to withdraw their forces from the neighbourhood of the Canal. At 1 p.m. yesterday, we were advised that the matter was under most urgent consideration by the United Kingdom Cabinet. At 1.30, we learned that Great Britain and France had delivered what was in effect an ultimatum to both Egypt and Israel calling for an answer within twelve hours. . . .

I now proceed to say something about two questions which will

---

1 The text of this resolution is as follows:

'*The Security Council*,
Noting that the armed forces of Israel have penetrated deeply into Egyptian territory in violation of the General Armistice Agreement between Egypt and Israel.
Expressing its grave concern at this violation of the Armistice Agreement,
1. *Calls upon* Israel immediately to withdraw its armed forces behind the established armistice lines;
2. *Calls upon* all Members:
   (a) to refrain from the use of force or threat of force in the area in any manner inconsistent with the purposes of the United Nations;
   (b) to assist the United Nations in ensuring the integrity of the armistice agreements;
   (c) to refrain from giving any military, economic or financial assistance to Israel so long as it has not complied with this resolution;
3. *Requests* the Secretary-General to keep the Security Council informed on compliance with this resolution and to make whatever recommendations he deems appropriate for the maintenance of international peace and security in the area by the implementation of this and prior resolutions.'
S.C.O.R., Eleventh Year, Supplement for October, November, and December 1956, Doc. S/3710.

present themselves to the minds of honourable members in relation to the actions of Great Britain and France.

First, is the United Kingdom at fault in not having engaged in a pre-consultation with the other British Commonwealth countries? Our answer to this question is that she was not at fault at all. The circumstances were those of great emergency. Hostile armed forces were rapidly approaching each other, and extensive combat was imminent. As I have said, in that combat vital interests in the passage of the Canal were quite likely to suffer the most serious damage. The Canal is an international waterway with a guaranteed freedom of passage for the ships of all nations, but that guarantee would cease to have much value if the Canal itself were put out of action by becoming part of a theatre of active war. There was literally no time to be lost if any action was to be taken to keep the combatants out of the Canal area, and afford it proper protection.

Effective consultation—and I say 'effective' because a mere 'form of consultation' would have been quite useless—would plainly have occupied considerable time and the urgent position might have fallen into irretrievable disaster. In our opinion, therefore, Great Britain, whose Canal and other Middle East economic interests are so vast, is correct in proceeding upon her own responsibility. We are not living in an academic world. The normal processes of consultation should always be followed wherever possible, but there are instances like the present one in which events move too fast for normal processes.

The second great question that arises is as to the propriety of the action taken by Great Britain and France. Upon this point the Government of Australia believes that the action was proper. . . . It was quite clear that the procedures of the Security Council were such that even assuming that some resolution could be carried, the Canal would have been involved in war long before any United Nations' action would become effective. Great Britain and France, therefore, decided that they would, so to speak, 'hold the pass'. Their purpose, as they plainly state, was to have the Israeli and Egyptian forces withdrawn from the Canal for a distance of 10 miles on either side so that the operation of the Canal would not be menaced. Their action, so considered, was a police action taken in a state of great emergency and was in fact calculated to keep the combatants apart and to enable counsels of moderation thereafter to prevail.

6. Extracts from speech by the Leader of the Opposition, Dr. Herbert Evatt, in the House of Representatives, 1 November 1956[1]

. . . Apparently the Prime Minister does not mind the fact that Australia was not even consulted by the United Kingdom Government despite the interests of this country in the Middle East, which the Prime Minister emphasised so eloquently only a few weeks ago when speaking of the Suez Canal dispute. At that time, he said that we must act in concert with the United Kingdom. Now, he does not worry about that. He goes even further and says that the action of the United Kingdom in issuing this ultimatum was right. President Eisenhower says it was wrong.

It was not only the British Commonwealth of Nations that was not consulted. President Eisenhower first read of this ultimatum in the newspapers of his own country, although there was ample time to notify him beforehand. . . .

. . . A country might, in certain circumstances or emergency, act on its own initiative, even without notice to other countries. But in this instance there was daily contact between the representatives of the three great powers—not only between the European representatives in the respective capitals, but also at the very seat of the United Nations. But other countries were not informed. I say they were not informed because Britain and France did not want them to know of it until action had been taken. That is the whole essence of the matter. . . .

. . . The Prime Minister talks about the facts of life. Well, here is an illustration of them. The Prime Minister prepared his statement, of course, before recent events had been reported in this country, but one of those events is this: It is not merely an intangible ultimatum that is being carried into effect; force is being used. Individuals, if only a few people, according to the evening press, have been killed by bombs which come from British or French sources. Those are the facts of life. Is not the life of an innocent Egyptian bystander and his family just as important to the Prime Minister as the life of any man in the world?

Government supporters interjecting,

Mr. Speaker: Order! . . .

Dr. Evatt: I say that certain events are happening, and innocent

[1] Australia, H.R. Deb. 1956 (Second Period), pp. 2060–5.

Rcs

people are being killed because of the action of the British and French Governments. It is disgraceful to think that this should occur without the authority of the United Nations.

Can anyone forget the extraordinary speech that the Prime Minister made a few weeks ago? [Part II, III, Document 5, *supra*]. He enunciated a new theory about the use of force in international affairs, and there would have been no need for Sir Anthony Eden to consult the Prime Minister of Australia in order to ascertain his views, because he would have learnt of them from the speech that the Prime Minister made recently. The right honourable gentleman has postulated the theory that you can use force without the authority of the United Nations if you think it is right that it should be used. That is what the Prime Minister said in connexion with the Suez Canal dispute. That is an absolutely intolerable and illegal doctrine. . . .

We repudiate and disaffirm the principle stated by the Prime Minister in his speech on that occasion, in which he expressed the bald view, the good old rule, or the simple plan that they should take who have the power and that they should keep who can. That is the rule of the nineteenth century—the policy of the gunboat diplomat. You go in and try to get what you want. If the people concerned are not acquiescent and do not recognise your superiority, you use armed force. I tell him that in the opinion of the people of this country that doctrine ought to be dead. . . .

I say that the action taken by Great Britain and France will not succeed. The shocking thing is that the Prime Minister, on behalf of the Government, and presumably of the Australian people, welcomes what has been done. The Government, after having hesitated and supported the action of the majority of the members of the United Nations Security Council, now turns in the opposite direction. It is setting an example which, I submit, will be resented by the Australian people. I believe that the proper course is to discuss this matter in the United Nations General Assembly and to see whether Great Britain and France can be persuaded to withdraw their operation of force, which is directed against Egypt. That would be a courageous thing to expect them to do, but it should be done. I consider that, if the governments of Australia and other countries such as Canada and New Zealand were not consulted, they should now bring all their persuasive powers to bear on Great Britain and France, not for the purpose of attacking and discrediting the United Nations, but for the purpose of supporting it when vital clauses of the Charter are under consideration. . . .

7. Extracts from speech by the Permanent Representative to the United Nations, Dr. Ronald Walker, to the General Assembly, 1 November 1956[1]

When the Secretary of State of the United States addressed us earlier this evening, he said that he spoke with a heavy heart. I must say that I heard his statement with a heavy heart.

In Australia we believe that the strength of the United Nations rests principally on two foundations: on the one hand, the participation of the wide range of members throughout the world, and, secondly, on the close friendship and close co-operation of the United States, the United Kingdom and France. It is with very heavy heart that we recognize the division of opinion, the very deep division of opinion, that has developed regarding the practical measures that should be taken at this time to deal with the tragic situation in the Middle East.

Moreover, our own relations with the United States are so close and so friendly that our hearts are heavy as we find ourselves in opposition to a resolution proposed by the United States, particularly as we know in our hearts that the objectives sought by the United States are those that we would seek ourselves in connexion with this grave problem. I wish I could say that I were as sure that all those who speak or vote in favour of this resolution share profoundly the same objective of restoring the peace in the Middle East. . . .

The main feature of this resolution[2] to which I wish to bring attention is the first operative paragraph and the second preliminary paragraph. The second preliminary paragraph reads: 'Noting that armed forces of France and the United Kingdom are conducting military operations against Egyptian territory'. The first operative paragraph reads as follows: 'Urges as a matter of priority that all parties now involved in hostilities in the area agree to an immediate cease-fire and as part thereof halt the movement of military forces and arms into the area.'

This paragraph, taking priority, as it were, over the other

---

[1] G.A.O.R., First Emergency Special Session, 1956, pp. 27–28.
[2] For the full text of the resolution, see below, p. 317, n.2.

operative parts of the resolution, seems to be clearly directed against the action which has been taken by France and the United Kingdom. It reflects a judgment which the Assembly is asked to make on that action.

We have heard a statement by the representative of the United Kingdom this evening, and we have yet to hear a statement from the representative of France. We have also heard statements in the Security Council, and they have been available in the press of the world. The declared objectives of the United Kingdom and France in this matter are not to wage war but to prevent some of the consequences of war and to prevent the development of the conflict between Israel and Egypt and, in particular, to safeguard the lives and property of the nations that are using the Suez Canal. Their objectives have been stated to be to prevent the Suez Canal itself from becoming a battle line between Israel and Egypt, a development which would deny the use of this international waterway to the whole world. Their objective is to interpose forces between the Egyptian and Israeli forces.

The United Kingdom and France have declared solemnly that this is temporary emergency action, to be terminated when peaceful conditions are restored. They are not setting out to impose a solution to the problems of the Middle East by force but to establish conditions necessary for a peaceful solution of those problems. Their objective is not to supersede the action of the United Nations but to reinforce the limited measures that the United Nations is able to take in present circumstances. The United Kingdom and France have declared that these emergency measures are not to be directed against Egypt's sovereignty or territorial integrity.

As I said in the Security Council yesterday [Document 3], the record of the United Kingdom and France in this Organization is such as to deserve a fair consideration of their statements. Their record is such that I believe they have a right to have their solemn assurance accepted by Members of this Organization.

For these reasons, a resolution drafted in the terms that have been placed before us is not acceptable to the Australian Government and its delegation.

8. Extracts from speech by the Prime Minister, Mr. R. G. Menzies, in the House of Representatives, 8 November 1956[1]

... We in Australia realise that the great bulk of our overseas trade, which is vital to our own economic existence, passes through the Suez Canal in one direction or the other. The Western European powers, including Great Britain, depend upon a free and open Suez Canal for the vital industrial ingredient, to wit, oil, of their own industrial life and employment. Under these circumstances, should the two great Suez Canal shipping powers, Great Britain and France, have stood aside and pretended that a war in the Suez Canal Zone was no concern of theirs? They would have been bent on economic suicide if they had thought so, or said so, or acted so. Were they to believe that the United Nations could and would promptly and efficiently deal with this matter, not only by words and resolutions, but by deeds? If they had done so, resolutions would have been passed in the General Assembly at any rate but there is no reason to believe that anything would have happened. ...

These two great powers, therefore, concluded that action was necessary if the Suez Canal was to be kept free and open and out of a zone of war. That is why Great Britain and France developed their military activities in the Middle East. They have, I believe, been well justified in the result. It is just because they took strong action that the United Nations itself has been galvanised into action. They made it perfectly clear that their object was and is to separate the belligerents, to get a peaceful settlement of disputes and to preserve the Canal. If, as a result of this, both Israel and Egypt have declared a 'cease-fire' and if the United Nations itself is prepared to put in an effective military force to replace the police action of Great Britain and France, we will all very willingly believe that practical action has been taken by the world organisation. But at the same time, it must not be forgotten that there will always be the threat of conflict around the Suez Canal if the outstanding issues are not really settled. It must, therefore, not be thought that an international force will have exhausted its function when hostilities are ended. It must continue its function until the outstanding questions between Israel and Egypt have been settled on a basis acceptable by both, and the future of the Suez

[1] Australia, H.R. Deb., 1956 (Second Period), pp. 2115–19.

Canal as an international waterway, insulated from the politics of any one nation, has been assured. . . .

Some casual but biased observers have suggested that we have merely 'toed the line'. This is, of course, nonsense. We have not, if I may say so, lacked the capacity for expressing our views, though we have at all times expressed them as British people. But I would think badly of myself and my colleagues would think badly of themselves, if we remained silent or neutral under circumstances in which the Government of the United Kingdom has been assailed for taking action which we regard as both practical and courageous. I think that already it is being realised more and more that taking a firm course in matters like the Egyptian conflict is not a means of provoking war but of averting war.

. . . A good deal of apprehensive talk has occurred about the differences which have been manifested over this Egyptian matter between some of the countries of Europe and some of the countries of Asia. In particular, honourable members will not have failed to notice that some of our Asian friends have protested strongly against Anglo-French action in Egypt, but have had little or nothing to say about the murderous activities of the Soviet Union in Hungary. These are matters which it is considered wise politics never to mention. But a time comes when this rule should be broken. There could be no greater tragedy in the world than for it to become settled doctrine that the great nations of Asia and the great European and neo-European nations have conflicting interests, and that they must, therefore, accept conflict about them as inevitable. We, in Australia, do not believe that, in world matters, the interests of India must be in conflict with those of Australia or the interests of Asia in conflict with those of Europe. Statesmanship requires that we should all swiftly bring ourselves to an understanding that the world is one, and that ordinary human beings all around the world have similar interests and the same dignified and human ambitions. . . . The freedom and integrity of the Suez Canal are of just as much importance to the villager of Pakistan or India as to the ordinary citizen of Australia or the wage-earner of Great Britain or France. The freedom of the Canal, therefore, has a universal quality, the significance of which is not altered by the pigment of the skin or the geographical locality of the Canal users. If we are to settle these problems by lining ourselves up in favour of a European bloc, or in favour of an Asian bloc, if actions taken by Egypt are to be regarded in Arab communities as good simply because Egypt is an Arab community, then the world will be committing itself to a dispute to which there can be no end except in bitterness and destruction. In dealing with such a matter, we must try to look objectively at the merits and at the common good of all; we will initiate the suicide of mankind if we substitute bigotry for judgment, or seek to revive racial hatred under the guise of instituting the brotherhood of man. . . .

9. Extracts from speech by the Leader of the Opposition, Dr. Herbert Evatt, in the House of Representatives, 8 November 1956[1]

. . . The attitude adopted by the Prime Minister today [Document 8] is a repetition of the niggardly and disloyal attitude that he adopted earlier to the United Nations, of which Australia is a member. . . .

I do not wish to be misunderstood for a moment, and I shall repeat that in referring to Great Britain in this crisis, one knows perfectly well that the Government does not, or at any rate it is by no means certain that it does, represent the majority of the people. . . .

The Prime Minister says that it is wrong to associate that deep penetration [by Israeli forces] of Egyptian territory with the decision of Great Britain and France to act by force against Egypt. I think that they are clearly associated and by that I am not saying—the Prime Minister mentioned this allegation—that it was a pre-arrangement; but it is perfectly clear from the statement of Sir Anthony Eden, which the Prime Minister read in this House last week, that the British Ambassador was instructed to go to the Israeli Government in Tel Aviv, and that he got from it the information that there was mobilisation of a very special character. He got the assurance of the Government of Israel that there would be no attack on Jordan. That meant, obviously, that there would be an attack on Egypt and, from the very first . . . it is perfectly clear that the British and French move was timed to coincide with that deep penetration of Egyptian territory by the forces of Israel. Is that not plain? One only has to look at it.

Honourable members who know much more of this aspect of the question than I do, will realise that the combined forces of two

[1] Australia, H.R. Deb., 1956 (Second Period), pp. 2120–5.

nations, France and Great Britain, operating from a base like Cyprus, must have had the operation planned weeks, and perhaps months, before. Split-second timing was necessary to carry that out with such efficiency. It was not suddenly arranged. It was ready to go when the Prime Minister spoke to this House in September, because Sir Anthony Eden would not give an undertaking to the Labour Party in the House of Commons that he would not use force. . . .

. . . The force was ready. . . . The question was one of getting British and French forces ready to make their attack against Egypt, with particular reference to the Suez Canal. That was surely the original intention, but for some time Britain and France were dissuaded, and then darker forces and evil advisers got to work, and in the crisis affecting Israel someone said, 'This is the time to make our attack.' Just imagine the position, the horror of it. . . .

I want the Prime Minister to go back to a question which he answered the other day with regard to what used to be called the British Commonwealth but is now called the Commonwealth of Nations. We always think of it, correctly, as the British Commonwealth, because Great Britain is the centre of that Commonwealth. The Prime Minister said that Britain was perfectly justified in not consulting or even informing the other members of the British Commonwealth. Would anybody accept that? Nobody protests more against action of that kind than the Prime Minister on the eve of Commonwealth conferences. He wants a better organisation when he gets there. He wants a scheme of consultation. Mr. Curtin wanted the same thing. Indeed, Mr. Curtin and I advocated it. If that does not happen very quickly to meet this emergency, the British Commonwealth will not be a commonwealth at all. That is one of the urgent problems that have to be grasped. The Labour movement is taking action of its own in Great Britain, Australia and New Zealand, so that, as members of the British Commonwealth, we can meet together to try to work out a solution to these problems. The governments of the various nations should do the same thing. Lack of consultation is one great disaster that is threatening us to-day. . . .

. . . In any event, the United Kingdom did not need to consult the Prime Minister of Australia or this Government. The Prime Minister laid down the law in favour of force as against the law; and in favour of the legality of direct action by force in connexion with Egypt over the Canal dispute. Therefore, Sir Anthony Eden would know that, whatever he did, he would be supported by the Australian Government.

Of course, the tragic figure of the Minister for External Affairs (Mr. Casey) during this situation needs only to be mentioned without being described. He is everywhere but at the point where decisions are being made. He should be at the General Assembly when the General Assembly is meeting. But he is not; he is in Great Britain or in Canada. Similarly, when decisions were being made here, he was away. When decisions were being made in Egypt or in England, he was here. I believe that he understands far better than the Prime Minister what the obligations of the United Nations are and that he would not have been—and he should not have been—a party to the great blunders that have been made. . . .

I ask the Government to review the action it has taken, to make a contribution to the international force in the Middle East and to support the United Nations. It is vitally important that Great Britain and France should observe and obey the further direction given by the United Nations. They say that they will leave the area as soon as the peace force arrives. Let the peace force arrive as quickly as possible! Those countries have done enough harm. . . .

I again emphasise that we are not opposing in any way the people of Great Britain. The people of Great Britain are our kinsmen and I believe that in this matter they take the same view as that expressed by the Labour Party here. . . .

IV. NEW ZEALAND

1. Extracts from statement by the Prime Minister, Mr. S. G. Holland, 30 October 1956[1]

Clearly, Israel's position as a newly created State surrounded by hostile neighbours is a difficult one for a small country to sustain. Nevertheless, the New Zealand Government cannot but regard with the utmost concern what appears to be a complete disregard of the existing Armistice Agreement between Israel and Egypt. . . .

New Zealand has always taken the view that acts of planned retaliation along the frontier are self-defeating in that they merely intensify and extend the range of reprisal and counter-reprisal. We have always emphasized, too, that acts of aggression cannot be condoned.

[1] *E.A.R.*, vol. VI, no. 10, October 1956, p. 2.

The implications of Israel's action for the future of peace in the Middle East are most serious and the Government is giving close attention to the situation.

2. Extracts from statement by the Prime Minister, Mr. S. G. Holland, 1 November 1956[1]

While several features of the present situation are disturbing, I have full confidence in the United Kingdom's intentions in moving forces into the Canal Zone. The United Kingdom Government has given an undertaking that its operations are designed solely to protect the Suez Canal and to halt fighting between Israeli and Egyptian forces and that this emergency police action is intended to be of limited duration. . . .

It is nevertheless a matter of grave concern that a situation should have arisen in which there are serious differences of viewpoint between the United Kingdom and one of her principal allies, the United States. The Government is receiving full and frequent reports from its representatives in London, Washington, and other Commonwealth capitals. It will continue to give the closest attention to developments, and is prepared to make whatever contribution it can through the United Nations or in other directions to secure a lasting solution of Middle East problems— an objective towards which the Commonwealth and the United States have been working in close unity for many years. . . .

3. Extract from speech by the Permanent Representative to the United Nations, Sir Leslie Munroe, to the General Assembly, 2 November 1956[2]

The gravity of the situation which faces this special session of the General Assembly leads me to speak with a deep sense of responsibility. I speak as the representative of a country which has always given devoted support to the objectives of the United Nations Charter, which is a fellow-member with the United Kingdom in the Commonwealth, with all that implies, and which is closely and enduringly associated with the United States, not only by formal treaty arrangements in the Pacific and South East Asian areas but also by ties of respect and friendship.

[1] *E.A.R.*, vol. VI, no. 11, November 1956, p. 2.
[2] G.A.O.R., First Emergency Special Session, 1956, p. 34.

My Prime Minister yesterday issued a statement [Document 2] in which he expresses full confidence in the intentions of the United Kingdom in moving forces into the Suez Canal zone. He notes that the United Kingdom Government has given an undertaking that its operations are designed solely to protect the Suez Canal and to halt the fighting between Israel and Egyptian forces, and are intended to be of limited duration. It is my Prime Minister's hope that there will in due course be wider understanding of the motives of the present action as explained by the Prime Minister of the United Kingdom.

My Government is confident that the measures taken by France and the United Kingdom constitute an emergency action of a limited nature designed to deal with a situation of great political danger, a situation likely to deteriorate still further unless checked by drastic action.

May I add one word in particular about the position of the United Kingdom, with which our ties are never closer than in times of stress and danger. My Government does not accept any charge or imputation of insincerity in the motives of the Government of the United Kingdom. Still less does it accept the charge of British complicity in the Israel attack upon Egypt, a charge which was, in our opinion, convincingly answered by the representative of the United Kingdom in the Security Council and again in the Assembly.

With these considerations in the forefront of our minds, my delegation has carefully examined the United States draft resolution. Let me say at once that we do not feel able to support proposals which, as so many speeches reveal only too clearly, gravely reflect upon those on whom now rests a principal responsibility for the restoration of peace and order in the Middle East.

The United States resolution is at present the only resolution before us. It raises issues which are at least as important as and perhaps more difficult than any on which my Government has been required to take a decision since the foundation of the United Nations. As I read the resolution, it appears to raise more questions than it answers. Who, for example, is to take steps to reopen the Suez Canal and restore secure freedom of navigation? The most serious defect in the resolution, it seems to my delegation, lies in the absence of any proposal for dealing effectively with the situation in the Middle East as a whole, a situation which has been so gravely aggravated by the action of the Egyptian Government in seizing the Canal.

The Prime Minister of my country has been at pains to emphasize that the New Zealand Government cannot but regard with the utmost concern what appears to have been complete disregard by Israel of the existing Armistice Agreement between Israel and Egypt. There is no doubt about our concern on that point. There is equally no doubt about our concern over the various past events which are part of the whole unhappy story contributing to the incursion of Israel forces into Egypt.

I should have liked, moreover, to have heard some response to the constructive suggestions of the United Kingdom in regard to the desirability of a conference to negotiate a lasting settlement of the problems which have thrown the Middle East into turmoil. There is obviously not time tonight for my delegation, or for many others, to obtain instructions based on the present text of the United States draft resolution. Despite the undoubted urgency of the situation, it would, I believe, be proper and justifiable for the Assembly to adjourn before a vote is taken. If, however, we are obliged to cast a vote tonight, my delegation will regretfully be constrained, for the reasons I have given, to oppose it.

May I say, finally, that New Zealand's interest in this issue is a vital but not a selfish one. Our sole aim here and elsewhere will be to do what we can to contribute to a solution of Middle Eastern problems which will be swift, which will be effective, and which will be lasting. Whatever our vote here tonight, let us not regard our work in this place as finished. . . .

4. Statement by the Leader of the Opposition, Mr. Walter Nash, 7 November 1956[1]

The Party recognizes the inability hitherto of the United Nations to maintain the terms of the armistice in the Middle East and to compel compliance with its decisions.

We appreciate to the full the magnificent contribution that Britain has previously made to world peace and the extension of democratic government.

We regret, however, the action of the United Kingdom, in co-operation with France, without consulting the other members of the British Commonwealth, and without reference to the United

[1] *The Dominion* (Wellington), 8 November 1956.

Nations, in taking aggressive action against Egypt. Equally do we regret the actions of Israel, Egypt and Jordan in launching attacks upon one another.

We condemn aggression by any country that violates the territory or independence of others. We deplore the brutal, ruthless attack of Russia on the citizens of Hungary in their struggle for independence.

It has always been the policy of the New Zealand Labour Party to support the United Nations as the only hope for maintaining world peace. We agree with the Canadian resolution to establish a United Nations force and we support every possible step to strengthen the United Nations Organization. We urge the United Nations to consider the establishment of a permanent force to be in readiness to enforce decisions made by the Security Council or the General Assembly.

The Labour Party believes that the only way in which peace can be assured in the areas bordering on Israel is for a United Nations force to be responsible for maintaining order. We therefore support the enlistment, under agreed conditions, of a New Zealand force to participate in the task of separating the combatants and ensuring that peace in the Middle East is re-established and maintained. We welcome the cease-fire agreement.

Finally, we reaffirm our opinion that all international disputes should be settled by negotiations and a peaceful solution found under the auspices of the United Nations.

V. SOUTH AFRICA

1. Statement by the Minister of External Affairs, Mr. Eric Louw, 31 October 1956[1]

I heard about the British-French ultimatum to Egypt for the first time on the radio last night. I immediately telephoned the Prime Minister to ask if he had been officially informed or consulted but he knew nothing about it either.

This morning I asked him again if he had received any direct information. He said he had not. All we have had so far is a telegram from our High Commissioner in London summarizing what Sir Anthony Eden said in the Commons yesterday [I, Document 4, *supra*].

It is quite likely that we shall be informed this morning. But there has been no consultation at all.

[1] *The Star* (Johannesburg), 31 October 1956.

2. Statement by the Permanent Representative to the United Nations, Mr. Donald Sole, to the General Assembly, 1 November 1956[1]

All speakers have emphasized the gravity of the situation which we have been discussing. It was no less true that all delegations faced a grave decision in determining how they would vote on the resolution which was before us. It was a decision to be taken not by delegations but by Governments. South Africa has the closest of ties with the United Kingdom and France. It also has cordial relations with both Egypt and Israel. The resolution submitted by the United States delegation became available only a few hours ago. My Government obviously had no time whatsoever to consider it in all its far-reaching implications, many of which have been referred to in the statement made by the previous speaker, the representative of Canada. In these circumstances, it would have been neither right nor proper for me to record South Africa's vote either for or against the United States resolution. My delegation therefore abstained.

3. Extract from statement by the Minister of External Affairs, Mr. Eric Louw, 2 November 1956[2]

. . . The ultimatum had come as a great surprise to South Africa, which was only notified at 10.30 the following morning [1 November]. This would seem to indicate a major change in policy, for in the past stress has been laid on consultations with other Commonwealth countries on matters of importance. The action of the British Government has relieved South Africa of responsibility in the present crisis.

4. Statement by the Leader of the Opposition, Mr. J. G. N. Strauss, 2 November 1956[3]

Judging from Press reports, the military measures taken by the British and French Governments are more in the nature of a local, though intensive, police action and the question of South Africa's direct participation does not therefore arise.

The United Party, however, would urge our Government not to turn a blind eye to events, and to take the earliest possible opportunity of informing the country of their attitude to the present crisis.

[1] G.A.O.R., First Emergency Special Session, 1956, p. 36.
[2] *Cape Times*, 3 November 1956.   [3] Ibid.

We do not share in the harsh condemnation directed at Britain and France. It is quite clear that, if the Suez area had become a battleground between Israel and Egypt, the Canal would have been closed to all shipping for a considerable period.

The action of Britain and France may well succeed in opening the Canal to safe passage much sooner than if no action had been taken.

While we condemn aggression, we know that the aggressor is not always he who fires the first shot. The provocation suffered by Britain, France and Israel from President Nasser has been extreme.

Egypt has shipped arms to, and encouraged violent action among, Arab rebels in French North Africa, and has declared it her aim to destroy the State of Israel and to drive British influence from the Middle East. Even we in the Union have to take cognizance of his threats. He had taken upon himself the role of leader of the movement to win 'Africa for the Africans' and drive the Europeans from Africa. In all these aims he has consistently had the communist nations on his side.

President Nasser has showed himself to be a man who cannot be trusted, while his action in taking over the controlling body of the Suez Canal is a clear indication of his lack of regard for international obligations. One cannot negotiate with nationalism gone berserk.

Meanwhile South Africa will once more have to take up her historic role of being the most important nation on the main trade and supply route between East and West. We trust that the Union Government will not shirk its duty in this respect.

5. Statement by the Minister of External Affairs, Mr. Eric Louw, 5 November 1956[1]

The Cabinet today fully discussed the serious situation in the Middle East in the light of everything that has happened since the nationalization of the Suez Canal Company by the Egyptian Government, until the landing today of the Anglo-French military forces.

When the Suez Canal Company was nationalized, the Union Government clearly stated its position, namely, that no South African ships make use of the Canal, and that the Union is thus not a 'user country' in the sense in which that term has been

[1] *Cape Times*, 6 November 1956.

employed in connection with the Suez issue. Furthermore, only a
small percentage of the Union's export trade passes through the
Suez Canal.

The fact that South Africa does not have a direct interest in the
nationalization of the Suez Canal Company has since been fairly
generally admitted, and is also understood. The absence of direct
interest was also confirmed when South Africa and, amongst
others, Canada, was for that reason not invited to attend the
'18 country' conference.

In the circumstances mentioned, the Union Government feels
that it is not involved in the hostilities which are now taking place
in the Suez area.

The Government is, however, deeply concerned about the
maintenance of peace in the Middle East, and wishes once more to
express the earnest hope that a solution will be found—or that, in
any case, the hostilities will remain limited and localized.

The South African Railways and Harbours Administration will
continue to provide all possible assistance and facilities to ships
which are now obliged to make use of the route around the Cape of
Good Hope.

VI. India

1. Statement by the Ministry of External Affairs, 31 October 1956[1]

The Government of India have learned with profound concern
of the Israeli aggression on Egyptian territory and the subsequent
ultimatum delivered by the United Kingdom and France to the
Egyptian Government which was to be followed by an Anglo-
French invasion of Egyptian territory. They consider that a
flagrant violation of the United Nations Charter and opposed to
all the principles laid down by the Bandung Conference. This
aggression is bound to have far-reaching consequences in Asia and
Africa and may even lead to war on an extended scale.

The Government of India are conveying their views to the
Governments concerned and earnestly trust that even at this late
hour this aggression will be halted and foreign troops withdrawn
from Egyptian territory. They hope that the world community as
represented in the United Nations will take effective action to this
end.

[1] *F.A.R.*, vol. II, no. 10, October 1956, p. 150.

2. Telegram from the Prime Minister, Shri Jawaharlal Nehru, to the Secretary-General of the United Nations, Dag Hammarskjold, 31 October 1956[1]

We have been profoundly shocked by recent developments in the Middle East and more particularly by the Anglo-French invasion of Egypt after their rejection of the Security Council resolution moved by the United States. It is clear and admitted that Israel has committed large-scale aggression against Egypt. Instead of trying to stop this aggression, the U.K. and France are themselves invading Egyptian territory. This is not only an affront to the Security Council and a violation of the U.N. Charter but also likely to lead to the gravest possible consequences all over the world.

Egypt, which has suffered from the Israeli aggression, has in addition to suffer grievously by the Anglo-French invasion of her territory. The argument that this invasion is meant to protect the Canal and to secure free traffic has no force as the first result of this invasion is for this traffic to cease.

In view of the disastrous consequences of this invasion of Egyptian territory, I earnestly trust that the United Nations will take strong steps in this matter to prevent the world plunging into war and demand immediate withdrawal of all foreign troops from Egypt. The procedures of the U.N. must be swifter than those of invasion and aggression.

In sending you this message, I am not only reflecting the unanimous views of my Government and people but also I am sure of large numbers of other peoples.

3. Extracts from speech by the Prime Minister, Shri Jawaharlal Nehru, 1 November 1956[2]

It has been rather difficult for me during the last day or two even to think of much of the changes in India because my mind was filled with forebodings, with apprehension of what is taking place in what is called the Middle East region. . . . Some extraordinary things have happened in the last two or three days. First was the sudden invasion of Egyptian territory by Israel. Remember, there was no declaration of war. It was a sudden, unheralded invasion of Egyptian territory in large numbers. It was a breach of

[1] *F.A.R.*, vol. II, no. 11, November 1956, pp. 174–5.
[2] *The Hindu*, 2 November 1956.

Scs

the Armistice. It was a breach of the rules of the U.N. Charter. It is clear, naked aggression.

Now our sympathies in these disputes have been with Arab countries in the past. We have expressed them so but at the same time we have not been in any sense opposed to Jews or Israelis. We have felt that the Arabs have had a raw deal and that must be ended. Even so, we have avoided thinking of Israel or Jews as people to be wiped out. So it is with no feeling of enmity or hostility to Israel that I have addressed myself to this question. But the fact remains that apart from what has happened in the past, in the present case it is clear and naked aggression by Israel on Egypt and every member of the U.N. should try to halt, stop and resist the aggression. What, however, had happened, was that suddenly England and France have issued ultimatums to Egypt to clear off the Canal region—say ten miles away—and agreed to plant their forces to protect the Canal. The Canal has been functioning all this time under Egyptian control. If anything has happened in the Canal, it came from the Israeli invasion. But Egypt is being asked to withdraw. . . .

After fairly considerable experience in foreign affairs, I cannot think of a grosser case of naked aggression than what England and France are attempting to do, backed by the armed forces of the two great powers. I deeply regret to say so, because we have been friendly with both the countries and in particular our relationship with the U.K. has been close and friendly ever since we attained independence. I realize also that the U.K. has made many liberal gestures to other countries and has been a force for peace, I think, for the past few years. Because of this my sorrow and distress are all the greater at this amazing adventure that England and France have entered into.

It is clear that Egypt has not done in the last few days anything against England and France. Israel commits aggression, admittedly, everyone admits it. England and France say that in order to protect the Suez Canal they are coming and asking Egypt to get away. It is the most extraordinary argument that I have ever heard. . . . Israel is the invader and Egypt is made to suffer.

You will remember the U.S. Government brought a resolution in the Security Council two days ago asking all the parties to stop the fighting and not take any military measures. It was a fair resolution and yet Britain and France rejected it and vetoed it.

The Soviet Union brought a resolution, a simpler one, calling upon Israel and Egypt not to fight. Even this has been rejected by England and France. It passes comprehension what has come to this world.

In the middle of the 20th century we are going back to the predatory method of the 18th and 19th centuries. But there is a difference now. There are self-respecting independent nations in Asia and Africa which are not going to tolerate this kind of incursion by the colonial powers. Therefore I need not say that in this matter our sympathies are entirely with Egypt.

4. Statement by the Permanent Representative to the United Nations, Mr. Arthur Lall, to the General Assembly, 1 November 1956[1]

In this hour overcast with clouds of war and aggression, I rise to make these brief remarks on behalf of the Government of India because my Government remains convinced that the horror which is upon us can be arrested and a major catastrophe averted if immediate and effective action is taken by this emergency special session of the General Assembly, which has been called by vote of the Security Council and for which supporting requests have been made by as many as forty-nine Members of the United Nations, that is, by well over half of its total membership.

Let me say immediately that I do not propose at this late hour to engage in a discussion on the remarks of my colleagues of France and the United Kingdom, except to say that if they would only re-read parts of their speeches side by side with the Charter, they would be amazed at the wide gulf between the two sets of documents and would find no support or sanction whatsoever for their self-imposed role of policemen in the Middle East.

At this very moment the people and the Government of Egypt are the victims of a triple invasion, and our deepest feelings go out to them in sympathy. They are being subjected to the brutal facts of war. Their sovereignty is being violently curtailed and their territory is being occupied by the armed might of a neighbour and of two powerful countries. There has been released over Egypt a manifestation of the law of the jungle instead of the law of peace and the law of nations as enshrined in our Charter. Thus, on the territory of Egypt, is mockery being made of the Charter of the

[1] G.A.O.R., First Emergency Special Session, 1956, pp. 30–31.

United Nations, and there the organs of the United Nations are being affronted by aggression and invasion. It is this appalling state of affairs which confronts us and which demands, through our united effort, an immediate remedy, a remedy which will directly replace folly and inhumanity by peace and justice.

There cannot, there must not, be the slightest delay in applying such a remedy, and that is why in our opinion we must waste no time in proceeding to a vote on the draft resolution before us. For the United Kingdom and France, in this moment when we are in total disagreement with the actions of their Governments and completely repudiate the course on which they have embarked, we have feelings.

Those feelings reflect our conviction that those two countries are capable of an entirely different and immeasurably superior quality of international action. We believe in this moment of clouded vision they will wish us to remind them of their own capacity to seek solutions by peaceful means which they themselves have brought to bear with success in other situations which, if handled as they are handling this one, could have become equally grave and equally perilous to the peace of the world. We believe that the interests, not only the best interests but the basic interests of France and the United Kingdom cannot be served by the course they have chosen. That course is self-defeating and destructive for them as much as for anyone else. We in fact ask France and the United Kingdom to apply now those standards of objectivity and good sense which they would expect of other nations, and much more strongly does all this apply to Israel, a State that owes its very existence to the United Nations and which should most appropriately, therefore, be the foremost in its adherence to the Charter. . . .

There has been much talk of the protection of vital interests, but we ask ourselves: how can the protection of the vital interests of some so blatantly crush the vital interests of others, and how can this protection be achieved by a course which sows the seeds of increasing strife and war? We do not ask for the recalling of those vital interests, but we do ask that those countries which have deluded themselves into the view that their present actions will protect their interests recognize immediately how futile is the course they are following.

Their argument is that their invasion is meant to protect the

Suez Canal and to ensure free passage, but this argument has no force, for the first result of this invasion is that traffic in the Canal has ceased. This has actually happened as a result of the military action of France and the United Kingdom, and these countries have added to their other follies their disregard of the Constantinople Convention of 1888. As was to be expected, this violent approach to the safeguarding of vital interests is thus plunging the world into chaos.

It is for this reason that we demand of the nations concerned that they immediately seek to pursue their vital interests only through the measures allowed to them by the Charter and sanctioned by all codes of civilized and humane behaviour. It is with these feelings and with a deep sense of urgency that we ask this Assembly to act immediately and to adopt the draft resolution before it.

This draft resolution calls for the very minimum that is required at this stage. The aerial and sea bombardment and other operations of this character against Egypt must stop forthwith. The armed forces of Israel, France and the United Kingdom must leave Egyptian territory and waters immediately and this Assembly, which has met here to fulfil the objectives of peace, must remain in session until this purpose has been achieved. We must, one and all, be determined that there will be no repetition of 1935 or 1939. We must remember that the engines of war today are immeasurably more destructive than they were before. We must rescue ourselves from the vistas of horror which face us, lest there be no returning for mankind to the paths of peace.

Only determined and immediate action will now suffice us. Again, as Prime Minister Nehru said in his message to the Secretary-General of the United Nations yesterday [Document 2], the procedure of the United Nations must be swifter than those of invasion and aggression. Thanks to the initiative of the United States, we have before us a draft resolution which, for the most part and with the interpretation given to its first operative paragraph by the representatives of Colombia and Ecuador, fulfils the immediate requirements of the situation. It decides, furthermore, that we remain in emergency session pending compliance with the resolution. We strongly urge this Assembly, including the parties mainly concerned, to accept the draft resolution by unanimous vote as a first step towards settling the issues involved.

5. Extracts from statement by the Permanent Representative to the United Nations, Mr. Arthur Lall, to the General Assembly, 3 November 1956[1]

Less than forty-eight hours ago in this very hall we adopted a resolution which was moderate in tone, which was restrained, which was constructive, and which attempted not to worsen the situation in any way but to do something to arrest the damage to human life and property and the repercussions of that damage which were then beginning. It is a matter of the gravest disappointment to us that that resolution, moderate though it was, statesmanlike though it was, and though it was introduced by the leading Western Power, has not been complied with by the Western Powers concerned. This is a matter of gravest concern because the result of non-compliance has led to a steady worsening of the situation.

Fighting has intensified. We note that there has been a cablegram from the United Nations representatives in Cairo informing us of the intensification of attacks. We also note that the representative of the United Kingdom has published a contradiction of this particular report, but the statement he has given us in itself indicates that there has been an intensification of the fighting. It does not give my delegation much comfort to be told that there has been a switch from air to army targets. After all, are army targets always isolated from places of civilian habitation; and, after all, are not army persons only human beings in uniform? Is not human life concerned in this case also? We do not see that this contradiction brushes aside the main fact, which is that the situation is worse than it was two days ago, that there is an intensification of the evils which confronted us when we met here two days ago.

Furthermore, we are now faced definitely with just those events which we were told were the reasons why certain Western Powers had embarked on their expedition in Egypt. The very situations have arisen which the Western Powers sought to avert by going into this operation in Egypt. I refer to the closing of the Canal and to the cutting of oil pipelines. This was just the sort of thing that we were told the Western Powers wished to avert, and here, as a result of their operations, just those things are taking place.

[1] G.A.O.R., First Emergency Special Session, 1956, pp. 53–54.

Though all this is happening, and though the picture is growing darker, at the same time I think it is also our duty to bring to the notice of this Assembly that efforts are being made behind the scenes to arrive at some settlement, at some arrangement which would bring to an end what is being perpetrated in Egypt. It is too early to say that those efforts will succeed. We hope they will succeed. We hope that the good sense that was voiced here almost unanimously two nights ago will prevail and that practical arrangements can be devised in the very near future to put an end to the hostilities in Egypt. . . .

But what of the immediate situation? What of the fighting that is going on? In the opinion of the delegation of India and in the opinion of the delegations of almost all the Asian and African countries represented here, something must be done immediately to try to arrest the situation that exists in Egypt today. In view of this extremely urgent fact, in view of the need to stop the hostilities in Egypt, nineteen delegations in this chamber have drawn up a draft resolution. . . .

. . . This is an attempt to make an appeal to the parties which are engaged in hostilities immediately to cease those hostilities. This authorizes the Secretary-General to let us have an urgent report on his efforts, in conjunction with the parties concerned, to put an end to hostilities and to obtain withdrawal of troops from Egypt.

This, surely, is the least we can do tonight. Surely, this Assembly, which passed a resolution almost unanimously not forty-eight hours ago, wishes to adhere to the terms of that resolution, wishes to call again on the parties to comply with that resolution and wishes to draw their attention, through our words here and through those of other speakers, to the worsening of the situation in Egypt, to its repercussions on the rest of the Arab world, to its reverberations in areas outside the Middle East.

If the situation is allowed to continue, it will soon not be one which can be met by resolutions which will affect just those parties that are now on two sides of the Suez Canal. If we allow the situation to continue, we will soon be faced with a situation which will require resolutions dealing not with a few hundred square miles of territory but with many tens of thousands of square miles of territory. It is therefore imperative that we stop this situation, arrest it at once. We therefore call again upon the parties concerned to comply with the resolution which we adopted by so large a majority, on the initiative of the United States of America, not two days ago. It cannot be the intention of the sixty-four countries which voted for that resolution to allow the situation to remain in mid-air, because there is no mid-air in a situation like this: there is a terrible fall-out from a situation of this kind, and the area of contamination and war will spread rapidly. . . .

6. Extract from speech by Shri C. Rajagopalachari, ex-Governor-General, 4 November 1956[1]

The resolution of the United Nations having been flouted— more time has elapsed after the resolution was communicated to Britain than the British Government's ultimatum gave to Egypt before opening fire on her—it remains for the Government of India to consider why the Republic of India should remain in the Commonwealth.

Things should not be done in a hurry or in passion. But sometimes crucial issues have to be answered at the right moment and should not be postponed.

We are in the Commonwealth to stabilize an area of peace, and seeking to widen it progressively in the world. An association based on that principle loses its justification when the principal partner in that association so obstinately persists in aggression and flouts a resolution of the General Assembly of the United Nations passed with such unanimity and earnestness of purpose.

India may not find it easy to wind up her affairs with the Commonwealth, but that will be the very measure of the value of our protest. Our parting from the Commonwealth will not mean ending of diplomatic relations, but only will signify our complete non-co-operation with her [sic] attitude towards world peace.

India's lead in this matter will signify much especially if followed by Ceylon and Pakistan, and the totality of effect will go to strengthen the resolution of the United Nations Assembly and give moral consolation to Egypt on whom such sudden wrongful damage has been inflicted and who stands so greatly in need of such moral support. . . .

[1] *The Hindu*, 5 November 1956.

7. Message from the Prime Minister, Shri Jawaharlal Nehru, to the Chairman of the Council of Ministers of the U.S.S.R., N. A. Bulganin, 8 November 1956[1]

I am grateful to you for your message which was handed to me by your Ambassador this afternoon. I had a talk with your Ambassador later and explained to him our attitude and approach to the subject of your letter.

You are aware that we have been deeply distressed at the Israeli invasion of Egypt and the aggression of the United Kingdom and France against Egypt. We have expressed in the United Nations and elsewhere our views about this aggression in clear and unequivocal language. We have further deplored that the resolutions of the United Nations General Assembly have been ignored. All our sympathy has been with Egypt in her hour of deep crisis when her independence is endangered and an attempt is made to impose political and other decisions upon her by superior might. This aggression involved dangers not only to Egypt but to other countries which may not be strong enough to resist it. For us in Asia this turn of events has come as a great shock. If countries which are militarily weak are to be threatened by more powerful countries, then we revert to the rule of brutal might and the law of the jungle. All our declarations of peaceful co-existence and respect for integrity and independence of nations, non-aggression, non-intervention, and mutual respect, which we have repeated so often, have no meaning left and the world reverts to international barbarism.

The one relieving feature of this deplorable situation is that the United Nations General Assembly and world opinion generally have condemned aggression and have earnestly sought a way to put an end to it. Some progress has been made to that end and it may well be that recent resolutions of the General Assembly may lead not only to cease-fire but also to withdrawal of the forces that have committed aggression on Egyptian territory. If that end is achieved, it will be easier to deal with the problem so that the independence of Egypt may be assured and the issues that have been raised solved peacefully. I have just been informed that the United Kingdom Government are demanding immediate withdrawal of Israeli troops from Egyptian territory, apart from any general settlement in the Middle Eastern region.

[1] *F.A.R.*, vol. II, no. 11, November 1956, pp. 173–4.

As you know, Mr. Chairman, we in India are resolutely opposed to war and we do not think that war solves any problems. Even if it appears to do so for the moment, it creates far more difficult problems. War today is too terrible to contemplate and humanity has rebelled against such a prospect. Your own country has taken a lead in the campaign for peace.

While, therefore, we entirely agree with you that aggression of all kinds must be put an end to, we feel strongly that any steps that might lead to world war would be a crime against humanity and must be avoided. It is indeed partly for this reason also that we have condemned the aggression on Egypt.

We feel that it is possible, even at this late stage, to rescue peace from the fog of war that threatens to suffocate it. In this task, your great country can play a great part. To the extent that we in India can help, we shall certainly do so with all our will and strength.

I agree with you fully that the situation is serious and delay may well lead to disaster. But I earnestly hope that there will be measures to bring back and ensure peace rather than to enlarge the circle of war and disaster.

I thank you for telling me that you will be good enough to send me information about Hungary. As you know, developments there have caused us much concern.

8. Extracts from speech by the Prime Minister, Shri Jawaharlal Nehru, in the Lok Sabha, 16 November 1956[1]

... As the House knows, India had viewed with grave apprehension the policy of the U.K. and French Governments after the nationalisation of the Suez Canal Company. In particular, the massing of troops and aircraft for the purpose of military operations in Egypt appeared to us to be a reversion to past colonial methods and an attempt to coerce Egypt by show of armed might. Indeed, it was stated by responsible statesmen in the United Kingdom and France that the regime in Egypt must be changed and, in particular, the Head of the State and of the Government of Egypt should be removed. We had hoped, however, that ... more peaceful methods would be adopted to solve this dispute. The starting of military operations against Egypt by the United Kingdom and France and, more particularly, the bombing of parts of Cairo city and other parts of Egypt came, therefore, as a profound

[1] Lok Sabha Deb., 1956, Part II, vol. IX, coll. 261–7.

shock not only to people in India but also to large numbers of people in other countries including the United Kingdom. This appeared to be a flagrant case of aggression by two strong powers against a weaker country with the purpose of enforcing their will, even to the extent of changing the Government of that country. . . .

The story of the past three and a half months, ever since the nationalisation of the Suez Canal Company, is full of tragic drama, and events have happened which I would have thought could not possibly occur in this modern age. I find it a little difficult to deal with this record of unabashed aggression and deception. The explanations that have been given from time to time, contradict one another and exhibit an approach which is dangerous to the freedom of Asian and African countries and to world peace itself. It has brought misery and disaster, hatred and ill-will, with no gain whatever, and, in addition, we live now under the threat of possible world war.

During all the controversies since the nationalisation of the Suez Canal Company, Egypt has conducted herself with a large measure of propriety and forbearance. Without the least justification, Egypt was attacked not only by Israel but also by the United Kingdom and France. Whether there was any previous consultation between the aggressor countries, I do not know. But it is obvious that their plans fitted in, and the Anglo-French attack helped Israel's aggression and was itself helped by it. Egypt, the victim of Israeli aggression, was attacked immediately after by the armed forces of the United Kingdom and France. It was only the widespread indignation of peoples not only in Asia and Africa but also in Europe and America and the action taken by the United Nations that put some check on this aggression. But it appears to me that the cease-fire having taken place, there is a tendency to complacency and to allow matters to drift. Indeed, there has even been some attempt made to minimise and justify this utterly unprovoked and brutal attack on Egypt. Attention has been diverted to some extent to the grave and distressing occurrences in Hungary.

Even as we were distressed by events in Egypt, we viewed with grave concern and distress events in Hungary. It is possible that what happened in one of these countries produced its reactions in the other, and both created a very serious international situation. But it is well to remember that though both deserve serious attention, the nature of each differed from the other. Neither can be held to justify the other. . . .

The world appears now to be in the grip of the fevered psychology of war, and I am reminded of the months preceding the last great war. I am convinced that it is not by war and violence that these problems will be settled or freedom established. I am convinced that colonialism, whatever new look it may put on, can revert to its old brutal self, and the only remedy is for it to give place to freedom.

The world stands facing great danger, and it may be that the little wars we have had are only a first round and bigger conflicts lie ahead. In particular, the ambitions of strong nations imperil weaker countries. The only hope lies in the United Nations, representing the world community, succeeding in putting an end to the law of force and substituting for it a more civilised method of dealing with problems. Today, the choice lies between the hydrogen bomb and the Panch Sheel.

VII. PAKISTAN

1. Statement by the Permanent Representative to the United Nations, Mr. Mir Khan, to the General Assembly, 1 November 1956[1]

I speak on a point of order. As I understand it, we have perhaps about twenty more speakers and perhaps there will be many more inscribed to speak as we go on. I also see that a draft resolution has been tabled by the United States; . . . I particularly have my eye on paragraph 3 which reads as follows: 'Recommends that all Members refrain from introducing military goods in the area. . . .' The draft resolution also urges that the cease-fire becomes effective immediately. I myself have a great deal to say in this matter. I have to say that we feel in great pain while our brothers in faith in Egypt are being killed. I have to say how angry we feel against the aggression committed by Israel against Egypt. I also have to say how unwarranted we feel the attack made by the United Kingdom and France against Egypt—the United Kingdom and France, with whom my country has such traditional and sacred ties.

We belong to the Commonwealth. We are associated with the United Kingdom in that wonderful association of countries which work for peace. With France we have traditional and cultural

[1] G.A.O.R., First Emergency Special Session, 1956, pp. 81–82.

friendship. However, I see that while we speak and perorate here, there is damage being done to valuable property; there are people being killed. With this in mind, I make a formal proposal that we take up this draft resolution for consideration and voting here and now, without prejudice to the right of any Member to speak after this resolution has been voted. I move that as a formal proposal, and I request the President to consult the Assembly on this proposal.

2. Statement by the Prime Minister, Mr. H. S. Suhrawardy, 3 November 1956[1]

Ever since the invasion of Egyptian territory by Israel and the subsequent action of Britain and France, feelings in this country have been understandably running high.

The Security Council resolution sponsored by the U.S.A. which was aimed at bringing about a cessation of hostilities and removal of all foreign troops from Egyptian soil has been vetoed by Britain and France and there has been bombing of military targets in Cairo and elsewhere in Egypt by British and French aircraft as a prelude to landing British and French forces to occupy the Suez in defiance of the United Nations.

These developments have created a very grave situation. Not only have these events incited and helped Israel in her aggressive designs on Egyptian territory, but the violation by these two Powers of Egyptian sovereignty and territory by the use of force, in disregard of the appeals of the other members of the United Nations, has shocked world opinion and placed the very concept of that world organization in jeopardy.

So far as Pakistan is concerned, the fact that a Muslim country towards which Pakistan has always entertained fraternal feelings should be the victim of such aggression has further exercised public opinion. What is happening in Egypt today constitutes a threat to the entire Muslim world.

The Government of Pakistan unreservedly condemn this aggressive action. They will continue to endeavour by every means within their power to bring about an early peaceful settlement of this conflict. With that end in view, I personally have been in almost constant touch with the principal Powers concerned. At our instance, a meeting of Prime Ministers and Foreign Ministers of Turkey, Iran, Iraq and Pakistan will shortly take place in

[1] *Dawn*, 4 November 1956.

Teheran, where this question will be further considered with a view to appropriate action to bring about a peaceful solution of this dispute.

Our people may rest assured that their Government are not an idle spectator of this drama, nor will they rest content with mere denunciation of the action of the aggressors. We have been exploring every avenue possible to ensure that wiser counsels prevail and this tragic episode is brought to an early end.

If the efforts we are making should unfortunately fail, we shall certainly proceed to consider whatever further measures may be possible to bring about a settlement which would ensure that Egyptian sovereignty and territorial integrity are fully respected.

During the last two or three days public opinion in Pakistan has been gravely perturbed on this issue. Demonstrations have taken place in Karachi, Dacca, Lahore, Peshawar and elsewhere in the country. While, on the whole, these demonstrations have been peaceful, there have also been certain ugly incidents.

I would earnestly ask my people to remain calm and refrain from violence. I appeal to law-abiding people themselves to intervene for the sake of Pakistan and its prestige to stop lawless elements from indulging in violence wherever they find them doing so.

On occasions such as this, certain elements are inclined to take advantage of the situation and to resort to lawlessness. We must on no account allow such elements to exploit this situation for their own nefarious ends.

The Government can deal with this grave international situation effectively only so long as the people stand peacefully behind it on this issue.

There should be respect for diplomatic immunities which is expected of all civilized nations, whatever may be the provocation. Whatever happens, we must maintain peace.

I am confident that in this matter the Government will have the willing co-operation of the people of all shades of public opinion.

3. Extracts from speech by the Prime Minister, Mr. H. S. Suhrawardy, 3 November 1956[1]

One aspect of your demonstration today makes me happy because it shows that in you, my children, there is the deep consciousness (*chsas*) that if there is any tyranny (*zulum*) or aggression

---

[1] *Dawn*, 4 November 1956. (The speech was delivered in Urdu.)

anywhere in the world you immediately react and stand up against it. I have already spoken in condemnation of British and French use of force against Egypt and I indicated the policy of my Government in this matter [Document 2]. . . .

But it cannot be done in a day. We shall now meet in Teheran and decide our further action. There are other Islamic countries in the world also with whom we must take counsel together. Your demands are also our demands (*loud cheers*).

You will remember that I was the first to speak out and condemn Israel as the aggressor immediately on my return from China.[1] I ask you to wait, and I assure you that you will be pleased when you find that you have now a Government in Pakistan who can face a situation like this fearlessly and take a bold stand on an issue of this nature. I ask you to leave it to us.

I must tell you, however, what I saw in China. I found the people there so devoted to their country that they are all the time working for it to make it better and happier and more prosperous. They seldom bother with what goes on outside. Here in Pakistan, we worry about the whole world but do not care if Pakistan goes to ruin.

Therefore, I would ask you first to see to it that Pakistan becomes strong and respected in the world abroad. Until you have a country which is respected outside, and until we gather sufficient strength ourselves, what influence can we have on international affairs?

A voice: We are first Muslims and then Pakistanis.

The Prime Minister: Who denies that? But if anyone says you must be Pakistanis first who is snatching away your 'Mussalmani'?

4. Extracts from speech by the Permanent Representative to the United Nations, Mr. Mir Khan, to the General Assembly, 4 November 1956[2]

On November 1, the Assembly met in emergency special session under a dark cloud, the effects of which on the international political climate were beyond calculation. We adopted a resolution calling upon the combatants on Egyptian soil to cease fire and retire from Egyptian territory. That resolution was adopted with

---

[1] At Karachi Airport on 30 October 1956, Mr. Suhrawardy declared: 'There can be no question, however, that Israel by its latest act of aggression has disturbed the peace of the region.' *Dawn,* 1 November 1956.

[2] G.A.O.R., First Emergency Special Session, 1956, pp. 81–82.

unprecedented speed and support. It has so far not been heeded, as may be seen from the reports which we have just heard. This is also true of the resolution adopted by the Assembly less than twenty-four hours ago. It therefore seems to me that the cloud which cast its shadow on this Assembly on Thursday still hangs over us tonight.

Pakistan is against colonialism and imperialism of all kinds and condemns aggression wherever it takes place. That must be evident from the vote we cast in the Assembly earlier this evening.

The sequence of events which commenced with the invasion of Egyptian territory has caused profound shock and grief to the people of Pakistan. . . .

As I stand on this rostrum, I do so more in sorrow than in indignation, because the time for indignation has passed. There was indignation when Israel marched its troops into Egyptian territory. It was a blow not only against the Egyptian people and Egyptian sovereignty, but it was also a blow against peace in the Middle East and possibly in the whole world. That what the United Nations created by mandate as a small State should so far forget itself as to strike a blow at the very basic purposes and foundations of that Organization was an occasion for indignation, but unfortunately the people of the world were not allowed much time to remain indignant. There came to the assistance of Israel two of the world's big Powers, and they started a pounding of Egypt from the air to which Egypt's defences were no match.

Today, as I stand before this Assembly, I feel in my heart black, profound and engulfing sorrow. Sorrow for the Egyptian people, their dream of progress and their dream of a better future. Sorrow for the nations which, by their action, have given a setback to humanity's efforts to evolve a peaceful method and machinery for solving disputes and disagreements. Above all, I feel sorrow for the world Organization. If the disregard for its edicts, of which we have seen evidence during the last three days, continues, I am afraid that I cannot beguile myself about what the future of this Organization is going to be.

The only silver lining which I see in the cloud of which I spoke at the outset is that the position is not yet irretrievable. Even now there is time to undo the wrong that has been done if the invaders desist from their plan. Let there be an immediate cease-fire, and let there be a complete withdrawal of foreign forces from the soil of

Egypt. This, alas, will not bring back the Egyptian dead, but the parties responsible for the despoliation of Egypt should make amends by fitting together the fragments of the dreams of the people of Egypt.

The least they can do, and the least this Assembly can do, is to see that Egypt, apart from the repair and restoration of damage, receives all necessary assistance in its long-term dream and plan for building a better future for its people.

5. Statement by the Premiers of the Four Muslim Powers of the Baghdad Pact, 8 November 1956[1]

(1) As a result of the negotiations between H.M. the Shah and the Pakistani President, the Imperial Government [of Iran] invited the Premiers and Foreign Ministers of Pakistan, Turkey and Iraq to visit Teheran to consider the situation created by (a) Israeli aggression against Egypt and the occupation of Egyptian territory by Israel, and (b) military intervention by Britain and France in Egypt.

(2) The Premiers of Persia, Turkey, Pakistan and Iraq considered the critical situation created in this zone and unanimously reached the following decisions:

(a) They condemned the Israeli act of aggression against Egypt, and deemed it necessary for the Israeli forces to return immediately to the cease-fire line and for all Egyptian prisoners to be repatriated.

(b) On the basis of the U.N. recommendations concerning the regrettable armed intervention of the British and French forces, they decided to ask the British and French Governments to end hostilities immediately, to withdraw their forces from Egypt, and to respect fully the sovereignty, independence and territorial integrity of Egypt.

(c) To ensure a permanent peace in this zone, the four Governments stated that the Israeli-Arab dispute must be settled immediately on the basis of the 1947 resolution of the United Nations.

(d) The four Governments state that the Suez Canal problem must be settled under the auspices of the U.N. by negotiations with Egypt. Freedom of navigation must be safeguarded, fully respecting Egyptian sovereignty.

[1] *Pakistan Horizon*, vol. IX, no. 4, pp. 209–11.

(3) The four Governments were pleased that the British Ambassador at Teheran had stated that the joint action by the four Governments and the British Government on November 5 affected the decision for a cease-fire order on November 6. They welcomed this decision and expressed the hope that there would be no delay in the implementation of the U.N. resolution for the establishment of an International Force.

(4) The four Governments appreciated the prompt action of America in the U.N. for safeguarding peace and supporting U.N. principles.

(5) They confirmed their promise to respect the principles of the U.N. Charter, and declared that aggression, in whatever manner, is resented by them. The independence and sovereignty of all nations must be respected and safeguarded.

(6) The four Governments were informed of the efforts made jointly or individually by them to safeguard peace. Emphasizing that their alliance has no aggressive nature, they reiterated their desire to strive for the peace and security of this zone. They expressed their distaste for the false statements made by some circles for the purpose of hampering peace and security in this zone.

6. Statement by the Prime Minister, Mr. H. S. Suhrawardy, 10 November 1956[1]

I am returning from Teheran where I have just had the privilege of having the most interesting and useful conversations with His Imperial Majesty the Shahinshah of Iran, and the Prime Ministers of Iran, Iraq and Turkey, with which countries we are, as you know, bound by the most close and friendly ties.

As you will perhaps remember, the President of Pakistan was on a visit to Iran at the gracious invitation of his Imperial Majesty, when a situation of grave international consequence arose in the Middle East. The two Heads of State very wisely decided at once to summon for consultations the Prime Ministers of Allied Powers whom the new crisis affected in a direct and dangerous manner.

These consultations have taken place and you will see from the joint communiqué [Document 5] the Four Powers reached the unanimous conclusions that the Israel aggression on Egypt must be condemned and that Israeli troops should be required to with-

[1] *Dawn*, 11 November 1956.

draw immediately to the Armistice Line and release all Egyptian prisoners.

The Four Powers moreover regarded the armed intervention of the Anglo-French forces as deplorable, and made it quite clear that these forces should also be withdrawn from Egyptian territory forthwith and that the sovereignty, integrity and independence of Egypt should be fully observed and respected.

The Four Powers expressed their considered view that the Suez Canal dispute should be settled through negotiations with Egypt under the auspices of the United Nations which should, *inter alia*, ensure free passage through the Suez Canal with full respect for Egyptian sovereignty.

The Four Powers further drew attention to the urgent need of solving once and for all the Palestine dispute between the Arab countries and Israel and considered the United Nations Resolution of 1947 to be the proper basis in negotiating such a settlement. The Palestine problem lies at the root of Middle East instability and is the most important problem to be solved.

Unfortunately, owing to misguided policies heretofore pursued by certain countries at the behest of countries who wish to keep this region in a state of ferment and turmoil, this problem has not only remained unsolved but has been aggravated.

Now a situation has been created which, if handled competently, may well render possible a permanent and just solution acceptable to the Arab people, since forces hostile to this concept have been weakened. Pakistan, tied as it is by ties of religion and culture with the Middle East, has always striven and will continue to do its best to bring about this happy result.

I trust it will now widely be realized that Israel is a real danger to the peace and tranquillity of this region and is not a 'martyr' but an aggressive nation.

You will be glad to know that before the issue of this joint communiqué, as a result of preliminary and urgent discussions between the Four Powers, a joint approach was made to Her Majesty's Government on November 5, which we have been told influenced the decision to cease fire at midnight on the following day.

The Four Powers have, since the issue of the joint communiqué, urgently represented to Her Majesty's Government that only by prompt action to implement the conclusions given in the joint communiqué would it be possible to prevent the conflagration from spreading and restore peace and stability to the Middle East. We now await anxiously, but with confidence, the quick acceptance by Her Majesty's Government of these proposals. I must however emphasize that we do not wish to see a situation created in this area of which Israel could take advantage to the detriment of Egypt's interests. It is for this reason that Pakistan was among the first to contribute to the International Police Force.

The stand taken by the Government of Pakistan in this matter has all along been clear. We have throughout unreservedly condemned Israeli aggression, deplored Anglo-French intervention, called for an immediate cease-fire and withdrawal of foreign troops from Egyptian soil, and taken the view that the dispute should be settled under the auspices of the United Nations through negotiation with Egypt.

This remains the attitude of the Government of Pakistan and we have in the last few days, in co-operation with our Allies, made the most urgent and anxious representations both in Karachi and through our diplomatic representatives abroad to bring this unfortunate conflict to an early end. Our representative at the United Nations has also been most active in encouraging and supporting measures taken in the United Nations to secure this objective.

We have taken the opportunity, while at Teheran, to re-emphasize the essentially defensive and non-aggressive nature of our Alliance. I must here point out that whatever action Her Majesty's Government thought fit to take was not given in its capacity as a member of the Baghdad Pact. On the contrary, the other members of the Baghdad Pact, realizing their duty to safeguard the peace and stability of the region and to uphold the principles of the United Nations Charter, have throughout, individually and collectively, done their very best to prevent hostilities from spreading and to bring the conflict to a speedy end.

In this way we have demonstrated our firm adherence to the cause of world peace, our abhorrence of aggression in any form, and our determination as free countries to pursue what we consider right.

7. Statement by the Prime Minister, Mr. H. S. Suhrawardy, 12 November 1956[1]

I find that the record of my airport interview of November 10 [Document 6] on my return from Teheran, which appeared in the Pakistan newspapers, is not quite correct. For instance, a question was put to me as to why I had made a distinction between Israeli 'aggression' and Anglo-French 'intervention' in Egypt. My reply was that as a member of the U.N., I was not prepared to differ from the United Nations.

The U.N. Resolution on the subject makes a distinction between Israeli aggression and Anglo-French intervention.

I tried to explain further that the United Nations' position might perhaps have been influenced by the plea which the U.K. and France had offered in justification of their action and which was that they had gone into Egypt in order to keep the Egyptians and the Israelis apart.

When someone asked me whether I accepted this plea of the British, I made it clear that we condemned the Anglo-French action, whatever might have been their motives.

It will thus be seen that this is generally the fate of verbal airport interviews.

8. Statement by the Prime Minister, Mr. H. S. Suhrawardy, 15 November 1956[2]

Let me tell you that I refuse to be isolated. So long as I am Prime Minister I will not do anything which will weaken Pakistan. We are living in difficult and dangerous times and we must have friends on whom we can rely. Our enemies are most anxious to isolate us and weaken us, so that we can fall a prey to their evil designs and neither be able to assert our rights nor to hold high our head among the nations of the world.

My main objective is to make more and more friends and not to loosen our ties. Grave dangers threaten the Muslim world. I would like the Muslim countries to be strong and to be united. Each of us has a long way to go in order to achieve this. The process will be slow. Most of the countries are weak and ill-armed and there are many conflicting interests which have to be resolved.

It is in pursuit of these objectives that I am making another visit to the Middle East countries.

[1] *Dawn*, 13 November 1956.   [2] Ibid., 15 November 1956.

I would, therefore, beg those of our people who desire that we should cast ourselves adrift from all our ties to pause and think. Let us to the utmost extent of our power help our brothers in distress. Let us uphold their prestige, but let us not in that process so weaken ourselves that we are no use to anyone and our words and efforts carry no weight.

VIII. CEYLON

1. Message from the Prime Minister, Mr. S. W. R. D. Bandaranaike, to the General Assembly of the United Nations, 31 October 1956[1]

On behalf of the Government and people of Ceylon I wish to express my sense of shock and perturbation at developments in the Middle East. I consider that there has been no adequate justification for the invasion of Egyptian territory by Israel and for the action taken thereafter by Britain and France. I consider the situation one of the utmost gravity to the whole world and press most strongly that Israeli, British and French forces be immediately withdrawn from Egyptian territory and the situation prevented from deteriorating into one that is bound to bring calamity and disaster to the whole world. I appeal to you in the name of humanity to do everything in your power to achieve this object.

2. Speech by the Permanent Representative to the United Nations, Mr. R. S. S. Gunawardene, to the General Assembly, 1 November 1956[2]

We meet today in a situation of extreme danger to international peace. The territory of a member nation has been subjected to armed attack by three Powers. Two of these are permanent members of the Security Council charged with the responsibility of maintaining peace in the world. All three Powers as Members of the United Nations have pledged themselves to uphold in spirit and in letter the principles embodied in the Charter of the United Nations. As a member of the Commonwealth, an association of nations dedicated to the cause of peace and international harmony, it is with profound sorrow that I express my Government's strong dissent and disapproval of the actions of a fellow member of the Commonwealth with whom we have traditional ties of friendship.

[1] G.A.O.R., First Emergency Special Session, 1956, p. 5. [2] Ibid., pp. 4–5.

It is to me a matter of profound grief that I should have to perform this duty. I have always been a stout upholder of the concept of the Commonwealth as one of the greatest contributing forces fostering international peace and the principles of liberty and justice.

Events are moving fast and time is very short. It is not for us to debate at tortuous length the intricate antecedents of this grave situation. The fact remains that aggression has taken place and that this must be halted immediately. The territorial integrity of Egypt has been violated and Egypt has been subjected to invasion and aerial bombardment. What seems to be deplorable is the fact that two members of the Security Council, instead of joining in the efforts to halt the initial aggressor, have themselves committed a further act of aggression—all this in the name of peace.

It is our strong belief, shared I have no doubt by all right-thinking people, that the United Nations is the proper organ for the settlement of disputes among nations. It is not for individual members of that body to arrogate to themselves the right to intervene with military force. In this case such intervention occurred at the very moment that the Security Council was earnestly engaged in the pursuit of a peaceful settlement. We have heard the phrase 'police action' with regard to internal matters within a sovereign State. We are now asked to accept in the name of 'police action' a wanton violation of a sovereign nation's rights.

This action on the part of three Members sworn to uphold the Charter constitutes a serious threat to the effectiveness of the United Nations. As a small nation we look to the United Nations as the guardian of peace, but if superior military strength is wilfully and, in our opinion, unjustifiably used, as in this case, by individual nations in the exercise of their fancied rights, the future of this Organization will be in serious jeopardy.

The events of the last few days have demonstrated the tremendous weight of world opinion that has been brought to bear against the reckless use of force. We refuse to believe that there is any nation so devoid of conscience and responsibility that it can continue to defy the moral judgment of the world.

In the view of many responsible nations the action of the United Kingdom and France is construed as a continuation of the tradition of colonialism. It does not require much foresight to see that no nation, however powerful, can turn back the clock in Asia and

Africa, and resist the inexorable march of events. The nations of resurgent Asia and Africa are determined to exercise their sovereign rights in conformity with the principles of the United Nations Charter. It is naïve to assume that this process of change can be arrested by the crushing of a leader or of leaders. The spirit of Asia and Africa can never be crushed.

It would be both expedient and dignified for the Colonial Powers to accept with grace the change. In this connexion, I have no hesitation in applauding the United Kingdom Government for its timely recognition of the claims for freedom and self-determination in India, Pakistan, Burma and Ceylon. This act earned for them the admiration and respect of the whole world. I would appeal to the United Kingdom even at this late stage not to jeopardize this well-earned good will in the pursuit of an objective that is fraught with disaster. May I also address a similar appeal to the Governments of France and Israel.

In the tragic situation that has developed, my Government feels that the least that should be done is that (1) there should be an immediate cease-fire, and (2) all military forces should be withdrawn to their own territory. We feel also that this Assembly should continue to be in session until these objectives are achieved.

At this stage may I take the liberty of delivering a message sent to this Assembly by my Prime Minister, the Honourable Mr. S. W. R. D. Bandaranaike . . . [Document 1].

I wish to express also my Government's appreciation of the positive stand taken by the United States in its earnest search for a peaceful solution of this problem in spite of its traditional friendships. May I express the sincere hope that the deliberations of this august body will restore peace and harmony to a trouble-worn world.

PART IV

THE COMMONWEALTH AND THE UNITED NATIONS EMERGENCY FORCE

Peace is far more than ceasing to fire. . . . Are we to go through all this again? . . . My own Government would be glad to recommend Canadian participation in . . . a United Nations force, a truly international peace and police force.

Mr. Lester B. Pearson, 2 November 1956

COMMENTARY

I

THE origins of the idea that a force of the United Nations might somehow be used to end the Egyptian imbroglio are still in dispute. Sir Anthony Eden has claimed it as his own. In his Memoirs, he refers to his suggestion, offered in the debate in the House of Commons on 1 November, 'that a United Nations force should eventually be associated with the Anglo-French police action', an 'idea . . . taken up in the General Assembly the next day by Mr. Lester Pearson and others'.[1] These were the words used by Sir Anthony Eden at that time:

> Effective action to re-establish peace will make easier an international solution of the many problems which exist in that area. Of course, we do not delegate to ourselves any special position in that respect. On the contrary, we would welcome and look for the participation of many other nations in bringing about a settlement and in upholding it.
>
> Israel and Egypt are locked in conflict. . . . The first and urgent task is to separate these combatants and to stabilize the position. That is our purpose. If the United Nations were then willing to take over the physical task of maintaining peace in that area, no one would be better pleased than we. . . .[2]

While the British Prime Minister was speaking in these terms at Westminster, the Canadian Cabinet, meeting for the first time since the crisis had moved, forty-eight hours earlier, into its new and foreboding phase, was discussing the idea of a United Nations force, not (it is clear from the sequence of events) on referment from London but on its own initiative. Word of Sir Anthony Eden's statement did not reach Mr. Pearson until he was about to leave Ottawa for the United Nations after the meeting of his Government on the morning of 1 November.[3]

The plan formed at the meeting of Canada's Cabinet was that Mr. Pearson should go to New York to attempt to persuade the General Assembly to prevail upon the United Kingdom and France to place their forces at the disposal of the United Nations—

---

[1] *Full Circle*, pp. 535–6.
[2] United Kingdom, H.C. Deb., 1956, vol. 558, col. 1649.
[3] Graham Spry, 'Canada, the United Nations Emergency Force, and the Commonwealth', *International Affairs*, vol. 33, no. 3, July 1957, p. 297.

'not to give U.N. respectability to the Anglo-French intervention but to change its character and make it serve different ends'.[1] This idea Mr. Pearson very soon abandoned after reaching the United Nations and sampling the mood of anger, alarm, and fright which, while strongest among the Afro-Asian delegations, was not confined to them. Many delegates already favoured branding Britain and France as aggressors under the Charter. To ask them to entrust these same countries with the task of police enforcement on the United Nations' behalf was clearly out of the question. Accordingly, Mr. Pearson dropped his original plan and began casting about for a more realistic alternative. More than anything else, time was at a premium: he strove mightily to obtain it.

All during the first critical night of November 1–2 [an American observer has written] while the . . . General Assembly was exploding with indignation and demanding a cease-fire, Mr. Pearson, in his private contacts, argued for delay. Make haste slowly, he counselled; divide the Assembly's pressure, exerting some on Egypt as well as Britain, France and Israel; make use of the 'incentive of fear' to pry from Egypt and her friends the first step toward a peace settlement. Pearson held an impromptu conference with . . . Dulles on the floor of the Assembly while the debate was in progress. Dulles was sympathetic to the idea of a U.N. force, but did not think it wise to delay the pending cease-fire resolution, which he had put forward.[2]

This cease-fire resolution, Mr. Pearson subsequently informed the Canadian House of Commons, had been introduced by the United States.

without very much consultation or very much opportunity for consideration. . . . We also felt that it was inadequate for the purpose . . . because it did not recognize the background, the previous problems which had brought about this situation, and made no provision for the absolute necessity for a peace settlement. Nor did it make provision for a United Nations police force to supervise and secure the cessation of hostilities. We were anxious not to give our support at that first meeting of the Assembly to a resolution which might seem to bring the fighting to an end but to do nothing else, or even to recognize the importance of doing something else.[3]

Shortly after his arrival at the United Nations, the Canadian Secretary of State for External Affairs was shown the text of the

---

[1] William R. Frye, *A United Nations Peace Force* (New York, 1957), pp. 1–2.
[2] Ibid., pp. 2–3.   [3] Canada, H.C. Deb., 1956 (Special Session), p. 56.

United States cease-fire resolution, and promptly put his name down on the speakers' list. But already twenty-one delegates had signified their intention to address the Assembly. Under strong pressure from the United States delegation debate on its resolution was closed off before all those wishing to speak to it had the chance to do so. Only after the vote had been taken and the resolution carried by a majority of 64 to 5, with 6 abstentions, was Mr. Pearson able to explain to the Assembly why his delegation had abstained. 'Peace', he declared, 'is far more than ceasing to fire. . .'; and he went on to state that in his Government's view, the cease-fire resolution should have authorized 'the Secretary-General to begin to make arrangements with member governments for a United Nations force large enough to keep these borders at peace while a political settlement is being worked out'. He hoped that the idea might still be considered and carried out. To such a force, Canada would be glad to contribute.[1]

This suggestion—this 'brilliantly tactical stroke', as a Canadian observer later described it[2]—was made early in the morning of 2 November. Soon afterwards, at 4.20 a.m., 'too exhausted to go any further', the Assembly adjourned, although not before hearing Dulles endorse the principle of Mr. Pearson's proposal. At 4.30 a.m. the Canadian Secretary of State for External Affairs and the Secretary-General of the United Nations conferred on the proposed U.N. force. Dag Hammarskjold was reported to have been 'wary' of the idea at this stage.[3]

As the morning of 2 November passed, it became increasingly more obvious that the call to cease their fire would be ignored by the combatants. Members of the United States delegation sought out Mr. Pearson to inform him that their country would support a United Nations force if he would formally propose that it be created. 'Pearson was willing but first wanted to consult Ottawa.'[4] An hour or so later,

Mr. Pearson sat down with Mr. Hammarskjold and Andrew W. Cordier, Executive Assistant to the Secretary-General, for lunch. The

---

[1] See below, pp. 317–20.

[2] Maxwell Cohen, 'The United Nations Emergency Force: A Preliminary View', *International Journal*, vol. XII, no. 2, Spring 1957, p. 110.

[3] Frye, op. cit., p. 4. See also Joseph P. Lash, *Dag Hammarskjold* (New York, 1961), p. 84: 'With a British and French invasion force approaching the coast of Egypt, the Secretary-General seriously doubted that the moment was propitious for talking about a lasting peace settlement.'

[4] Lash, op. cit., p. 85.

Canadian explained his plan in greater detail. It was urgent for many obvious reasons, he said, including the fact that a police-force was one of the British-French conditions for a cease-fire. . . .

Hammarskjold . . . was still doubtful about the practicality of the Pearson idea. From which countries would the troops come? he wanted to know. Would there be a response? The idea must not start out with fanfare and then flop; it could backfire badly. The risk of acting must be weighed against the risk of not acting. . . . However, Pearson was able to convince Hammarskjold that the obstacles could be overcome; and the rest of the meal was spent planning an approach to the General Assembly and deciding immediate next steps.[1]

After these consultations, Mr. Pearson returned to Ottawa to discuss with his colleagues in the Cabinet whether introducing a resolution at the Assembly calling for the creation of a United Nations police force 'would be a desirable thing to do'. The Government, Mr. Pearson later told the House of Commons, 'were anxious to keep in touch with our friends in Washington and our friends in London. . . . As soon as it was decided here [in Ottawa] . . . that this might be a useful and helpful Canadian initiative under certain circumstances we cabled London and Washington at once and asked them what they thought about the idea. . . . Then . . . I returned to New York.'[2]

Armed with the approval of his Government (and perhaps of the Governments of the United Kingdom and the United States, although it may have been too early for their replies to have been received in Ottawa before the General Assembly reconvened on 3 November), Mr. Pearson began the job of rounding up support from other delegations:

Pearson had put a proposed first step into writing in the form of a draft resolution endorsing the idea of a police force and asking an Assembly committee to work out the details. This draft, in a form revised by the United States to give Hammarskjold the executive responsibility, had been shown to as many influential delegates as Pearson and his aides could collar in U.N. corridors and lounges.

This process of lobbying continued during the debate that night [3–4 November], Pearson masterminding the process from his seat in the Assembly Hall. A key move was a contact made with Arthur S. Lall, the permanent representative of India, head of the Indian delegation, pending the arrival from New Delhi of . . . Krishna Menon. Lall, on

1 Frye, op. cit., p. 5.
2 Canada, H.C. Deb., 1956 (Special Session), pp. 57–58.

behalf of India, accepted the Pearson draft and promised to support it. Pearson in turn indicated support for a new and strengthened cease-fire resolution which India and a group of other Afro-Asian nations were offering. . . . Pearson accepted the arrangement, even though the Indian resolution represented considerable pressure on Britain, France and Israel, and such pressure was not entirely popular at home [in Canada]. Canada could go along [with the strengthened cease-fire resolution] if the Indians—and consequently the 26-nation Afro-Asian bloc—would accept the U.N. force. Attaching the police force as a rider to the pressure was precisely the strategy Pearson had himself been urging.[1]

That night of November 3 and 4 [Mr. Pearson recalled three weeks later in the Canadian House of Commons]—and the session went on all night—tempers were rather high. The talk was strong and the danger of a rash—as we would have thought it—condemnation of the United Kingdom and France as aggressors was very real. The situation was deteriorating and the communists were working feverishly and destructively to exploit it.

In these circumstances and having, as I have said, canvassed the situation carefully with our friends and having studied Sir Anthony Eden's speech, we moved this resolution concurrently with the 19-power Asian-Arab resolution. . . . It was a very short resolution[2], and it asked the Secretary-General merely to submit, within 48 hours, something we had been unable to do anything about for ten years, namely, a plan for setting up an emergency international United Nations police force with the consent of the governments concerned. . . .[3]

The Canadian resolution was adopted at 2.17 on Sunday morning (4 November) by a vote of 57 to 0 with 19 abstentions (including France and Egypt, all of the Soviet bloc and, from the Commonwealth, the United Kingdom and Australia). The Afro-Asian resolution was carried by a vote of 59 to 5 with 12 abstentions, Canada voting, according to Mr. Pearson's arrangement with the Indian delegation, in its favour.

II

There now had to be improvised, within the forty-eight-hour

[1] Frye, op. cit., p. 6.
[2] 'The General Assembly, bearing in mind the urgent necessity of facilitating compliance with the resolution of 2 November 1956, requests, as a matter of priority, the Secretary-General to submit to it within forty-eight hours, a plan for the setting up, with the consent of the nations concerned, of an emergency international United Nations force to secure and supervise the cessation of hostilities in accordance with the terms of the aforementioned resolution.' G.A.O.R., 1st Emergency Special Session, 563rd Meeting, 3 November 1956, Doc. A/3276.
[3] See below, p. 330.

period specified in the Canadian resolution, a plan for the projected United Nations force. 'We had something to do with this', Mr. Pearson subsequently informed the House of Commons at Ottawa, 'because we were the sponsors of the resolution and had a certain obligation in connection with helping the Secretary-General carry it out. We started to work.'[1] Seven hours after the Assembly's endorsement of his resolution, Mr. Pearson met in the Secretary-General's office with Dr. Francisco Urrutia of Colombia, Mr. Arthur Lall of India, and Mr. Hans Engen of Norway. They discussed together the provisions of the working paper that Dag Hammarskjold had somehow been able to prepare in the meantime. They agreed, as a first step, that the General Assembly should be asked to authorize the creation of a United Nations Command; it was to be placed under the direction of Major-General E. L. M. Burns, the Canadian Army officer who since 1954 had been Chief of Staff of the United Nations Truce Supervision Organization charged with keeping the peace between Arab and Israeli. Staff officers were to be recruited immediately; troops would come later, though not (it was stipulated at the outset) from permanent members of the Security Council (and hence not from Britain and France). A report incorporating these principles was presented to the General Assembly. It was adopted, shortly after midnight on the morning of 5 November, by the same vote that had carried the Canadian resolution the previous day.

The way was now opened for putting a United Nations force together, and for deciding what, exactly, should be done with it. The second problem had to be solved first, for governments, asked to contribute troops, would naturally want to know what they would be doing. Its solution had been entrusted by the Assembly to the Secretary-General and his advisers; and in working it out they had necessarily to take into account the wishes of the Assembly. But which of the conflicting views expressed in that unruly body were they to take as their guide? To Egypt, the proposed force was acceptable as a device to clear the aggressors out of Egyptian territory; that done, its job was done, and there was no further reason for a United Nations presence. To Israel, France, the United Kingdom, and Australia, the force was a necessary inconvenience, occupying in their stead the Sinai Desert and Suez Canal area until such time as the safety and stability of these

[1] See below, p. 331.

regions could be assured and a settlement of the Suez Canal dispute worked out to their satisfaction. And in between national attitudes towards the proposed force ranged from the enthusiastic support of its Canadian innovator to the frankly suspicious views of Afro-Asian nations fearful of inroads upon their newly acquired sovereignties.[1]

To the complexities involved in compromising between these very different attitudes to the satisfaction of all concerned in the very short time available there were now added two aggravating factors. British and French troops landed at Port Said; and the Soviet Union spoke ominously of intervention and, in letters to the Prime Ministers of Britain and France, of 'rocket weapons' to rain upon London and Paris in the event that the invaders were not immediately withdrawn. 'The situation at the United Nations', Mr. Pearson recalled, 'immediately began to deteriorate. Things became very tense. The Security Council was called into emergency session and refused to consider a Soviet proposal for Soviet and United States intervention.' There were 'rumours that there would be a determined demand by the Arab and Asian members of the Assembly to brand the United Kingdom and France formally as aggressors under the Charter and to invoke sanctions against them.'[2] It was amidst these exceptionally harrowing pressures that the Secretary-General and his advisers, working between 5 November and the early morning of the following day, produced what Dag Hammarskjold called his 'second and final report' to the Assembly on the proposed United Nations force. It was

[1] 'There is a paradox in the attitude of Asians and Africans towards United Nations forces which will have to be resolved if the United Nations is to become a stronger instrument in troubled areas. While they call for the United Nations to step in and do all sorts of things they want, they are suspicious of United Nations interventions in which white men are deeply committed. Bitter experiences, reinforced by the facts and myths of the anti-colonialist creed, have given them a phobia about marines and paratroopers who land in weak countries. Even though Arabs and Asians on the whole welcomed the expedition of U.N.E.F. in 1956, it was necessary to argue hard to convince them that this was not just a scheme to bring back the old imperialism under a new flag. (The problem of Canadian diplomats during that strenuous period was to persuade the Arabs on the one hand that they were not acting merely as agents of what the Arabs considered to be sinister British intentions and to persuade the British on the other hand that U.N.E.F. could carry out more effectively the mission of pacification and protection which they had taken upon themselves.)' John Holmes, 'The United Nations in the Congo', International Journal, vol. XVI, no. 1, Winter 1960–1, pp. 5–6. The author of this passage knows whereof he speaks, having been a representative of Canada at the United Nations throughout the Suez crisis.

[2] See below, p. 331.

placed before the Assembly on the morning of 6 November.

In it were laid down certain principles and procedures by which the proposed force was to be governed. Of these, the most important for the future was the provision that the permanent members of the Security Council should be excluded from the list of possible contributors, thus ruling out Anglo-French participation and, no less significantly, the participation of the Soviet Union. A second significant feature was its emphasis on the need to keep control of the United Nations Emergency Force (U.N.E.F.) vested in the General Assembly, which meant in practice the Secretary-General and his advisers. 'If the force is to come into being with all the speed indispensable to its success', the Report declared, 'a margin of confidence must be left to those who will carry the responsibility for putting the decisions of the General Assembly into effect.' The force itself was to be a police force, not a fighting force: 'more than an observers' corps, but in no way a military force temporarily controlling the territory in which it is stationed.' While the cease-fire was being established, it was 'to enter Egyptian territory, with the consent of the Egyptian Government, in order to help maintain quiet during and after the withdrawal of non-Egyptian troops, and to secure compliance with the other terms established in the Resolution of November 2nd. . .'. The force was to be as politically neutral as possible. 'There is no intent in the establishment of the force to influence the military balance in the present conflict and thereby the political balance affecting efforts to settle the conflict.'[1]

The Assembly accepted these guiding principles. At the Secretary-General's suggestion an Advisory Committee was established, with himself as chairman, consisting of representatives of Brazil, Canada, Ceylon, Cambodia, India, Norway, and Pakistan. Its purpose was to help him bring into being and into position the kind of force which the Assembly, by adopting his second Report, had agreed should be brought into being.

III

Word of the proposed United Nations force reached its future commander at his Jerusalem headquarters at about 11.30 p.m. on 4 November. Dag Hammarskjold's message set out the proposal in the very tentative form that it had reached at that stage, and asked General Burns for his views about the numbers and types of

---

[1] G.A.O.R., Doc. A/3302, First Emergency Special Session, 1956.

troops which would be required for the job, and what kind of equipment they would need. General Burns dispatched a 'lengthy reply' the following morning.

My suggestions as to organization [he has since candidly stated] were based on certain premises as to the political situation, some of which proved untenable, as things developed in the General Assembly. ... I stated that I thought the force should be so strong that it would be in no danger of being thrust aside, pushed out, or ignored, as the U.N. Military Observers had been in Palestine. ... I thought such a force, in view of the strength of the armed forces of Israel and Egypt, would have to be about the size of a division, with a brigade of tanks, and attached reconnaissance and fighter-aircraft units—the whole organized as an operational force capable of fighting. I suggested that contingents should not be less than battalion size, as a force made up of many smaller units of different nationalities would be difficult to control, from the administrative as well as the tactical viewpoint.[1]

What General Burns wanted, and what General Burns got, turned out to be very different things.

When the various nations eligible began to offer contributions to the force [he has since written] these turned out mostly to be odd-sized units, usually smaller than the normal battalions I had requested. ... The composition of the force had many of the disadvantages, from the military viewpoint, which I had hoped to avoid by my 'specifications'. While, by and large, the United Nations Emergency Force accomplished its purpose in spite of its peculiar composition, I still feel that a stronger and more coherently organized force might have been a better instrument for the execution of U.N. policies.[2]

However, as General Burns realized better than anyone else, one had to judge what emerged not by the canons of the Staff College but as Dr. Johnson did the woman preacher. It was not only that a 'force capable of fighting' was out of the question: the very different kind of operation then being planned in New York was at the outset in danger of being refused permission to enter Egyptian territory. Egypt's delegate at the United Nations had pointedly abstained from voting upon either the resolution of 4 November authorizing the Secretary-General to draw up plans for an international force or the resolution of 5 November endorsing the plans he had drawn up. Where did Egypt stand? One who

[1] Lieut.-Gen. E. L. M. Burns, *Between Arab and Israeli* (Toronto, 1962), p. 188.
[2] Ibid., p. 190.

attempted to find out was Canada's newly appointed Ambassador, Herbert Norman, who sought out President Nasser and spent two hours with him arguing for the admission of U.N.E.F.[1] Another was General Burns himself who, on 8 November, flew to Cairo from Jerusalem for an interview with various Egyptian authorities, including the Foreign Minister, Dr. Mahmoud Fawzi. Dr. Fawzi told General Burns that he thought there would be no objection to U.N.E.F.'s presence, provided that its composition and activities were acceptable to the Government of Egypt.

This was agreement in principle only. In theory, the Secretary-General's terms of reference placed the determination of the composition of U.N.E.F. in the hands of the Assembly—that is to say, his own; in practice, the composition of U.N.E.F. had to meet with the approval of the host country. Among the first to offer to contribute to the Force had been Canada, Colombia, Norway, and New Zealand. The Secretary-General and his advisers decided that the New Zealand offer, in view of New Zealand's support of the United Kingdom's action, would be less of an asset than a liability; that Canada and Colombia would be eminently acceptable; and that other Scandinavian contributions were both desirable and available. General Burns was instructed by the Secretary-General to ascertain from the Egyptian Government whether it would object to the participation in the proposed force of Canada, Colombia, Denmark, Finland, Norway, and Sweden.

I got a considerable shock [General Burns has recorded] when, at my meeting with Dr. Fawzi, he intimated that it might not be acceptable to Egypt that a Canadian contingent should form part of the United Nations Emergency Force. I pointed out the leading part which Mr. Pearson had played in the General Assembly in the creation of the force, promoting a policy contrary to the views of not only the United Kingdom but also the other Dominions, and one which would be greatly to the advantage of Egypt. Fawzi said he understood Canada's independence in foreign policy and Mr. Pearson's helpful efforts very well; but the trouble was that Canadian soldiers were dressed just like British soldiers, they were subjects of the same Queen—the ordinary Egyptian would not understand the difference, and there might be unfortunate incidents. He said this was not a 'firm answer' to the Secretary-General's proposal that Canadians should form part of the force, but wanted me to pass the thought on to Mr. Hammarskjold. I said I hoped no such

[1] Sidney Katz, 'What Kind of Man was Herbert Norman?', *Maclean's Magazine* (Toronto), 28 September 1957.

regrettable decision as to exclude Canadian participation in the force would be taken. For one thing, I should naturally not be able to act as commander in such a case. . . .[1]

As the Canadian Government had already made arrangements to send a battalion to Egypt (an advance party of thirty-four men was already in Naples, awaiting Egyptian permission to enter), word of Cairo's stand was conveyed immediately to it, *via* a telephone call to Mr. Pearson from United Nations headquarters at New York. 'The Egyptian authorities', Mr. Pearson later reported he had been told, 'were concerned about the possibility of Canadian troops being mistaken for United Kingdom troops and that incidents might take place especially if the proportion of Canadian troops to the total were high as would be the case if the Canadian infantry battalion had arrived at that time'.[2]

Genuine or not, there was some reason for the Egyptian Government's misgivings. More than one country, even in recent years, had misunderstood, or found it useful to misunderstand, the relationship between the United Kingdom and other members of the Commonwealth; the uniforms worn by British and Canadian troops were similar if not identical (General Burns himself soon found it expedient to exchange his Canadian khaki for a uniform of special and distinguishing design[3]). But, whatever their validity, the Egyptian objections were doubly unacceptable to the Canadian Government. From the outset it had opposed the principle that any single government should be allowed to dictate the composition of the Force; Egypt, while within its rights in refusing it admission, had no right to determine its national components. Mr. Pearson put this forcibly to the Secretary-General in discussions with him in New York on 13 November. 'On this matter', he afterwards informed the House of Commons, 'we would negotiate only with him, . . . although we recognized, of course, that it was right and proper that he should discuss these matters with Egypt in order to avoid, if possible, subsequent difficulties.'[4]

But there was a further, if unstated, reason for the Canadian

---

[1] Burns, op. cit., p. 198.   [2] See below, p. 335.

[3] 'The design was basically the same as the uniform worn by the security officers (guards) of the United Nations, at New York H.Q. and in missions abroad. . . . At first I felt somewhat self-conscious in this new garb, after forty years of khaki, but soon got used to it.' Burns, op. cit., p. 238.

[4] See below, p. 336.

Government's concern. It had already become apparent that the policy it had been following since 2 November, so far from commanding universal support among the Canadian people, had aroused a storm of controversy and criticism which, if its strength could not be accurately gauged, could not be overlooked. At the same time the movement of Canadian men and material to Halifax, from where they were to embark for the Middle East on the aircraft carrier H.M.C.S. *Magnificent*, had taken place in a blaze of publicity which the Government, to say the least, had done little to discourage. The political repercussions of the spectacle of these troops, so ostentatiously dispatched, returning dispiritedly to barracks because Colonel Nasser had refused them admission were distasteful to imagine, let alone to experience. Whether or not this aspect of the situation was put before the Secretary-General, it helped to account for Canada's insistence that neither Nasser's nor any one else's government was to be allowed to say what the composition of U.N.E.F. ought to be. Dag Hammarskjold informed Mr. Pearson that he believed Egypt's objections could be overcome, and advised him to instruct his Government to proceed on this optimistic assumption. Arrangements were accordingly made for the R.C.A.F. to fly 300 troops to join the advance party at Naples, and for the *Magnificent*, with the remainder of the infantry battalion on board, to follow, arriving at Naples around the end of November.

Intensive negotiations with the Egyptians ensued. General Burns was in the thick of the fight. He first discussed the issue of Canadian participation with President Nasser on the afternoon of 9 November:

The President rehearsed the arguments against inclusion of the Canadians in the force which I had previously heard from Dr. Fawzi. He said it would be easy for agitators to incite some of the people against troops that looked like the British; there might be nasty incidents, and he did not wish to become embroiled with Canada and the United Nations. I said I thought this was a risk which should be taken. I did not voice my impression, however, that the real reason Canadians were not desired was the fear that Canadian policy, while so far favourable to Egypt in the General Assembly, might later veer to the 'Western' if not the British stand in regard to the control of the Canal.[1]

---

[1] Burns, op. cit., p. 200.

A second interview between the two men took place on 12 November. 'We talked for some time . . ., but we both only repeated previous arguments. He was friendly and agreeable in his attitude throughout, and appeared very willing to co-operate, apart from the composition of the force. He said that he hoped Mr. Pearson would understand his position, and that there was no prejudice against Canada.'[1] It was only after these discussions that General Burns learned, to his dismay, that the unit selected by the Canadian Government for participation in U.N.E.F. was the Queen's Own Rifles—a name, however glorious, unlikely to make its acceptance by Egypt any easier. 'There is no regiment in the Canadian forces that I respect more than the Queen's Own', he wrote afterwards, 'but it did seem an unlucky chance, in view of the Egyptian argument, that out of Canada's six regular infantry regiments this one had been selected.'[2]

Another intermediary working on behalf of the Canadian Government in this matter was Mr. Krishna Menon. On 15 November, he told Mr. Pearson that he and other spokesmen for the Indian Government had exerted pressure in Cairo, with the result that the Egyptian Government was now prepared to allow Canada to provide air transport and a field ambulance unit for U.N.E.F. The same concession was reported by the Secretary-General, who had travelled to Cairo to take up this and other problems connected with the Force with President Nasser. Dag Hammarskjold was told by Mr. Pearson, however, that the Canadian Government 'could not accept the principle that any one government could determine what contribution or whether any contribution would be made by a member state in connection with the United Nations force'.[3] The Secretary-General thereupon resumed negotiations. On 17 November he was able to report further progress. He and the Egyptian Foreign Minister had drawn up and agreed upon the following statement:

Canada is welcome as a country from which elements of the U.N.E.F. may be drawn. It is felt that the most important contribution that could be given at the present stage from that country would be air support in

---

[1] Ibid., p. 204.
[2] Ibid., p. 209. The names of the two regiments next in line were not much more propitious: they were the Princess Patricia's Canadian Light Infantry and the Black Watch.
[3] See below, p. 336.

the transport of troops from Italy and for the current functioning of the U.N.E.F. in Egypt.[1]

The concluding words of this statement provided the Canadian Government with a means of overcoming the grave political embarrassment in which Colonel Nasser's obduracy threatened to place it. For in the phrase 'for the current functioning of the U.N.E.F. in Egypt' could be discerned Cairo's willingness to admit administrative troops which could be substituted for the Queen's Own Rifles on a greater scale and therefore more acceptably than could air transport personnel and an ambulance unit. One further obstacle remained. As matters stood, the conclusion could hardly be avoided that Colonel Nasser, objecting to the Queen's Own Rifles but acquiescing in administrative forces, had in fact dictated the nature of the Canadian contribution. What then became of the principle that this was a matter for the Secretary-General, in consultation with the contributing country, to decide? To the rescue came U.N.E.F.'s Commander.

While in New York [16 to 19 November] I had several conversations with Mr. L. B. Pearson, . . . and before I left wrote him a letter stating that as a result of my studies and discussions with military representatives at U.N. headquarters, I had concluded that the most urgently required and valuable contribution which Canada could make at the time would be the air transport squadron of the R.C.A.F. and administrative elements for the Force. It may be thought that this was merely making the best of an awkward situation created by President Nasser's obduracy in refusing to accept a Canadian infantry battalion. However, in the light of subsequent experience, I feel that Nasser's refusal was a blessing in disguise, for the administrative and supporting troops Canada provided then and subsequently were absolutely essential, and the Force could not have operated without them. It was not feasible for other contributing nations to furnish technical and administrative troops of the kind needed, as was proved by the lack of response to the request for such contributions which had been sent out . . . on November 10.

I would have been extremely glad to have a Canadian infantry battalion in the Force; but if the choice had been between the infantry battalion and the technical troops I should have been obliged to take the latter. . . .[2]

The Canadian Government, naturally enough, was relieved at the outcome although, as Mr. Pearson stated at the time, it regretted

---

[1] Quoted in Frye, op. cit., p. 30.   [2] Burns, op. cit., pp. 215–16.

the change of plan. 'I think the course we took was the right course, and it was considered the right course by the United Nations officials concerned.'[1]

By a number of Canadian newspapers and, it may be presumed, by a considerable number of Canadians, it was not considered the right course at all. *The Globe and Mail* of Toronto, critical of the Government's Middle Eastern policies throughout the crisis, greeted the news that Colonel Nasser would not play host to the Queen's Own Rifles with an editorial entitled 'The U.N. Police Farce'.[2] *The Calgary Herald*, also a persistent critic, remarked of Mr. Pearson's part in the creation of U.N.E.F. that 'all it did was to provide a face-saver for the Soviet's puppet-dictator in Egypt and preserve his power for another day—a rank disservice in the cause of peace'.[3] During the four-day special session of Parliament begun on 26 November, the amendment to the Address moved by the Conservative Opposition accused the Government of having 'placed Canada in the humiliating position of accepting dictation from President Nasser'.[4] However, the idea of the International Force was not attacked from any quarter in the House of Commons. Neither was the principle of Canada's contribution, although an Independent Member, Mr. Fernand Girard, warned that 'le rôle préponderant que le Canada joue dans cette force policière peut paraître honorable, mais il risque de faire de nôtre pays la principale victime si la situation tourne mal', and another Independent, Mr. Raoul Poulin, thought Canada's contribution 'exagérée'.[5] 'Exagérée' or not, it was certainly 'préponderant'. By the end of 1956, U.N.E.F. consisted of some 5,500 men, of whom 1,100 were Canadians. And so it continued for the next five years.

IV

Of the eight members of the Commonwealth, only two made no offer of men or material to the United Nations Emergency Force. South Africa remained aloof, expressing neither interest nor opinion. Australia expressed both. On 3 November, Mr. Menzies stated that his Government awaited 'the concrete development of Mr. Pearson's helpful proposal' with 'most sympathetic interest'.[6]

---

[1] Canada, H.C. Deb., 1956 (Special Session), p. 64.
[2] *The Globe and Mail*, 21 November 1956.
[3] *The Calgary Herald*, 17 November 1956.
[4] Canada, H.C. Deb., 1956 (Special Session), p. 92.
[5] Ibid., pp. 91, 92.   [6] See below, p. 346.

On 8 November, he told the House of Representatives that the Force should be 'an effective military force'. He pointed out that as Egypt had just broken off diplomatic relations with Australia, it would be highly unlikely that Australia would be a welcome contributor to the Force; but 'if the proposal is to constitute a military establishment which will facilitate the making of a permanent settlement in the Middle East, Australia will certainly be not unwilling to make such quick practical contribution as it can'.[1] In these statements may be detected the restraining hand of official draftsmen; on his own, Mr. Menzies was his usual blunt self. In a speech at Canberra on 12 November, he criticized the exclusion of Britain and France from U.N.E.F., which he described disparagingly as 'a force to consist of people from Portugal and Colombia and little groups and bits and pieces'; he also criticized the United Nations—'Lake Success—or whatever it is called'— for 'imposing an indignity in the name of some barren technical argument upon the Great Power from which we derive . . .'.[2] His Minister of External Affairs, speaking (some said allowed to speak) for the first time since the Anglo-French ultimatum on his country's behalf, was understandably more moderate when he addressed the General Assembly on 21 November[3] and again on 26 November.[4] On 4 December, Mr. Casey expressed his belief 'that the decision of the British Government to withdraw their forces from Egypt is a wise decision, if only as a means of bringing about normal relations between the British and American Governments'.[5] Two months later, Sir Percy Spender, speaking in the general debate of the General Assembly, reverted to Australia's earlier criticism by a point-blank refusal to accept the basic principle for the United Nations Force in Egypt enunciated in the Secretary-General's report—namely, that its presence and operation were at the discretion of the host Government.[6] At least one Australian commentator found this intervention less than helpful. 'It is to be feared', he wrote, 'that the discussion initiated by Sir Percy Spender . . . was designed less for the strengthening of U.N. authority than for scoring off the Egyptians.'[7]

New Zealand (it will be recalled), while abstaining from voting

---

[1] See below, p. 347.   [2] See below, p. 347.   [3] See below, pp. 348–9.
[4] See below, pp. 349–50.   [5] See below, p. 350.   [6] See below, pp. 351–2.
[7] Geoffrey Sawer, 'Problems of Australian Foreign Policy, June 1956–June 1957', *Australian Journal of Politics and History*, vol. III, no. 1, November 1957, p. 2.

on the resolution creating the United Nations Force, had been one
of the first to offer to contribute to it.[1] On 9 November, in a letter
to the Secretary-General, the offer was made precise. New Zealand
was ready to contribute 300 men for duty as 'land forces' on the
'clear understanding' that the Force would be 'para-military' and
not a fighting force.[2] Confident that their offer would be accepted,
the New Zealand Government proceeded to make arrangements
for enlistment. On 14 November the Prime Minister announced
that the force would be 'a volunteer body specially enlisted on
terms similar to those applying to the New Zealand Special Air
Service squadron' then serving in Malaya.[3] To this unusual call to
arms more than enough responded. Both as a means of extricating
the Mother Country from an untenable position and as a means of
realizing the concept of collective security to which New Zealanders had for long been strongly attached, the Emergency Force
was enthusiastically supported by New Zealand opinion. The
Minister of External Affairs, Mr. Macdonald, had spoken in the
General Assembly on 21 November of 'the admirable and imaginative proposal of the Minister of External Affairs of Canada'.[4] It
was therefore hard and humiliating to learn that New Zealand's
offer was not accepted by the United Nations planning group.
The Secretary-General had with great tact and consideration
explained to Mr. Macdonald the need for U.N.E.F. to be a
balanced force, and the priority then attaching to technical and
administrative units; but it was clear enough that the important
disqualifying factor was the inability on the part of the Egyptian
Government to discern any significant difference between the
attitude of New Zealand and the attitude of the United Kingdom
throughout the crisis. And so, as a New Zealand commentator
ruefully remarked, 'a country which for eleven years had regarded
itself as a model small nation in the field of international security
was excluded from a force composed almost entirely of small
nations'.[5] The Prime Minister accepted the situation with a good
grace (if it be asked what else could he do, there was always Mr.
Menzies' very different demeanour), and on 4 December, in a
statement approving the United Kingdom's decision to withdraw

---

[1] See below, p. 352.   [2] See below, pp. 353–4.
[3] See below, p. 354.   [4] See below, pp. 354–6.
[5] J. K. Cunningham, 'New Zealand as a Small Power in the United Nations',
Political Science (Wellington), vol. 9, no. 2, September 1957, p. 36.

from Egypt, he announced that 'our own offer to help still stands'.[1]

V

All three Asian members of the Commonwealth responded to the Secretary-General's call for contributions to the United Nations Force on 4 November: Pakistan eagerly, Ceylon less eagerly, India with reservations.

Pakistan's offer, conveyed in a letter to the Secretary-General on 5 November by its Permanent Representative to the United Nations,[2] provided Mr. Suhrawardy with one of his very few opportunities to oblige his Western allies without alienating his public. President Nasser, however, professing his distaste for Pakistan's activities as a member of the Baghdad Pact, refused to accept a Pakistani contingent as a part of U.N.E.F. The United Nations command, which had been content to accept without protest Egypt's refusal of New Zealand's offer, were unhappy at the rejection of Pakistan's. It was not only that 'their unit included an excellent military band',[3] or even well-trained and disciplined soldiers of just renown; the inclusion of contingents of both Pakistan and India in a United Nations Force dedicated to peaceful solutions in disputed territory might conceivably have been turned to advantage in the solution of the Kashmir problem. No doubt it was on this account disliked by India, though perhaps not to the extent later believed by Pakistani commentators, one of whom accused India of spreading hostile propaganda in Egypt to the effect that 'Pakistan welcomed the invasion of Egypt by England' and that 'prayers were said for Queen Elizabeth in the mosques in Pakistan'.[4]

Early in December it was proposed to bring U.N.E.F.'s strength, then at about 4,500 men, up to about 6,000, and as the most suitable contributors for this purpose the United Nations at New York had selected Brazil, Indonesia, and Pakistan. As always, Egypt's consent was practically essential. No objection was raised to Indonesia, that to Brazil was overcome, but to a Pakistani contribution Egypt remained adamantly opposed. Coming as it did hard on the heels of Colonel Nasser's refusal to see Prime

---

[1] See below, pp. 357–8.   [2] See below, p. 367.   [3] Frye, op. cit., p. 23.
[4] Quoted in Hafeez-ur-Rahman Khan, 'Pakistan's Relations with the U.A.R.', *Pakistan Horizon* (Karachi), vol. XIII, no. 3, Third Quarter 1960, p. 222.

Minister Suhrawardy in Cairo, Egypt's continued opposition to Pakistan's participation in U.N.E.F. added insult to injury, and feelings against Egypt (and against India, as the real or imagined instigator of Egyptian protests) ran very high in Karachi. 'Whether we like it or not', commented the newspaper *Dawn*, 'whether we believe it or not, and however much it may hurt our national pride, Nasser himself now proves the painful fact that he is not Pakistan's friend and even spurns the sympathy, support and helping hand which we so sincerely and spontaneously proffered.'[1]

Ceylon's offer of a contribution to U.N.E.F. was made only a day after Pakistan's, on 6 November, but, as its Prime Minister subsequently informed the House of Representatives, it was made not on its own initiative but 'in pursuance of a request urgently made by the Secretary-General to our representative at the U.N. which was transmitted here [to Colombo]'.[2] It is possible that S. W. R. D. Bandaranaike would not have found it so imperative to stress that the idea of Ceylon's contribution was not his Government's idea had the United Nations found it possible to make use of the infantry company of 150 men promised by Mr. Gunawardene in a letter to the Secretary-General on 9 November.[3] Ceylon's rejection, unlike Pakistan's, was due not to Egypt's unwillingness to accept its troops but rather to Ceylon's inability to provide the United Nations command with the specialized personnel it required at that time.

Of the three Asian members of the Commonwealth, India viewed the project of resorting to an international police force with the greatest suspicion and with the most reservations. Such a view was entirely consistent with India's previously expressed attitudes. Late in 1950 the Indian representative at the United Nations was almost alone among non-communist delegates in objecting to the so-called Uniting for Peace resolution, declaring that it 'gave the impression that the United Nations was more concerned with perfecting its enforcement machinery than with promoting international co-operation and good will'; India had opposed the effort to implement the resolution and in reply to the inquiry by the Collective Measures Committee of the General Assembly as to what, if anything, the Government of India was doing by way of implementation had cabled tersely: 'Indian Government feel that creation U.N. force would not assist at present in creation proper

---

[1] Quoted ibid., p. 220.   [2] See below, p. 373.   [3] See below, pp. 371–2.

psychological atmosphere preservation of peace. . . .'[1] To this position India adhered throughout subsequent sessions. During the Ninth Session (1954), a new report of the Collective Measures Committee asked that the burden of peace enforcement be more equitably distributed among the members and that enforcement machinery be placed effectively at the disposal of the General Assembly. Mr. Krishna Menon supported the opposing Soviet view that the Security Council was the appropriate place for such activities, adding the distinctively Indian refinement that such activities were not conducive to the relief of international tension. It was, he declared, 'unwise to emphasize the military aspects of the United Nations when all the world was longing for peace'. His delegation 'wished to place before the Committee the alternative of collective peace to collective measures'.[2] Underlying this generalized concern lest collective measures threaten the peace of the world it was not hard to discern the fear lest they threaten the Indian position in Kashmir.

With such a record of unbroken hostility to the idea of an international police force, the Indian Government was obviously not to be looked to during the early days of the formation of U.N.E.F. as a pillar of strength and support. And yet, as has been noted, it was India's support for Mr. Pearson's resolution of 4 November which largely assured its passage in the General Assembly. This fortuitous development was due to an accident of personality. Had Mr. Krishna Menon rather than Mr. Arthur Lall been representing his country during the first crucial forty-eight hours of negotiations, the outcome might have been very different.

On Monday, two days after the Pearson resolution had been adopted overwhelmingly, Mr. Krishna Menon strode into a private meeting of the Afro-Asian bloc and started resisting the whole idea of a police force. There was a shocked silence; then someone pointed out to him that virtually everyone in the room, including the representative of India, had already voted in favour of the idea, and that it was being implemented. Krishna Menon subsided. . . .[3]

The fact that Mr. Menon, *en route* from New Delhi to New York, was delayed a day or so on the journey is in the view of a Canadian

---

[1] Quoted in Ross N. Berkes and Mohinder S. Bedi, *The Diplomacy of India* (Stanford, California, 1958), pp. 6–7.
[2] Quoted ibid., p. 9.   [3] Frye, op. cit., pp. 6–7.

at that time representing Canada at the United Nations 'one of the fortunate accidents of history'.[1]

Once on the scene, however, Mr. Krishna Menon very quickly made his mark upon events. It was at his insistence that Mr. Lall's letter to the Secretary-General of 6 November bristled with so many reservations and understandings: the Force was not to be regarded as a successor to the Anglo-French invaders; it was to function with the consent of the Egyptian Government; it was to be balanced in its composition; above all, it was to be temporary.[2] In the event the results proved more impressive than the reservations. Indian troops were offered and accepted. From 15 September 1957 the Indian component in U.N.E.F.—27 officers and 930 other ranks—was larger than any other except the Canadian.

Mr. Krishna Menon made things difficult for the Canadian delegation by his reference to the United Kingdom's treatment of the Canadian proposal for the creation of the United Nations Force as a continuation of British policy.[3] As one of Canada's principal motives in taking the initiative in bringing the Force into being had been to extricate its Commonwealth partner and N.A.T.O. ally from what it regarded to be a dangerous, untenable, and humiliating position, no Canadian spokesman was able to dissociate his country's policy from that of the United Kingdom in the manner Mr. Menon seemed to desire; nor, given the divided state of Canadian public opinion, would it have been expedient to do so. The difference between Canadian and Indian attitudes towards U.N.E.F. continued to be emphasized by Mr. Menon from time to time; he referred to it in a speech in the Lok Sabha on 26 March 1957. This feature of India's diplomacy could hardly have seemed helpful to the Canadian Government; it certainly did little to strengthen that 'Indo-Canadian *entente*' whose existence an Indian scholar later identified and extolled.[4]

There was, however, a difference; and it derived from something more fundamental than Mr. Krishna Menon's peculiar temperament and outlook upon the world. Following the evident success of U.N.E.F. in the Middle East, it was proposed to create a permanent or 'stand-by' police force for United Nations service

[1] John Holmes, 'The United Nations in the Congo', *International Journal*, vol. XVI, no. 1, Winter 1960–1, p. 14.
[2] See below, pp. 361–4.   [3] See below, pp. 364–5.
[4] M. S. Rajan, 'The Indo-Canadian *Entente*', *International Journal*, vol. XVII, no. 4, Autumn 1962, pp. 358–84.

anywhere it could be used. Canada enthusiastically supported the proposal. India no less strongly opposed it. The Indian view prevailed.

VI

The entry of the United Nations Emergency Force into Egypt placed the occupying powers in a highly anomalous position. If the blue-helmeted troops of U.N.E.F. were policemen, what were then the British and the French? To the United Kingdom Government, they were vigilantes. The substitution of U.N.E.F. for the Anglo-French force was a changing of the guard. One form of 'police action' was replacing another, volunteers by more properly instituted authority. Methods necessarily differed, but the motive in each case was to restore and keep the peace. Privately, no doubt, sourer versions might be heard. Frustration produced by abandoning a dangerous enterprise on the threshold of apparent military success passed easily over into bitter criticism of the United Nations—criticism which might with more reason have been directed against Messrs. Eisenhower and Dulles, or indeed against Sir Anthony Eden himself, but which fitted more easily into the tradition of antipathy towards the United Nations and all its works which long before the celebrated strictures of Lord Home marked public and official attitudes in the United Kingdom. But the British Government, at any rate, was not now in any position to resort to the kind of criticism by which Mr. Menzies relieved his feelings in Canberra. It was only several years later that Sir Anthony Eden disclosed what may be presumed to have been his real sentiment on the subject by choosing as his title for that chapter in his memoirs dealing with the early days of U.N.E.F. the word 'Myopia'. For the time being, British spokesmen could only wish U.N.E.F. well; they went so far, indeed, as to take credit for its creation. In any event, they insisted, there was no reason why relations between their forces and those of U.N.E.F. should not be as cordial and co-operative as those of any partners in a common cause.

To other governments there was every reason why not. If U.N.E.F. was the sheriff, the Anglo-French forces, so far from being vigilantes, were the rustlers. The apparatus of justice at the disposal of the international community was admittedly imperfect, but it was not so malformed as to require the deputizing of outlaws to redress their own crimes. Such was the opinion of most of the Afro-Asian members of the United Nations, and a fair number of the Latin American members as well. Among Commonwealth countries it was tenaciously held by India, Mr. Krishna Menon, as we have seen, taking particular pains to prevent U.N.E.F. from being regarded as taking over from where the 'aggressors' left off. But to view U.N.E.F. as responsible for arresting the wrong-doers was, if anything, less realistic than the opposing conception of partnership. If U.N.E.F. was not allied with the Anglo-French forces in the common task of pacification, it was certainly not designed or equipped to deal with them as policemen are supposed to deal with law-breakers.

In fact, none of the usual analogies came close to the existing situation. (Analogies are seldom helpful in political analysis.) If the Anglo-French forces, whether as vigilantes or as outlaws, were to retire from the scene, it would be by their own volition and under their own power, and not as the result of coercion from a force many times more physically feeble than they. Relations between the invaders and those charged with overseeing their departure proved in the event to be unusually harmonious, with one or two exceptions. The nature of the relationship was set even before the first units of U.N.E.F. had entered Egypt. As soon as the cease-fire went into effect, the Egyptian authorities complained that it was being violated by the Anglo-French forces. The United Nations decided to send a number of military observers to investigate these complaints, but no transport from Jerusalem to Port Said could be found. 'We eventually arranged', General Burns has recalled, 'for the observers and their vehicles to be taken in a British landing ship, H.M.S. *Striker*, from Haifa to Port Said, where they arrived on November 13. This was the first act of co-operation by the Anglo-French forces, and was a considerable encouragement.'[1] When in New York, General Burns had conferred with the British Foreign Secretary, who promised, subject to approval by the Cabinet, to make available to U.N.E.F. mechanical transport and other stores out of the supplies of the United Kingdom forces in Egypt. 'Thus, when I got to Port Said and met General Keightley, commander of the Allied force, among the first items of business was seeing what vehicles and stores the "enemy" could furnish to the United Nations Emergency Force

[1] Burns, op. cit., p. 195.

which, while not supposed to drive the invaders out, at least was politely to usher them out. . . . The stores and particularly the M.T. vehicles so obtained were invaluable to U.N.E.F. in the early days, and the force would have been in great difficulties without them'.[1] Additional stores—petrol, oil, and lubricants, medical supplies and rations—were made available a little later on.[2]

Not everything went as smoothly as this. Sir Pierson Dixon's letter to the Secretary-General of 6 November, conveying his Government's readiness to order a cease-fire, put forward its proposal that Anglo-French forces should at once begin the work of clearing the Suez Canal of the sunken ships then obstructing passage through it. This proposal was unacceptable to the General Assembly's majority and therefore to the Secretary-General. Here the United Kingdom believed it could negotiate from strength. The Queen's Salvage Fleet, Mr. Selwyn Lloyd informed the Secretary-General's Executive Assistant on 12 November, was not only the best in the world, it was the only salvage fleet that could do the job.[3] In that he was mistaken. The United Nations Secretariat turned to General Raymond A. Wheeler, a retired United States Army Engineer, and a noted salvage technician; under his direction an improvised fleet, consisting of thirty-two vessels mobilized from six Western European nations (other than Britain and France) and Yugoslavia, began the work of clearing the Canal. This operation did not improve Anglo-American relations.[4] Even here, however, a mutually beneficial *modus vivendi* was worked out. 'Eventually it was agreed that about ten of the Anglo-French vessels would remain, that their naval crews would be dressed as civilians and would be protected by small detachments of "civilian guards" recruited from the Swedes and Finns of U.N.E.F.'[5]

The timing of the Anglo-French withdrawal also created difficulty. The presence of British and French forces in Egypt offered their Governments the possibility, however faint, of influencing whatever settlement might be reached with President Nasser.

---

[1] Burns, op. cit., p. 212.   [2] Ibid., p. 224.   [3] Lash, op. cit., p. 96.
[4] 'The British and French seemed to blame General Wheeler for the rejection of their offers to help in the clearance of the Canal, and for a while he was the target for a good deal of abuse in the British Press. . . . General Wheeler, with whom I came to be on very friendly terms, remarked to me that he didn't like this at all, as he had always got on very well with the British in India. . . .' Burns, op. cit., p. 236.
[5] Ibid., p. 236.

Though committed to withdrawal, they were committed as yet to
no date by which to withdraw; and they were determined to
postpone it until such time as conditions judged by them to be
satisfactory had been imposed by the arrival and deployment of
U.N.E.F. and by simultaneous political negotiation. Egypt,
however, supported by most of the Afro-Asian states, insisted that
the 'aggressors' had no right to suggest terms, still less to try to
impose conditions. Withdrawal had to precede settlement.

The argument was carried to the floor of the General Assembly.
The British Foreign Secretary, 'struggling', in Sir Anthony Eden's
words, 'to inject some sense of values',[1] attempted a scholastic
distinction between 'immediate' withdrawal and 'instantaneous'
withdrawal; the former, to which his Government had agreed, did
not mean the latter. The British and French troops would go
when, as trustees for the international community, they judged
that their going would not cause a further deterioration. As
evidence of their good faith, the British Government was prepared
to withdraw one battalion; but the remainder would linger until
such time 'as the United Nations Force becomes effective and
competent to discharge its functions'.[2] By Mr. Selwyn Lloyd's
protestations of good faith the Assembly remained largely un-
moved. Mr. Krishna Menon argued strenuously on behalf of the
Government of India against the British Government's attempt to
describe the invasion of Egypt 'as though it were a service to the
world. The moment we permit this halo to get around it, to portray
an act of aggression as an act of morality, we shall be unable to
take any corrective action.'[3] An amendment offered by the Foreign
Minister of Belgium put these opposing points of view to the test
of an Assembly vote by proposing that the Afro-Asian draft resolution enjoining 'withdrawal forthwith' upon Britain and France be
replaced by a less peremptory formula allowing those Governments discretion as to timing.[4] On the afternoon of 24 November,
23 delegations voted for the Spaak amendment, 37 against it, 18

---

[1] *Full Circle*, p. 565.   [2] See below, pp. 312–15.   [3] See below, p. 364.
[4] The Spaak amendment 'Notes that, according to the information received,
one-third of the French forces has been withdrawn, the United Kingdom
Government has decided to withdraw one infantry battalion immediately, and
Israel has withdrawn a part of its troops, and considers that France, the United
Kingdom and Israel should expedite the application of the resolutions of 2 and
7 November in the spirit in which they were adopted, particularly with regard to
the functions vested in the United Nations forces'. G.A.O.R., Eleventh Session,
1956–7, I, Doc. A/L. 215.

abstained. The United Kingdom, Canada, Australia, New Zealand, and South Africa supported the amendment. India, Pakistan, and Ceylon opposed it. The Afro-Asian resolution demanding 'withdrawal forthwith' was passed immediately afterwards by a vote of 63 to 5, with 10 abstentions. The United Kingdom, flanked by Australia and New Zealand, voted with France and Israel against it; Canada and South Africa abstained; India, Pakistan, and Ceylon voted in its favour. Rebuffed on so great a scale, the British Government, on 3 December, communicated to the Secretary-General its decision to continue withdrawal of its forces from Port Said 'without delay', according to a time-table to be worked out with the United Nations Commander.[1]

General Burns opened negotiations with General Keightley, the Anglo-French military commander, on the timing of the withdrawal two days later.

This was the first of the occasions [General Burns has written] on which, after withdrawal had been accepted in principle, I was instructed to press the commanders of the invading forces to pull back more speedily. I found I had very little bargaining-power to enforce or persuade to such a withdrawal. . . . The arguments for withdrawal were political arguments, and the ultimate sanctions forcing the withdrawals, though very real, were certainly not provided by U.N.E.F. The negotiations were essentially political, not military, and it was for politicians to carry them on, not myself as commander of a force having defensive powers only, and limited defensive powers at that.[2]

Fortunately for U.N.E.F., the political decision to withdraw, with as much speed and with as little loss of face as possible, was never in doubt after 3 December. General Keightley told General Burns that he 'wanted to leave as quickly as possible', and that only technical reasons made it impossible for the withdrawal to take place in less than a fortnight.[3] That meant 19 December. In fact the final withdrawal was completed on 22 December. General Burns thus describes the Anglo-French departure:

. . . I went up to Port Said, where I met General Stockwell [Commander 2nd Corps, Allied Force] in the small perimeter which the British were holding near the waterfront just by the Casino Palace Hotel. . . . It was a beautiful bright day, and all was quiet in the city. There was not much to talk about; General Stockwell thanked me for the harmonious way

---

[1] See below, pp. 315–16.   [2] Burns, op. cit., pp. 229–30.   [3] Ibid., p. 229.

in which the unprecedented operation had been carried out. I thanked him for the co-operation given to U.N.E.F., especially in the sale of vehicles and stores. . . . From time to time jet fighters screamed over-head, at low altitude, to remind the Egyptians that the Allied air power was still ready to counter any resumption of hostilities. But all was very quiet as the British regular soldiers went about the final tasks of em-barkation (and there was little that remained to be done) with the quiet precision which denotes well-trained and disciplined troops.

It made a great show of military power, and yet—they were getting out. Attending their departure was the tiny half-organized and lightly armed United Nations Emergency Force. It was an historic day, per-haps marking the last time that an Empire (here two Empires in associa-tion) would seek to impose its will by force on a weaker nation. It marked the end of an epoch. . . .[1]

DOCUMENTS

I. UNITED KINGDOM

1. Extracts from speech by the Permanent Representative to the United Nations, Sir Pierson Dixon, to the General Assembly, 1 November 1956[2]

Before I enter into the substance of the matter for which this emergency session of the General Assembly has been called, I feel bound to point out . . . that the procedure under the 'Uniting for Peace' resolution of the General Assembly has, in our view, been improperly invoked on this occasion. . . .

Her Majesty's Government in the United Kingdom has never-theless decided to attend this session for an important reason. It is because it believes that the United Nations can and should do what it can to make effective contributions to the present grave situation in the Middle East. . . .

. . . The world has not been able to rely on the United Nations for the collective security which the Organization was designed to provide. Least of all, in view of the intransigence of the parties and the cynical misuse of their veto power by the Soviet Union, could we expect swift effective action from the United Nations in an emergency in the Middle East.

---

[1] Ibid., pp. 238–9.
[2] G.A.O.R., First Emergency Special Session, 1956, pp. 5–8.

It is hard to say these things, but I fear they are true. It is precisely because of this unhappy limitation in the effective powers of the Security Council to deal with such an emergency that the United Kingdom and French Governments were compelled to intervene at once, as they were fortunately in a position to do. It was through no wish of ours that a situation arose in which we were compelled to act independently of the United Nations. Indeed, as soon as the news of the Israeli action reached us here in New York on the afternoon of 29 October, I took immediate steps with my United States and French colleagues to make it clear that the Council should be seized of this situation at once. We did not however consider that the course of action proposed by the United States, without consultation with Her Majesty's Government, could effectively achieve the twin objectives of separating the belligerents at once and of safeguarding free passage through the Canal.

It was in these circumstances that we were obliged to impose our negative votes in the Security Council. The action which we and the French Government have taken is essentially of a temporary character and, I repeat it, designed to deal with a unique emergency. Our intervention was swift because the emergency brooked no delay. It has been drastic because drastic action was evidently required. It is an emergency police action. The situation is not dissimilar to that which obtained at the time of the North Korean invasion. On that occasion the Member of the United Nations which had forces on hand and was in a position to intervene at once courageously did so.

By a happy chance—and I mean the absence of the Soviet representative from the Security Council on that occasion—the Council was able to endorse the United States action. The same fortunate choice was not ours.

I cannot, however, believe that the United States would not, in any case, have acted, and rightly so, in the circumstances. . . .

. . . We believe that the United Nations now has a unique opportunity to bring peace to the Middle East. It is our hope that the emergency action we have taken to protect the Canal, to terminate hostilities, and to separate the belligerents will result in a settlement which will prevent such a situation from arising in the future. We must speedily work for a settlement of the whole Middle East question which takes account of the legitimate interests of the Arab countries as well as those of Israel.

I am not making any precise proposals—it would be inappropriate on such an occasion—but I should like to throw out the suggestion that one method of achieving this would be to convene a suitably constituted conference to consider how best to promote a permanent settlement.

I realize that there may at this moment be a temptation for this Assembly to take no effective action but merely to call upon all parties to cease hostilities and withdraw, but I must solemnly state—and I say this with great emphasis—that if that were the only action which the United Nations was prepared to take at this time of crisis we would merely revert to the continuation of the chaos in the Middle East which we have endured in the last eight years. We should thus inexorably be drawing nearer to the time when the growing threat of war became a reality.

The first urgent task is to separate Israel and Egypt and to stabilize the position. That is our purpose. If the United Nations were willing to take over the physical task of maintaining peace in the area, no one would be better pleased than we. But police action there must be, to separate the belligerents and to stop the hostilities.

In my sober submission, all Members of the United Nations should earnestly bend their efforts to bring about a lasting settlement which can replace the Armistice Agreements which have now proved to be too fragile for their task of preserving peace and order in the Middle East.

2. Letter from the Permanent Representative to the United Nations, Sir Pierson Dixon, to the Secretary-General, Dag Hammarskjold, 3 November 1956[1]

I have the honour, on instructions from Her Majesty's Government in the United Kingdom, to communicate the following. This communication is made in response to the resolution A/3256 adopted by the General Assembly on 2 November 1956 (resolution 997/ES-I), during the emergency special session.

'1. The British and French Governments have given careful consideration to the resolution passed by the General Assembly on 2 November. They maintain their view that police action must be carried through urgently to stop the hostilities which

[1] G.A.O.R., First Emergency Special Session, Annexes, Agenda item 5, Doc. A/3269.

are now threatening the Suez Canal, to prevent a resumption of
those hostilities and to pave the way for a definitive settlement
of the Arab-Israel war which threatens the legitimate interests
of so many countries.

'2. They would most willingly stop military action as soon as
the following conditions could be satisfied:

(a) Both the Egyptian and the Israeli Governments agree to
accept a United Nations Force to keep the peace;

(b) The United Nations decides to constitute and maintain
such a Force until an Arab-Israel peace settlement is reached
and until satisfactory arrangements have been agreed in
regard to the Suez Canal, both settlements to be guaranteed
by the United Nations;

(c) In the meantime, until the United Nations Force is constituted, both combatants agree to accept forthwith limited
detachments of Anglo-French troops to be stationed between
the combatants.'

I request you to be so good as to circulate this note immediately
to all Members of the United Nations.

3. Extract from speech by the Permanent Representative to the United Nations, Sir Pierson Dixon, to the General Assembly, 3 November 1956[1]

. . . I can add very little at this time to what I said in my letter
this morning [Document 2], and I can say only this: we believe
that police action of the most urgent kind is called for, and that is
why we, with the French, have stepped in. As I have said many
times, we have done so on a purely temporary basis, and the sooner
the United Nations can take over from us the more we shall
welcome that, particularly since our action has been so much
misunderstood and criticized, even by our friends.

. . . This idea is reflected in the draft resolution, which has just
been adopted and which was advanced by the Secretary of State
for External Affairs of Canada. I found the central idea in that
resolution of great interest, but, as Mr. Pearson knows, my delegation did not see the text of his draft resolution before he read it
out. I have not had the time to give it a full and complete study,
nor have I been able to refer it to my Government. Indeed, at first

---

[1] G.A.O.R., First Emergency Special Session, 1956, p. 72.

sight it does seem to me that in some respects the resolution goes too far, and in others not far enough. It was for that reason that I was not able to vote for the resolution, but, equally, I was able to abstain. . . .

4. Statement by the Secretary of State for Foreign Affairs, Mr. Selwyn Lloyd, in the House of Commons, and extracts from ensuing debate, 5 November 1956[1]

With your permission, Mr. Speaker, and that of the House, I will make a further statement on the Israel-Egypt situation.

Since the House met on Saturday the General Assembly of the United Nations, meeting in emergency Special Session, has passed three resolutions.

The first was sponsored by a number of Asian and African states. This called for a cease-fire, the halting of the movement of military forces and arms into the area and the withdrawal of all forces in the area behind the armistice lines. It authorised the Secretary-General to obtain compliance.

The second resolution was sponsored by Canada [II, Document 2, *infra*]. It requested the Secretary-General to submit within 48 hours a plan for the setting up, with the consent of the nations concerned, of an emergency international United Nations force to secure and supervise the cessation of hostilities in accordance with the terms of the cease-fire resolution of 2nd November.

In a telegram received yesterday morning the Secretary-General of the United Nations drew the attention of Her Majesty's Government to these resolutions and requested all parties to bring a halt to all hostile military actions in the area by 8 o'clock Greenwich Mean Time yesterday. Her Majesty's Government had already invited the French Ministers to come to London for consultations. They informed Mr. Hammarskjold of this fact and explained that it was not possible to give him a definite answer to his message within the time limit which he had stipulated. As a result of their consultations with the French Government they sent a telegram to the Secretary-General very early this morning. This read as follows:

The Governments of the United Kingdom and France have studied carefully the resolutions of the United Nations General Assembly passed on 3rd and 4th November.

[1] United Kingdom, H.C. Deb., vol. 558, coll. 1956–62.

They warmly welcome the idea which seems to underlie the request to the Secretary-General contained in the resolution sponsored by Canada, and adopted by the Assembly at its 563rd meeting, that an international force should be interpolated as a shield between Israel and Egypt pending a Palestine settlement and a settlement of the question of the Suez Canal. But according to their information neither the Israeli nor the Egyptian Government has accepted such a proposal. Nor has any plan for an international force been accepted by the General Assembly or endorsed by the Security Council.

The composition of the staff and contingents of the international force would be a matter for discussion.

The two Governments continue to believe that it is necessary to interpose an international force to prevent the continuance of hostilities between Egypt and Israel, to secure the speedy withdrawal of Israeli forces, to take the necessary measures to remove obstructions and restore traffic through the Suez Canal, and to promote a settlement of the problems of the area.

Certain Anglo-French operations, with strictly limited objectives, are continuing. But as soon as the Israeli and Egyptian Governments signify acceptance of, and the United Nations endorses a plan for an international force with the above functions the two Governments will cease all military action.

In thus stating their views, the United Kingdom and French Governments would like to express their firm conviction that their action is justified. To return deliberately to the system which has produced continuing deadlock and chaos in the Middle East is now not only undesirable but impossible. A new constructive solution is required. To this end they suggest that an early Security Council meeting at the ministerial level should be called in order to work out an international settlement which would be likely to endure, together with the means to enforce it.

This message to the Secretary-General crossed a telegram from him informing Her Majesty's Government of the passing of a third resolution. This referred to the Canadian resolution, which I have already described, and to a preliminary report from the Secretary-General on the plan to set up an emergency International United Nations Force. It called for the establishment of United Nations command to secure and supervise the cessation of hostilities in accordance with all the terms of the earlier cease-fire resolution. It appointed General [E. L. M.] Burns, Chief of Staff of the United Nations Truce Supervision Organisation, Chief of Command on an emergency basis.

It authorised General Burns immediately to recruit from the Observer Corps of the Truce Supervision Organisation a limited number of officers who shall be nationals of countries other than those having permanent membership of the Security Council and further, in consultation with the Secretary-General, to undertake the recruitment directly from various Member States other than the permanent Member States of the Security Council the additional number of officers required. Finally, it invited the Secretary-General to take such administrative measures as may be necessary for the prompt execution of the actions envisaged in this resolution.

Her Majesty's Government abstained from voting on this resolution. [*Hon. Members:* 'Shame!'] They fully approved the principle of an International United Nations Force. But although the steps called for in this latest resolution might be considered to be a beginning, they are not in themselves likely to achieve the purposes set out in our message to the Secretary-General. We do not know that hostilities between Israel and Egypt have ceased or that they will not be resumed. The measures to be taken under the latest resolution could not be sufficient to ensure that.

It is the policy of Her Majesty's Government to ensure that the Israeli forces withdraw from Egyptian territory. We have also told the United Nations that we believe it is necessary to secure the speedy withdrawal of Israel forces. But we cannot ensure that the Israelis withdraw from Egyptian territory until we are physically in the area to keep the peace, to give the necessary guarantees and to prevent a repetition of the events of the past few years.

There must also be immediate means on the spot to take the necessary measures, as I have said, to remove obstructions and restore navigation through the Suez Canal, and to promote a settlement of the problems of the area.

It will, of course, be a matter for the Security Council, if our proposal for an early meeting at ministerial level is accepted, to consider what part the United Kingdom and France should play in achieving all the objectives to which I have referred. Meanwhile, Her Majesty's Government believe that the Anglo-French forces, once they are established in the area, will be the best guarantee that these purposes will be effectively and speedily achieved.

Mr. Gaitskell: The House has heard with astonishment the statement of the Foreign Secretary on why Her Majesty's

Government abstained from voting on the Resolution of the United Nations General Assembly setting up an international force. Is the Foreign Secretary aware that the Resolution was sponsored by Canada, that New Zealand has already said she is prepared to contribute troops in this force?—[*An Hon. Member:* 'So are we.'] On the contrary, the Government have just explained that they could not support this proposal.

Is the right hon. and learned Gentleman also aware that the United States Government, whilst accepting the proposal that the force should be composed of the troops of countries which are not permanent members of the Security Council, has nevertheless made it plain that she will make available aircraft and supplies for this international force? Can we clarify a little more the attitude of Her Majesty's Government to the international force proposal?

Therefore, the second question I ask is this: is it not the case that in the first Resolution, again introduced by Canada, the purpose of this force was defined as to secure and to supervise the cessation of hostilities in accordance with the terms of the Resolution of 2nd November; that is to say, the cease-fire, the withdrawal of the combatants to within their own territories and the absence of any other intervention by any other party? If that is the case, why did Her Majesty's Government, in replying to the Resolution passed by the Assembly, suggest that the idea of the international force was not merely to secure and supervise the cessation of hostilities but also to secure a final settlement of the Suez Canal problem?

Hon. Members: Why not?

Mr. Gaitskell: Is the Foreign Secretary aware that by imposing that particular implication the effect is to confirm in the minds of the whole world that the real reason for British and French intervention here was not to separate the combatants but to seize control of the Canal? Is he further aware that if Her Majesty's Government insist that the purpose of the international force under the United Nations must be to deal with the Suez Canal problem, they will effectively sabotage the whole idea of that force?

Mr. Lloyd: With regard to the question of the Resolution, the right hon. Gentleman has asked why the representative of the United Kingdom abstained. First, we could not vote for a proposition which excluded detachments of the forces of the permanent members of the Security Council from this international force. [*Hon. Members:* 'Why not?'] The reason for this is quite simple—

that there has to be some reality about the situation. One has to consider the people who are able effectively to contribute those forces.

If hon. Members opposite still think that a few officers—because that is all that the Resolution amounts to—rather fewer than there have been in the Truce Supervision Organisation already, are going to solve this problem, they are quite mistaken. This Resolution may be a beginning but it will not solve this problem unless the international force is constituted to a much greater extent than is envisaged by the Resolution.

The Leader of the Opposition made a second accusation that we had done wrong in suggesting that the problems of the Suez Canal should be included to be settled whilst the international force was there. I should have thought he would have understood that the blockade on the shipping of a certain country going through the Canal—[*Interruption.*]—is one of the matters a settlement of which we have to try to get out of this situation. As for the talk of people being humbugs or hypocrites—[*Interruption.*]—those words apply to the people who for the past four years have consistently urged forcible action to deal with this matter.

Mr. Gaitskell: Are we to understand from what the right hon. and learned Gentleman has just said that when Her Majesty's Government, in their reply to the Secretary-General of the United Nations, made reference to the settlement of the Suez Canal problem, all that they had in mind was ensuring the free passage of Israeli ships?

Mr. Lloyd: Certainly not. We had in mind all problems affecting the free and open transit through the Canal guaranteed by the 1888 Convention.

Mr. [Walter] Elliot: Is it not a thousand pities that this positive, constructive proposal for an international police force, which may well be the key to the problem, should not receive much more objective treatment than apparently it has received up to now? In particular, the many points—major points, but still points of detail—such as the composition of the international force, could well be examined with an open mind, because all of us would agree that there should be reality in this business and not merely imagination. It is also very desirable that the four belligerents should not find themselves actively engaged in a police force, if that proves to be possible.

Surely, the Leader of the Opposition is at fault in suggesting that the six principles which were voted by the United Nations in its consideration of this very question should be left out altogether when the international police force is being considered. Therefore, from all these points of view, I beg that it might be possible for the House to examine this—it may be the only solution for the problem—without the terrible heat which seems to be creeping into it just now.

Mr. Gaitskell: For my part, I entirely agree that we should examine this proposal objectively. I assure the right hon. Member for Kelvingrove (Mr. Elliot) that I was endeavouring to get clarification of the Government's view on it. I must, however, also say this and I put it in the form of a question. Is the Foreign Secretary aware that it is vitally important to distinguish between the setting up of an international force to deal with the Arab-Israeli question and an international force to impose on Egypt a solution of the Suez Canal problem? Does the Foreign Secretary realise that it is because I feel that this distinction is vital that I am pressing the difference between the two proposals?

Has the right hon. and learned Gentleman any information about where the proposed international force would be stationed? Is he, for instance, in agreement with the Prime Minister of New Zealand, who is reported as saying today that he presumed that the force would be stationed on the border between Israel and Egypt?

May I also ask the Foreign Secretary—again, pursuing what the right hon. Member for Kelvingrove described as an objective examination—why Her Majesty's Government, in their first reply on this matter, made their consent to—and, indeed, their acceptance of—the United Nations Assembly Resolution conditional on Israel and Egypt both accepting the idea of an international United Nations force, whether they still adhere to that view, and whether, if Egypt accepts, as she has done, it is the view of Her Majesty's Government that nothing further can be done about this unless Israel also accepts? Could the Foreign Secretary enlighten us on these points?

Mr. Lloyd: On the first point, with regard to the question of imposing a settlement of the Suez Canal issue, there is no question of imposing a settlement. I should have thought that every sensible person would have agreed that these things having happened, it was wiser that there should be a settlement of all those matters before the international policemen were removed. With regard to where the international force is to be stationed, I think that that would be a matter for the force commander himself to discuss with the Governments concerned.

Thirdly, with regard to the matter of acceptance by Egypt and Israel, I should have thought that as a practical proposition it would very much facilitate the development of this idea if those countries did agree to accept the international force, as it would have very much facilitated matters had both countries accepted our request of last Tuesday.

Mr. Gaitskell: Is it still the case that, as stated on Saturday, Her Majesty's Government's attitude to the proposal for an international force depends upon the acceptance by Israel and Egypt of this proposal?

Mr. Lloyd: I think that for practical purposes that is bound to influence us.

Mr. [Stephen] McAdden: Will my right hon. and learned Friend urge upon our representative at the United Nations that in any future discussion on this subject, the test as to the composition of this United Nations force should not be whether those prepared to participate are small Powers who cannot provide the forces or large Powers who can, but should be the willingness of the nations concerned to accept United Nations observers and a police force in their own country?

Mr. Lloyd: There is something in what my hon. Friend says.

5. Letter from the Permanent Representative to the United Nations, Sir Pierson Dixon, to the Secretary-General, Dag Hammarskjold, 6 November 1956[1]

I have the honour, on instructions from Her Majesty's Government in the United Kingdom, to inform you that they have received and most carefully considered the communication which you addressed to them yesterday evening and your second and final report to the General Assembly which was issued this morning.

I have been instructed to convey to you at once the following message from Her Majesty's Government:

[1] G.A.O.R., First Emergency Special Session, Annexes, Agenda item 5, Doc. A/3306.

Her Majesty's Government welcome the Secretary-General's communication, while agreeing that a further clarification of certain points is necessary.

If the Secretary-General can confirm that the Egyptian and Israeli Governments have accepted an unconditional cease-fire, and that the international Force to be set up will be competent to secure and supervise the attainment of the objectives set out in the operative paragraphs of the resolution passed by the General Assembly on November 2, Her Majesty's Government will agree to stop further military operations.

They wish to point out, however, that the clearing of the obstructions in the Suez Canal and its approaches, which is in no sense a military operation, is a matter of great urgency in the interests of world shipping and trade. The Franco-British force is equipped to tackle this task. Her Majesty's Government therefore propose that the technicians accompanying the Franco-British force shall begin this work at once.

Pending the clarification of the above points, Her Majesty's Government are ordering their forces to cease fire at midnight GMT unless they are attacked.

I request you to be so good as to circulate this reply to all Members of the United Nations.

6. Extracts from speech by the Secretary of State for Foreign Affairs, Mr. Selwyn Lloyd, to the General Assembly of the United Nations, 23 November 1956[1]

. . . In my speech this morning I want to cover three points: first of all, the United Nations Emergency Force; secondly, the withdrawal of British troops; and thirdly, the clearance of the Suez Canal.

First of all, so far as the United Nations Emergency Force is concerned, over the past few years the United Nations, whether in the General Assembly or in the Security Council, has completely failed, so far as the Middle East is concerned, either to keep the peace or to procure compliance with its own resolutions, or to pave the way for final settlement. I am not criticizing; I am stating a fact which is very well known to all of us. And that fact is the greatest reason, the principal reason, for what took place on 29 and 30 October.

I deny emphatically the allegation that Her Majesty's Government in the United Kingdom instigated the Israeli attack or that there was agreement between the two countries about it. The

[1] G.A.O.R., Eleventh Session, 1956–7, I, pp. 258–9.

United Kingdom and French Governments decided to intervene to prevent the spread of hostilities, to stop the conflagration from spreading. We wished to put, as rapidly as possible, a protective shield between the combatants, and that was a situation which really brooked of no delay. And that, in fact, was what was achieved.

Whatever may be thought of our actions or our motives, out of the painful discussions regarding them there has come the idea of a United Nations Force, the idea that the United Nations should act through an international force. The idea was first mooted by Sir Anthony Eden in his speech before the British Parliament on 1 November 1956, when he said that if the United Nations were willing to take over the physical task of maintaining peace in the area, then no one would be better pleased than the British. That statement of the Prime Minister's was immediately repeated in the General Assembly by Sir Pierson Dixon [Document 1]. Mr. Pearson, the Canadian Secretary of State for External Affairs, referred to it at the same session, and he introduced the draft resolution for establishing the Force. After that, the concept of an international force gained rapid acceptance. Many nations have offered contingents. The Secretary-General and his staff have worked untiringly at the detailed arrangements, and a rapid start has been made in bringing in advance contingents of the Force to Egypt.

We are doing what we can to help. In response to requests made on behalf of the Secretary-General, arrangements were made for a Norwegian-Danish company to enter Port Said. We have agreed that the main body of the Yugoslav contingent should disembark at Port Said and be assisted in transit. We have agreed to provide, if wanted, military transport for the Indian infantry battalion which will form part of the Force, the necessary vehicles for the Norwegian medical company, and some medical supplies and food for the Force itself. In other words, it is our declared purpose to co-operate to the best of our ability with the Force and with those who are seeking to make arrangements for it.

With regard to the tasks of the Force, we have noted the Secretary-General's report and, in particular, the annex to it. We understand this to mean that the Force will carry out all the tasks laid upon it in accordance with the resolutions of 2, 6 and 7 November. We have great confidence in the Secretary-General and we believe that he and the General Assembly will in good faith see to it that the Force is effective and competent to carry out those tasks. On that basis we have agreed to withdraw our forces.

The action we took was of a restricted, temporary character designed to meet an emergency. It was not directed against Egyptian sovereignty or Egyptian independence. Therefore we wish to withdraw as soon—I say again, as soon—as the United Nations Force is in a position to assume effectively the tasks assigned to it. One of those tasks, of course, is to ensure that hostilities are not resumed.

It is our desire that the Force should be in that position as soon as possible. However, it does take a little time to organize the command arrangements for the Force, to integrate a sufficient body of its units, to make the necessary arrangements for supply and command and control, so that it will be a Force and not just a hotchpotch of military units. It is my hope that this Force will be a credit to the United Nations. It will be under close scrutiny; and I say that unless you give General Burns and his officers time to organize it, you will bring the United Nations itself into disrepute. You do not want this Force to be laughed at.

When the word 'immediately' was put into resolution 1002 (ES-I), many representatives expressed the view that it did not mean 'instantaneously', that there had to be a relationship between the withdrawal of the forces referred to in the resolution and the arrival and functioning of the United Nations Force. I think that was recognized in many of the speeches which were made on 7 November. . . . The representative of Canada said: '. . . we give the same interpretation to the word "immediately" that has been given by others, that is "as quickly as possible". In our minds, there is a relationship, implicit in the word "immediately", between the withdrawal of the forces referred to in the resolution and the arrival and the functioning of the United Nations Force.' I think that others spoke in similar terms. I believe that to act otherwise would bring discredit rather than credit upon the United Nations. Nevertheless, as an indication of the sincerity of our intentions, we have given immediate orders that one battalion should be withdrawn as quickly as possible.

I hope that members of the Assembly will realize that, in taking up this position on withdrawal, the United Kingdom Government has asked the British people to endorse an act of faith. We believe

—and, whether you agree with us or not, we believe it sincerely—that we have stopped a small war from spreading into a larger war. We believe that we have created the conditions under which a United Nations Force is to be introduced into this troubled area to establish and maintain peace; and we believe that thereby we have given the Assembly and the world another opportunity to settle the problems of the area. We believe that we have brought matters to a head, to a crisis, that we have cast down a challenge to world statesmanship, the statesmanship of this Assembly, to achieve results. We think that there is in this a great test for the United Nations and for the Powers on whose continued support the United Nations ultimately depends.

We are, therefore, prepared to make this act of faith because we believe that the United Nations has the will to ensure that the United Nations Emergency Force will effectively and honourably carry out all the functions laid down for it in the Assembly resolutions. But, should our faith prove to have been misplaced, should all this effort and disturbance have been for nothing, should the United Nations fail to show the necessary will-power to procure the lasting settlement required, then indeed there will be cause for alarm and despondency.

That is our position with regard to this question of withdrawal: it will take place as soon as possible, as the United Nations Force becomes effective and competent to discharge its functions. . . .

7. Note Verbale from the Permanent Representative to the United Nations, Sir Pierson Dixon, to the Secretary-General, Dag Hammarskjold, 3 December 1956[1]

The Permanent Representative of the United Kingdom of Great Britain and Northern Ireland to the United Nations presents his compliments to the Secretary-General and has the honour to make the following communication on behalf of Her Majesty's Government in the United Kingdom:

1. Her Majesty's Government and the French Government note that

(a) An effective United Nations Force is now arriving in Egypt charged with the tasks assigned to it in the Assembly resolutions of 2, 6 and 7 November.

---

[1] G.A.O.R., Eleventh Session, Annexes, Agenda item 66, Doc. A/3415.

(*b*) The Secretary-General accepts the responsibility for organizing the task of clearing the Canal as expeditiously as possible.

(*c*) In accordance with the General Assembly resolution of 2 November free and secure transit will be re-established through the Canal when it is clear.

(*d*) The Secretary-General will promote as quickly as possible negotiations with regard to the future regime of the Canal on the basis of the six requirements set out in the Security Council resolution of 13 October.

2. Her Majesty's Government and the French Government confirm their decision to continue the withdrawal of their forces now in the Port Said area without delay.

3. They have accordingly instructed the Allied Commander, General Keightley, to seek agreement with the United Nations Commander, General Burns, on a time-table for the complete withdrawal, taking account of the military and practical questions involved. This time-table should be reported as quickly as possible to the Secretary-General of the United Nations.

4. In preparing these arrangements the Allied Commander will ensure:

(*a*) That the embarkations of personnel or material shall be carried out in an efficient and orderly manner;

(*b*) That proper regard will be had to the maintenance of public security in the area now under Allied control;

(*c*) That the United Nations Commander should make himself responsible for the safety of any French and British salvage resources left at the disposition of the United Nations salvage organization.

5. In communicating these conclusions Her Majesty's Government and the French Government recall the strong representations they have made regarding the treatment of their nationals in Egypt. They draw attention to the humane treatment accorded to Egyptian nationals in the United Kingdom and France. They feel entitled to demand that the position of British and French nationals in Egypt should be fully guaranteed.

II. CANADA

1. Speech by the Secretary of State for External Affairs, Mr. L. B. Pearson, to the General Assembly of the United Nations, 2 November 1956[1]

I rise not to take part in this debate, because the debate is over. The vote has been taken. But I do wish to explain the abstention of my delegation on that vote.

It is never easy to explain an abstention, and in this case it is particularly difficult because we are in favour of some parts of this resolution, and also because this resolution deals with a complicated question.[2]

Because we are in favour of some parts of the resolution, we could not vote against it, especially as, in our opinion, it is a moderate proposal couched in reasonable and objective terms, without unfair or unbalanced condemnation; and also, by referring to violations by both sides to the armistice agreements, it puts, I

---

[1] G.A.O.R., First Emergency Special Session, 1956, pp. 35–36.
[2] The text of the Resolution is as follows:

'*The General Assembly*,
'*Noting* the disregard on many occasions by parties to the Israel-Arab armistice agreements of 1949 of the terms of such agreements, and that the armed forces of Israel have penetrated deeply into Egyptian territory in violation of the General Armistice Agreement between Egypt and Israel of 24 February 1949,
'*Noting* that the armed forces of France and the United Kingdom of Great Britain and Northern Ireland are conducting military operations against Egyptian territory,
'*Noting* that traffic through the Suez Canal is now interrupted to the serious prejudice of many nations,
'*Expressing its grave concern* over these developments,
'1. *Urges* as a matter of priority that all parties now involved in hostilities in the area agree to an immediate cease-fire and, as part therefore, halt the movement of military forces and arms into the area;
'2. *Urges* the parties to the armistice agreements promptly to withdraw all forces behind the armistice lines, to desist from raids across the armistice lines into neighbouring territory, and to observe scrupulously the provisions of the armistice agreements;
'3. *Recommends* that all Member States refrain from introducing military goods in the area of hostilities and in general refrain from any acts which would delay or prevent the implementation of the present resolution;
'4. *Urges* that, upon the cease-fire being effective, steps be taken to reopen the Suez Canal and restore freedom of navigation;
'5. *Requests* the Secretary-General to observe and report promptly on the compliance with the present resolution to the Security Council and to the General Assembly, for such further action as they may deem appropriate in accordance with the Charter;
'6. *Decides* to remain in emergency session pending compliance with the present resolution.' G.A.O.R., First Emergency Special Session, Supplement No. 1, Resolution 997 (ES–I).

think, recent action by the United Kingdom and France—and rightly—against the background of those repeated violations and provocations.

We support the effort being made to bring the fighting to an end. We support it, among other reasons, because we regret that force was used in the circumstances that face us at this time. As my delegation sees it, however, this resolution which the General Assembly has thus adopted in its present form—and there was very little chance to alter that form—is inadequate to achieve the purpose which we have in mind at this Assembly. These purposes are defined in that resolution of the United Nations under which we are meeting—resolution 377(V), uniting for peace—and peace is far more than ceasing to fire, although it certainly must include that essential factor. This is the first time that action has been taken under the 'Uniting for Peace' resolution, and I confess to a feeling of sadness, indeed, even distress, at not being able to support the position taken by two countries whose ties with my country are and will remain close and intimate; two countries which have contributed so much to man's progress and freedom under law; and two countries which are Canada's mother countries.

I regret the use of military force in the circumstances which we have been discussing, but I regret also that there was not more time, before a vote had to be taken, for consideration of the best way to bring about that kind of cease-fire which would have enduring and beneficial results. I think that we were entitled to that time, for this is not only a tragic moment for the countries and peoples immediately affected, but it is an equally difficult time for the United Nations itself. I know, of course, that the situation is of special and, indeed, poignant urgency, a human urgency, and that action could not be postponed by dragging out a discussion, as has been done so often in this Assembly. I do feel, however, that had that time, which has always, to my knowledge, in the past been permitted for adequate examination of even the most critical and urgent resolution, been available on this occasion, the result might have been a better resolution. Such a short delay would not, I think, have done harm, but, in the long run, would have helped those in the area who need help most at this time.

Why do I say this? In the first place, our resolution, though it has been adopted, is only a recommendation, and its moral effects would have been greater if it could have received a more unanimous vote in this Assembly—which might have been possible if there had been somewhat more delay.

Secondly, this recommendation which we have adopted cannot be effective without the compliance of those to whom it is addressed and who have to carry it out. I had ventured to hope that, by a short delay and in informal talks, we might have made some headway, or at least have tried to make some headway, in securing a favourable response, before the vote was taken, from those governments and delegations which will be responsible for carrying it out.

I consider that there is one great omission from this resolution, which has already been pointed out by previous speakers—more particularly by the representative of New Zealand, who has preceded me. This resolution does provide for a cease-fire, and I admit that that is of first importance and urgency. But, alongside a cease-fire and a withdrawal of troops, it does not provide for any steps to be taken by the United Nations for a peace settlement, without which a cease-fire will be only of temporary value at best. Surely, we should have used this opportunity to link a cease-fire to the absolute necessity of a political settlement in Palestine and for the Suez, and perhaps we might also have been able to recommend a procedure by which this absolutely essential process might begin.

Today we are facing a feeling of almost despairing crisis for the United Nations and for peace. Surely that feeling might have been harnessed to action or at least to a formal resolve to act at long last and to do something effective about the underlying causes of this crisis which has brought us to the very edge of a tragedy even greater than that which has already taken place. We should then, I think, have recognized the necessity for political settlement in this resolution and done something about it. And I do not think that, if we had done that, it would have postponed action very long on the other clauses of the resolution. Without such a settlement, which we might have pushed forward under the incentive of fear, our resolution, as I see it, may not make for an enduring and real peace. We need action, then, not only to end the fighting but to make the peace.

I believe that there is another omission from this resolution, to which attention has also already been directed. The armed forces of Israel and Egypt are to withdraw, or, if you like, to return to the armistice lines, where presumably, if this is done, they will once again face each other in fear and hatred. What then? What then, six months from now? Are we to go through all this again? Are we to return to the *status quo*? Such a return would not be to a position of security or even to a tolerable position, but would be a return to terror, bloodshed, strife, incidents, charges and counter-charges, and ultimately another explosion which the United Nations armistice commission would be powerless to prevent and possibly even to investigate.

I therefore would have liked to see a provision in this resolution —and this has been mentioned by previous speakers—authorizing the Secretary-General to begin to make arrangements with member governments for a United Nations force large enough to keep these borders at peace while a political settlement is being worked out. I regret exceedingly that time has not been given to follow up this idea, which was mentioned also by the representative of the United Kingdom in his first speech [I, Document 1, *supra*], and I hope that even now, when action on the resolution has been completed, it may not be too late to give consideration to this matter. My own Government would be glad to recommend Canadian participation in such a United Nations force, a truly international peace and police force.

We have a duty here. We also—or, should I say, we had—an opportunity. Our resolution may deal with one aspect of our duty —an urgent, a terribly urgent, aspect. But, as I see it, it does nothing to seize that opportunity which, if it had been seized, might have brought some real peace and a decent existence, or hope for such, to the people of that part of the world. There was no time on this occasion for us to seize the opportunity in this resolution. My delegation therefore felt, because of the inadequacy of the resolution in this respect, that we had no alternative in the circumstances but to abstain in the voting.

I hope that our inability to deal with those essential matters at this time will very soon be removed and that we can come to grips with the basic core of this problem.

2. Speech by the Secretary of State for External Affairs, Mr. L. B. Pearson, to the General Assembly of the United Nations, 3 November 1956[1]

The immediate purpose of our meeting tonight is to bring about as soon as possible a cease-fire and a withdrawal of forces, in the area which we are considering, from contact and from conflict with each other. Our longer-range purpose, which has already been referred to tonight and which may ultimately, in its implications, be even more important, is to find solutions for the problems which, because we have left them unsolved over the years, have finally exploded into this fighting and conflict.

In regard to this longer-range purpose, important resolutions have been submitted this evening by the United States delegation. We value this initiative, and our delegation will give the resolutions the examination which their importance deserves and will, I hope, make its own detailed comments concerning them later.

So far as the first and immediate purpose is concerned, a short time ago the Assembly passed, by a very large majority, a resolution which is now a recommendation of the United Nations General Assembly. And so we must ask ourselves how the United Nations can assist in securing compliance with the terms of that resolution from those who are most immediately concerned and whose compliance is essential if that resolution is to be carried out.

How can we get from them the support and co-operation which is required, and how can we do this quickly?

The representative of India has just read to us, on behalf of a number of delegations, a very important resolution which deals with this matter.[2] In operative paragraphs 2 and 3 of that resolution,

---

[1] G.A.O.R., First Emergency Special Session, 1956, pp. 54–55.
[2] The text of this draft resolution, submitted by India for nineteen delegations, reads as follows:

'*The General Assembly,*
'*Noting with regret* that not all the parties concerned have yet agreed to comply with the provisions of its resolution of 2 November 1956,
'*Noting* the special priority given in its resolution to an immediate cease-fire and as part thereof to the halting of the movement of military forces and arms into the area,
'*Noting further* that the resolution urged the parties to the Armistice Agreements to withdraw all forces behind the armistice lines, to desist from raids across the armistice lines into neighbouring territory, and to observe scrupulously the provisions of the Armistice Agreements,
'1. *Reaffirms* its resolution of 2 November 1956 and once again calls upon the parties immediately to comply with the provisions of the said resolution;
'2. *Authorizes* the Secretary-General immediately to arrange with the parties

certain specific proposals are made with a view to setting up machinery to facilitate compliance with the resolution.

I ask myself the question whether that machinery is adequate for the complicated and difficult task which is before us. I am not in any way opposing this resolution which we have just heard read. I appreciate its importance and the spirit in which it has been put forward. But I do suggest that the Secretary-General be given another and supplementary—not conflicting—but supplementary —responsibility: to work out at once a plan for an international force to bring about and supervise the cease-fire visualized in the Assembly resolution which has already been passed.

For that purpose my delegation would like to submit to the Assembly a very short draft resolution which I venture to read at this time. It is as follows:

The General Assembly, bearing in mind the urgent necessity of facilitating compliance with the Resolution (A/3256) of November 2, requests, as a matter of priority, the Secretary-General to submit to it within forty-eight hours a plan for the setting up, with the consent of the nations concerned, of an emergency international United Nations force to secure and supervise the cessation of hostilities in accordance with the terms of the above resolution.

I would assume that during this short period the Secretary-General would get into touch with, and endeavour to secure co-operation in the carrying out of the earlier resolution from, the parties immediately concerned—whose co-operation, I venture to repeat, is essential—as well as endeavouring to secure help and co-operation from any others whom he thinks might assist him in his vitally important task.

This draft resolution which I have just read out, and which will be circulated shortly, has an added purpose of facilitating and making effective compliance with the resolution which we have already

concerned for the implementation of the cease-fire and the halting of the movement of military forces and arms into the area and requests him to report compliance forthwith and, in any case, not later than twelve hours from the time of adoption of this resolution;

'3. *Requests* the Secretary-General, with the assistance of the Chief of Staff and the members of the United Nations Truce Supervision Organization, to obtain compliance of the withdrawal of all forces behind the armistice lines;

'4. *Decides* to meet again on receipt of the Secretary-General's report referred to in operative paragraph 2 of this resolution.'

passed on the part of those whose compliance is absolutely essential.

It has also the purpose of providing for international supervision of that compliance through the United Nations, and, finally, it has as its purpose the bringing to an end of the fighting and bloodshed at once, even while the Secretary-General is examining this question and reporting back in forty-eight hours.

If this draft resolution commended itself to the General Assembly—and I suggest that it is not in conflict with the draft resolution which has just been read to us by our Indian colleague—and if it were accepted and accepted quickly the Secretary-General could at once begin the important task which the draft resolution gives him.

I apologize for adding to his burdens in this way, because they have already been added to in the immediately preceding draft resolution, but we know that he can carry burdens of this kind both unselfishly and efficiently.

Meanwhile, during this period of forty-eight hours, we can get on with our consideration of and decision on the United States draft resolution and other draft resolutions before the General Assembly which deal with this grave and dangerous situation which confronts us, both in relation to its immediate as well as its wider and perhaps even more far-reaching results.

3. Letter from the Secretary of State for External Affairs, Mr. L. B. Pearson, to the Secretary-General of the United Nations, Dag Hammarskjold, 4 November 1956[1]

With respect to the proposed international United Nations Force referred to in resolution A/3276 (resolution 998 (ES-1)), approved by the General Assembly on 4 November 1956, the Canadian Government has decided to make an appropriate contribution, the details of which will be communicated to you shortly, subject to the required constitutional action which will be put in motion without delay.

4. Statement by the Prime Minister, Mr. L. S. St. Laurent, 6 November 1956[2]

To comply with the resolutions of the United Nations, the

---

[1] G.A.O.R., Doc. A/3302, Annex 1, 1st Special Emergency Session, 1956.
[2] E.A., vol. 8, no. 11, November 1956, p. 326.

Canadian Government has agreed to make an offer of a Canadian contingent to the Emergency International United Nations Force for the Middle East. This proposal is subject to adjustment and/or rearrangement after consultation with the United Nations Commander. Arrangements have already been made for a group of Canadian officers to be available today for consultation with the U.N. Commander in New York as soon as he arrives.

It is proposed to offer a Canadian contingent of battalion strength, augmented by ordnance, army service corps, medical and dental detachments to ensure that the battalion group is self-contained and can operate independently from a Canadian base. The size of the contingent is expected to be over 1,000 men.

Canada will be prepared to have this force lifted by the R.C.A.F. to the Middle East.

It is proposed to provide this contingent with a temporary mobile Canadian base for the first phase of its policing operations. The Canadian Government is prepared to use H.M.C.S. *Magnificent* for the purpose of transporting vehicles and stores to the Middle East and for use as a temporary mobile Canadian base for rations, medical supplies, ammunition, fuel and limited accommodation stores. H.M.C.S. *Magnificent* will also provide a small hospital to accommodate the sick and injured in the force; accommodation for a force headquarters; and communications between the force and Canada.

5. Extract from speech by the Secretary of State for External Affairs, Mr. L. B. Pearson, to the General Assembly of the United Nations, 7 November 1956[1]

My delegation, representing the Canadian Government, gives strong support to the draft resolution which has just been put before us in Document A/3308. We also wish to endorse the report [A/3302 and Add. 1 to 7] of the Secretary-General, which is related to this draft resolution. In doing so, I should like to echo the words of gratitude and appreciation which have just been spoken by the representative of Denmark for the tireless and effective work done by the Secretary-General, without which we would not have this report today. I should also like to state that my Government is proud to have been able to offer a contribution to

---

[1] G.A.O.R., First Emergency Special Session, 1956, pp. 93–94.

this United Nations Force and has taken steps as a matter of urgency to organize that contribution.

The purpose of this draft resolution—and it is vital—is to complete the process of setting up an emergency United Nations Force for the purposes set out both in this and in the earlier resolution which we have approved and which deals with this subject. With the acceptance of this resolution—and surely it can be unanimously supported—the ending of hostilities can be confirmed and safeguarded and the work of peace-making begun on a solid United Nations foundation.

In a sense, that work has begun. But a great deal remains to be done, of course, before it is finished. I think that this is a moment for sober satisfaction, but certainly not one for any premature rejoicing. And yet, whilst appreciating the necessity for being realistic about these things, it is hard not to rejoice at the thought that we may have been saved from the very edge of catastrophe—and saved, let us not forget, not by threats or bluster but by the action of the United Nations. If we can draw the necessary conclusions from the manner of our escape—if we have escaped—and if we can act on them, perhaps we will not in the future have to come so perilously close to disaster again. I repeat, however, that much remains to be done even in the first stage, which is now under way. The organization of a United Nations Force from other than the permanent members of the Security Council is bound to be a task of greater complexity and difficulty. We are breaking new ground, but I feel sure that we can reap a rich harvest from that ground in terms of peace and security in the area concerned and indeed, I hope, in wider terms as well.

We must now pass on to the greater and perhaps even more difficult task of reaching a political settlement which will be honourable and just and will provide some hope of security and progress for millions in that part of the world who have not known them in these troublous and distracting years. . . .

6. Order-in-Council, 20 November 1956[1]

Whereas by a resolution dated November 4, 1956, the General Assembly of the United Nations established a United Nations Command for an emergency international force to secure and supervise the cessation of hostilities in the Middle East;

[1] *The Globe and Mail* (Toronto), 21 November 1956.

And whereas member nations were invited to contribute self-contained battalion groups to which Canada agreed, as a result of which preparations were made to assemble such a force for dispatch in part by air and in part by H.M.C.S. *Magnificent*;

And whereas the United Nations commander has now indicated that the most valuable and urgently required contribution that Canada could make to the force at present would be to supply an augmented transport squadron of the R.C.A.F. and administrative and technical elements of the army contingent to help in organizing the administration at the base of the force in Egypt;

And whereas the United Nations commander has also advised that the dispatch of the battalion should now be deferred until consideration of the detailed requirements of the force permits him to determine where and when the battalion can best be used.

Therefore, His Excellency the Governor-General in council, on the recommendation of the Minister of National Defence, is pleased, hereby, to make the following order:

Authority is hereby given for the maintenance on active service of officers and men of the Royal Canadian Navy, the Canadian Army, and the Royal Canadian Air Force, not exceeding 2,500 in number at any one time, as a part of or in immediate support of an emergency international force organized by the United Nations to secure and supervise the cessation of hostilities in the Middle East.

7. Extract from speech by the Secretary of State for External Affairs, Mr. L. B. Pearson, to the General Assembly of the United Nations, 23 November 1956[1]

. . . There is very strong, enthusiastic support in my country for this Force, but only as a United Nations Force, under United Nations control, and as an effective and organized force which can do the job that has been given to it and which, if it can do that job, may be the beginning of something bigger and more permanent in the history of our Organization: something which we have talked about at United Nations meetings for many years, the organization of the peace through international action. Therefore, it is important that this Force should be so constituted and so organized that it will be able to do the work that it has been given to do and thereby set a precedent for the future.

[1] G.A.O.R., Eleventh Session, 1956–7, vol. I, p. 268.

It is also important that the principles on which the Force is to operate are sound. What are these principles? They have been laid down for us in the Secretary-General's report. The Force must be fully independent, in regard to its functions and its composition, of the political situation of any single member. The United Nations alone controls it and is responsible for it.

I agree, of course, that the Force—I am not talking about individual elements in the Force but of the Force as such—in the circumstances and on the basis of which it was set up, could not operate in the territory of a country without the consent of that country. That is why we are happy that Egypt has given that consent in principle and I am sure that we all agree that, in giving that consent to the constructive and helpful move, no infringement of sovereignty is involved. It is rather an example of using national sovereignty to bring about peace and security and a political settlement through United Nations action.

The control of this Force, then, is in the hands of the United Nations and must remain there. Otherwise it would not be a United Nations Force but it would be merely a collection of national forces, each under the control of its own Government and serving in another country with the consent of and under conditions laid down by that country. That, I am sure, would be unacceptable to most of the Governments of this Assembly.

Having said that, however, I do agree that the Secretary-General should certainly consult with the Government of the country in which the Force is serving, on all matters of any importance that affect it; also, as we understand it, the Force is to remain in the area until its task is completed, and that would surely be for the determination of the United Nations itself. . . .

8. Extracts from broadcast interview of the Secretary of State for External Affairs, Mr. L. B. Pearson, by the United Nations correspondent of the Canadian Broadcasting Corporation, Mr. Charles Lynch, 25 November 1956[1]

Mr. Lynch: . . . Could you tell us, Sir, has the United Nations Emergency Force worked out as you hoped it would when you first suggested it?

[1] *S. and S.*, No. 56/34.

Mr. Pearson: It was less than three weeks ago when the resolution was introduced setting up this Force. During that very short time far more has been accomplished than any of us could have reasonably expected, although there are a great many difficulties to overcome yet. But what has been done I think has been really quite amazing. . . . The Forces, from six or seven countries, including Canada, are on the spot now. Offers have been received from another nineteen or twenty which have not been accepted in the sense that they have been incorporated in the force, largely because work of preparation for the absorption of forces has not yet been completed. On the whole, however, an amazing amount of work has been done.

Mr. Lynch: We have heard a lot here [United Nations headquarters in New York], and I believe in Ottawa as well, about these supposed Egyptian objections to Canadian participation in the Force; that the Egyptians feel that the Canadians are too British for their tastes. Can you pin that one down?

Mr. Pearson: I know a great deal of interest has been aroused in that question, and it is quite true that our own participation in the Force at the moment is not as we expected it to be. A fortnight ago when the Canadian offer was made to the Secretary-General it was of an infantry battalion, as you know. And that was accepted very gratefully by the Secretary-General and the Commanding Officer, who had been appointed by then and who is a Canadian (which would have had some bearing on Egyptian objections). And we were told at that time that we would be performing a very useful service if we could move that regiment down to the sea coast, down to Halifax where it could be shipped on the *Magnificent*, and steps were taken to do that at once. Then, as you know, the Secretary-General went to Cairo. He there discussed a great many things about the Force, its functions and composition, with the Egyptian Government. That is quite understandable because after all this Force has to serve on Egyptian territory, and though I for one, and a good many others also, don't admit that the Egyptian Government could have a veto over the composition of the Force, I think the Secretary-General is very wise in consulting them and trying to get their co-operation. And when he did consult, he found that there was a reluctance on the part of the Egyptian Government to have such a large part of the infantry Force at the beginning consisting of Canadians.

Because the Egyptian Government thought it would create misunderstandings in Egyptian public opinion, which wasn't able easily to distinguish between various members of the Commonwealth, the Secretary-General, who has the decision in this matter, was impressed by this point of view in respect of the immediate functioning of the Force. When he came back to New York, he discussed it with us and with General Burns who was here then. By that time General Burns had decided that the most important thing was to get his headquarters organized and his service troops out there—signallers and that kind of thing—and air transport, not only air transport for the Force, but the air component for the Force generally. And therefore he asked us if we would supply those units at once with the infantry to come along later when he felt it was possible to absorb them. By that time there shouldn't be any difficulty on anybody's part. I want to make it quite clear, however, that the participation of Canadians in this Force has been accepted by the Egyptian Government itself, that the Egyptian Government does not veto the participation of any units in this Force and that this is quite clear with the Secretary-General. The question of when the Canadian infantry units come forward will be determined by the advice we get from the Commanding Officer sent on by the Secretary-General.

Mr. Lynch: Do you think it was unfortunate that a regiment with the name 'The Queen's Own Rifles' which might be calculated to set the Egyptian hair standing on end, was chosen as the Canadian Regiment?

Mr. Pearson: Perhaps, but that is an honourable name for a Canadian regiment and, of course, while it may have led to some temporary misunderstanding, we are not likely to change the names of our regiments for purposes of that kind.[1]

Mr. Lynch: Canada seems to be playing a role in this matter of the Force out of all proportion to her population. Do you think she can carry it off and can we expect the Canadian role to continue on this level?

Mr. Pearson: I think we will be happy to participate in this force to the extent of our ability. I think Canadian opinion is behind this decision of the Government. This is an imaginative and important move on the part of the United Nations and Canada,

---

1 See above, p. 287, n. 2.

which has been interested in the idea of a United Nations Police Force for many, many years, and has made previous proposals precisely to that end, will want to do her full part. . . .

9. Extracts from speech by the Secretary of State for External Affairs, Mr. L. B. Pearson, in the House of Commons, 27 November 1956[1]

. . . When our United Nations force resolution was introduced, . . . the 19-power Asian-Arab resolution had already been introduced, which reaffirmed the earlier United States resolution which had been carried by this time and which insisted on a cease-fire and a withdrawal of troops, and which asked the Secretary-General to report within 12 hours on the compliance with that injunction. That night of November 3 and 4—and the session went on all night—tempers were rather high. The talk was strong and the danger of a rash—as we would have thought it—condemnation of the United Kingdom as aggressors was very real. The situation was deteriorating and the communists were working feverishly and destructively to exploit it.

In these circumstances and having, as I have said, canvassed the situation carefully with our friends and having studied Sir Anthony Eden's speech [Part III, I, Document 7, *supra*], we moved this resolution concurrently with the 19-power Asian-Arab resolution which was an attempt to get British, French and Israeli forces out of . . . Egypt. It was a very short resolution, and it asked the Secretary-General merely to submit, within 48 hours, something we had been unable to do for ten years, namely a plan for setting up an emergency international United Nations police force with the consent of the governments concerned. If we had not put in that phrase 'with the consent of the governments concerned' we might not have been able to secure a majority for our resolution. As it was, the resolution passed unanimously. . . . Steps were taken immediately by the Secretary-General to report back what he was able to do in 48 hours in the setting up of this force to supervise and secure a cessation of hostilities in accordance with the earlier resolution of November 2, one of which was to ensure freedom of navigation in the Suez Canal.

We obtained 57 votes as sponsors for the resolution. There

[1] Canada, H.C. Deb., 1956 (Special Session), pp. 51–65.

were 19 abstentions. Nobody voted against us. The United Kingdom and France did not find it possible to vote for that resolution at that time, but they have indicated, both privately and publicly, their great appreciation of the initiative which resulted in its being adopted and they have also stated their support for it since then. At the same time—and this is related to the first resolution—the Asian-Arab resolution was put to the vote and carried by a large majority, 59 to 5 opposed. . . . Canada voted for that resolution asking for a cease-fire and a withdrawal of the forces from Egypt. . . . Then on November 4 we started to work, and we had something to do with this because we were the sponsors of the resolution and had a certain obligation in connection with helping the Secretary-General carrying it out. We started to work on organizing a United Nations police force or at least to form the basis of the organization and report back in 48 hours.

As it happened the Secretary-General, who has played a magnificent part throughout all these difficult days, was able to make a first report within 24 hours. Offers of contributions to the force began to come in within that 24-hour period. That Sunday night when we were working on the establishment of the force the United Kingdom and French ground forces landed at Port Said. The situation at the United Nations immediately began to deteriorate. Things became very tense. The Security Council was called into emergency session and refused to consider a Soviet proposal for Soviet and United States intervention because the matter was before the United Nations Assembly. Then in the midst of rumours of Russian intervention, rumours that there would be a determined demand by the Arab and Asian members of the Assembly to brand the United Kingdom and France formally as aggressors under the Charter and to invoke sanctions against them, the Assembly met on Tuesday morning, November 6. It had before it the Secretary-General's final report on the organization of the United Nations force. At that time he was able to report progress with regard to the composition of the force. He was able to lay down certain principles and functions for that force but not to go into detail, for two reasons. He did not have enough time, in the first place; and in the second place if we had attempted to do it in detail, we would still be arguing about what those functions should be. There was however one important detail, namely that the force should exclude contingents from the permanent members of the Security Council. The significance of that detail is obvious.

A draft resolution was drawn up supporting this report and authorizing the Secretary-General to go ahead on that basis, to discuss participation with other governments. It set up also an advisory committee of seven members of the Assembly[1] to help in this task. . . . It is interesting to note in passing that four members of that Committee are members of the Commonwealth of Nations. While we were trying to get this resolution through and get it through quickly and with a big majority—it was finally passed unanimously—another resolution, in the atmosphere of the fighting that was going on at that time in Suez, was introduced demanding the immediate withdrawal of forces, and that the Secretary-General should report that this had been done in 24 hours. Both these resolutions were being considered together. . . . The resolution demanding immediate withdrawal was passed by a vote of 65 with only one opposed, Israel, and with 10 abstentions. The United Kingdom and France did not oppose that resolution, they abstained on it. We voted for that resolution after having stated our interpretation, which was accepted by a good many other delegations, of the word 'immediate'. If that interpretation had not been stated and accepted by many we would not have voted for it. By 'immediate' we said we had in mind that the United Kingdom and French forces would withdraw from Egypt as soon as the United Nations forces had been moved there and were operating satisfactorily. By getting our United Nations force resolution through and by accepting this Arab-Asian resolution of withdrawal, which had in it no element of sanctions, we were able to reject extreme demands which were being made, and which would have led us into grave danger indeed. . . .

There were only two more resolutions subsequent to the one I have just mentioned. The one last Saturday asked for withdrawal once again. We did not support it because we felt that the withdrawal had begun. We had confidence in the good faith of the British and French when they told us that the withdrawal would be completed. We felt at that time that to support another resolution of withdrawal would be to assimilate the position of the British, French and Israelis to that of the Russians in Hungary. . . .

[1] Brazil, Canada, Ceylon, Colombia, India, Norway, and Pakistan.

Then the final resolution carried Saturday night approved an *aide mémoire* which gave the Secretary-General further authority to organize the United Nations police force. By a very important paragraph in that resolution he was told to get ahead with the clearing of the Suez Canal. In spite of efforts by Soviet and certain Arab-Asian countries to hold up the work on political grounds, he has now authority to go ahead with the vitally important work.

Now, Mr. Speaker, we have the United Nations force in being, and I am sure the House . . . would like me to say something about the functions, operations and composition of that force, and Canada's contribution to it. . . .

The function of this force which is now in being is to secure and supervise the cessation of hostilities . . . and carry out its task in accordance with directions received from the United Nations, not from any one member of the United Nations. The force—and it is interesting to recall that the resolution authorizing this force was passed not much more than three weeks ago—is now in being in Egypt where it will be stationed, or any place else where the United Nations considers it necessary to be stationed. . . . At the present time the headquarters of the force is along the Suez, but it may of course be moved.

It is not a fighting force in the sense that it is a force operating under, say, Chapter VII of the United Nations Charter, which deals with enforcement procedures. It is not a United Nations fighting force in the sense that the force in Korea was; it is operating under a different chapter of the Charter dealing with concilia-tion procedures. . . . It is not the purpose of this force to be used in fighting operations against anybody. . . .

This force will stay in Egypt until the United Nations decides that its functions are discharged, or, of course, until the govern-ments participating in the force withdraw their contingents. It must, of course, not infringe on the sovereignty of the government of the territory in which it is operating. . . . But the exercise of that sovereignty in the case of the government of Egypt where the force is operating now must be qualified by the acceptance by Egypt of the resolution of the United Nations concerning the force. Egypt has already agreed to the admission of this United Nations force to its territory; and it seems to me to be obvious, because it is not an enforcement action of the United Nations under Chapter VII of the Charter that every effort should be made by the Secretary-General of the United Nations, and by the United Nations itself, to secure and maintain the co-operation of the Egyptian Government in the functioning of this force, and the co-operation of the other Governments concerned, including the Government of Israel.

But this does not mean, as I understand it—and I assure you, Mr. Speaker, this has been made very clear in meetings of the advisory committee—that Egypt or any other government can determine by its own decision where the force is to operate, how it is to operate or when it must leave. Furthermore, the right of Egypt to consent to the admission of a United Nations force to its territory does not imply the necessity of consent to the admission of, or the right to reject, separate units or elements of that force. That is a stand, Mr. Speaker, which the Canadian representative on the advisory committee has taken. . . .

The first resolution dealing with this force was passed in the United Nations Assembly on November 4. We had already said by the time that resolution was passed . . . that we were in favour of it and that we would recommend a contribution to it. The day after the resolution was passed I met the Secretary-General as the sponsor of the resolutions and discussed with him the question of putting some United Nations troops into the area at once. He considered it to be a matter of the most immediate urgency. So I said I was authorized to state that the Canadian Government was willing to participate, and later in the day I wrote a formal communication to him to that effect, saying that we had decided to make an appropriate contribution subject to the required constitutional action being taken in Canada.

The next day I also talked with the Secretary-General about the force and he was then also emphatic, for the obvious reason that the situation seemed to be deteriorating, that we must proceed quickly. We discussed the nature of our contribution that afternoon, I by telephone with my colleagues in Ottawa, when the question of a battalion came up. . . .

General Burns was asked to come to New York, and those countries that had already announced their desire to contribute were asked to send military advisers to New York to discuss the problem with the Secretary-General, his staff and General Burns. The Canadian Department of National Defence sent three officers down immediately and the next day, Tuesday, November 6, the Prime Minister announced that Canada would offer, and I quote: 'Subject to adjustment and/or rearrangement after consultation with the United Nations commander' a self-contained battalion group with H.M.C.S. *Magnificent* as a temporary mobile base [Document 4].

The consultations which we had had in New York up to that time led us to believe that would be a most welcome contribution and we were urged to press ahead with it. The Secretary-General told me he was most anxious for us to get our battalion to a place where it could be embarked without delay.

General Burns reached New York a little later than we expected because he had to go to Cairo en route. The possibility then was mentioned that one country might provide all the administrative and air support at least in the initial stages. General Burns had found that difficulties were already developing because the infantry that had arrived, mostly from the Scandinavian countries and also from Colombia, were reaching the base without the necessary services and there was no headquarters organized to receive them.

These reports were sent by me to Ottawa. I returned to discuss them with my colleagues over the week-end, and while I was in Ottawa the Secretary-General through his executive assistant phoned me on Saturday, November 10, about another difficulty that was developing . . . , namely that the Egyptian authorities were concerned about the possibility of Canadian troops being mistaken for United Kingdom troops and that incidents might take place especially if the proportion of Canadian troops to the total force were as high as would be the case if the Canadian infantry battalion had arrived at that time.

We in New York, and indeed in Ottawa on advice from New York, felt that these difficulties would be overcome, and in discussing them again with the Secretary-General he once again asked us to make no changes in our plans pending further discussions and he hoped satisfactory arrangements could be made. So the Government went ahead with the arrangements as originally contemplated.

. . . I think it was on Tuesday, November 13, when back in New York from Ottawa that I had another talk with the Secretary-General in relation to the new difficulties which had occurred. I emphasized to him at that time that we felt it absolutely essential to the success of this effort that neither Egypt nor any other country should impose conditions regarding the composition of the force. I told him that on this matter we would negotiate only with him, the Secretary-General, although we recognized, of course, that it was right and proper that he should discuss these matters with Egypt in order to avoid, if possible, subsequent difficulties.

Nevertheless, on that Tuesday I asked him again about composition in view of the developing difficulties, and whether we should proceed with our plans for moving the regiment. The Secretary-General said . . . that he hoped we would go right ahead with our plans.

He also discussed with me the question of composition on the next day, Wednesday. Then later we had a meeting of the Advisory Committee on the matter. . . . Following that the Secretary-General flew to Cairo. He left New York in the hope that these difficulties would all be cleared up before he had returned. As we were having diplomatic discussions about them and as it seemed that these discussions might end in a satisfactory way, we did our best, I quite admit, to discourage any premature publicity about difficulties. . . .

During that week-end when General Burns had reached New York and the Secretary-General was in Cairo I was in touch with the Secretary-General by telephone and cable through our Embassy [in Cairo]. I stated to him that I had had word about his discussions with the Egyptians; that while I appreciated the difficulties which had arisen and while naturally we wanted to help the Secretary-General, already so overburdened with problems, in any way possible, nevertheless we could not accept the principle that any one government could determine what contribution or whether any contribution would be made by a member state in connection with the United Nations force. I am glad to say that the Secretary-General has taken the same position.

. . . As a result of these discussions the Secretary-General had sent a communication to me from Cairo which I shall put on the record:

The question of when and where ground troops shall be used can best be considered when the U.N.E.F. can assess its needs at the armistice line. The present situation seems to be one where it is not a lack of troops for the immediate task but of possibilities to bring them over and maintain their lines of communications.

That was a message from the Secretary-General, not from the Egyptian Government. He also emphasized that in sending it neither he nor anyone else was laying down conditions for Canadian participation because he felt that that would be improper. On his return and after further discussion with General Burns it was agreed that for the time being we should concentrate on getting these other forces to Egypt and hold the infantry battalion in reserve. General Burns himself said he agreed that it was even more important at the present moment to have an air transport headquarters, administration units, signals, engineers, army service, medical units and forces of that type; which were later to be sneered at by some excitable persons as constituting a typewriter army, something that will not I think commend itself to the members of these very gallant Canadian regiments.

We agreed then to this change in plans, although regretting it. . . . I ask whether we could or should have proceeded otherwise. . . . To have made no offers or to have made no plans, to have held back our offer until everything was cleared up; to have permitted no movement of troops of any kind, would I think have left us open to criticism, to the charge that we were dragging our feet in connection with a proposal which we ourselves had put forward. . . . It did not seem to be the time . . . or the occasion for national pique or peevishness or sneering at this new United Nations force as being Nasser's farce. It seemed to me that the situation was far too serious for that. What was required from every member of the United Nations was to back up the United Nations force to the best of its ability, after receiving the best advice it could. After receiving such advice from the United Nations itself, we took that course, and as a result there is now a United Nations force which within three and four weeks of the resolution authorizing it now includes on the spot . . . 1,700 troops of which 20 per cent. or 350 are Canadians . . . now working together under the United Nations blue flag for peace. . . .

10. Statement by the Acting Prime Minister, Mr. C. D. Howe, 10 December 1956[1]

A request has been received from the United Nations, on recommendation of Major-General Burns, the U.N. Commander in the Middle East, for the immediate provision by Canada of

[1] *The Crisis in the Middle East*, pp. 27–28.

additional maintenance, support and communications personnel for the U.N. Emergency Force.

Major-General Burns states that the detachments of similar troops which have already been sent from Canada have filled a most important and essential role in the rapid build-up of the United Nations Emergency Force and now, as additional infantry units arrive in the Middle East from other countries, there is developing a need for more signals, servicing and maintenance units.

He has accordingly requested, and the U.N. has authorized, the despatch by Canada of a signals squadron of approximately 150 all ranks, a R.C.E.M.E. [Royal Canadian Electrical and Mechanical Engineers] workshop of about 150 persons and two transport platoons of all ranks.

In addition to these Army personnel, General Burns has asked that upon completion of the airlift between Naples and Egypt, which is now being carried out by the R.C.A.F., an air component for communications and observations be established by Canada in the Middle East as part of U.N.E.F. The number of R.C.A.F. personnel involved in this operation will be between 250 and 300.

In addition to the above-mentioned Army and R.C.A.F. contributions, there will be a small number of Canadian officers employed on the staff of General Burns' headquarters.

When the foregoing personnel are despatched to the Middle East the Canadian Army and Air Force component of the U.N.E.F. will comprise over 1,000 service personnel.

The Canadian Army component as stated above, with vehicles and equipment necessary for them to perform their tasks, will be despatched in H.M.C.S. *Magnificent* before the end of the month.

Because of Canada's comparatively favourable position among the nations contributing forces to the U.N.E.F. it is apparent that requirements for the support of elements so necessary to round out and weld the U.N.E.F. into an effective and efficient force can best be supplied by Canada. The number of these specialists to be provided by Canada has now reached the point where we are about in balance, so far as numbers are concerned, with the other contributing nations. It is desirable from the U.N. point of view to preserve this balance and as a result it now appears doubtful whether an infantry unit will be required from Canada. For this reason it has been decided to return, at least for the time being, the

1st Q.O.R. [Queen's Own Rifles] to their home station. This will be done during the next few days. The members of this unit are to be commended for the speed and efficiency with which they prepared themselves for overseas service and for their exemplary conduct during those weeks of waiting in Halifax, and we all regret that changes in United Nations plans have not made it possible for the battalion to proceed overseas as originally planned.

11. Extracts from speech by the Secretary of State for External Affairs, Mr. L. B. Pearson, in the House of Commons, 14 January 1957[1]

. . . Recently the [General] Assembly took a very important step indeed in extending its functions into the field of security after the Security Council itself became powerless in that field through the exercise of the veto. I refer, of course, to the emergency force which was set up to supervise and secure a cessation of hostilities. Now, Mr. Speaker, the immediate value of this force which now numbers, incidentally, about 5,500 of whom over 1,100 are Canadians, in respect of the specific emergency which brought it into being has I think been well established. Its continuing value in helping to bring about and maintain peaceful conditions and security in the area in which it operates remains, of course, to be proven. I myself think it should be of great value for this purpose also, provided it remains genuinely international in control, composition and function, and providing also that its limitations are recognized, especially that it is a voluntary organization which must act strictly within the terms of resolutions which are only morally binding and which must be passed by two-thirds of the Assembly in each case. But even within these limitations the United Nations force can, I think, play an important part in bringing about an honourable and enduring political settlement in the Palestine and Suez area. . . .

While the political climate of the Middle East is maturing toward the time when conditions will be more appropriate for a comprehensive settlement, it is essential, I think, for the countries of the region, and indeed for us all, that there should be no return to the former state of strife and tension and conflict on the borders; that security should be maintained and, indeed, guaranteed. I suggest that for this purpose there will be a continuing need during

[1] Canada, H.C. Deb., 1957, I, pp. 175–7.

the period until a political settlement is achieved for the stabilizing international influence that the emergency force is now exercising. And this essential stabilizing role might well require the continuing presence of a United Nations force along the boundary between Egypt and Israel; perhaps also for a time in the Gaza strip and, with the consent of the states involved, along the borders between Israel and her other Arab neighbours, though that of course would require a further resolution from the United Nations Assembly. . . .

Looking further ahead, the experience of the United Nations in respect of the Suez crisis, especially the necessity for hasty improvisation, underlines, I think, the desirability and the need of some international police force on a more permanent basis. We have recognized this need in the past. We have expressed that recognition at the United Nations and elsewhere as recently as in the General Assembly before the recess and we have done all we could to translate that necessity into reality, but for one reason or another it has never been possible for the United Nations, except in the special and limited cases of Korea and the Middle East, to have armed forces at its disposal. . . .

This present emergency force in the Middle East is a unique experiment in the use of an international police agency to secure and supervise the cease-fire which has been called for by the General Assembly. Why should we not, therefore, on the basis of this experience—the experience we have gained by the operation and establishment and organization of this force—consider how a more permanent United Nations machinery of this kind might be created for use in similar situations as required?

What the United Nations now would seem to need for these limited and essentially police functions is perhaps not so much a force in being as an assurance that members would be prepared to contribute contingents when asked to do so, to have them ready and organized for that purpose; with some appropriate central United Nations machinery along the lines of that which has already been established for this present emergency force.

The kind of force we have in mind would be designed to meet situations calling for action, intermediate if you like, between the passing of resolutions and the fighting of a war, and which might incidentally have the effect of reducing the risks of the latter. It would not, however, as I see it, be expected to operate in an area where fighting was actually in progress; it would be preventitive and restoratory rather than punitive or belligerent.

It is not possible to determine in advance what would be required in any emergency, but surely members through the proper legislative processes could take in advance the necessary decisions in principle so that should the occasion arise the executive power could quickly meet United Nations requests for assistance which had been approved by it. In doing so we would be making at least some progress in putting international action behind international words.

12. Extracts from speech by the Secretary of State for External Affairs, Mr. L. B. Pearson, in the House of Commons, 22 March 1957[1]

. . . Our position, Mr. Speaker . . . is that the United Nations should be associated to the maximum possible extent in the administration of the Gaza strip. . . . The problem . . . is to find an acceptable balance in the administration of Gaza between the practical position which the United Nations must take and the legal position of Egypt under the Armistice Agreement. Any such 'suitable balance' in particular must give the United Nations that control of internal security in the Gaza strip necessary to enable it to carry out effectively its operations and responsibilities on the demarcation line. So far as the Canadian Government is concerned, U.N.E.F. could not be expected to discharge effectively its duty of preventing raids and incursions and maintain peaceful conditions along that line if it were not in a position to carry out observations or investigations and to exercise necessary control in the strip itself. A satisfactory agreement to this effect is a fundamental prerequisite of the effective continuance of U.N.E.F.'s role on the demarcation line. If no such agreement is made and kept, there will not only be renewed trouble between Israel and Egypt but the continued operation of U.N.E.F. will be prejudiced. Certainly it would be difficult for Canada to continue to participate in the Force under conditions, and we hope those conditions will not materialize, in which it would not be able to discharge satisfactorily the responsibilities given to it by the United Nations Assembly. . . .

[1] Canada, H.C. Deb., 1957, I, pp. 2589–90.

13. Extracts from interview of the Secretary of State for External Affairs, Mr. L. B. Pearson, by Mr. Blair Fraser and Mr. Lionel Shapiro, May 1957[1]

Mr. Fraser: . . . What is your view of the status of U.N.E.F. in Egypt right now? Must it get out whenever Nasser wants it to?

Mr. Pearson: I would say no. And we've made this position of ours as clear as we could in parliament here, to the Committee of Seven of which I am a member—sort of the Executive Committee of U.N.E.F.—and to the Assembly itself. We feel that Egypt had the right to be consulted and to agree to the entry of an international force, but having given that consent as she did, she has no right to control the Force, to order it about, to tell the Force when it shall leave. If Egypt is dissatisfied with the operation of the Force, or if anybody else is dissatisfied, or if Egypt wants the Force to withdraw, feels its work is completed, Egypt should make its views known to the Secretary-General who would take it up with the Committee of Seven and then it would go to the full Assembly, and until the Assembly had decided the Force would carry on.

Mr. Fraser: Do I understand you correctly to say that if President Nasser tomorrow decided he didn't want the Force in Egypt any longer the Force would *not* leave within a reasonable period?

Mr. Pearson: You put the question in very difficult practical terms. The position I stated is, I think, theoretically sound. But there are several governments participating in the Force—who don't accept our position and say that any time Nasser wants them to leave they'll go—India particularly. So it's a difficult question.

Mr. Fraser: The practical answer then is that the Force must get out when Nasser decides it must.

Mr. Pearson: I'm afraid that having regard to the views of some of the members of the Force and having regard for the practical difficulty of the position, the Force couldn't operate constructively on Egyptian territory with the active opposition of the government of Egypt. But, it is one thing to say that, and another to admit the right of Egypt to take that position. . . .

Mr. Fraser: On the other hand, you do feel that right now President Nasser would not like to see the Force go.

[1] 'Where Canada stands in the World Crisis', *Maclean's Magazine* (Toronto), 6 July 1957.

Mr. Pearson: Oh, I don't think so.

Mr. Fraser: Well, doesn't that give us a bargaining position?

Mr. Pearson: Yes, it does.

Mr. Fraser: On that point, isn't it a fact that a few weeks ago Nasser was calmed down a good deal in his soundings-off by the Canadian hint, which I think came from you—that if . . .

Mr. Pearson: It wasn't a hint. . . . It was a very forthright statement!

Mr. Fraser: . . . that we jolly well would get out if he didn't behave himself?

Mr. Pearson: We made that statement the day Hammarskjold arrived in Cairo.

Mr. Fraser: And that made him behave himself.

Mr. Shapiro: Doesn't it remain, then, the fact that time is running out on us? I mean, our bargaining position, which is probably the last one we will have, is running out?

Mr. Pearson: Yes. Our bargaining position will decrease as time goes by. . . .

Mr. Fraser: Do you think that U.N.E.F., the United Nations police force, or something like it, should become a permanent thing?

Mr. Pearson: I don't think of a U.N. police force in the terms that were in the minds of the people who drew up the Charter. They thought of the U.N. preserving the peace, policing the peace by overwhelming force. If anybody wanted to start trouble the U.N. force would move in. That postulated the working together of the great powers and was provided for in the Charter through the Security Council, and the Security Council was given certain powers to enforce the peace. Now that's all gone and there is no point in talking about it as long as the world is divided. But a U.N. police force in the sense of an organization created with a headquarters in New York and with governments pledged to contribute up to a certain amount to that police force, which would be ready to go into action at a moment's notice to put an end to brush fires, or to get between the combatants and stay there until the danger of new fighting was over—that kind of police force, it seems to me, makes sense. It really would be a sort of extension and a perpetuation of the present police force we have in Egypt and in Gaza.

Mr. Fraser: How could it be permanent when the personnel of the force has to be adapted to each particular situation?

Mr. Pearson: It wouldn't be very difficult if the will to do it were there. Take our own position. We would earmark, as we were willing to earmark in 1950 under the Uniting for Peace resolution, a certain number and kind of troops, available for United Nations service to carry out United Nations resolutions, subject, of course, always to the consent of our parliament. Other countries would make the same kind of offer and there would be a headquarters in New York which would know at any one time what it could call on. If we needed, say, three thousand men now to go out to Jordan under a U.N. resolution, the military director in the Secretary-General's staff in New York would know that so-and-so had offered in advance to send a thousand infantrymen and that other governments had done the same. It would all be organized in advance. The troops would have been trained for this kind of work. The staff would have been created and the force would be on the way in a few days.

Mr. Fraser: But the operative point is: will the consent of the country to which it is sent be necessary?

Mr. Pearson: In theory it would have to be. But let's take Jordan again. Supposing the situation collapsed there and various countries knew that Jordan was collapsing. But there was great rivalry as to who would move in. It might be possible by a large vote of the United Nations, although it wouldn't perhaps be strictly speaking legal under the Charter, for the United Nations to take action even against the wishes of a state which was disappearing to prevent confusion and trouble.

Mr. Fraser: You are taking an extreme case now.

Mr. Pearson: Normally you would have to have the consent of the state in which the U.N. force was operating. If the force was to go to a frontier between two conflicting states and get between the forces, you'd have to have agreement from both those countries. In other words, you would have an armistice between the conflicting forces and you would also have an agency to police the armistice. . . .

14. Extract from the Nobel Peace Prize Lecture, by Mr. L. B. Pearson, 11 December 1957[1]

. . . Certainly the idea of an international police force effective against a big disturber of the peace seems today unrealizeable to

[1] Lester B. Pearson, *Diplomacy in the Nuclear Age* (Toronto, 1959), pp. 103–4.

the point of absurdity. We did, however, take at least a step in the direction of putting international force behind an international decision a year ago in the Suez crisis. The birth of this force was sudden and it was surgical. The arrangements for the reception of the infant were rudimentary, and the mid-wives—one of the most important of whom was Norway—had no precedents or experience to guide them. Nevertheless, U.N.E.F., the first genuinely international police force of its kind, came into being and into action.

It was organized with great speed and efficiency even though its functions were limited and its authority unclear. And the credit for that must go first of all to the Secretary-General of the United Nations and his assistants. Composed of the men of nine United Nations countries from four continents, U.N.E.F. moved with high morale and higher purpose between national military forces in conflict. Under the peaceful blue emblem of the United Nations it brought, and has maintained, at least relative quiet on an explosive border. It has supervised and secured a cease-fire.

I do not exaggerate the significance of what has been done. There is no peace in the area. There is no unanimity at the United Nations about the functions and future of this force. It would be futile in a quarrel between, or in opposition to, big powers. But it may have prevented a brush fire from becoming an all-consuming blaze at the Suez last year, and it could do so again in similar circumstances in the future. . . .

III. AUSTRALIA

1. Statement by the Prime Minister, Mr. R. G. Menzies, 3 November 1956[1]

As I pointed out in my statement to Parliament [Part III, III, Document 5, *supra*], the action taken by the United Kingdom and France was and is police action; the only quick and practical means of separating the belligerents and protecting the Canal.

Without such police action the Israel-Egypt position might by now have become completely out of hand with spreading consequences.

If the United Nations Assembly accepts the idea of a United Nations police force in and around the Canal, and the Security Council adopts it and acts on it, the object of protection of the Canal will have been achieved.

[1] *C.N.I.A.*, vol. 27, no. 11, November 1956, p. 735.

Zcs

Meanwhile it is clear that the United Kingdom and France cannot withdraw. Police action, to be effective, must be continuous while the danger exists.

I fear there is much confusion about the position of Egypt. That the author of the Suez Canal confiscation and the promoter of anti-British and anti-Israel activities in the Middle East should now be represented as the innocent victim of unprovoked aggression is, of course, both wrong and absurd.

We will await the concrete development of Mr. Pearson's helpful proposal for an international police force with most sympathetic interest.

2. Extract from speech by the Permanent Representative to the United Nations, Dr. Ronald Walker, to the General Assembly, 7 November 1956[1]

. . . the Australian Government, in considering the documents submitted within the last few hours, has felt some doubts regarding the effectiveness of some of the arrangements proposed and the enduring nature of the international police force which it is proposed to establish. There is some question in our minds as to whether it will remain in the area long enough effectively to secure the Egyptian-Israel situation along the lines that are necessary for the establishment and maintenance of peace and, we hope, a permanent settlement. Nevertheless, despite the sort of doubts that may arise in our minds as we examine some parts of this proposal as set out in the document before us, it is the view of my delegation and of my Government that it is a matter of urgency to proceed with the establishment of an international force and for the force to take up the duties which the Assembly is entrusting to it. . . .

3. Extract from speech by the Prime Minister, Mr. R. G. Menzies, in the House of Representatives, 8 November 1956[2]

. . . It would be rather a strange circumstance if the properly armed and equipped troops of the United Kingdom and France should be replaced by a force of no military consequence without adequate supply and backing. It should be an effective military force. At present, we in Australia do not know whether it is desired

[1] G.A.O.R., First Emergency Special Session, 1956, p. 114.
[2] Australia, H.R. Deb., 1956 (Second Period), p. 2116.

that we should contribute to it. It appears that the introduction of such a force, under United Nations rules, must be with the consent of the nation whose territory is to be entered, and as our Minister in Egypt has just received notification that diplomatic relations with Australia are cut off, it may be that Australia will not be included in a general approval. I do not know. . . . All I need say at present, on behalf of the Government, is that if the proposal is to constitute a military establishment which will facilitate the making of a permanent settlement in the Middle East, Australia will certainly be not unwilling to make such quick practical contribution as it can. . . .

4. Extracts from speech by the Prime Minister, Mr. R. G. Menzies, 12 November 1956[1]

. . . I wonder whether there would be a United Nations force in Egypt if the British and French had stayed at home? I wonder if they could have got there fast enough, because the Israelis are magnificent fighters and the Egyptians have a considerable turn of speed? [But] they are to be ordered out. They are not to be allowed to participate in an international force—a force to consist of people from Portugal and Colombia and little groups and bits and pieces. But not Great Britain and France. I would have thought that around their forces could have been built a U.N. force which would have power, authority, and could have got this Canal going and could have prevented President Nasser from having a feeling that his great enemies have been 'warned off the course'. But for some reason, intelligible only at Lake Success—or whatever it is called—intelligible only in the United Nations, Great Britain and France are not to be in this. . . .

Does anyone suppose that we are helping world peace by imposing an indignity in the name of some barren technical argument, upon the great Power from which we derive and upon the other great Power of the Western European defence system? This, to me, is sheer bedlam nonsense. I cannot understand it. . . . Britain's action, I personally say—and I will say it if I am the only one left to say it—was brave and correct. . . .

[1] *Sydney Morning Herald*, 13 November 1956.

5. Extracts from speech by the Minister of External Affairs, Mr. R. G. Casey, to the General Assembly of the United Nations, 21 November 1956[1]

. . . Of the Middle East situation, it is essential to try to see things in perspective. It is clearly wrong to judge any one particular international incident by itself and to ignore all that preceded it. To use an analogy, when one man assaults another, there is usually something that went before. When the alleged aggressor comes before a magistrate, evidence is collected to establish the background, as to whether there was provocation and, if so, the nature of it, and whether or not the alleged aggressor was justified in taking the law into his own hands, for instance by reason of the police being either absent or unable to provide the necessary protection. This analogy is relevant to our consideration of the Middle East situation. . . . It would be utterly wrong to picture President Nasser as a man who had been walking along quietly, minding his own business, when suddenly somebody hit him. The plain fact of the matter is that it was Egypt that upset the balance in the Middle East without any prior consultation with other countries in the region or with others having interests in the region. . . .

Before I speak of the longer-term objectives in the Middle East, let me say something about the immediate situation which is still critical and which could still flare up again at any moment. It is plainly most urgent that all haggling over the admission of the United Nations Emergency Force should cease and that the Force should take up its duties at once and in such numbers that it inspires confidence and creates some stability in the area. It seems to me quite absurd that there should be a long argument about what contingents are acceptable and what are not acceptable. Such objections, if pursued, make a farce of the very idea of a United Nations operation and a United Nations force. . . .

. . . We must get the United Nations Emergency Force performing the function for which it was intended; otherwise we risk a possible recrudescence of fighting in the area. Passions are still high, and there must be created what is in effect a demilitarized zone between Egypt and Israel, which it will be the particular function of the United Nations Emergency Force to achieve. . . .

One lesson of all this is that our machinery for maintaining

[1] G.A.O.R., Eleventh Session, 1956–7, I, pp. 197–200.

international stability in the Middle East—and probably elsewhere—was defective and that, if peace is now to be preserved, we
must try to evolve machinery that is more in line with realities.
But machinery is not the whole answer; it is not even at the real
root of the problem. Many of our troubles in the Middle East stem
from the refusal of Egypt and others to recognize the right of the
State of Israel to exist. . . .

I believe that an important part of our duty at this present
Assembly is to insist on watertight arrangements for the physical
security of the countries concerned in the Arab-Israel dispute.
Only when both sides are convinced that there is nothing to hope
for or to fear from the use of force, will they be prepared to make a
real peace. As an immediate measure, I think that the expansion
this year of the United Nations observer group under General
Burns was a good move. I would like to see more widespread use of
United Nations observers and their complete acceptance by each
side. We might also well consider the establishment of adequate
demilitarized zones in all the areas of particular tension. . . .

As we look out upon the world, we in Australia are impressed by
the fact that there are still great mental barriers between the
peoples which were until recently under colonial administration
and the countries that previously ruled them. That is understandable. I certainly do not attempt to defend all that was done by
colonial rulers in other parts of the world in the past. . . . For our
part we in Australia sincerely offer our friendship to all peoples on
a basis of complete and frank equality. We are determined to
continue to work with our friends, in Asia and elsewhere, for the
strengthening of mutual security, and for the improvement of
understanding and of social and economic conditions. . . .

6. Extract from speech by the Minister of External Affairs, Mr. R. G. Casey, to the General Assembly of the United Nations, 26 November 1956[1]

. . . I would like to inform the Assembly of the attitude taken
by the Australian Government towards possible Australian
participation in the United Nations Emergency Force. This
preliminary attitude has already been communicated by me to the
Secretary-General. Speaking in the Australian Parliament on
8 November, the Prime Minister of Australia, Mr. Menzies, said:

[1] G.A.O.R., Eleventh Session, 1956-7, I, p. 318.

It should be an effective military force. At present we in Australia do not know whether it is desired that we should contribute to it. It is probably too soon for anybody to have worked out what its constitution is to be and how it is to be used, and in what particular respect individual nations should take part in it. All I need say at present, on behalf of the Government, is that if the proposal is to constitute a military establishment which will facilitate the making of a permanent settlement in the Middle East, Australia would not be unwilling to make such quick, practical contributions as it can.

It is common knowledge that the Secretary-General has encountered considerable difficulty in making arrangements for an effectively organized force to enter Egypt. We hold strongly that the composition and functions of the United Nations Emergency Force, and its area of operations, can be determined only by the United Nations. If the United Nations can succeed, as it must, in establishing an adequate and effectively organized international force, as a shield between the belligerents, it will, in effect, bring about the demilitarization of the potential area of conflict between Israel and Egypt—an area which extends from the Suez Canal to the Israel border, and whose future limits must be determined by the United Nations in the light of circumstances and developments.

No doubt, there will be questions to be settled in the future, but the most urgent requirement is to complete the organization of the Force and to get it into position. . . .

7. Statement by the Minister of External Affairs, Mr. R. G. Casey, 4 December 1956[1]

I believe that the decision of the British Government to withdraw their forces from Egypt is a wise decision, if only as a means of bringing about normal relations between the British and American Governments. I say this with knowledge of the situation in Washington and in the United Nations over recent weeks.

The restoration of the closest possible relations between Britain and America is essential and urgent and the withdrawal of British forces from Egypt is an essential preliminary to this in the minds of the American Administration.

---

[1] *C.N.I.A.*, vol. 27, no. 12, December 1956, p. 830.

8. Extracts from speech by the delegate to the United Nations, Sir Percy Spender, to the General Assembly, 1 February 1957[1]

. . . I would not find it possible to accept, in the wide and general terms in which it is expressed, the proposition advanced by the Secretary-General in the first sentence of sub-paragraph 5 (b), namely, that: 'The use of military force by the United Nations other than that under Chapter VII of the Charter requires the consent of the States in which the force is to operate.'

. . . If the Secretary-General's proposition is of universal application—and if I have given it a wider meaning than the words of the Secretary-General would justify or than he intended, I shall be very happy to be corrected—then it would seem to me to follow that 'consent' is a continuing concept, and, if that is so, consent may be discontinued or consent, having been given, may at any time be withdrawn unless the party concerned has by agreement otherwise precluded itself from so doing. Nor would the matter stop here. Consent must, so it would seem, if the Secretary-General's proposition is correct, cover every phase of the operation at every point of time. It would mean that no plans could be worked out for the discharge by the United Nations Force of its functions, since it would reside solely within the power of the State concerned as to the specific area or areas in which such Forces should operate, when, for how long, and in what circumstances.

In other words, the functions and operations of the United Nations Force could be determined by the State concerned entirely in its own interest and so render completely inoperative all efforts of the Force to discharge effectively its functions.

It seems to me to follow from the proposition enunciated by the Secretary-General that, unless agreement already given by Egypt to the operation of the United Nations Emergency Force preclude it, the Force presently operating within the borders of Egypt is entirely subject to Egypt's ultimate control. Egypt could, on that basis, at any time withdraw its consent wholly or in part or in relation to any area or operation. If this were the position, then we certainly would have a situation where the United Nations Force, which, as the Secretary-General states, in sub-paragraph 5 (*b*) of his report, should not 'serve as a means to force settlement, in

[1] G.A.O.R., Eleventh Session, 1956–7, p. 1037.

the interest of one party, of political conflicts or legal issues recognized as controversial', could in fact be used by Egypt to that end.

. . . If I may say so, with due deference to the view expressed by the Secretary-General, this is how, taken to its logical conclusion, his proposition could work out, and I am unable to accept it unless I am forced to do so by clear language of the Charter. I am not unaware [sic: not aware?] of any such language. . . .

IV. New Zealand

1. Extract from speech by the Permanent Representative to the United Nations, Sir Leslie Munroe, to the General Assembly, 3 November 1956[1]

My delegation welcomes the initiative of the Canadian delegation. It was with great regret that I was unable fully to support the form in which its draft resolution was put forward.

My delegation, for reasons which I stated at the 562nd meeting of this Assembly [Part III, IV, Document 3, *supra*], was obliged to vote against the resolution which was then adopted. Clearly we were, as a result, unable to support a text which referred to the urgent necessity for complying with that resolution, namely, the resolution which we had already voted against. Accordingly, also, we were unable to support a text which referred to the cessation of hostilities in accordance with the terms of the aforementioned resolution—namely the one which we had previously voted against.

Nevertheless, I am authorized to state that the New Zealand Government is prepared to support the establishment of a United Nations force, on acceptable terms, to assist in the establishment of peace and order in the Middle East. Not only that; I have been specifically authorized tonight to state that New Zealand is prepared to contribute to a force of the nature envisaged. . . .

2. Statement by the Permanent Representative to the United Nations, Sir Leslie Munroe, to the General Assembly, 4 November 1956[2]

I wish to explain to the Assembly why on behalf of my Government I abstained on the draft resolution which has just been carried. I abstained with regret because, as the members of the Assembly know, my Government has already offered to contribute

---

[1] G.A.O.R., First Emergency Special Session, 1956, p. 76.   [2] Ibid., p. 89.

to the international force which has been proposed and, subject to satisfactory conditions, is still prepared to do so. But I must say that I am concerned that in a matter of such moment as this, the debate should be concluded so quickly. I realize the urgency of the matter, but obviously there is need for delegations representing Governments at a great distance, as mine is, to get instructions. At least, that is what I would have thought. My Government is a responsible government. There are important matters in this resolution which require some time to examine and which it would be entirely proper for the Government of New Zealand to discuss with other Governments.

Under these circumstances, while thoroughly approving in principle the proposal for an international force, and having shown in practice that my Government is ready to contribute to that force, I found to my regret that it was necessary for me to abstain in the vote.

3. Letter from the Permanent Representative to the United Nations, Sir Leslie Munroe, to the Secretary-General, Dag Hammarskjold, 7 November 1956[1]

I have the honour, upon instructions, formally to confirm my statement in the General Assembly at the 563rd meeting [Document 1] concerning the readiness of the New Zealand Government to contribute to the emergency international United Nations Force.

I shall inform you of the nature of the proposed New Zealand contribution as soon as possible.

4. Letter from the Permanent Representative to the United Nations, Sir Leslie Munroe, to the Secretary-General, Dag Hammarskjold, 9 November 1956[2]

I have the honour to refer to my letter of 7 November (A/3302/Add. 8) [Document 3] concerning a New Zealand contribution to an emergency international United Nations Force.

I have now been asked to inform you that my Government contemplates a contribution consisting of units of land forces. The total strength of these units would be up to approximately 300 men. The contribution would be made upon the determination

---

[1] G.A.O.R., Doc. A/3302/Add. 8, First Emergency Special Session, 1956.
[2] G.A.O.R., Doc. A/3302/Add. 12, First Emergency Special Session, 1956.

of acceptable conditions, and on the clear understanding that the New Zealand contingent would be of a para-military character and not an operational entity.

5. Report of statement by the Prime Minister, Mr. S. G. Holland, 14 November 1956[1]

The Government has given authority to the Army to begin recruiting for the New Zealand contributions to the United Nations force to preserve peace in the Middle East.

The Prime Minister, Mr. Holland, said today the force would be a volunteer body specially enlisted on terms similar to those applying to the New Zealand Special Air Service squadron now serving in Malaya. A final decision on the number included in the force for the Middle East has not yet been made, but it will be approximately 300.

Mr. Holland confirmed that the Chief of the General Staff, Major-General C. E. Weir, who last Friday attended a special meeting of Cabinet, had submitted alternative plans for an effective contribution by New Zealand in the terms of her original offer to the United Nations.

He said these plans included the possibilities of the New Zealand contribution being called on to act as a self-contained unit or as part of a larger unit. The Government would not be in a position to make a decision on these alternatives until the wishes of other Governments were known.

'There has been a constant flow of information from the United Nations on the arrangements for the International Force and the steps that have been taken in New Zealand have been in conformity with the progress made in the overall plan. The Government, however, has held the view throughout that any force sent from New Zealand in such circumstances as those should be a volunteer force. It has also been contemplated that the International Force should enter on its task in stages, according to the circumstances of the nations which are making contributions to it.'

6. Extracts from speech by the Minister of External Affairs, Mr. T. W. Macdonald, to the General Assembly of the United Nations, 21 November 1956[2]

. . . The concern of New Zealand with the maintenance of peace

---

[1] *E.A.R.*, vol. VI, no. 11, pp. 11–12.
[2] G.A.O.R., Eleventh Session, 1956–7, I, pp. 212–13.

in the Pacific area is a direct one. However, history has also given tragic proof of the importance of the Middle East to our security and communication. Twice in recent times New Zealanders have been called upon to stake their lives and resources in its defence against aggression. Eighty thousand of our soldiers out of a population of less than 2 million fought in the Middle East in the Second World War for the victory which made possible the founding of this Organization.

What has happened in the Middle East in recent weeks is of the greatest importance to us. New Zealand's attitude towards Anglo-French intervention in the recent fighting between Egypt and Israel has already been made clear. From the outset it has had full confidence in the intentions underlying the action taken with France by the United Kingdom. There have been other times when the United Kingdom, virtually alone, has acted in the world interest against odds even heavier than the weight of adverse opinion. Time will show, we believe, that in this case, too, action was taken in the general interest rather than in pursuit of narrow ends. And we are not without hope that, high as the immediate costs may have been, the long-range results both for this Organization and for world peace may yet prove salutary.

We consider it a gain that the extent of Soviet penetration in the Middle East, the magnitude of its supply of arms to the area, and the malevolence of its intentions should have been unmistakably exposed. If any Middle East nation, proud as they all are to have thrown off Western control, chooses now to assist the entry into the Middle East of Soviet imperialism, then it will do so with full knowledge of the risks to which it exposes itself and the world.

It is a gain that the situation should have provided the stimulus for the creation of a United Nations Force, perhaps the first step towards investing the United Nations with the practical means to make its decisions effective. And it is a gain that it should have at last been brought home that it is time—perhaps the last opportunity—for this Organization to stop backing away from the hard realities and difficulties of the Middle East. For make no mistake, this Organization has backed away. . . .

Certainly in the past fortnight there has been no disinclination on the part of the majority in the Security Council or the General Assembly to deal with the situation created by the Israel attack on Egypt and the Anglo-French intervention. Certainly, too, there has been no failure on the part of the General Assembly to respond to the admirable and imaginative proposal of the Minister of External Affairs of Canada for the creation of a United Nations Emergency Force. . . . Nevertheless, I am not alone, I think, in detecting already a reluctance in some quarters to extend our work from the study of effects to the study of causes, and to accept the responsibility from which we have retreated in recent years. It was with this in mind that the New Zealand representative, Sir Leslie Munroe, proposed on 2 November 1956 [Part III, IV, Document 3, *supra*] that the whole problem of Arab-Israel relations should be fully and effectively considered at the present Assembly. . . .

In the view of my delegation, this Assembly should now frame recommendations on the Palestine problem and should at the same time decide what obligations it is prepared to assume in order to give them meaning. It is obvious that the situation which will prevail when the United Nations has a force in the Middle East capable of taking over from the forces of the United Kingdom and France will not be static but dynamic. It is obvious too that unless steps are taken to make that situation better, it will get worse.

7. Report of statement by the Prime Minister, Mr. S. G. Holland, 26 November 1956[1]

Cabinet agreed on 26 November after considering reports from the Minister of External Affairs, Mr. Macdonald, to the suspension of further recruiting for the United Nations Emergency Force, and to defer any decision to call men into camp.

This was announced by the Prime Minister, Mr. Holland, who said that Cabinet had given very careful study to the reports of the Minister of External Affairs on his discussions in New York with the Secretary-General, and also to a letter from Mr. Hammarskjold, the Secretary-General of the United Nations.

'It now seems clear', said Mr. Holland, 'that, at the present time, the New Zealand offer is not likely to be taken up.' No arrangements for calling men into camp would therefore be made before the New Year at the earliest, but the selection and medical examination of those who were volunteers would proceed.

Mr. Holland said that the Secretary-General had renewed his expression of sincere gratitude for the New Zealand offer. He had

---

[1] *E.A.R.*, vol. VI, no. 11, pp. 12–13.

pointed out that at this early stage of an entirely new type of operation considerable practical difficulties were confronting the
United Nations and the situation was also one of great delicacy.
The Secretary-General had had to act as quickly as he could to
produce a balanced composition of contributions from the assistance immediately forthcoming. He and the Commander of the
Force (Major-General E. L. M. Burns) were at present giving
priority to the administrative difficulties which had to be faced as
the first units of the Force moved into Egypt.

'It seems possible that the Emergency Force as constituted at
present will be sufficient to undertake the first stage of its operations; Mr. Hammarskjold had in his communication to the New
Zealand Government expressed the view that it would be unwise
to consider making additions to the Force until the units now
represented had been moulded into an integrated whole and until
greater clarity prevailed concerning the proper size of the Force.'

Mr. Holland concluded by expressing the appreciation of the
government at the excellent response made to the call for volunteers for a New Zealand contingent. The numbers were more than
sufficient for the purpose. He thanked all those who had volunteered and wished to assure them that, should further discussions,
which the Secretary-General wishes to hold, necessitate the
despatch of a New Zealand Contingent, men who had volunteered
would be given ample time before being called into camp. The
situation in the Middle East was, to say the least of it, in a fluid
state, and it was difficult to foresee the outcome at this stage.

8. Extracts from statement by the Prime Minister, Mr. S. G. Holland, 4 December 1956[1]

Now that the United Nations has organized and landed a
substantial part of its Emergency Force, the United Kingdom are
in my view justified in handing over to the Secretary-General
responsibility for the attainment of the objectives of British policy
in the Middle East. It is now possible for Britain to withdraw its
forces without creating a dangerous vacuum in the area. . . .

New Zealand has never wavered in its support of Britain because
it has never wavered in its belief in the sincerity of Britain's
motives. Briefly, these were to bring about a cessation of hostilities
between Israel and Egypt, to prevent a local conflict spreading into

[1] *E.A.R.*, vol. VI, no. 12, pp. 2–3.

a wider conflagration and to ensure at the same time the maintenance of the Suez Canal as an international waterway free to all
nations.

It is my earnest hope, and I am sure it is shared by all New
Zealanders, that British confidence in the United Nations is not
misplaced, and that it will in fact complete the work of pacification
and settlement in the Middle East. This will be a testing time for
the United Nations, and I am sure that the New Zealand people
will give it full support. Our own offer to help still stands. . . .

V. South Africa

1. Statement by the Minister of External Affairs, Mr. Eric Louw, to the General Assembly of the United Nations, 24 November 1956[1]

I am addressing myself to the amendment moved by the Foreign
Minister of Belgium.[2] I wish to state the attitude of the Union of
South Africa on that amendment in relation to the issues raised by
the original draft resolution.

It will be recalled that throughout the debates on the situation in
the Middle East and Suez during the first emergency session the
South African delegate refrained from participating in the substance of the discussion and also abstained in the voting. The line
taken by my delegation was in accordance with the attitude of the
South African Government when the Suez Canal Company was
nationalized by the Government of Egypt.

South Africa is deeply concerned with the maintenance of peace
in the Middle East, which is the gateway to Africa. We are, however, not a user of the Suez Canal, and for that reason we did not
regard ourselves as involved either in the nationalization of the Suez
Canal Company or in the subsequent events, including the landing
of the British-French military force with the express intention
of protecting the Canal. Now, however, we are faced with the

---

[1] G.A.O.R., Eleventh Session, 1956–7, I, p. 299.
[2] The Belgian amendment, introduced by M. Spaak on 24 November,
proposed replacing paragraphs 1 and 2 of A/3385, Rev. 1, by the following:
'*Notes* that, according to the information received, one-third of the French
forces has been withdrawn, the United Kingdom Government has decided to
withdraw one infantry battalion immediately, and Israel has withdrawn a part of
its troops, and considers that France, the United Kingdom and Israel should
expedite the application of the resolutions of 2 and 7 November in the spirit in
which they were adopted, particularly with regard to the functions vested in the
United Nations force.'

realities of a new situation, which cannot be ignored. British-French forces are in occupation of Port Said, Israel forces hold the Sinai peninsula and a United Nations force is moving into the area of conflict for the purpose of securing and further supervising the cease-fire order by the General Assembly and performing other tasks and functions entrusted to it by the Assembly.

The General Assembly has before it the twenty-Power draft resolution of the Asian and North African countries which, although now slightly amended, still calls upon the British-French forces to withdraw forthwith and immediately. In response to that draft resolution, the British Foreign Minister has given to the Assembly [I, Document 6, *supra*] specific and unequivocal assurances that the British-French forces will be withdrawn as rapidly as the United Nations Emergency Force is in a position to carry out the tasks assigned to it by the General Assembly.

The South African delegation is prepared to accept those assurances, and is of the opinion that the Assembly should equally accept them. We doubt the wisdom of adopting a further draft resolution, such as is contained in document A/3385/Rev.1. We do not believe that the adoption of such a text or even of the amended draft resolution will contribute to a practical or constructive solution of the difficulties, and may, in fact, aggravate them.

Without expressing any views regarding the action taken by the General Assembly in sending an Emergency Force to the Suez Canal area, or on the question whether it is likely to have the desired effect, the South African delegation is none the less not prepared to doubt the assurances given by the British Foreign Minister. Having regard for the realities of the present situation, we consider the amendment moved by the Belgian delegation to be a great improvement on the revised draft resolution. Therefore, without prejudice to South Africa's attitude in regard to the resolution mentioned in the Belgian amendment, we are prepared to support that amendment.

VI. INDIA

1. Letter from the Permanent Representative to the United Nations, Mr. Arthur Lall, to the Secretary-General, Dag Hammarskjold, 6 November 1956[1]

I have the honour to confirm the agreement of the Government

[1] G.A.O.R., Doc. A/3302/Add.4, First Emergency Special Session, 1956.

of India, in principle, communicated to you today by Mr. Krishna Menon, to participate in the United Nations international emergency Force to be organized in accordance with the resolutions of the General Assembly of 4 and 5 November 1956 with a view to secure and supervise the cessation of hostilities in accordance with the terms of the resolution of the General Assembly of 2 November 1956.

In conversation with Mr. Krishna Menon on 5 November 1956 you were good enough to set out for his information the conditions and circumstances of participation by Member States in such a Force. These were communicated to the Government of India who have accepted them as the context and conditions in which they are invited to participate. You will also recall that the formulation of them as in the enclosure [see below] was confirmed by you this morning to Mr. Krishna Menon.

We would be glad to know what immediate steps are now required to implement our acceptance in principle. The Government of India are prepared to use their best efforts for the implementation of their participation as soon as the details are settled and on the assumption that the General Assembly will approve the plan.

[Enclosure]

United Nations Emergency Force
1. The emergency Force is set up in the context of the withdrawal of Anglo-French forces from Egypt and on the basis of the call to Israel to withdraw behind the armistice lines.
2. The Force is not in any sense a successor to the invading Anglo-French forces, or in any sense to take over its functions.
3. It is understood the Force may have to function through Egyptian territory. Therefore, there must be Egyptian consent for its establishment.
4. The Force is a temporary one for an emergency. Its purpose is to separate the combatants, namely, Egypt and Israel, with the latter withdrawing as required by the resolution.
5. The Force must be a balanced one in its composition.
6. The agreement would be in principle and the position in regard to actual participation is reserved until the full plan is before us.

It is understood that the size of Indian participation is about a battalion strength.

It is also understood that transport, including airlift and all facilities, will be provided by or through the United Nations.

2. Statement by the Government of India, 6 November 1956[1]

The Government of India have received an invitation from the Secretary-General of the United Nations to associate themselves with the International United Nations Force for the cessation of hostilities and the maintenance of the old Armistice line between Egypt and Israel.

This invitation is supported by the Egyptian Government. The Government are reluctant to send their armed forces outside India. But in view of the grave crisis and with a view to help in some peaceful solution, they have agreed, in principle, to send a limited armed force.

They have, however, made it clear that their active participation can only be decided upon after seeing the final plan which must have the consent of the Egyptian Government. Also the force should be set up in the context of the withdrawal of the foreign forces which have entered Egyptian territory.

The Government cannot agree to any such force being sent if this is in any sense a recognition of aggression.

3. Extracts from speech by the delegate to the United Nations, Mr. V. K. Krishna Menon, to the General Assembly, 7 November 1956[2]

. . . My Government desires me to state before the General Assembly—because I have just come from India—its general reaction and attitude towards the present difficulties.

Our country, our people and our Government, indeed our whole part of the world, have been shocked by the developments that have taken place in relation to Egypt. We desire to state without any superlatives that we regard the action of Israel as an invasion of Egyptian territory, and the introduction of the forces of the United Kingdom and France as an aggression without any qualification. We have said this from the very beginning.

We received this news not only with a sense of shock but also with a great deal of sadness. I say 'with a great deal of sadness' for two reasons, first because of the great physical harm that is being inflicted upon the people of Egypt, and secondly because those engaged in a part of this aggression, namely the Franco-British

[1] *The Hindu* (Madras), 7 November 1956.
[2] G.A.O.R., First Emergency Special Session, 1956, pp. 116–18.

alliance, are countries and people with whom we have very close kinship. That is particularly so in the case of the United Kingdom. It is therefore more in sorrow than in anger that we approach this problem. During the last several weeks of great anxiety, we have tried to exercise our influence by way of persuasion to change the course of these developments. I think that I should leave this aspect at this stage, stating the minimum that is required. . . .

Since we are now embarking on a very serious phase of activity on behalf of the United Nations, and since those countries that would be contributing troops or in other ways contributing to the maintenance of the international police force would be taking on responsibilities, it is essential that clarification should be sought and made and interpretations given.

I am directed by my Government to deal with this matter in that way. Two or three days ago, when this question was discussed between the Secretary-General and our Government, he very kindly gave us his view of the context and conditions in which any participation of countries in the International United Nations Force would take place. My delegation has formulated these conditions [Document 1] and further discussed them with the Secretary-General, and we understand that they represent the accurate set of conditions and circumstances in which such a force would function; and we would like this to be put into the record. . . .

I have been instructed by my Government to communicate to the Secretary-General—which I have done—that the Government of India would be willing to participate in the United Nations Force contemplated by his report in the context of the conditions that I stated and would be willing, when the arrangements are made final and the consent of the Assembly received, to send officers immediately to enter into consultations with the Secretary-General in regard to the details. Of course, when we say 'immediately', we are taking into account the fact that communications between India and this country have now been somewhat interrupted by the damage done to the airfields in the Middle East.

The Government of India has been able to accede to the position that its contribution would be something in the order of a battalion in strength. I am instructed to say that transport facilities, including airlift, would have to be provided through the United Nations, because it is not possible for us to transport this body of troops and equipment on such short notice without outside assistance. It will be possible to implement such agreement as we may make in a very short period, not more than ten days from the day of agreement, although it may be possible to send advance bodies beforehand. I have already made this communication to the Secretary-General and it is among the papers that have been circulated.

I want to draw the attention of the Assembly to the Secretary-General's report which is contained in document A/3302. My delegation is glad to note that it is specifically stated in paragraph 4 that the authority of the United Nations Commander would be so defined as to make him fully independent of the policies of any one nation. . . .

Now we come to paragraph 8 of the report and I should like the Assembly to look at the last two sentences of paragraph 8. Here the difficulty is not perhaps so much one of substance, but in a matter of this kind, in view of the not always happy experience that my Government has had in dealing with similar problems in Korea and in Indo-China, clarification beforehand is essential in order that work may proceed without impediments, and also so that we may not have difficulties with other participants. Paragraph 8 says the following:

It follows from its terms of reference that there is no intent in the establishment of the Force to influence the military balance in the present conflict and, thereby, the political balance affecting efforts to settle the conflict. By the establishment of the Force, therefore, the General Assembly has not taken a stand in relation to aims other than those clearly and fully indicated in resolution 997 (ES/1) of 2 November 1956.

I confess I am a little perplexed by this statement. I do not know quite what its implications would be. If it means that the United Nations Force is not intended to support the parties in the aggression or to intervene militarily, then I understand it. But if its meaning is that the occupation forces would remain where they are and therefore that their military balance would not be affected then of course it is totally contrary to its purpose. I draw attention to this because as the statement stands it is a little perplexing.

The Secretary-General will bear with me. We have had difficulties of this kind—I would not say without number, but quite a number—both in Indo-China and in Korea, of a very serious character. My Government would have grave apprehensions in undertaking military obligations which are subject to the interpretation of foreign office lawyers of different countries afterwards. Therefore, I think that some clarification of this paragraph would greatly relieve our minds.

The Secretary-General has told us that this is the second and final report, but I suppose that that is only a procedural description, because it goes on to say, in paragraph 11: 'However, the general observations which are possible should at this stage be sufficient.' Therefore, my delegation wants to be assured that there is no finality about this report in the sense that it is a kind of army manual in regard to these forces. There, again, it is not because we want to be punctilious, but because we have been once bitten, twice bitten, and are three times shy at the moment. . . .

4. Extracts from speech by the delegate to the United Nations, Mr. V. K. Krishna Menon, to the General Assembly, 26 November 1956[1]

. . . The invasion of Egypt by the United Kingdom Government stands on a par with the attack on Alexandria in 1880 and the occupation of Egypt thereafter. This must be clearly understood, and we keep on reiterating these things because in the last week, in the United Kingdom, in this country and in this Assembly and elsewhere, there has been an attempt to describe this action as though it were a service to the world. The moment we permit this halo to get around it, to portray an act of aggression as an act of morality, we shall be unable to take any corrective action.

Mr. Lloyd further told us [I, Document 6, *supra*] that the action we were taking in regard to the United Nations Emergency Force had been first conceived by the Prime Minister of the United Kingdom, that the idea had been repeated in this Assembly by Sir Pierson Dixon, and that thereafter it had been put before the Assembly by the Foreign Minister of Canada. I must say that we are rather taken aback by this. We had accepted, we continue to accept the actions of the Canadian Government as taken in good faith, as measures arising from themselves, not as part of the policies of the two aggressor countries. Mr. Lloyd's statement treats the Canadian proposal as though it were part and parcel of

[1] G.A.O.R., Eleventh Session, 1956–7, I, pp. 330–3.

Anglo-French foreign policy. The Canadian Government can make its own explanations and defend itself.

So far as our Government is concerned, the proposal for setting up an emergency force was a conception—there is nothing unusual about it—put forward by the Foreign Minister of Canada as one of the ways to bring about a cease-fire. . . .

. . . To suggest in any way that the withdrawal is dependent upon the judgment of the United Kingdom and French Governments as to the competence of the United Nations Emergency Force, is again to seek to usurp the powers of this Assembly. Who are these two Governments to make their own judgments? They can no more make judgments about the action of the Assembly by themselves than we can—and we do not claim that right. Therefore, whether this United Nations Emergency Force is competent, is a matter for General Burns on the one hand and the Secretary-General on the other. Constitutional responsibility for it rests in the Assembly, and my delegation denies the right of the Governments of the United Kingdom and France to appropriate to themselves the right to say that this Force is competent for any one purpose or another.

But the position becomes much worse when we go into the substances of this competence. Competent for what? Competence to perform the duties that the Anglo-French invaders were supposed to be attempting? In other words, the view expressed in the statement before this Assembly is that the United Nations Emergency Force is a continuation of the invading forces. It is to perform the part of putting what is called a protective shield between the combatants, of staying there for the solution of various problems, of preventing conflicts in the sense they understood it—and therefore hallowing the aggression.

I hope that the Assembly will at no time lend itself to a position where by various and devious methods it is called upon to give its blessing to invasion. . . . I submit that the competence of the United Nations Emergency Force is entirely a matter for the United Nations. The Governments of France, the United Kingdom and Israel will contribute in that judgment in the proportion of their power here, namely, each as one sovereign State amongst seventy-nine nations. That is our position. . . .

. . . Our understanding of . . . the basis on which the United Nations Force is organized, is that there can be no violation of

Egyptian sovereignty. It is the sovereign right of every Government to admit to its territory whom it likes and to refuse to admit those whom it does not like. It is equally the prerogative of the Assembly to determine the composition of the Force. . . .

5. Extracts from speech by the Minister without Portfolio, Shri V. K. Krishna Menon, in the Lok Sabha, 26 March 1957[1]

. . . I now come to the more specific issues which have been raised during the debate. The first of these is the Middle East. . . . This is a matter in which Canada and we are commonly concerned. Incidentally, while the Prime Minister has spoken fully about the Commonwealth relations, I may as well draw the attention of the House to this fact that there were two Commonwealth states— Canada and India—that stood against Britain and Australia and others in regard to the situation in the Middle East in the United Nations. So, there is no question of inhibition in regard to judgments or policies. . . .

. . . When India agreed to participate in the United Nations Emergency Force, it made it very clear and categorical, and the assurance was given, which was endorsed by the United Nations, that the forces would not at any place be asked to violate the sovereignty of Egyptian territory; they would not take up the functions of invading forces; they would not be armies of occupation. We and the Canadian Government, according to the speech made in the House there [II, Document 12, *supra*], have slight differences in this matter in that the Canadian Government apprehends that when the United Nations Emergency Forces are placed on the Armistice line, they must feel assured that there would be no attacks from the Egyptian side. So far as we are informed and so far as our knowledge goes, there is no need for this apprehension. . . . Mr. Lester Pearson has said that the armistice agreement is the basis on which these territories rest and that is our position also. But to put in foreign forces in this [Gaza] strip, which is *de facto* Egyptian territory but whose status *de jure* is unsettled, would be a violation of the Armistice Agreement of 1949. Our country has always said that we would not take over the powers of occupation. Over and above all that, what we are all looking for is a peaceful settlement of this matter and if it were possible, the

[1] Lok Sabha Deb., 1957, Part II, vol. I, coll. 800–2.

establishment of at least as much of non-conflict on this border which may lead gradually step by step towards a wider settlement of the question that involves the two countries and the other Arab states. Therefore, the functions of the United Nations forces are of a neutral character. Any police functions, except as requested by the Egyptian Government, would involve them in trouble. . . .

VII. PAKISTAN

1. Letter from the Permanent Representative to the United Nations, Mr. Mir Khan, to the Secretary-General, Dag Hammarskjold, 5 November 1956[1]

I have the honour, upon instructions from my Government, to inform you with reference to resolution A/3290 adopted on 5 November 1956 at the first emergency special session of the General Assembly (resolution 100 (ES-1)) that the Government of Pakistan offers to contribute a contingent of Pakistan armed forces for the emergency international United Nations Force.

2. Speech by the delegate to the United Nations, Begum Shaista Ikramullah, to the General Assembly, 23 November 1956[2]

Pakistan has once again co-sponsored a draft resolution [A/3385] demanding the immediate withdrawal of the invading forces from Egypt. It is imperative that this should be done immediately to restore the shattered confidence of the world in the principles of the Charter. The attack on Egypt has destroyed the belief that has been laboriously built up that the era of aggressive force is on the wane. The shock has been greater because one of the parties to this aggression has been the United Kingdom, which since the last war has been one of the foremost countries that had worked for a new morality in international affairs and which, by gracefully accepting the liquidation of an empire and welcoming in its stead the free co-operation of like-minded nations which is the Commonwealth, seemed to have abjured force. Therefore, it is imperative and in the interest of the United Kingdom, that this unfortunate reversion to imperialistic tactics be rectified immediately and that the United Kingdom, together with France and Israel, withdraw their troops forthwith from Egypt. And it is incumbent upon the

[1] G.A.O.R., Doc. A/3302, Annex 6, First Emergency Special Session, 1956.
[2] G.A.O.R., Eleventh Session, 1956-7, I, pp. 270-1.

United Nations to see that this is done. The United Nations has, by overwhelming majorities on 2 and 7 November, directed that this should take place. If the United Nations declarations are to have any force and meaning in the world, the Organization should see that its decisions are implemented and complied with.

For the last few years a feeling of disillusion had been growing amongst the smaller nations of the world regarding the United Nations. They had begun to feel that this Organization that came into being with such high hopes and such faith, born out of bitter suffering and great trial, was after all nothing better than holy alliances for the unholy purposes of the past and that it was almost futile to hope that justice regardless of power politics could be had at the hands of the United Nations; but by taking at last a bold and prompt action in the case of Israel and British-French aggression, the United Nations has redeemed itself. It has restored the faith of the small peoples of the world in its integrity. That the United Nations could condemn the action not only of its protégé Israel, but of two of the permanent members of the Security Council, has generated a new wave of hope, and peoples and nations have once again begun to look to it for justice.

If now the United Nations fails to see that its resolutions are complied with, it would lose its new-found strength. It must see that its resolutions do not join the archives with the other un-implemented resolutions of the Security Council and the General Assembly. The time has come when it must make clear that the United Nations resolutions are not mere pious declarations, but are meant to be obeyed and applied without fear or favour.

My delegation has co-sponsored a draft resolution asking for withdrawal of foreign troops from Egypt because we are against all foreign troops and troops of occupation anywhere, under any pretext by anybody. We condemn aggression and suppression of liberty equally in Egypt and in Hungary, and in Algeria and in Kashmir. We are opposed to the last-ditch stand of the waning imperialism of Europe and we are equally determined to oppose the nascent rise of imperialism in Asia. As we have said before from this rostrum, we are against imperialism of all types and colours, white or red, black or brown. Our objection to imperialism is deep and sincere and real, and springs out of a genuine hatred of domination and a genuine love of liberty, out of faith and determination that all small nations of the world have the right, and shall have the right to exercise the right, to decide their fate, and determine the destiny of their own country according to their own choice. Our condemnation of aggression is not determined by any 'isms'. It is not coloured by other considerations except those of abstract and absolute justice and morality.

3. Extracts from statement and answer to question by the Minister of Foreign Affairs and Commonwealth Relations, Mr. Malik Firoz Khan Noon, 7 December 1956[1]

. . . We must now create a U.N. police force as is already laid down in the Charter. But the first person who will oppose the creation of such a police force will be Pandit Jawaharlal Nehru, because it does not suit his policies. . . . He is definitely out to destroy smaller countries, and cannot support the creation of an international police force which will give protection to these small nations.

But I have no doubt in my mind that if this international permanent police force is not created, the U.N.O. will be a complete failure, like the old League of Nations. I hope sincerely that the United Nations Assembly will create such a force immediately. . . .

Q. Do you think that this force should have the right to move into a country against the will of that country?

A. Yes, in keeping with the Charter. . . .

VIII. CEYLON

1. Extracts from speech by the Permanent Representative to the United Nations, Mr. R. S. S. Gunawardene, to the General Assembly, 3 November 1956[2]

. . . We are truly grateful to the representative of the United States for his laudable objective in trying to secure an over-all settlement of the problem, but we do not normally think of the disposition of property when a patient is dying, is gasping for breath. Egypt is a patient gasping for breath, and we cannot ask Egypt to sign its last will before we are satisfied that Egypt is in a sound condition and able to take a reasonable course of action.

In the position in which Egypt is placed today, I say that our first duty is to see that Egypt is in a condition to exercise its

---

[1] *Dawn* (Karachi), 8 December 1956.
[2] G.A.O.R., First Emergency Special Session, 1956, p. 65.

sovereign rights. When that position has been created, then it will be the time to consider how best the Suez Canal dispute should be decided and how best the Israel-Arab dispute should be disposed of. I freely grant that a solution must be found. This cannot go on for long, but at the same time the present moment certainly is not the occasion for the consideration of that problem. The problem before us is simply how we can secure a cease-fire, how we can secure a cessation of hostilities, and how we can put Egypt into a *status quo ante.* That is exactly the position in which we are placed.

The draft resolution introduced by the representative of India,[1] which Ceylon and other countries of the Asian group had the privilege and honour of co-sponsoring, gives the Secretary-General at least some time—it may be twelve hours, it may be eighteen hours, just as the Secretary-General would require—to see if something more can be achieved during that period to bring these nations to reason, at least reason as it is accepted by the rest of the world. Sometimes when people embark upon a course of action, they find it difficult to retreat, and obstinacy sometimes becomes a virtue. In the present circumstances, I think that these three nations should be guided by the weight of world public opinion and by the moral judgment passed by this august body only the other day.

The Secretary-General, who is well known for his tact, impartiality, ability and experience in handling this kind of intricate problem, will perhaps find time, a little breathing space, to see whether other methods are possible.

I hope that this Assembly will find no difficulty in accepting the draft resolution sponsored by the nineteen Asian countries. It merely requests the Secretary-General, with the assistance of the Chief of Staff and members of the United Nations Truce Supervision Organization, to obtain compliance with the request for the withdrawal of all forces behind the Armistice lines. In other words, this Assembly would decide to give them a second chance to reconsider the position they have taken in the light of the discussions that are taking place this evening.

I see no difficulty in accepting at the same time the draft resolution submitted by the representative of Canada [II, Document 2, *supra*]. One does not conflict with the other. Far from conflicting, this resolution would help the first resolution. The carrying out of the objectives of the draft resolution introduced by the Asian countries

---

[1] For the text of the 19-power resolution, see p. 321, n. 2.

would be greatly helped by the draft resolution introduced by the representative of Canada.

. . . I hope that this Commission, this police force—I do not know how to describe it—will not include representatives from Israel or the United Kingdom or France. That goes without saying. So long as this task is undertaken by States which are not parties to the dispute, to this unfortunate situation, then I think that this august Assembly should also be able to accept the draft resolution introduced by the representative of Canada. . . .

2. Letter from the Permanent Representative to the United Nations, Mr. R. S. S. Gunawardene, to the Secretary-General, Dag Hammarskjold, 6 November 1956[1]

With respect to the proposed international United Nations Force referred to in resolution A/3276, approved by the General Assembly on 4 November 1956 (resolution 998 (ES-1)), the Ceylon Government will be glad to make an appropriate contribution, the details of which will be communicated to you shortly, subject to the required constitutional action which will be put in motion without delay.

3. Letter from the Permanent Representative to the United Nations, Mr. R. S. S. Gunawardene, to the Secretary-General, Dag Hammarskjold, 9 November 1956[2]

I confirm the offer I made of Ceylon's participation in the emergency international Force under United Nations command (A/3302/Add. 3) [Document 2]. The Ceylon contingent will be composed of an infantry company and a small administrative staff totalling 150 men under the command of an Infantry Major. This contingent will be equipped with personal weapons and transport for one infantry company. It will include two motor-cycles, three jeeps and five 3-ton vehicles. It will be ready to move in approximately two weeks from now.

I shall be glad if you will inform me whether my Government's offer is accepted. If it is accepted I trust that arrangements for transport from Colombo of the men and equipment would be made by you.

---

[1] G.A.O.R., Doc. A/3302/Add. 3, First Emergency Special Session, 1956.
[2] G.A.O.R., Doc. A/3302/Add. 15, First Emergency Special Session, 1956.

I presume that logistical support for this contingent will be provided by you.

4. Extracts from speech by the Prime Minister, Mr. S. W. R. D. Bandaranaike, to the General Assembly of the United Nations, 22 November 1956[1]

. . . I must tell the Assembly that it is my view, and the view of my colleagues, the other Asian Prime Ministers, that the position is still extremely delicate and dangerous. We do not feel that there is any occasion for undue complacence. I am glad that, substantially, a cease-fire has taken place. But the withdrawal has not yet taken place. A United Nations Emergency Force is already in the process of being established in Egypt in order to carry out the decisions of the United Nations in supervising a withdrawal of these forces. I say this, and I say it with all seriousness, that as long as foreign troops—be they Israel, United Kingdom or French —continue to remain on Egyptian territory, the position is one that is fraught with the greatest danger and one that may bring about results leading to a third world war. I wish to say that those forces must be withdrawn now, without any delay. I wish to say that I think it would be very unwise to follow some principle of a phased withdrawal, a withdrawing of those forces in numbers according to the numbers of the United Nations Emergency Force who enter: for example a hundred United Nations troops going into Egypt and a hundred being withdrawn; two hundred United Nations troops going in and two hundred being withdrawn. There can be no greater mistake than that.

The moment that even a token United Nations force is established on Egyptian territory, it will be sufficient occasion, in the interests of us all, for Israel forces to be withdrawn behind the armistice line, and United Kingdom and French forces to be withdrawn from Egyptian territory. I cannot conceive that either Egypt or Israel would make an assault upon forces of the United Nations. I just do not believe it. So that it is really not required for a large force of the United Nations to be present before those forces are withdrawn. The first and the most vital thing is a withdrawal of forces from Egyptian territory now, as early as possible. If that does not happen, even if under the guise of 'volunteers'— and we know what 'volunteers' mean—other countries, in order to

[1] G.A.O.R., Eleventh Session, 1956–7, I, p. 232.

secure the observance of the decisions of the United Nations, take steps, I fear that the results may be very far-reaching and all our efforts so far be swept away in a moment. . . .

5. Extract from statement by the Prime Minister, Mr. S. W. R. D. Bandaranaike, in the House of Representatives, 11 December 1956[1]

. . . I had a talk with the Secretary-General. What he told me was that every country was prepared to send infantry but that it was not possible to maintain such a force without any other supporting service. He asked us to give if possible an armoured unit—signallers, engineers, and so on. I said that that would be rather difficult for us to supply with our small, rather rudimentary force, whereupon he informed me that at present there were sufficient numbers of infantrymen but the U.N. force was in the process of being built up and that we would be informed as early as possible so that we could send a certain number of infantrymen if necessary. But it was a very specialised type of service that he urgently wanted. I understood from the Prime Minister of India that even he was unable to offer units of all types so as to comprise a cohesive whole. All the countries wanted to send infantrymen. There was indeed a plethora of such offers. But what was really wanted were specialists in order to assist in the build-up of the Force. It was difficult for us to supply armoured units, and so on.

I might also add that this offer was in pursuance of a request urgently made by the Secretary-General to our representative at the U.N. which was transmitted here. We did not make the offer on our own. The request came from the U.N. and must have gone to so many other countries. It was in pursuance of that request that we made this offer.

[1] Ceylon, H.R. Deb., 1956–7, vol. 27, col. 809.

PART V
COMMONWEALTH RECKONINGS

It was because Canada values the Commonwealth, because the world needed the Commonwealth, and because we wanted no nation to leave this association, that Canada was so concerned last November. . . .

Mr. L. S. St. Laurent, 16 May 1957

There is enough breakage in this world not to add to it. . . .

Shri Jawaharlal Nehru, 7 December 1956

We cannot carry opposition and venom for all time.

Mr. H. S. Suhrawardy, 22 February 1957

COMMENTARY

I

SIR ANTHONY EDEN was compelled by illness to leave the United Kingdom after three weeks of incessant diplomatic and political activity following the Anglo-French ultimatum. He sought temporary refuge and respite in Jamaica, but these did not restore his health. Soon after his return to Britain, illness compelled his resignation as Prime Minister on 9 January 1957. Mr. Holland extended an invitation to convalesce in New Zealand; it was gratefully accepted. A farewell message issued from shipboard on 18 January stated once more, and for the last time, Sir Anthony Eden's unshaken conviction that 'the difference between the West and Egypt has not been colonialism; it is the difference between democracies and a dictatorship'.[1]

This curious comment (of all comments possible, it was noted at the time, 'this was assuredly the most irrelevant'[2]) aids at least in understanding Sir Anthony Eden's final and tragic act of statecraft. That act 'seemed to many like a desperate striving after heroic virtues with which Nature had not endowed him; it seemed out of character even to those who approved it. The experience of Suez seemed to carry his mind back to the days of his youthful glory when he tried to rally the world against Mussolini, in whose role he now saw Nasser'.[3] Here was a parallel as fateful as it was inept, for it caused the Prime Minister of the United Kingdom to embark upon and hold to a course which more than a few of his countrymen thought criminal and a great many more than these mistaken. But to as many more again, perhaps to more than that, 'the last phase of Sir Anthony's political career [appeared] as the most striking instance of personal courage in their memory, and there are those who suspect that history, while not congratulating the Government on its competence, may see its Suez policy as a worthy attempt to uphold a vital national interest against the overwhelming force of prejudiced allies and factious political opponents'.[4] History, however, has yet to deliver its verdict.

Sir Anthony Eden's resignation left the Conservative Party

---

[1] *Full Circle*, p. 584.   [2] *The Round Table*, vol. 47, 1956–7, p. 200.
[3] Ibid.   [4] Ibid.

without a leader and the United Kingdom without a Prime Minister. In these circumstances it fell to the Sovereign to select a successor from among the obvious candidates, of whom there were on this occasion two. The Queen, guided by the counsel of Sir Winston Churchill and Lord Salisbury, chose Mr. Harold Macmillan rather than Mr. R. A. Butler. Her choice, in retrospect, appears both wise and inevitable; at the time it caused surprise.

There were contrary interpretations of the significance of the new Prime Minister's succession.

To some, Mr. Macmillan's appointment instead of Mr. Butler meant a victory for the Right and presaged a tough policy designed to rescue as much as possible from the wreck and to maintain as much independence as was possible in the pursuit of British interests; to others, Mr. Macmillan seemed to have been chosen to perform a graceful somersault and to lead the country in a gay and elegant manner towards the acceptance of the status of a second-rate power. When two opposite views of the nature of a statesman's task prevail within his own party . . . he will be careful, if he is wise, to postpone the day when what he is after becomes unmistakably clear. This is what Mr. Macmillan is doing.[1]

But his immediate task upon becoming Prime Minister required no concealment. He had to restore, with all possible speed and efficiency, the bonds of Commonwealth and trans-Atlantic unity.

The Bermuda Conferences of March 1957 brilliantly began the restoration of both. From 21 March to 23 March, Mr. Macmillan met with the President of the United States. Their discussions were conducted, according to their brief communiqué, 'with the freedom and frankness permitted to old friends in a world of growing interdependence'. As tangible evidence of interdependence, President Eisenhower undertook to provide the United Kingdom with ballistic missiles of intermediate range, and Mr. Macmillan to receive them. The rift of the previous months was not to be allowed permanently to impair the relations of the United States with its British ally, and the task of reconciliation was considerably helped by the fact that General Eisenhower and Mr. Macmillan had worked closely together during the Second World War. The Bermuda meeting was received with approbation by the British Press and public, which perceived that it had effectively eased the strain of the immediate past.

[1] *The Round Table*, vol. 47, 1956–7, p. 269.

A day or so after the President's departure, Mr. Macmillan played host to three members of the Canadian Government. The meeting of Messrs. St. Laurent, Pearson, and Howe with the British Prime Minister could not properly be described as a reconciliation, for throughout the crisis the British and Canadian Governments had kept in amiable contact, notwithstanding their differences of approach. 'Our relations with the United Kingdom', Mr. Pearson recalled the following summer, 'remained close and friendly throughout all the difficult days and hours of last autumn. Our delegations [at the United Nations] were in close touch and even when we disagreed we talked things over. I have spent hours with them trying to see how we could work things out together.'[1] For their part, the members of the British group at Bermuda 'went out of their way to say that there was great appreciation in Britain for all that Lester B. Pearson . . . had done for the Mother Country in the United Nations, "especially in the last month" '.[2] This aura of good feeling did not entirely dispel an evident disagreement between the British and Canadian leaders over the place and priority of the United Nations in the foreign policies of their respective nations. For Canada, the Suez crisis had if anything vindicated the importance attached by its Governments since 1945 to the world organization; for the United Kingdom it had no less confirmed deeply seated suspicions. Dispatches from Tucker's Town reported that the different evaluations of the United Nations became evident soon after the Anglo-Canadian discussions began, and that both sides held fast.[3] But the disagreement was not allowed to mar the harmony of the proceedings. To 'the value of the family relationship between the peoples of the Commonwealth', the Anglo-Canadian communiqué dutifully paid tribute, as it did to the 'special relationship' of the two Governments which, it insisted, 'will always enable them to work together effectively with a constructive purpose'.[4] The full restoration of the special relationship was certainly not impeded by the United Kingdom's agreement at Bermuda (announced in an Annexe to the communiqué) to purchase from the Canadian Government's agency, Eldorado Mining and Refining Ltd., uranium to the value of about $120 millions over the next five years.

[1] See below, pp. 433–4.
[2] 'Canada, Britain differ over U.N.', *New York Times*, 26 March 1957.
[3] Ibid.   [4] See below, p. 427.

All this Mr. Macmillan achieved without in any way repudiating his predecessor, to whom, on 15 May 1957, he paid a warm and moving tribute.[1] The tribute was strangely unreciprocated; a reviewer of Sir Anthony Eden's Memoirs noted that Mr. Macmillan was the only Minister or official serving under their author to whom they do not offer an appropriate word of recognition or gratitude.[2]

The opponents of Sir Anthony Eden's policies had predicted that every sort of disaster would result from them. Their most dire prophecies were confounded. The Commonwealth did not disintegrate, though disintegration might have followed had the resort to force against Egypt been allowed to continue. The trans-Atlantic bridge from London to Washington, though severely damaged, was restored with remarkable speed. The prestige of the United Kingdom, though shaken, was not greatly set back for long. Even Anglo-Egyptian relations, which might have been expected to remain in disrepair for the duration of Conservative rule, had to most outward appearances returned to normal only two or three years after Egyptian civilians had been machine-gunned in the streets of Port Said.

In the United Kingdom itself, there was a tendency, understandable enough, to treat the Suez crisis as an unpleasant, but momentary, aberration in Britain's affairs, and to forget the episode as quickly as possible. A by-election in Chester on 15 November offered the voters of that constituency the opportunity to pass judgement upon the Government. Speakers in the campaign reported that if Suez was on everybody's mind, it was far from being on everybody's tongue. The Conservative candidate remarked that 'what the electors are most interested in is me', and won the seat by a substantial, if sharply reduced, majority (6,348, as against the Conservative majority of 11,002 in the General Election of 1955), which really proved nothing at all. In the campaign for the General Election of 1959, candidates for all three parties found that raking Middle Eastern embers did them little good and less credit, and for the most part were content to leave them alone. 'The Conservatives ignored [Suez] because it had caused internal party strife; Labour was hesitant to stress the

---

[1] See below, p. 415.
[2] Martin Wight, 'Brutus in Foreign Policy: The Memoirs of Sir Anthony Eden', *International Affairs* (London), vol. 36, no. 3, July 1960, p. 300.

issue for fear that its criticisms made it seem "unpatriotic".'[1]
Aneurin Bevan was one of the very few to place the Suez issue at
the centre of his campaign. 'In the eyes of the world', he declaimed
on one occasion, 'the guilt was the guilt of Macmillan and his
friends. But if, on October 8, you vote for the Tories once more,
then the guilt will be yours.'[2] That was a burden which 13,749,830
British voters were evidently prepared to shoulder.

The Government was able to resist without difficulty the efforts
of members of the Labour Party to appoint a commission of
inquiry into its Suez policy of the kind that in previous years had
investigated the Jameson Raid or the Gallipoli Campaign; one
may surmise that there were not a few among the Opposition just
as thankful that there was no further exposure of its melancholy
failure. Four years afterwards a British commentator could write
of the Suez expedition that it already seemed 'remote and rounded
off, a failure of high drama, but small historical effect, less consequential internationally than the Mexican expedition of Napoleon III, less consequential domestically than the failure to relieve
Gordon at Khartoum'.[3]

Had it then no impact at all upon the nation which was the
principal actor in the drama? The shock of Suez upon the British
public was severe, but it was traumatic in its nature. It is the way
with trauma to lie buried in the sub-conscious, not emerging until
many years after the event and then affecting behaviour in unexpected ways. More than one student of recent British foreign
policy have seen Suez as a watershed, a great dividing of that
period when it was possible for a Prime Minister with a hold on
reality to pursue, independently of others, a British grand strategy,
from that period, ever afterwards, when it was not. Some have
argued that the failure of Suez set Britain's course for Europe:

Suez was without doubt Britain's postwar 'moment of truth'. In a
blinding lightning flash her essential weakness was revealed, militarily
and politically. The cosy concept of the 'interlocking circles' was cruelly
exposed. Faced with a choice between her ally Britain and the goodwill
of the Afro-Asian bloc, America unhesitatingly chose the latter. Other
than France, there was no support for the Suez operation from the
Continent of Europe. Even in the Commonwealth majority opinion condemned the British action in no uncertain terms. World disapproval

[1] D. E. Butler and Richard Rose, *The British General Election of 1959* (London, 1960), p. 64, n. 1.
[2] Quoted ibid., p. 54.   [3] Wight, op. cit., p. 309.

would not perhaps have mattered if we could have achieved our ends despite it. . . . But we could not. . . . We pulled out.

This humiliating defeat had a cathartic effect on British public opinion. One section took refuge in an angry xenophobia, burying their heads more fiercely in the sand. But most people felt, however confusedly, that a decisive turning point had been reached in Britain's relations with the outside world, necessitating fundamental changes of policy. And it was instinctively to Europe rather than the Commonwealth that people turned.[1]

Analysis of this kind is only superficially convincing. It is not only that it is at present (January 1963) far from certain that Britain has chosen Europe in preference to the Commonwealth, or, indeed, that these are in fact the alternatives. Nations, even less than individuals, gain insight into their situations by some 'blinding lightning flash', whether on the road to Damascus or on the road to Suez. That Britain was no longer a great power in the classical sense of being able to impose its will by force upon other powers was something that, six years after Suez, the British people and their Government were still unable to concede.[2] Suez was but a stage, albeit an important stage, in that process by which a nation, once great, readjusts itself to the fact that it is no longer as great as it was. The process is both protracted and painful. It began, for Britain, not in 1956 but in 1946, or even in 1919; and its end is not yet in sight.

II

A convention of the Canadian constitution requires Parliament to meet to approve a Government's decision to send troops abroad on active service, and a special session of Parliament was accordingly convened to approve the dispatch of Canadian forces for

[1] Michael Shanks and John Lambert, *Britain and the New Europe: The Future of the Common Market* (London, 1962), p. 27.

[2] As when, in December 1962, a former U.S. Secretary of State uttered some plain words about the place of the United Kingdom in the second half of the twentieth century. 'Great Britain', Mr. Dean Acheson declared, 'has lost an empire and has not yet found a role. The attempt to play a separate power role— that is, a role apart from Europe, a role based on a "special relationship" with the United States, a role based on being the head of a "Commonwealth" which has no political structure or unity or strength—this role is about played out.' *The Times*, 11 December 1962. For this assessment, Mr. Acheson was placed alongside those who had previously miscalculated Britain's position in the world —Philip of Spain, Napoleon, and Hitler—and placed there not by some reckless don or irresponsible journalist, but by the Prime Minister of the United Kingdom.

service with the United Nations Emergency Force in the Middle
East. Opinion in Canada (as recorded in the Commentary to Part
III) was deeply and even passionately divided on the merits of
the United Kingdom Government's action in Egypt, and on the
Canadian Government's response to its action. The debate in
Parliament took place at a time when public controversy was at its
fiercest stage. The debate lasted four days (26 November through
29 November); it was unusually intense; and it illuminated the
spectrum of national opinion with remarkable clarity.[1]

Nor did the debate subside when Parliament adjourned. 1957
was a general election year in Canada, and the record of the
Government during the Suez crisis attracted a good deal of attention on the hustings. Little was heard, either from Opposition
speakers or from audiences, rebuking the Government for its part
in creating the United Nations Emergency Force. But about
Canada's voting record at the General Assembly Liberal cam-
paigners heard a good deal, most of it sharply critical. There is
reason to believe that the Conservatives, reluctant to be drawn into
discussion on a subject with which the Government, through Mr.
Pearson, was so admirably equipped to deal, decided at the outset
not to press the issue, or at least to press it only at those points at
which the Government was clearly vulnerable. The most obvious
of these was Mr. St. Laurent's incautious reference to the 'supermen of Europe'; and of this, in its most unfavourable interpretation, Conservatives tirelessly reminded their audiences throughout
the campaign. In his first major election speech, delivered in
Toronto (the heart of old-fashioned Tory imperialism), Mr. John
Diefenbaker declared: 'In the tradition of this Party, we did and
do resent the British people being castigated and derisively condemned as those "supermen" whose days are about over.'[2] A week
or so later, speaking at Truro, Nova Scotia, Mr. Diefenbaker
departed from his prepared text:

The Liberals tell you they have brought peace to the world. Did they
do it? Did they do it by saying to Britain and the British people when
Britain had her back to the wall—I ask you this, do you agree with the
Prime Minister when he referred to those 'supermen' whose days are

[1] See below, pp. 416–25.
[2] Quoted in John Meisel, *The Canadian General Election of 1957* (Toronto,
1962), pp. 57–58.

about numbered? We do not believe that. We believe that the Commonwealth still has a responsibility for freedom in the world.[1]

Rhetoric of this kind so stirred his listeners that the Conservative leader resorted to it with increasing frequency and confidence throughout the remainder of his campaign. On 13 May, at Guelph, Ontario, in what was reported to be his 'most effective speech so far', they

responded with boos when Mr. Diefenbaker reminded them that during the Suez crisis last fall, Prime Minister St. Laurent had condemned the British and French as 'those supermen of Europe whose days are about over'. 'We believe', Mr. Diefenbaker added to cheers, 'that those "supermen" still have a great contribution to make if freedom is to be maintained.' And Canada, he added, which was going to take no stand on the Suez situation, found itself allied at the United Nations with Russia.[2]

At Richmond Hill, Ontario, towards the close of the campaign, Mr. Diefenbaker referred to the Liberal Government as 'the people who turned down our good friends the Israelis where, if we had left them alone, we would have had no trouble'; the Premier of Ontario, Mr. Leslie Frost, who had joined his colleague in federal politics on the Richmond Hill platform, expressed similar views.[3] Further west, however, Mr. Diefenbaker's reference to Middle East matters were less well received. In Minnedosa, Manitoba, there was less interest in Suez than in wheat; when he spoke soon afterwards in Portage la Prairie, in the same Province, he ingeniously if tenuously joined the two subjects: 'The Government had wasted no time in protesting to Britain over the Anglo-French invasion of Egypt. Why, then, hadn't it been just as punctual in protesting U.S. policies which could reduce wheat to an almost valueless commodity?'[4]

The Government could of course make a strong case for its part in the Suez crisis, and it did not try to dodge debate on the issue. The Leader of the Social Credit Party, Solon Low, came unwittingly to its assistance by intruding into the discussion a

---

[1] Quoted in J. H. Aitcheson, 'Canadian Foreign Policy in the House and on the Hustings', *International Journal* (Toronto), vol. XII, no. 4, Autumn 1957, p. 286.
[2] *The Globe and Mail* (Toronto), 14 May 1957.
[3] Ibid., 8 June 1957; *The Telegram* (Toronto), 8 June 1957.
[4] Quoted in Meisel, op. cit., p. 58.

palpably spurious contention—namely, that Mr. St. Laurent's telegram to Sir Anthony Eden of 5 November 1956 contained the threat that Canada might withdraw from the Commonwealth if the Anglo-French invasion were not called off. The prompt repudiation of this charge by the Prime Minister as 'ridiculous' and 'a downright lie' did not deter Mr. Low from expanding his accusation. It seems unlikely that the Conservatives watched his performance with approval, for none of their spokesmen was drawn into it. The Liberals, however, struck back strongly. Mr. Pearson defended his Middle Eastern policy in Victoria, British Columbia, on 15 May. A day later, in a speech at London, Ontario, Mr. St. Laurent vigorously rejected the charge that his Government had weakened Commonwealth ties. 'It was because Canada values the Commonwealth, because the world needed the Commonwealth, and because we wanted no nation to leave this association, that Canada was so concerned last November at the time of the Middle East crisis.'[1]

To what extent did dissatisfaction with the Government during the Suez crisis contribute to its unexpected defeat in June 1957? Post-election comment attributing the voters' desertion of the Liberal Party to their disapproval of its leaders' response to the Eden-Mollet adventure probably exaggerated the significance of the issue in determining the outcome of the campaign. It was true that eighty-two of the one hundred and twelve Conservative members in the new Parliament came from the predominantly British areas of Ontario and the Maritimes, and that—taking a closer look—the Conservatives increased their vote in Victoria, the most 'English' constituency in the Dominion, by more than 25 per cent. But this proves little if anything about the significance of Suez.

Ontario is no longer as predominantly British as it once was [a close student of Canadian voting behaviour has pointed out]. The swing towards the Conservatives was as great as elsewhere in the city in those Toronto constituencies where the recently enfranchised New Canadians and others of non-British origin could have effected the outcome. While the Conservative increase in Victoria was substantial, it was greater in four constituencies in the British Columbia region which are not as British as Victoria is held to be. In any case, there is no evidence to suggest that New Canadians reacted differently to Suez than older

[1] Quoted ibid., p. 58.

settlers. Those who have escaped from communism may have favoured a tough anti-Nasser policy.[1]

A study of the campaign in the populous constituency of York-Scarborough, most of whose 104,000 voters lived in Metropolitan Toronto, revealed that none of those asked for their opinions regarded any aspect of foreign policy as a significant election issue; if Suez was thought not to be crucial by the electors of York-Scarborough it was unlikely to be thought crucial elsewhere.[2] In French-speaking Quebec, where the Conservatives obtained only four of the sixty-six seats held by the Liberals since 1953, the Suez issue may have prevented a more severe defeat than the Government actually sustained; but it is hard to be sure.

Of more lasting importance than the place of Suez in the General Election of 1957 is that by being called upon to form a government the Conservative Party was able to regain its hold upon reality in foreign affairs, and only just in time. Twenty-two years in opposition, its rank-and-file dwindling, a succession of leaders cast aside, a woeful lack of expertise, all had combined to produce an outlook upon the world full of quirks and dangerous distortions. Deprived of time, incentive, and facilities for developing reasoned and well-informed positions on external policy, the Opposition indulged too much in thinking with its blood. The Conservative attacks upon the Government's Suez policies were not just rancorous, they were irresponsible, and in some cases downright foolish. It was no brash newcomer to the political scene but a seasoned and respected senior member of the Conservative Party who had accused the Government of Canada of 'knifing Britain and France . . . in the back'. Mr. Howard Green was soon given the opportunity to develop a less rabid and more reasoned *Weltanschauung*, and it was well for the nation that he proved himself to be the kind of man capable of benefiting by his opportunity. Reproached, when in office, for some *folie de jeunesse* when in opposition, Mr. Green, now Secretary of State for External Affairs, replied with engaging candour: 'I have learned a lot since then.' So, indeed, had the Government as a whole; and nothing that it had learned was more valuable than that there was more to

---

[1] Quoted in Meisel, op. cit., p. 225.
[2] Paul W. Fox, 'A Study of One Constituency in the Canadian Federal Election of 1957', *The Canadian Journal of Economics and Political Science*, vol. XXIV, no. 2, May 1958, p. 239.

Commonwealth relations than relations between Canada and the United Kingdom.

Suez also contributed, in a manner not altogether fortunate, towards a redefinition of Canada's role in world affairs. The initiative leading to the creation of the United Nations Emergency Force, in which Canadian diplomacy and personnel had played so large a part, encouraged the hope and expectation that henceforth it was the nation's destiny to participate on a regular and continuing basis in para-military peace-making of the kind so successfully attempted in the Middle East. In this notion there resided much wishful thinking. The brilliance of Mr. Pearson's performance in the Suez crisis obscured one of its essential features. It was a one-night stand, an exceptional turn staged amidst circumstances unlikely to recur. Beyond that, the tendency of Canadians to see their country as a sort of saintly mediator pursuing, in the Swedish manner, an antiseptic path among the sordid manœuvres of the powers, was to see it not as others saw it. Canadians had, and have, few qualifications for so unusual a mission. Upon the flimsy foundation of their claim to be free from what the Afro-Asian nations understand by 'colonialism' they reared expectations altogether too grandiose. Canada's treatment of its aboriginal population does not compare particularly favourably with the record of other Western powers in their colonial possessions; the franchise has been withheld from Eskimoes and Indians for nearly as long as from Algerians and Congolese, while less has been spent upon their welfare. Afro-Asian leaders having any interest in the matter understand this very well, even if they are too polite to point it out. Again, the prospects for successfully repeating the U.N.E.F. adventure (and it is conveniently forgotten by those who urge its repetition how close Canada's part in U.N.E.F. came to being an ignominious failure) are dimmed by the inescapable fact that Canadians have the wrong colour of skin for the job; this much, at least, the United Nations Operation in the Congo has disclosed. Finally, the national tradition is, on balance, a liability rather than an asset for para-military peace-making on a regular beat. In Latin-American disputes Canada is unlikely to develop a taste for intervention; in quarrels among Commonwealth nations there is a reluctance to express opinions, let alone intrude troops; while Canada's membership in N.A.T.O. is too easily exploited at the expense of whatever reputation it may have acquired as a disinterested mediator between East and West.

III

To Australia and New Zealand, the Suez crisis left a legacy of no mean value. It demonstrated the danger of their exposed position as appendages of South-East Asia ('the part of the world', as an Australian has observed, 'where people still fight'[1]); and it demonstrated, as well, the difficulty of conducting policies which, whatever abstract arguments might be mustered in their defence, ignored the feelings and aspirations of newly independent peoples, expectant and restless.

In New Zealand, the new emphasis of external policy was made easier by a change of government. The General Election held on 30 November 1957 brought the defeat (though by no great margin) of the National Party by the Labour Party. Mr. Walter Nash became Prime Minister and Minister of External Affairs, and his first major move in international matters was to undertake a tour of Asian countries. His visit early in 1958 to Ceylon, India, Malaya, Pakistan, the Philippines, and Thailand was intended to convey New Zealand's interest and sympathy in their affairs, and to erase their impression, to which New Zealand's Suez policies had lent credence, that his country would stand by Britain, 'right or wrong', to the end of time.

No such change of government eased Australia's conversion to a more generous and realistic appreciation of Afro-Asia's place in the modern world. Mr. Menzies remained Prime Minister; in January 1963 he and Mr. Nehru, alone of Commonwealth Prime Ministers at the time of Suez, still survived in office. Mr. Casey remained Minister of External Affairs until his resignation in January 1960, when he accepted a life peerage and entered the House of Lords at Westminster.[2] Many months before the Suez crisis, Mr. Casey had written that 'if the day ever arrives when the

---

[1] Donald Home, 'Australia Obsolescent', *The Spectator*, 28 December 1962.

[2] An event on which an Australian critic commented as follows: 'As Minister for External Affairs Lord Casey suffered in recent years particularly from occasional interventions by the Prime Minister which put him in eclipse and Australian foreign policy into some confusion. The most notable of these interventions occurred during the Suez crisis and it is commonly supposed that the coolness between himself and the Prime Minister which developed during this episode cost him the deputy leadership of his party.' L. C. Webb, 'Political Review', *The Australian Quarterly*, vol. XXXII, no. 2, June 1960, pp. 104–5.

Americans wrap themselves up in the Stars and Stripes and we British wrap ourselves up in the Union Jack, that would indeed be a tragedy for the world'.[1] That day had arrived, that tragedy occurred, in the hours following 30 October 1956; and to Australia's leaders the priorities were clear. In their view, the gravest damage had been sustained not in the West's relations with the emerging peoples of Asia and Africa, but in the relations of the United Kingdom and the United States. As Mr. Menzies put it:

> The worst result of all is that there is beyond question an . . . angry division of opinion among those who are, by tradition and practice, bound to be friends. . . . It needs to be made clear once more that the defence of the free world depends essentially not upon relations of the General Assembly but on the close and most friendly co-operation between the Great Powers of Western Europe and the United States.[2]

It was a bitter experience to discover that Australia lacked the qualifications to assist in the work of reconciliation. When Mr. Casey sought an interview with President Eisenhower in Washington, he was informed that it would serve no useful purpose and that it would not be granted to him. At this rebuff he did not conceal his dismay, remarking in a speech in New York City on 26 November that though he had 'devoted much of my time here to efforts to repair the breach, . . . I have not had any response to speak of. I have found difficulty in even discussing the matter.'[3]

Australia's isolated support of the British and French in a United Nations General Assembly that was overwhelmingly hostile—an isolation that some Australians thought splendid and others shameful—left a residue of bitterness which did not quickly disappear. Even Mr. Casey, who as a frequent participant in United Nations meetings might be expected to be more moderate in his views, was angry enough during the debate on foreign affairs in the House of Representatives on 11 April 1957 to blurt out, in reply to Dr. Evatt's accusation that the Government was 'guilty of a lack of persistent and continuous faith in the United Nations': 'Nobody takes any notice of it.'[4] Not too much significance should be attached to the remarks of statesmen goaded by opposition criticism into utterances of this kind (certainly no more

[1] R. G. Casey, *Friends and Neighbours* (Melbourne, 1954), p. 41.
[2] See below, p. 438.   [3] See below, p. 435.
[4] Quoted in Geoffrey Sawer, 'Problems of Australian Foreign Policy, June 1956–June 1957', *Australian Journal of Politics and History*, vol. III, no. 1, November 1957, p. 17.

to Mr. Casey's than to Mr. St. Laurent's 'supermen of Europe');
but statesmen under stress sometimes say illuminating things.
This was one of them. 'The truth is', an Australian commentator
has observed of Mr. Casey's incautious revelation, 'that the
Government, like most of the older Western powers, is unwilling
to depart from the power politics of military and economic group-
ings, conducted on the lines of the old diplomacy, in order to take
whatever risks are involved by relying on action solely or even
chiefly within [the] U.N.'[1]

Mr. Menzies made no attempt to conceal his distaste for the
methods and working of a United Nations wherein power and
authority, in defiance of the Charter, had moved from the Security
Council into the General Assembly—particularly a General
Assembly whose membership by the time of its first post-Suez
session had been augmented mainly by newly independent states
from Asia and Africa. 'I confess myself baffled', Mr. Menzies had
declared in Canberra on 30 November 1956, 'by the activities of
the General Assembly';[2] and on other occasions he spoke of what
he considered to be its shortcomings in a way that was derisory
and almost brutal. Perhaps the Prime Minister, who has taken
pride in his legal training, was in his unrestrained attacks upon the
new United Nations only applying the lawyer's old maxim: No
case—abuse plaintiff's attorney. That he had perhaps gone further
than the national interest would warrant he himself later realized
and admitted. 'My conclusion', he declared in a speech in London
on 8 July 1957, 'much contrary to what I said in my anger a few
months ago, is that the great and significant powers in the world
must treat the United Nations more seriously.'[3]

Mr. Menzies' admission did not signify as dramatic a conversion
in foreign affairs as many of his critics at home—and there were
many critics at home—would have liked. The countries that he
was prepared to admit to a roster of 'great and significant powers'
were not (one may confidently suppose) those that might have
been selected by Departments of External Affairs at Wellington
or Ottawa, certainly not at Delhi or Colombo. The notion that
neutrals, however influential or numerous, ought on occasion to be
cultivated at the expense of powers powerful by virtue of powerful
weapons of war was one to which Mr. Menzies remained wholly

---

[1] Quoted in Sawer, op. cit., p. 17.
[2] See below, p. 437.   [3] See below, p. 447.

unsympathetic. At the General Assembly in October 1960, the 'Menzies amendment' urging a Big Four summit meeting was overwhelmingly defeated by a vote of 45 to 5, with 43 delegations (including the New Zealand) abstaining. Mr. Menzies appeared anything but daunted by this outcome. 'I have learnt, perhaps, very little in my life', he declared to the House of Representatives, 'but I have learnt to know who are our friends.'[1] His friends, on this occasion, were Britain, France, the United States, and (for Ottawa, also, had by this time lost its touch) Canada.

In calling his book on Australian foreign policy *Friends and Neighbours* it was not Mr. Casey's intention to imply that friends were other than neighbours, still less that neighbours were other than friends. During his ten years as Minister of External Affairs he had made a conscious effort to try to make friends of Australia's neighbours. Under his guidance Australia embarked upon

a bipartisan good neighbour policy which left out no paragraph in the modern text-book on the creation of good will among nations. The diplomatic service in South-east Asia was greatly expanded. Friendship missions were sent, though not always received in the proper spirit. Officials exchanged visits at frequent intervals. Radio broadcasts were beamed in ever greater numbers to Asia in Asian languages. With official approval 'Meet Your Neighbour' campaigns were undertaken to familiarize Australians with their Asian guests. In 1956 a number of Australian-Asian Associations were created in the capital cities. The universities were beginning to introduce social science and language courses on Asia in their curricula. . . . They were beginning to compete seriously with British and American universities as the Mecca of Asian students.[2]

By 1956 these efforts, notwithstanding the considerable handicaps imposed upon the creation of Australian-Asian friendship by the 'White Australia' policies of the past and present, by Australia's membership in S.E.A.T.O., and by Australia's commitment to maintaining the *status quo* in West New Guinea, had begun to be repaid. 'A *rapprochement* between Australia and some countries of

[1] Quoted in D. C. Corbett, 'Problems of Australian Foreign Policy, July–December 1960', *The Australian Journal of Politics and History*, vol. VII, no. 1, May 1961, p. 7.
[2] Werner Levi, *Australia's Outlook on Asia* (East Lansing, Michigan, 1958), p. 177.

Asia had taken place with a rapidity and even a certain degree of intimacy which was truly amazing. . . .'[1]

Not the least remarkable feature of Australian policy during the Suez crisis and after was its readiness to cast all this aside. For Australia's good name in Asia was the first and paramount casualty of Mr. Menzies' activities between the time he sought to impose terms upon President Nasser in Cairo and the time when his representative voted with Britain and France in the United Nations. 'Australia by its recent actions', an Australian scholar resident in South-East Asia has written, 'has damaged or destroyed the reservoir of goodwill it had built up [there] during the post-war years.'[2] How could so precious an asset be so heedlessly squandered? The factor of personality is important, even crucial: had Dr. Evatt, or even Mr. Casey, been Prime Minister, Australia's policies would have been noticeably different, more like Canada's, or at least like New Zealand's. But it would be misleading as well as unfair to place all the blame (or praise) at Mr. Menzies' doorstep. A Prime Minister who consistently puts personal whim ahead of national interest, or who consistently confuses personal whim with national interest, is not likely to remain Prime Minister for long; and Mr. Menzies has demonstrated a remarkable durability as a Commonwealth Prime Minister. In acting as he did, during the Suez crisis, at the United Nations in the aftermath, and at Prime Ministers' Conferences since, Mr. Menzies, by so bluntly expressing his own views, has expressed something with which most Australian voters are able strongly to identify and of which they no less strongly approve. A keenly perceptive student of Australian society, himself Australian, has observed that the White Australia policy (perhaps the harshest manifestation of Australia's readiness to favour friends at the expense of neighbours)

is not just a matter of colour. It is concerned with the whole texture of life, with the preservation of a quality and a character which have developed without much conscious planning but which are now recognised as distinctive. To Australians it is manifestly in the interest of Australia to continue to be as it is now, only more so. Even Australian intellectuals who are dissatisfied with some aspects of the life around them are appreciative of other aspects which they know cannot be

---

[1] Levi, op. cit., p. 177.
[2] K. G. Tregonning, 'Australia's Imperialist Image in South-East Asia', *The Australian Quarterly*, vol. XXXIII, no. 3, September 1961, p. 51.

dissociated from these. They think that drive, energy, inventiveness and solidarity in trouble, together with a form of practical social equalitarianism, are good value for the price paid in anti-intellectualism, intolerance and intermittent official puritanism. All this can be summarised by saying that Australians as a body are proud of their way of living and intend to preserve it, even though the movement of world events might seem to threaten it.[1]

One might have expected that Australia's membership in the Commonwealth, so many of whose citizens now live in the new nations of South and South-East Asia, would exert a powerful countervailing force against a too exclusively European orientation in the Australian outlook. It was not powerful enough, evidently, to make much headway at the time of Suez. The fact of the matter was, and probably still is, that 'the Commonwealth' did not count for a great deal in the Australian scheme of things.

Apart from the fact that Australians use the phrase to describe their own Federal government, and so must make a considerable adjustment of mind to use it about anything else, it is fairly clear that Australians think either of the attachment to Britain, which is of long standing and fully comprehensible, or of the fortunate connection with the Asian member-nations, but not of the two as being part of a whole. Relations with Britain are one thing; relations with the Asian countries another.[2]

This is why concern for the future of the Commonwealth played so slight a part in Australian reckonings at the time of the Suez crisis. Mr. Casey's attitude when in Ottawa during the first week of November 1956 differed in this respect very markedly from that of his official hosts. 'If the Commonwealth means anything', he told reporters anxious to ascertain his reaction to rumours that the Commonwealth was on the point of disintegration, 'and it certainly does—and if it can't ride out this, I tremble for the future. The Commonwealth link would be very flimsy if it were broken up by this.'[3] But for Mr. Casey, the 'Commonwealth link' meant primarily the bond between Australia and Britain. This was, as he confidently predicted, a hardy thing which could withstand infinite buffeting; but his faith in the survival of the Commonwealth was based upon a different conception of Commonwealth than that which prevailed in Ottawa, or in the capitals of its Asian members.

[1] J. D. B. Miller, *The Commonwealth in the World* (London, 1958), p. 166.
[2] Ibid., p. 179.   [3] *The Globe and Mail* (Toronto), 6 November 1956.

IV

Of all Commonwealth member governments, only South Africa's refrained from passing judgement upon the merits of the Anglo-French action from 30 October to 6 November. For this attitude there were (as noted in the Commentary to Part III) reasons peculiar to the South African scene. Having refrained from taking a stand at the time, it was natural enough that it should refrain from the kind of assessment and appraisal undertaken in other Commonwealth capitals during the aftermath. 'Up till now', the Minister of External Affairs declared in the House of Assembly on 11 February 1957, 'I have not said one word about the British-French action, nor do I intend doing so. . . . It was not for me or for South Africa to interfere.'[1] Five years later, however, his new Prime Minister had come to a different conclusion. 'If Britain had been prepared to execute the policy which she had believed to be the correct one', Dr. Hendrik Verwoerd observed on 29 March 1962, 'and had not allowed herself to be dissuaded, I am sure there would be less unrest in the world, and especially in Africa, today.'[2]

The Nationalist Government was ready enough to express its opinion on other aspects of the crisis. One of these was the role of the United Nations. From its inception, South Africa's statesmen had nothing but distaste for what they felt to be the over-intrusive activities of the General Assembly; General Smuts himself had returned from its first session in 1946 in the conviction, as he told his Parliament, that it was nothing but a compound of 'emotion, passion and ignorance'.[3] For the next decade they were principally engaged in the losing struggle to plug the leaking dykes of domestic jurisdiction. By 1956, their patience, never in great supply, had all but disappeared. On 27 November, the Minister for External Affairs announced to the General Assembly that his Government had decided to withdraw from all but token participation in the work of the United Nations. In the litany of grievances which Mr. Louw recited as having led to this drastic decision, the United Nations' part in the Middle East crisis was mentioned only in passing. 'The adoption of temporary expedients', he declared, 'is not likely to provide a solution of the troubles, but may tend rather

---

[1] See below, p. 451.   [2] *The Times* (London), 30 March 1962.
[3] Quoted in J. E. Spence, 'Tradition and Change in South African Foreign Policy', *Journal of Commonwealth Political Studies*, vol. I, no. 2, May 1962, p. 148.

to aggravate the fundamental disease.'[1] Beyond this he did not go, neither diagnosing the disease nor prescribing a remedy. But the last thing in his mind was the creation of a permanent United Nations police force; on that, at least, he and Mr. Krishna Menon could see eye to eye. Rather, it was to purge and cleanse the organization by banishing its Afro-Asian members, or at least by restricting their influence which (since it was directed against South Africa) was regarded by the Nationalist Government as wholly unhelpful. 'The problems raised by Egypt's nationalization of the Suez Canal Company will not be solved by a thoroughly divided United Nations, in which the Bandung countries, generally acting in concert with the Communist States, are playing so important a role under the leadership of India. . . .'[2]

There was some disposition on the part of opposition members of the South African Parliament to criticize their Government for not having offered support for the United Nations Emergency Force. That it had been wise in not doing so supporters of the Government argued with great emphasis, pointing to the Canadian experience as a vindication of its wisdom in staying out. 'Our Government followed the sensible policy of keeping out of the Suez dispute and by doing so we were not humiliated in the way in which Nasser was enabled to humiliate Canada. . . .'[3]

It was hardly to be expected that the Nationalist Government would experience any concern about the effects of the Suez crisis upon the future cohesion of the Commonwealth. It had always been common ground between them and their political opponents within the Union that the admission of the Commonwealth's members in Asia, so far from adding to the strength and value of the association, seriously diminished its usefulness. It would indeed be hard to imagine a more acrimonious relationship, between either governments or peoples, than that which had prevailed between the Union Government and the Government of India since India became independent (the granting of Indian independence itself was thought by Smuts to have been 'an awful mistake'[4]). Had India decided to leave the Commonwealth as a protest against the United Kingdom's policies in Egypt, the Nationalist Government, consistent with this attitude, could only have

---

[1] G.A.O.R., Eleventh Session, 1956–7, I, p. 356.   [2] See below, p. 448.
[3] South Africa, House of Assembly Debates, vol. 93, col. 60. The speaker was Dr. J. H. O. Du Plessis.
[4] Quoted by Spence, op. cit., p. 144.

welcomed her departure. As it happened, India did not depart; but in the wake of Suez, South African-Indian relations took a turn for the worse. One of the advantages of the new importance of the Cape Route following the blocking of the Suez Canal, Mr. Louw declared on his return to South Africa in January 1957, was that 'South Africa was now in the position to take retaliatory measures on account of trade sanctions applied by India against South Africa for more than ten years. . . . The Union would be fully justified in refusing the facilities of South African ports to Indian vessels.'[1] Though he denied that his statement had been intended as a threat—such a threat, he said in the inevitable clarifying statement issued four days later, would 'obviously be futile'[2]—it was understandably regarded as such by India. The Indian Government apparently made no official representations to the Union Government on the subject, but Indian shipping companies instructed their vessels to by-pass South African ports. Mr. Louw chose to regard India's retaliation as a victory. 'My *ballon d'essai* succeeded admirably. Few South Africans will regret the fact that the Indians have now inconvenienced themselves.'[3] However that might be, there were not a few South Africans who regretted the fact that their Minister of External Affairs had, in their estimation, made a fool of himself. 'I think that that statement of the hon. Minister', declared Mr. S. F. Waterson, a former High Commissioner in London and a Minister in General Smuts' wartime administration, 'was one of the stupidest that I can conceive any responsible Minister of External Affairs making at the present time. I am afraid the Minister is a disciple of what one may call "knobkerrie diplomacy". That kind of diplomacy is always dangerous, but it becomes very dangerous indeed, as well as highly stupid, if you have not got a knobkerrie. . . .'[4]

A hardly more productive intervention was directed against the United Kingdom. There had appeared in *The Times* for 28 December 1956 an article strongly critical of the Union Government's disregard for civil liberties. The Minister of External Affairs, then in Rome, wrote immediately to the Editor protesting what he described as 'the campaign of hate that has been conducted against South Africa for the past eight years', a campaign 'the more surprising in view of the good will shown to Britain during the past

[1] *Johannesburg Star*, 18 January 1957.   [2] Ibid., 22 January 1957.
[3] Ibid.   [4] South Africa, House of Assembly Debates, vol. 93, col. 571.

eight years'. Mr. Louw sought to demonstrate that the servicing of British ships and shipping bound for Britain at the Union's ports was an illustration of that good will, involving, as he claimed it did, a real sacrifice of South African interests: the money earned 'by no means compensates for the dislocation of our harbours, the strain on our railways, and the adverse effects of certain aspects of South Africa's economy, such as a rise in the cost of foodstuffs to the local consumers'.[1] The British Government, like the Indian, made no public (nor, so far as is known, private) reply to this outburst, which again brought unofficial protest abroad and at home. 'This letter', declared Mr. Waterson, 'is a thoroughly badly drafted, irresponsible and clumsy document, and it is calculated to do far more harm than good to our foreign relations, for which that hon. Member is responsible.'[2]

It may be doubted whether Mr. Louw, or for that matter his colleagues, felt either remorse or regret at any of these reproaches. They went their way, as a certain statesman once admired by their party went his, 'with the assurance of a sleepwalker'; and their reaction to criticism, if any, was the reaction of the paranoiac. Of this disastrous combination their unhappy country seemed likely to be the victim.

V

The three members of the Commonwealth in Asia find themselves (as do the growing number of members of the Commonwealth in Africa) in the Commonwealth for reasons differing greatly from those of the so-called 'White Dominions'. Their membership is not the result of a protracted process of historical development into which the element of conscious choice entered hardly at all; it is rather the result of a deliberate weighing of its advantages and disadvantages. Two important consequences flow from this. First, having decided to enter the Commonwealth, they are all for using the Commonwealth. Having calculated that it is better to remain than to resign, they are naturally anxious to justify that decision by benefits of membership more tangible than participation in a mystique which must surely remind most of them of the Emperor's clothes. 'I am sure that most of our countries are . . . plain sober and realistic', remarked the High

[1] *The Times*, 7 January 1957.
[2] South Africa, House of Assembly Debates, vol. 93, col. 568.

Commissioner of Ceylon in the United Kingdom in 1960, 'and do not experience a spiritual ecstasy and elation as such at being in the Commonwealth. We are quite down-to-earth in being there and in wanting to be there. Our sentiments towards the Commonwealth are rational, not emotional.'[1] Secondly, and as a corollary, if at any time it seems to them that Commonwealth membership creates more difficulties than it confers benefits, they will not hesitate to withdraw.

With us [declared Mr. Julius Nyerere in 1961, when he was Prime Minister of Tanganyika] geography, pre-colonial history, and future prosperity, all demand that African unity must have priority over all other associations; and however great our belief in the Commonwealth, it is a fact that other parts of Africa have been under different colonial dominations, without the same historical links. . . . If we must give up membership of the Commonwealth in order to achieve unity with the ex-French or ex-Belgian states, Tanganyika can have no hesitation in doing so.[2]

It was therefore the natural reaction of the three Asian members of the Commonwealth to consider, during the aftermath of the Suez crisis, whether the events of the preceding months had made it expedient for them to terminate their membership.

Termination was strongly urged in India. Mr. C. Rajagopala-chari, the former Governor-General, doubtless spoke for many when he put the case for India's withdrawal from the Commonwealth in a statement issued before the cease-fire in Egypt.[3] But he spoke in the anger of the moment. More significant as an indication of enduring attitudes are the speeches of those who argued, in the course of the foreign affairs debate in the Lok Sabha in March 1957, that the Commonwealth no longer served the national interest. 'By remaining in the Commonwealth', the Member for Mysore declared, 'we are only encouraging colonial tendencies of a few powers. We are used as mere scapegoats or instruments to further their ends. . . . The United Kingdom has committed a great blunder. . . . It would be fitting to our national prestige and honour that we should cease to be a member of the Commonwealth of Nations.'[4]

[1] Speech by Mr. Gunasena de Soyza, *Journal of the Royal Commonwealth Society*, vol. III (New Series), no. 5, September–October 1960, p. 191.

[2] Julius Nyerere, 'For Commonwealth and/or African Unity', ibid., vol. IV, no. 6, November–December 1961, p. 254.

[3] See above, p. 256.   [4] See below, p. 456.

The Prime Minister declined to accept this counsel. He conceded—and it was a highly important concession—that he had felt, 'and for the first time I felt, the first time in these many years', that India's membership in the Commonwealth 'may some time or other require further consideration'. But that time had not come. There were two reasons for not breaking Commonwealth ties. First, 'our policies . . . are in no way conditioned or deflected from their normal course by that association'. Second, 'when there are so many disruptive tendencies in the world, it is better to retain every kind of association, which is not positively harmful to us, than to break it. Breaking it itself is a disruptive thing.'[1] These were not very stirring statements, and certainly Pandit Nehru's description of the Commonwealth as an association 'which is not positively harmful to us' held little to satisfy those who looked hopefully to Delhi to provide a reaffirmation of its romantic image. But then they expected too much. There was, after all, good reason for them to be satisfied by the Indian reaction to the crisis. As early as December 1956 Mr. Nehru had spoken without qualification against a motion in the Upper House which proclaimed that 'India's further continuation of her membership in the Commonwealth is inconsistent with the principles of Panch Sheel'.[2] And though he might justify India's continued membership by somewhat negative arguments, there were other, more compelling, justifications which he kept to himself. One of these was the importance of the Commonwealth association for influencing the policy of others. Of that there was ample evidence, some of which was doubtless disclosed to Mr. Nehru when he visited Ottawa in December 1956. There was also the Indian Prime Minister's appreciation of the fact that the policies of the British Government during the crisis were opposed by millions of British citizens and, in particular, by the Labour Party, whose spokesmen indeed appealed to India at the time of the Anglo-French invasion not to judge the nation by what so many of its own people regarded as a mistaken and even wicked adventure. And finally there was the persisting tradition of the Mahatma. Gandhi, writing in 1922, had a more profound understanding of the Commonwealth of Nations than did the British Government in 1956: 'India's greatest glory will consist not in regarding Englishmen as her implacable enemies . . . but in turning them into friends and

---

[1] See below, p. 455.   [2] See below, pp. 452-3.

partners in a Commonwealth of Nations in place of an empire
based upon exploitation of the weaker or undeveloped nations and
races of the earth and therefore finally upon force'.[1]

In Pakistan, demands that the Commonwealth connexion
should be broken off were even more forcibly expressed. No such
sentimental bond as existed between so many influential Indians
and those members and emissaries of the Labour Government
which had granted independence in 1947 prevailed among Pakistani politicians and their counterparts in the United Kingdom.
The Commonwealth connexion had indeed for long been a source
of grievance to Pakistan. A prime reason for membership had been
the expectation of acquiring Commonwealth support for Pakistan's
position over Kashmir; and as it became steadily more apparent
that that expectation would not be realized, Pakistan's enthusiasm
for the Commonwealth perceptibly diminished. Moreover, the
Anglo-French invasion of Egypt was seen in Pakistan as a far
more heinous offence than it appeared in India, for it could be
construed only as an assault upon, and an affront to, the entire
Muslim world for whose leadership Pakistan had always contended.

To be with England [argued Mian Mumtaz Mohammad Daultana, a
brilliant if erratic critic and an experienced politician whose last official
post had been that of Chief Minister of West Pakistan from 1951 to
1953] is to be without our own people. . . . This does not mean that we
must be sworn to the enmity of England. We have relations with China,
with Russia, with Australia, with South Africa; why not with Britain?
But it is not possible for us to involve ourselves in what amounts to a
family relationship with the very Power whose dagger is ever turned
towards the breasts of our nearest brothers. . . . Pakistan therefore has
no place in the family of the British Commonwealth of Nations.[2]

In February 1957, several speakers in the National Assembly
expressed a similar view. It was not enough to say, as Pandit
Nehru had said, that membership in the Commonwealth was
justified because it does not compromise the independence of the
member. 'If you do not get any benefit', declared Mr. M. A.
Khuhro of the Muslim League, 'then the correct course would be
to get out of it. . . . It is better to say "good-bye".'[3]

---

[1] Quoted in S. R. Mehrotra, 'Gandhi and the British Commonwealth',
India Quarterly, vol. XVII, no. 2, January–March 1961, p. 49.

[2] Mian Mumtaz Mohammad Daultana, 'Reflections on Pakistan's Foreign
Policy', *Dawn* (Karachi), 10 December 1956.

[3] See below, p. 459.

But Mr. Suhrawardy, defending his Government's decision to remain within the Commonwealth notwithstanding the iniquity of the British attack upon Egypt, was not content to rest his case upon the negative claim that membership merely meant independence. His view, like that of the British Minister who coined the phrase, was that it meant 'independence plus'. The influence of the Asian members of the Commonwealth, he insisted, had been in no small degree responsible for the British agreeing to a cease-fire in Egypt. 'I do not know why people consider that it is only the threat of enemies that counts and not the protests or pleadings of friends. . . .' And, he declared, 'if the member countries maintain their faith in the noble purpose underlying the concept of Commonwealth and we faithfully practise them, we are bound to exert strength, and moral influence, on international affairs. . . .'[1]

In Ceylon, also, the Opposition party had called for severing Commonwealth ties in retaliation against British policies in Egypt. Mr. Leslie Goonewardena derided the notion that membership afforded the Government any opportunity of influencing the behaviour of other members. 'It is quite possible', he argued, 'that for a big country like India . . . the advantages of remaining within the Commonwealth may outweigh its disadvantages, but . . . a small country . . . like Ceylon, whose interventions in Commonwealth affairs cannot reasonably be expected to exert much influence, . . . [is] likely to get all the disadvantages and very few of the advantages. . . .'[2] Such an analysis did scant justice to the power of small states in the modern world, and less to the power of Ceylon, which flowed in no small measure from its Commonwealth association. That association, Professor J. D. B. Miller has pointed out,

gives her stature which other small Asian states, such as South Viet Nam and Cambodia, do not possess. It eases relations with Britain, which might otherwise prove difficult when a radical government in Ceylon talks of nationalising tea plantations or ending British occupation of naval and air bases. Its main effect, however, is to give her an artificial but useful equality with India. . . . When the Prime Ministers come to London, a Bandaranaike or a Kotelawala will get almost as much of the spotlight as a Nehru; this is good for Ceylonese self-respect, and it may have some effect upon the relative positions of the two countries when they enter into diplomatic negotiations.[3]

---

[1] See below, p. 458.   [2] See below, p. 471.   [3] Miller, op. cit., p. 223.

No one understood all this better than S. W. R. D. Bandaranaike himself, who had experienced at first hand the easy access to the corridors of power available to the Prime Minister of a Commonwealth country. In the immediate aftermath of the Suez crisis, he had visited the capitals of three Commonwealth countries—the United Kingdom, Canada, and Pakistan—and had not hesitated to speak his mind in each. In London, he 'impressed [his] point of view upon Sir Anthony Eden': in Ottawa his conversations with the Canadian Ministers left him 'satisfied that Canada adopts a fairly impartial and objective attitude in dealing with these important problems'.[1] To his audience in Karachi he emphasized the influence that a state such as his own could exert if it went to the trouble of keeping its channels of communication open and in good repair.[2] Here was a Prime Minister determined to get something worth while out of the Commonwealth connexion, and succeeding in no mean degree.

Without exception, those of the Asian members of the Commonwealth who had spoken of resigning from it had in mind the resignation of their own country. None had suggested that when a member of a group brings by its own behaviour discredit upon the rest it is more appropriate to demand the resignation of the offender than to resign oneself. That the offender was in this case the United Kingdom itself undoubtedly made the demand for expulsion less realistic than it might otherwise have been; there still persists the notion that the United Kingdom is unique in being the indispensable member of the Commonwealth, without which it could not longer function. But it also reflected the lack of what may be termed a sense of proprietary concern for the Commonwealth. Its members, not experiencing collective consternation when the policies of any one of them threaten to bring all into disrepute, were more ready to resign than to band together for the protection and preservation of the whole.

It might be thought that the Commonwealth today is further than it ever was from the proprietary ideal; but there is evidence to the contrary. When, five years after Suez, the policies of another member affected the Commonwealth's good name, the consensus of the Commonwealth was that South Africa should resign. There are those who felt in March 1961, and who feel still, that the Prime Ministers' Conference should have been readier than it was to pay

---

[1] See below, p. 466-7.   [2] See below, p. 463-5.

the price of magnanimity; the Prime Ministers themselves (with the notable exception of Mr. Menzies) thought it much too high. The Commonwealth upon which they consciously set their course is one in which, not for the first time, the older members will be affected much by, and will have much to learn from, the new. It is one whose spirit has been best expressed by the Prime Minister of what, in 1961, was its most recent recruit. 'It seems to us', wrote Mr. Julius Nyerere, 'that the key is the common acceptance of certain standards of behaviour . . . a mutual trust and understanding of each other's basic purposes. . . . We have nothing in common with a state which denies the common humanity of all mankind.'[1]

DOCUMENTS

I. UNITED KINGDOM

1. Extacts from statement by the Foreign Office, 8 November 1956[2]

Her Majesty's Government fully understand the concern of their Baghdad Pact allies regarding the very grave situation which has arisen in the Middle East. It is their firm conviction that the action which has been taken to separate Egyptian and Israeli forces was the only way to prevent a general conflagration in the area.

Her Majesty's Government are also most appreciative of the initiative of the Governments of Iraq, Iran, Pakistan and Turkey. The views offered both individually and collectively [Part III, VII, Document 5, *supra*] by these Governments have weighed heavily in the decision to bring an end to military action in Egypt. They have listened with close attention to the friendly and constructive advice offered by the four Governments on the form of an early settlement and are happy to state that the proposals formulated by the Middle Eastern members of the Baghdad Pact are consistent with Her Majesty's Government's own views. As regards the settlement of the Palestine problem, Her Majesty's Government's views are well known. . . .

Her Majesty's Government believe that the opportunity is now

---

[1] Julius Nyerere, 'For Commonwealth and/or African Unity', loc. cit., p. 253.
[2] *Commonwealth Survey*, 13 November 1956, p. 964.

ripe for a general settlement of Middle Eastern problems which have defied solution for so long and in the settlement of which all members of the Baghdad Pact together will be able to play a valuable and constructive role. They believe further that through co-operation in this task the Baghdad Pact Alliance will be greatly fortified.

2. Extracts from statement by the Secretary of State for Foreign Affairs, Mr. Selwyn Lloyd, in the House of Commons, 3 December 1956[1]

. . . For the last ten years we have been living with a world-wide struggle going on between Communism and the free world. The introduction of nuclear weapons has made a global war unattractive to the aggressor. The Soviet, therefore, has used the methods of political subversion from within and military pressure from without.

The existence of the North Atlantic Treaty Organisation has halted the direct advance of Russia across Europe to the sea. But all the time there has been an open flank in the Middle East which Russia has been making a determined effort to turn. Certain factors have developed there to her advantage. There have been hostilities smouldering between Israel and the Arab States, and the United Nations has so far not been able to solve that problem at all. The situation has been deteriorating rather than improving. At the same time, Colonel Nasser has come to power with his plans for the aggrandisement of Egypt and the subjection to his domination of the material resources of the Arab countries. The seizure of the Suez Canal was part of that design. . . .

The Arab-Israel tension has afforded opportunity for Soviet mischief making. The large supply of Soviet arms to Colonel Nasser put him very much under Soviet influence. The Baghdad Pact gave a measure of security against direct Soviet penetration from the North, but the arming of Syria and Egypt was no doubt intended to turn its flank also.

Let there be no misunderstanding. The situation was deteriorating. It was one which sooner or later was likely to lead to war. The only doubtful question was the scope or extent of that war. A general conflagration in the Middle East would have been disastrous for many countries, not least our own. It was against that

[1] United Kingdom, H.C. Deb., 1956–7, vol. 561, coll. 878–83.

background that on 29th October major hostilities began between Israel and Egypt.

The French and British Governments decided immediately to intervene. We are quite sure that, by our timely action, we not only rapidly halted local hostilities, but forestalled the development of a general war throughout the whole Middle East and perhaps far beyond. . . .

Our second purpose was to interpose a force to prevent the resumption of fighting. . . .

Another vitally important result of our actions has been that the Russian designs have been exposed and dislocated. It is to be hoped that the free world will use the breathing space that we have provided to frustrate them altogether. . . .

. . . We have created conditions under which there can be hope of wider settlements. Of course, there will be heavy costs to bear, but they would have been far greater if our action had not been taken, and it is now for the United Nations and its member States to see that this opportunity is turned to good account.

3. Extracts from speech by the Member for Ebbw Vale, Mr. Aneurin Bevan, in the House of Commons, 5 December 1956[1]

. . . I have been looking through the various objectives and reasons that the Government have given to the House of Commons for making war on Egypt, and it really is desirable that when a nation makes war upon another nation it should be quite clear why it does so. It should not keep changing the reasons as time goes on. . . . When I have looked at this chronicle of events during the last few days, with every desire to understand it, I just have not been able to understand, and do not yet understand, the mentality of the Government. . . .

. . . We sent an ultimatum to Egypt by which we told her that unless she agreed to our landing in Ismailia, Suez and Port Said, we should make war upon her. We knew very well, did we not, that Nasser could not possibly comply? Did we really believe that Nasser was going to give in at once? Is our information from Egypt so bad that we did not know that an ultimatum of that sort was bound to consolidate his position in Egypt and in the whole Arab world?

[1] United Kingdom, H.C. Deb., 1956–7, vol. 561, coll. 1269–82.

We knew at that time, on 29th and 30th October, that long before we could have occupied Port Said, Ismailia and Suez, Nasser would have been in a position to make his riposte. So wonderfully organised was this expedition—which, apparently, has been a miracle of military genius—that long after we had delivered our ultimatum and bombed Port Said, our ships were still ploughing through the Mediterranean, leaving the enemy still in possession of all the main objectives which we said we wanted.

Did we really believe that Nasser was going to wait for us to arrive? He did what anybody would have thought he would do, and if the Government did not think he would do it, on that account alone they ought to resign. He sank ships in the Canal, the wicked man. What did hon. Gentlemen opposite expect him to do? The result is that, in fact, the first objective realised was the opposite of the one we set out to achieve; the Canal was blocked, and it is still blocked.

The only other interpretation of the Government's mind is that they expected, for some reason or other, that their ultimatum would bring about disorder in Egypt and the collapse of the Nasser regime. None of us believed that. If hon. Gentlemen opposite would only reason about other people as they reason amongst themselves, they would realise that a Government cannot possibly surrender to a threat of that sort and keep any self-respect. We should not, should we? If somebody held a pistol at our heads and said 'You do this or we fire', should we? Of course not. Why on earth do not hon. Members opposite sometimes believe that other people have the same courage and independence as they themselves possess? Nasser behaved exactly as any reasonable man would expect him to behave.

The other objective was 'to reduce the risk . . . to those voyaging through the Canal'. That was a rhetorical statement, and one does not know what it means. . . .

On 31st October, the Prime Minister said that our object was to secure a lasting settlement and to protect our nationals. . . . In the meantime, our nationals were living in Egypt while we were murdering Egyptians at Port Said. We left our nationals in Egypt at the mercy of what might have been merciless riots throughout the whole country, with no possibility of our coming to their help. . . .

On 1st November, we were told the reason was 'to stop hostilities' and 'prevent a resumption of them'. But hostilities had already been practically stopped. On 3rd November, our objectives became much more ambitious—'to deal with all the outstanding problems in the Middle East'. . . . After having outraged our friends, after having insulted the United States, after having affronted all our friends in the Commonwealth, after having driven the whole of the Arab world into one solid phalanx, at least for the moment, behind Nasser, we were then going to deal with all the outstanding problems in the Middle East. . . .

The next objective of which we were told was to ensure that the Israeli forces withdrew from Egyptian territory. That, I understand, is what we were there for. We went into Egyptian territory in order to establish our moral right to make the Israelis clear out of Egyptian territory. That is a remarkable war aim, is it not? In order that we might get Israel out, we went in. To establish our case before the eyes of the world, Israel being the wicked invader, we, of course, being the nice friend of Egypt, went to protect her from the Israelis, but, unfortunately, we had to bomb the Egyptians first.

On 6th November the Prime Minister said: 'The action we took has been an essential condition for . . . a United Nations force to come into the Canal Zone itself'. That is one of the most remarkable claims of all. . . . It is, of course, exactly the same claim which might have been made, if they had thought about it in time, by Mussolini and Hitler, that they made war on the world in order to call the United Nations into being. If it were possible for bacteria to argue with each other, they would be able to say that of course their chief justification was the advancement of medical science. . . .

. . . at the end of all these discussions . . . the war aim of the Government now becomes known. . . . It was a red peril all the time. It was Russia all the time. . . . I am not for one moment seeking to justify the Russian supply of arms to Egypt. I think it was a wicked thing to do and I think it is an equally wicked thing for us to supply arms. . . . It seems to me—and here I probably shall carry hon. Members opposite with me—that Nasser has not been behaving in the spirit of the Bandoeng Conference which he joined, because what he did was not to try to reduce the temperature of the cold war: what he did was to exploit it for Egypt's purposes. Therefore, Nasser's hands are not clean by any means. . . . We must not believe that because the Prime Minister is wrong Nasser is right. That is not the view of this side of the House.

What has deeply offended us is that such wrongs as Nasser has done and such faults as he has have been covered by the bigger blunders of the British Government. That is what vexes us. We are satisfied that the arts of diplomacy would have brought Nasser to where we wanted to get him, which was to agree about the free passage of ships through the Canal, on the civilised ground that a riparian nation has got no absolute rights over a great waterway like the Canal. . . .

It has been clear to us, and it is now becoming clear to the nation, that for many months past hon. Members opposite have been harbouring designs of this sort. One of the reasons why we could not get a civilised solution of the Cyprus problem was that the Government were harbouring designs to use Cyprus in the Middle East, unilaterally or in conjunction with France. . . . We have had all these murders and all this terror, we have had all this unfriend-ship over Cyprus between ourselves and Greece, and we have been held up to derision in all the world merely because we con-templated using Cyprus as a base for going it alone in the Middle East. And we did go it alone. Look at the result. . . .

The social furniture of modern society is so complicated and fragile that it cannot support the jackboot. We cannot run the processes of modern society by attempting to impose our will upon nations by armed force. If we have not learned that we have learned nothing. Therefore, from our point of view here, whatever may have been the morality of the Government's action—and about that there is no doubt—there is no doubt about its imbecility. There is not the slightest shadow of doubt that we have attempted to use methods which were bound to destroy the objectives we had, and, of course, this is what we have discovered. . . .

It has been proved over and over again now in the modern world that men and women are often prepared to put up with material losses for things that they really think worth while. It has been shown in Budapest, and it could be shown in the Middle East. That is why I beg hon. Members to turn their backs on this most ugly chapter and realise that if we are to live in the world and are to be regarded as a decent nation, decent citizens in the world, we

have to act up to different standards than the one that we have been
following in the last few weeks.

I resent most bitterly this unconcern for the lives of innocent
men and women. It may be that the dead in Port Said are 100,
200, or 300. If it is only one, we had no business to take it. Do hon.
Members begin to realise how this is going to revolt the world
when it passes into the imagination of men and women everywhere, and in this country, that we, with eight million here in
London, the biggest single civilian target in the world, with our
crowded island exposed, as no other nation in the world is exposed,
to the barbarism of modern weapons, we ourselves set the example.

We ourselves conscript our boys and put guns and aeroplanes
in their hands and say, 'Bomb there'. Really, this is so appalling
that human language can hardly describe it. And for what? The
Government resorted to epic weapons for squalid and trivial ends,
and that is why all through this unhappy period Ministers—all of
them—have spoken and argued and debated well below their
proper form—because they have been synthetic villains. They are
not really villains. They have only set off on a villainous course,
and they cannot even use the language of villainy.

Therefore, in conclusion, I say that it is no use hon. Members
consoling themselves that they have more support in the country
than many of them feared they might have. Of course they have
support in the country. They have support among many of the
unthinking and unreflective who still react to traditional values,
who still think that we can solve all these problems in the old ways.
Of course they have. Not all the human race has grown to adult
state yet. But do not let them take comfort in that thought. The
right hon. Member for Woodford (Sir W. Churchill) has warned
them before. In the first volume of his *Second World War* he writes
about the situation before the war and says this: 'Thus an Administration more disastrous than any in our history saw all its errors
and shortcomings acclaimed by the nation. There was however a
bill to be paid, and it took the new House of Commons nearly ten
years to pay it'. . . .

4. Extract from report of Sir Edwin Herbert, December 1956[1]

B. I have examined the series of instructions given by Ministers to the military forces as to the care to be taken for civilian life and property and the orders given and carried out for the purpose of implementing Ministerial instructions. I do not think it necessary to burden this report by setting out even a selection of the numerous signals I have seen but I think it right to emphasise the following points:

(i) Not one bomb was dropped on Port Said or its environs, all air strikes being low level attacks by rocket, machine gun and cannon of great accuracy and very localised.

(ii) The Naval bombardment was both localised to the beaches and reduced in intensity to 1/10 of the appropriate level.

(iii) Every possible warning was given by radio to the civil population. Indeed to one like myself, with a long experience of security in war-time, it came as a shock to find that the two over-riding considerations in the minds of all concerned were firstly to inform the enemy of our plans and secondly to do him as little damage as possible in the execution of those plans, which I think is a fair summary of the steps taken.

(iv) The steps taken to protect the interests of civilians in my opinion jeopardised British lives and increased British casualties.

C. No mathematical certainty based on unambiguous evidence as to Egyptian casualties can be obtained. Several pointers can, however, be given. They all point to a figure of some 650 deaths. ...

5. Extracts from speech by the Prime Minister, Mr. Harold Macmillan, in the House of Commons, 1 April 1957[2]

Before attempting to summarise the results of the Bermuda meeting, Mr. Speaker, I should like to say a few words about its general character.

The meeting took place not at my suggestion, but on the proposal of President Eisenhower. It was he, also, who suggested that it should take place on British soil. ... That there was need for

[1] Cmd. 47, p. 26.
[2] United Kingdom, H.C. Deb., 1956–7, vol. 568, coll. 37–39.

such a meeting is, I think, clear. There has been in recent months
the serious deterioration in Anglo-American relations. It was not
the purpose of the meeting to go back over the past, except so far
as the lessons to be learned might help us for the future. We had
no intention of crying over spilt milk, and certainly not of wallow-
ing in it. Nevertheless, both sides spoke with absolute frankness
and sincerity to each other, and this was wise, because it is only by
complete frankness that we can hope to rebuild and strengthen a
partnership so vital to the future of the world. . . .

. . . I should like to say a word about the meeting with the
Canadian Prime Minister. I am deeply indebted to Mr. St.
Laurent and his colleagues for their willingness to go to Bermuda
to suit my convenience. We had two excellent days of conversation
of that intimate character which we always have within the close
family of the Commonwealth. These discussions are always
informal in the sense of being off the record, but they are all the
more valuable, and they were particularly useful following on our
talks with the United States. The power and authority of the
Canadian Government are continually growing in international
affairs, and we are glad to recognise this. In general, I can perhaps
best sum up our discussion with a phrase which Mr. St. Laurent
used in public, which I think was a good one. He said: 'We found
that our anxieties were your anxieties'. . . .

6. Extract from speech by the Leader of the Opposition, Mr. Hugh Gaitskell, in the House of Commons, 15 May 1957[1]

I beg to move, 'That this House expresses its concern at the
outcome of the Government's Suez Canal policy, and deplores the
damage to British prestige and economic interests resulting
therefrom'.

. . . The dispute between us and the Government has never
been on the merits of Colonel Nasser's action in nationalising the
Canal. We condemned this as they did and as did many other
countries, and we condemned it not because of the act of nationalisation but because of the way it was done and because of our very
natural doubts about the future of the Canal. Our dispute with the
Government has throughout been simply on the question of how
this problem should have been dealt with. From the start we have

[1] United Kingdom, H.C. Deb., vol. 570, coll. 411, 421–5.

urged two things: first, that we should not act alone in this matter, and, secondly, that we should not use force contrary to the Charter of the United Nations. Both these points were perfectly clearly stated in the very first speech that I made on behalf of my party on 2nd August [Part I, I, Document 5, *supra*], and they were repeated again and again in the innumerable debates that took place in the summer and autumn.

We accepted the 18-Power proposals. We did not think them unreasonable. We urged that the Government should not regard them as an ultimatum to be accepted or not, but should be ready within the United Nations to negotiate with Egypt about them. We believe that by the end of October we were in sight of a settlement which, though not ideal, was reasonable.

It was at that point that the Government chose, using the excuse of the Israeli attack, to invade Egypt; and it is clear that they planned to do so when they learned of the possibility of the Israeli attack. What the consequences were is clear to all and cannot be denied—the blocking of the Canal, the cutting of the pipelines, the strain on the pound, the introduction of petrol rationing here, the check to industrial expansion, a tremendous blow to our reputation in the world and particularly in the Middle East and in Asia and, at the very least, a very serious breach between ourselves and America and grave misunderstanding in and a threat to the unity of the Commonwealth. Let me say one further thing about this. One of the worst features of this whole episode, I think, is that it was a really serious blow to that better relationship between the peoples of the East and the West which we had so carefully fostered during the previous ten years.

Some of those consequences are now past, but others still remain. I know there are those who would say—there are no doubt many of them on the Government benches—'If only we had been allowed to go through with it, if only the United States had behaved differently, if only the United Nations had behaved differently, and if only Russia had behaved differently.'

Hon. Members: If only the Opposition had behaved differently.

Mr. Gaitskell: Yes, and if only the Opposition had behaved differently.

Let us take these points that the United States might have behaved differently. Is it really suggested that after we had deceived our closest ally we should expect it to come to our aid in

doing something of which we knew from the start it would disapprove? The French Prime Minister admitted that the reason
why we never consulted the Americans was that we knew that they
would disapprove. Was it really to be supposed that the United
Nations would take lying down this flagrant violation of the
Charter, that it would be prepared to back us up? Did the Government really believe that there would be no two-thirds majority
against us?

As to Russia, is it really suggested that the Russians were not
going to react at all? Certainly, in the correspondence now published between Mr. Bulganin and Sir Anthony Eden it is made
perfectly plain by the Russians what was going to happen. It
really is absurd to suggest that these reactions are something which
need not have happened, that they are not part of the whole facts
of the world situation which should have been taken into account.

Now, what of the Opposition? There are those, I know, who talk
nowadays about us—I repeat, us—searching our consciences and
standing in white sheets. If those who talk this way are serious,
they utterly fail to understand the point of view of the Labour
Party. We could not possibly support a policy which was, in our
view, utterly wrong and utterly damaging to our national interests.
We could not support a policy which violated all the principles on
which our foreign policy had been based for the past ten years.

We wished to see, and still wish to see, our country not putting
itself above international morality but giving moral leadership to
the world. Nor can we accept, as Her Majesty's Opposition, that
we have to adopt as our slogan, 'My Government, right or wrong'.
We claim, too, that in refusing to support the Government's
action in going to war, and in opposing it, we did something at
least to preserve a bit of our reputation in the Middle East and in
Asia. And let hon. Members be quite clear about this. Whatever
attitude the Opposition had adopted in this matter would not have
made the slightest difference to the reaction of the United States,
or the reaction of the United Nations or the reaction of Russia.

I appreciate that, of course, there may be a certain inner logic
among those who have recently left the party opposite, by which
they say, 'We will go it alone, never mind what happens.' But,
while there is an inner logic there, it is utterly out of relation with
the realities of the external situation.

This notion that we can 'go it alone', and that we have the power to do so, regardless of the attitude of our allies and the rest of the Commonwealth, in the world as it is today is nothing short of lunacy. The mystery is how the Government themselves for a time followed this policy, for, leaving morality out of it altogether, if hon. Members wish, they will not deny that an appalling error of judgment was made.

It is not right that the responsibility for this should be laid to the door of one man. It is not the former Prime Minister, but the whole Government who are responsible—those who backed it throughout, those who had opposed it but had not the courage to resign, and those who backed it first and then withdrew their support. The present Prime Minister's role in these weeks remains somewhat obscure. Yet he is a man who could, perhaps more than any other member of the Government, have given sound advice in this matter. He had been Foreign Secretary, he had been Minister of Defence, and he was Chancellor of the Exchequer.

I wonder whether he warned the Cabinet of the time that there might be certain defence difficulties, which have since emerged. Did he tell them that he thought the Americans would welcome all this with open arms? In the light of his experience as Foreign Secretary, did he imagine and tell the Cabinet that the United Nations would give it their blessing? Did he, as Chancellor of the Exchequer, consider for a moment that perhaps, if this were to be done, there might be a little trouble with the pound, or was it only at the last that he came to realise this? Is it true that in the early stages he was the most flamboyant and romantic supporter of the whole plan? Perhaps we shall hear a little more about this point of view this afternoon.

So much for the past. As for the future, the lesson is clear enough. We must reaffirm our principles in the three essentials of our foreign policy—a united Commonwealth, support for the Charter of the United Nations, and the Anglo-American Alliance. We must not only reaffirm these, but we must base our foreign policy upon them too, for only in that way and by abandoning the path down which the Government strode last autumn, can we hope to regain some of our lost influence and prestige in the world. Only in this way can we recreate confidence between ourselves and our allies and unity in the Commonwealth; and both these things are essential if we are to achieve the lasting world settlement for which the sorely distracted peoples of the world are longing today.

7. Extract from speech by the Prime Minister, Mr. Harold Macmillan, in the House of Commons, 15 May 1957[1]

We cannot complain of the decision of the Opposition to put a Motion of censure on the Order Paper, or even at their attempt to exploit a political situation from which they may hope to draw some dividends, but I am bound to say that I am slightly surprised at the terms of the Motion.

I try to judge these matters as objectively as possible, but, with the best will in the world, I cannot agree that over the past seven or eight months the Leader of the Opposition has chosen as his single directing purpose the maintenance of British prestige. The incidents of last November have been debated over and over again, and the pros and cons of the Government's action have been exhaustively discussed. No doubt, there will be some Members in this long two-day debate who will wish to go over it once more, but I must say that I would not have thought that the Leader of the Opposition would be very anxious to recall some of the incidents of these discussions.

The broad opinion of the great majority of our fellow-country-men can, I believe, be stated quite simply. They have a full understanding of the crucial problem and the crucial challenge which confronted the Prime Minister and the Government of the day. They do not think that the Opposition did very much to help. They know that Sir Anthony Eden was actuated by the highest patriotic motives, based upon a longer experience and fuller understanding of world affairs than almost any man in this country. They have the deepest sympathy for him in the cruel illness which struck him down at the critical moment. They believe that if the decisions that he took, with the full, complete and unanimous support of his Government—

An Hon. Member: Why did he resign?

The Prime Minister: —with the full, complete and unanimous support of the Government—

Hon. Members: No.

The Prime Minister: —they believe that, if those decisions were misrepresented at home and misunderstood abroad, yet there is a growing body of opinion in every part of the world which is beginning to appreciate and understand the reasons for the Government's policy. They therefore believe that his actions and those of his colleagues will be fully justified by history. . . .

[1] United Kingdom, H.C. Deb., vol. 570, coll. 425–6.

II. Canada

1. Extract from speech by the Acting Leader of the Opposition, Mr. W. Earl Rowe, in the House of Commons, 26 November 1956[1]

. . . I would be ashamed to stand in this House if in the United Nations I had seen the United States voting as they did to drive Britain and France out of the Mediterranean area. Britain and France did agree they would vacate the area when there was an adequate police force, but to turn around now and tell them to get out and leave the area on which their lifeline depends represents a strange attitude. That is not good enough for the senior member of the British Commonwealth of Nations. Therefore, Mr. Speaker, I move on behalf of Her Majesty's loyal opposition, seconded by the hon. member for Vancouver-Quadra (Mr. Green):

That the following be added to the address:
That this House regrets that Your Excellency's advisers

(1) have followed a course of gratuitous condemnation of the action of the United Kingdom and France which was designed to prevent a major war in the Suez area;
(2) have meekly followed the unrealistic position of the United States of America and have thereby encouraged a truculent and defiant attitude on the part of the Egyptian dictator;
(3) have placed Canada in the humiliating position of accepting dictation from President Nasser;
(4) have failed to take swift and adequate action to extend refuge to the patriots of Hungary and other lands under the cruel Russian yoke.

2. Extracts from speech by the Prime Minister, Mr. L. S. St. Laurent, in the House of Commons, 26 November 1956[2]

. . . These gentlemen [opposite] who utter these high-flown phrases seem to forget that the nations of the world signed the Charter of the United Nations and thereby undertook to use peaceful means to settle possible disputes and not to resort to the use of force.

I have been scandalized more than once by the attitude of the larger powers, the big powers as we call them, who have all too

[1] Canada, H.C. Deb., 1956 (Special Session), p. 18.   [2] Ibid., pp. 22–23

frequently treated the Charter of the United Nations as an instrument with which to regiment smaller nations and as an instrument which did not have to be considered when their own so-called vital interests were at stake. I have been told, with respect to the veto, that if the Russians had not insisted upon it the United States and the United Kingdom would have insisted upon it, because they could not allow this crowd of smaller nations to deal decisively with questions which concerned their vital interests.

An hon. Member: Why should they?

Mr. St. Laurent (Quebec East): Because the members of the smaller nations are human beings just as are their people; because the era when the supermen of Europe could govern the whole world has and is coming pretty close to an end. . . .

. . . There was no gratuitous or other condemnation by Canada but there has been an expression of regret that certain members of the United Nations had felt it necessary to take the law into their own hands when the matter was before the Security Council; and there was an expression of regret that what took place in the Middle East was used as a screen to obscure the horrible actions, the horrible international crimes, that were being committed in mid-Europe at the same time. Events in the Middle East made it more difficult to marshal world opinion in unanimous and vigorous condemnation of what was taking place in Hungary at that very moment.

That is what we regretted. We feel that there can come out of this situation one that will be better than that which existed previously. It is our hope and it has been our objective to get all those in the western alliance . . . working together toward . . . a settlement of the Mid-Eastern situation that will be lasting. . . .

Mr. Diefenbaker: Would the Prime Minister allow a question? . . . Is the Prime Minister in a position to say whether he will reveal the communication that was sent to Sir Anthony Eden. . . ?

Mr. St. Laurent (Quebec East): I would be very happy to be able to reveal that correspondence. . . . I sent a message to Sir Anthony Eden dated November 21, asking him what would be the attitude of the United Kingdom Government in the face of such a request, because a somewhat similar request had been made in the House of Commons at Westminster in respect of confidential correspondence with Ceylon and the answer had been that that correspondence could not be published. So I wrote a letter which I might read into the record together with the answer I received to that communication . . . :

Message from the Right Honourable Louis S. St. Laurent to the Right Honourable Sir Anthony Eden:

A leading member of the official opposition has stated publicly that, when our Parliament meets in the near future, he proposes to ask for the tabling of one of the communications I addressed to you recently in reply to one of yours.

It is obvious that this correspondence between us could not be published piecemeal and that, if one of these confidential communications were published, they would all have to be published.

A similar question arose in our parliament some eight years ago about similar communications between Mr. Churchill and Mr. Mackenzie King regarding the international situation some seven years before that.

Mr. Attlee's government at that time took the view that such communications should not be published at any time because, as they said: 'Such telegrams are framed on the basis that they will not be published and the whole system of full and frank communication between His Majesty's Government would be prejudiced if telegrams of this nature had to be prepared on the basis that this rule might not eventually be observed.'

I would be glad to know what would be your attitude now and the attitude of the government of the United Kingdom with respect to these confidential communications which have recently passed between us.

I would like to read this message and your reply to it into our record of debates for future reference whenever similar requests for publication of confidential communications may arise.

Message from the Right Honourable Sir Anthony Eden to the Right Honourable Louis S. St. Laurent:

Thank you for your message and for consulting me about the possible publication of the confidential communications which have recently passed between us.

The United Kingdom Government's view on such publication remains identical with that expressed by their predecessors on the occasion you mention in your letter. It is, we feel, essential, if there is to be that full intimate and frank exchange of minds between Commonwealth governments on which alone policy can be based, that we should all of us be able to proceed on the basis that such correspondence shall be and remain confidential and shall not be published. That is the principle to which we in this country have consistently worked, and, as it happens, it was reiterated so lately as November 15, in answer to a parliamentary question in the House of Commons.

I am sure this is the only possible practice.

I am sorry, because statements have been made or at least have been reported to have been made, not only in this country but in the United Kingdom, suggesting that I had sent a blistering reply to Sir Anthony Eden. I am not free to disclose that correspondence. . . .

3. Extracts from speech by the Member for Vancouver-Quadra, Mr. Howard Green, in the House of Commons, 26–27 November 1956[1]

. . . Feelings on these questions raised by the Suez crisis, Mr. Speaker, are running very deep in Canada, far deeper I believe than the Government has the slightest conception. Listening to the Prime Minister I could not help but think he has been living in some other land altogether so far as public reaction to these issues is concerned, and particularly reaction to the attitude of the Canadian Government.

This attitude has come as a great shock to millions of Canadian people. In Vancouver the story broke in the headlines on October 31, and I must admit that even I was shocked, although the stand taken was just in line with the stand this government has been taking for the last 10 years. . . .

This afternoon the Prime Minister said that when the vote came up about the cease-fire, then Canada abstained. He did not explain that while the Minister for External Affairs abstained, in his speech the Minister showed very clearly that he was condemning the United Kingdom and France. The Prime Minister should have made that clear. . . .

Again . . . when the second resolution about the cease-fire was under discussion the Minister got up and said that this was all wrong, there had already been a resolution passed and the United Kingdom, France and Israel were complying with it. They had already taken steps to comply with that resolution and this second resolution should not be passed. Then the Canadian Government did not have the courage to get up and vote against it. Only the United Kingdom, France, Israel, Australia and New Zealand

[1] Canada, H.C. Deb., 1956 (Special Session), pp. 40–41, 49–50.

voted against that foolish and provocative resolution. The Canadian Government, representing the land of courageous people, did not have the backbone to get up and vote against that resolution; they were so busy currying favour with the United States. . . .

[November 27] . . . It was very strange yesterday that the Prime Minister of this country did not have one word to say in condemnation of the action of Egypt in blocking the Canal or her other actions. His whole attitude to Egypt is unbelievably soft. Yet the minute the United Kingdom and France moved, Canada rushed to condemn them.

The man on the street in Canada is asking today, and has asked ever since the Canadian Government took such action, why Canada took the lead in the attack on her friends. . . .

It so happens that hundreds of thousands of Canadians have been associated with the people of the United Kingdom and France. They know something about those two countries and from personal experience they know a great deal about their people. They know that no other two countries have ever done so much to preserve our way of life. In the first war at Verdun hundreds of thousands of young Frenchmen gave their lives. At the Somme the very flower of the United Kingdom was wiped out in order to preserve the democratic way of life. In the second world war France and the United Kingdom stood up against the aggressors when other nations who now talk a great deal were not sufficiently concerned to do the same thing. . . .

The Canadian people know, even if the Prime Minister does not, that the United Kingdom and France have never been aggressors and are not aggressors on this occasion. Yet the Prime Minister yesterday had the effrontery to compare the actions of the United Kingdom and France with the actions of Russia in Hungary.

Mr. St. Laurent (Quebec East): Mr. Speaker, I protest . . . against this statement, which is an entire misconstruction of what is the official record of the debate in this House. . . .

Mr. Speaker: The hon. member for Vancouver-Quadra has the floor.

Mr. Green: Actually the United Kingdom and France by their action—taken, I am sure, in the full realization of the tremendous risk involved—prevented a Russian-dominated Middle East, and prevented a major war. The United Nations could never have done it. The United Nations was taking no step to meet that situation. Now the United Kingdom, France and Israel are co-operating with the United Nations and they are being kicked in the teeth for their pains. . . .

4. Extracts from speech by the Secretary of State for External Affairs, Mr. L. B. Pearson, in the House of Commons, 27 November 1956[1]

. . . We are facing today a situation of gravity and danger, far too serious a situation to be dealt with from a purely partisan point of view. The hon. gentleman who has just taken his seat [Mr. Howard Green] talked about Canada being the chore boy of the United States. Our record over the last years, Mr. Speaker, gives us the right to say we have performed and will perform no such role. It is bad to be a chore boy of the United States. It is equally bad to be a colonial chore boy running around shouting 'Ready, aye, ready'. . . .

The debate in this House—and we have been meeting for only a few hours—has already shown that a very real difference on policy has developed between the Government and the Official Opposition. The speeches of the Acting Leader of the Opposition [Mr. W. Earl Rowe] and the hon. member for Vancouver-Quadra [Mr. Howard Green], who has just preceded me, have made that quite clear. The Official Opposition . . . now apparently support every move made by the United Kingdom and France in their intervention in Egypt after the attack on Egypt by Israel, an intervention brought about with army, navy and air forces after a 12-hour ultimatum. They claim, I have the right to conclude, that we as a government should have approved of those moves at once and should have backed up the United Kingdom and France at the United Nations even on those matters and on those resolutions where not a single member of the United Nations supported the resolution in question. . . .

Now, Mr. Speaker, we did not follow that particular line of policy in this matter, and I shall try to explain why. . . .

It is quite obvious—it was quite obvious by the summer—that there was no meeting of minds between Washington and London and Paris. . . . And, of course, the fault was not by any means entirely on the side of London and Paris, and no one on this side of the House has ever tried to take a one-sided view of this

[1] Canada, H.C. Deb., 1956 (Special Session), pp. 51–56.

situation. The vital importance of the Suez to Western Europe is perhaps not appreciated in Washington, and it might have been better appreciated there if this situation could have been related by them to the Panama Canal.

Now our own attitude in this matter was—and we expressed this attitude in the House of Commons and in a good many messages to the United Kingdom Government during the summer —that we did not stand aloof and indifferent, and we did appreciate the importance of this development not only to Western Europe but to Canada itself. Our attitude was that this question should be brought as quickly as possible to the United Nations and a solution attempted there; that at all costs there should be no division of opinion, no division of policy, between Washington and London and Paris on a matter of such vital importance, and that there should be no action taken by anybody which could not be justified under the United Nations Charter; otherwise the country taking that action, no matter how friendly to us, would be hauled before the United Nations and charged by the country against which the action had been taken. That is something that has happened, and it is something we tried to talk over with our friends before it happened.

It will be recalled that eventually the matter was taken to the Security Council of the United Nations, and it will also be recalled that not long before the use of force by Israel against Egypt certain principles for a settlement of the Suez question had been agreed on at the Security Council. One of those principles which had been accepted by Egypt at that time, was that the Canal should be insulated from the policies of any one nation, including Egypt. Therefore at that particular moment, through these conversations at the Security Council, and what is more important through conversations going on in the Secretary-General's office, we had some hope that an international solution might be reached which might be satisfactory to all concerned.

At that time, and I am speaking now of a period of only a week or two before the attack by Israel took place, we had no knowledge conveyed to us of any acute deterioration of the situation, nor did we have any knowledge or information about anything which could be called a Russian plot to seize Egypt and take over the Middle East. At that moment, and against that background, the Israeli Government moved against Egypt. . . .

I admit—and I am sure all members in this House must admit—the provocation which may have prompted this move. We in the Government tried to understand that provocation; nevertheless we did at that time, and do now, regret that the attack was made at that time and under those circumstances. Then, as the House knows, the United Kingdom Government and France intervened in the matter on the ground, so they claim, that it was necessary to keep the fighting away from the Suez Canal and thereby keep the Canal open. They wished, so they said in Paris and in London, to keep a shield between the opposing forces. . . .

Well, to carry out that purpose, as we know, the French and British Governments sent an ultimatum to Egypt and to Israel, a 12-hour ultimatum that was accepted by Israel whose forces at that time had come within ten miles of the Suez Canal, but was rejected by Egypt which had been asked to withdraw its forces beyond the Suez Canal and following that rejection the United Kingdom and French forces intervened by air and later on the ground.

At that time far from gratuitously condemning the action, the Canadian Government said through the Prime Minister, and indeed through myself, that we regretted the necessity for the use of force in these circumstances; and these circumstances, I confess, included an element of complete surprise on our part at the action taken.

There was no consultation—and this has been pointed out—with other members of the Commonwealth and no advance information that this very important action, for better or for worse, was about to be taken. In that sense consultation had broken down between London and Paris on the one hand, the Commonwealth capitals and—even more important, possibly—Washington on the other.

Nevertheless, instead of indulging then or since in gratuitous condemnation we expressed our regret and we began to pursue a policy, both here by diplomatic talks and diplomatic correspondence, and later at the United Nations, which would bring us together again inside the Western Alliance and which would bring about peace in the area on terms which everybody could accept. . . .

It was certainly a matter of urgent and distressing importance, especially to a Canadian, and I expressed this also in public at the

United Nations, that the United States should be on one side of this issue and the United Kingdom and France, our two mother countries, on the other. We were especially distressed at this because there were people down in New York, and they are still there, who are gleefully exploiting this division.

Having mentioned the breakdown of consultation, I think it would only be fair to add that this breakdown of consultation and agreement was not the fault exclusively of the United Kingdom and France over the preceding months. No other member, indeed, no member of the Western Alliance, is free of some responsibility, and particularly the United States of America, which is the major and most powerful member of that group. Therefore we felt and we still feel that this is no time nor is this an occasion on which to adopt an attitude of superior virtue or smug complacency over the righteousness of our own position. . . .

Then also, and this was a matter which was very much on our minds, we were anxious to do what we could to hold the Commonwealth together in this very severe test. It was badly and dangerously split. At one stage after the fighting on land began it was on the verge of dissolution, and that is not an exaggerated observation. . . .

There are those in this country and there are some whose views have been expressed in this House who feel that we should have automatically supported the United Kingdom and France, either because of the ties of friendship, indeed of kinship, with the countries concerned, or because they were convinced the United Kingdom and France were right in the course adopted and in the methods followed. Those who feel that way will be disappointed at the action we have taken. We thought it was the right action for a Canadian delegation to take.

It was an objective attitude, it was a Canadian and an independent attitude. Believe me, the Arab and Asian countries, including the Asian members of the Commonwealth, were watching us, as they were watching others, very carefully to see if our policy was based on those considerations I have mentioned or whether we were just following automatically any other power. If we had given any evidence that would have justified the impression that we were supporting without reservation the United Kingdom and France in all their tactics and attitudes towards this matter we would not have been of any help to our friends subsequently, nor would we have been able to play the part which we at least tried to play. . . .

Our purpose was to be as helpful to the United Kingdom and France as we possibly could be. Believe me, that attitude has been appreciated in London even if it has not been appreciated by my hon. friends opposite. Far from criticizing us in private or in public in London or Paris for our gratuitous condemnation of their course we have had many expressions of appreciation for the line we have been trying to follow, and which has been helpful in the circumstances to the United Kingdom and France. . . .

5. Extract from speech by the Secretary of State for External Affairs, Mr. L. B. Pearson, in the House of Commons, 14 January 1957[1]

. . . In actual practice, there have been over the last 10 years or so since World War II very few international occasions when we have not been on the side of Great Britain; the centre of our Commonwealth. But the rarity of dissenting occasions stems not from our automatic acceptance of the policies of Great Britain but from the fact [that] in the vast majority of international questions our interest and hers have happily been almost invariably identical. When that does not happen we, of course, regret it deeply and we do our best to reconcile our differences without delay and without recrimination. We experienced such regret indeed to the point of distress when we differed, not perhaps in objectives but in methods and procedures, with the United Kingdom on certain occasions at the United Nations Assembly meeting last autumn in connection with the Suez crisis. The Commonwealth was indeed deeply split on that issue and our relief was therefore correspondingly great, a relief shared in full measure by the Asian members of the Commonwealth, where the separation pressures were most intense, when this danger to the Commonwealth was removed by the Anglo-French decision to accept the cease-fire resolution of the United Nations Assembly. So the Commonwealth association remains strong and close. The friendly, informal and frank exchange of views in a sincere effort to reach agreement on all matters of common concern goes on, and the Commonwealth continues to play its invaluable and constructive role in today's

[1] Canada, H.C. Deb., 1957, vol. I, pp. 173-4.

troubled world; a role for which the whole world has reason to be grateful.

Mr. [Gordon] Churchill: Will the hon. gentlemen permit a question? What nations of the Commonwealth would have left the Commonwealth had the British and French not abided by the resolution of the United Nations?

Mr. Pearson: There is evidence, strong evidence, which I and others have received, to suggest that if the fighting in Egypt between Anglo-French and Israeli forces and Egyptian forces had continued and if the United Nations Assembly cease-fire resolution had been repudiated or rejected, the pressures in regard to separation from the Commonwealth in certain Asian members of the Commonwealth would have been so great that it would have been indeed very difficult to resist them. We have had evidence to that effect both from New Delhi and from Karachi.

Mr. Churchill: Has that not been denied by both Ceylon and India?

Mr. Pearson: This has been questioned, I believe, in Ceylon, including the Prime Minister. Mr. Speaker, I am giving my opinion on the basis of information which I have received from the highest authorities in the government of India. I am not suggesting, Mr. Speaker—and in my earlier statement on this I think I made it clear in the House I did not suggest—those pressures affected what we sometimes call the old members of the Commonwealth, but they certainly did affect those new members which, as I have just said, constitute four-fifths of the population of the Commonwealth.

It seems to me that this Commonwealth association, which all its members wish to preserve, to be of enduring value must strive for the widest possible areas of agreement between its members. It seems to me also that the limits of such areas, though not often expressed, may be pretty clearly discerned. Whether or not we speak of it, there are certain fundamental things that unite the governments and peoples of the Commonwealth: freedom, personal and national; parliamentary democracy and the supremacy of the individual over the state. There is also a certain basis of morality in political action to which Commonwealth members are by tacit consent expected to adhere. Such a basis can easily be disregarded, on the other hand, by those who do not share our Commonwealth beliefs and our ways of doing things. . . . It is more important than ever for us at this time to strengthen within the Commonwealth our will to work together in defence of these principles; for very significant events are now about to occur in the Commonwealth, as significant perhaps as those which took place 10 years ago when India, Pakistan and Ceylon became members. . . .

6. Extracts from communiqué of the meeting of Ministers of the Governments of the United Kingdom and Canada, 26 March 1957[1]

. . . In the course of their talks the Canadian Ministers noted with satisfaction the results of the Anglo-American meeting of March 21 to 23. . . . The Ministers discussed the steps which might be taken to secure acceptable settlements of the short-term problems in the Middle East. They also reviewed a number of less immediate political and economic questions in this area. . . .

The Ministers recognized the need to strengthen and improve the working of the United Nations as an instrument for improving peace with justice. . . .

This meeting has again demonstrated the value of the family relationship between the peoples of the Commonwealth, and the close and continuing co-operation between the Governments of the United Kingdom and Canada.

While there may from time to time appear to be differences in their approach or reaction to international developments, the two Governments are confident that their special relationship will always enable them to work together effectively with a constructive purpose.

7. Extracts from interview of the Secretary of State for External Affairs, Mr. L. B. Pearson, by Mr. Blair Fraser and Mr. Lionel Shapiro, May 1957[2]

Mr. Fraser: Mr. Pearson, the commonest thing we hear said about Canadian foreign policy . . . is that . . . Canada hasn't got a foreign policy, that we're just a tail to the American kite. What is your answer to this?

Mr. Pearson: Well, my short answer would be that it is not true. The results over the last ten years would show, I think, that

[1] *The Times* (London), 27 March 1957.
[2] 'Where Canada stands in the World Crisis', *Maclean's Magazine* (Toronto), 6 July 1957.

we did have a foreign policy as much as any middle power or
smaller power can have a foreign policy where events are dominated by giants, and that in the application of that foreign policy
we have had some influence on international developments.

Mr. Fraser: The charge came up most particularly of recent
months over the Suez crisis last fall. Formerly it was a perfectly
good answer to say that we follow the Anglo-American line; but
events knocked the hyphen out of Anglo-American last October
and November and the Americans went one way and the British
went another. Some people in Canada say that we deserted the
British by failing to support them as Australia and New Zealand
did, but those who take the opposite view feel that we could have
been a little stronger on the other side. Actually, since the chips
were down, since our friends and exemplars were differing, we
were unable to do anything—we abstained.

Mr. Pearson: No. That I don't accept. If we had automatically
followed British policy in the Middle East—and it would have had
to be pretty automatic because we didn't have much time to learn
about it—then we certainly would be open to the accusation we
had no foreign policy. The fact that we did not follow the British
policy did not mean that we weren't anxious to work with the
British. Nor did it mean that we just swung over to the other
extreme and automatically followed the American policy. When
the British and the Americans disagree we're in a dilemma; it's a
great dilemma in Canadian foreign policy. But, if, when they do
disagree and break and we follow one or the other, that doesn't
mean, surely, that we have no foreign policy of our own. If we
agree with a greater power are we to refuse to follow it just because
it is a greater power?

Mr. Fraser: I, personally, was one of those who agreed with
what the government thought at the time of Suez but felt that it
would have been better if the government had said more openly
exactly what it did think.

Mr. Pearson: Why did we abstain? Incidentally, we weren't the
only people who abstained and, in other votes, both the British
and Americans abstained when we voted for or against. But on this
occasion, on the big vote, whether there would be a cease-fire or
not, we abstained. Three Commonwealth countries voted for,
three voted against; Canada and South Africa abstained. We
explained our abstention in the following terms: that the vote had

been rushed; it was too important to take so quickly; while we were, in principle, in favour of a cease-fire and would not vote against the United Nations' cease-fire, at the same time we thought that a cease-fire should be accompanied by other measures which would secure the cease-fire and also provide for a settlement of disputes which brought about the fighting in the first place. We had no opportunity to plead that case and try to amend the resolution which we supported in principle. Therefore, if we couldn't vote against, we didn't want to vote for it because it was inadequate, we thought. So we had to abstain. Now, does that mean that we had no policy on that occasion?

Mr. Shapiro: . . . In view of what I call the revolution that took place last fall in our values, what are the problems in conducting Canada's foreign policy that now arise?

Mr. Pearson: There is no change so far as I can see in kind, but there is a change in degree. The United States for the last ten years, by the compulsion of force and events, has been the leader of the Western coalition, and we have had to adjust our policies to that development. So have the British and so have other people. All this came to a head in a very dramatic way last autumn. It didn't begin last autumn, but it came to a head when it became quite clear that no longer could the British or the French—and if they couldn't nobody else on our side could—take independent action in a part of the world where power, economic and military, was essential to bring that action to a success without, at least, the tacit support of the United States. Now, that has underlined the development you have mentioned, and hasn't caused it. I think you have to take that into account in Canadian policy, probably to an even greater extent than we did in the past—and so must the British and the French. We have to readjust ourselves to this new situation where the United States is even more dominant in the free world than it ever was before.

Mr. Shapiro: But doesn't this give Canadian policy a little more independence and latitude? Before last November the Western world had a more or less common policy, especially on the great issues. The United States, Great Britain and France voted more or less the same way at the United Nations. We had no large problem there, but we have a large problem now. We, at least, have got to have a certain independence.

Mr. Pearson: Yes, that is true—a certain independence.

Unfortunately, but perhaps inevitably, no country has independence in foreign policies these days, not even the United States. What I think you are suggesting is that, as the facts of American power become more obvious, we should be careful not to yield too easily to those facts and be very, very careful that American power is being used in the right way and try to influence it in the right direction. I think we tried to do that last autumn at the United Nations. We worked as hard as we possibly could with the Americans to get their support for a resolution which would lay down in detail specifically the arrangements we should follow for the withdrawal of foreign troops and Israeli troops from Egypt. We weren't successful. But we did our best to modify their attitude. We also told them, and in no uncertain terms, that if they supported a resolution of sanctions against Israel, we would have to break with them because we would not support it in those circumstances. I think we modified their approach.

Mr. Shapiro: Our policy was described the other day as 'schizophrenic'—and it was a natural description; it was not said in derogation at all. We apparently agreed with the United States on the morality and the unwisdom of the Anglo-French action at Suez. We also agreed with France and Britain on the fact that there were great provocations. So out of this schizophrenia we devised the United Nations' Emergency Force. We were torn between both sides.

Mr. Pearson: Yes. And when we are torn between both sides we instinctively try to find some kind of a solution on which the British and Americans can agree. As I have often put it, we have to keep in step even though we are standing on our own feet. We have to keep in step with these two people, and we can't do that if they don't keep in step with each other. . . .

Mr. Fraser: We started out by talking about Canada's attitude towards Anglo-American differences, and this is another Anglo-American difference, isn't it? The British are saying now, and have been saying for the last six months, that the Americans are unrealistically and foolishly passing the buck to the United Nations. The British say that when the Americans say this is a matter for the United Nations to settle, they are really just making an ineffectual attempt to abdicate power which they alone can wield, or which at any rate the great powers alone can wield. They're saying that the United Nations, being in essence merely a debating society, is just too soft to be the cornerstone of anybody's foreign policy. Without suggesting that I accept this view, could I ask for your comment on it?

Mr. Pearson: I have heard a good deal about that, and we talked about it in Bermuda. It seems to me that both the British and the Americans are influenced in their attitude to this particular problem by their experiences of the last six months. The United Kingdom didn't have a very happy experience with the United Nations Assembly, to say the least. You would expect them to have a certain feeling of disillusionment and disappointment about the United Nations because of that experience. The United States, on the other hand, found that the United Nations was a most valuable and important international agency to use for this emergency and perhaps as an escape from some of the urgency for immediate national decisions. They think the United Nations, for that reason, is a most important and valuable organization. Now, being a Canadian, I naturally find myself halfway between these views. I think it is folly to use the United Nations as an escape for making national decisions. But I think it is equal folly to say that the United Nations Assembly is now under the control of a lot of Asians and Arabs with no sense of responsibility and we must extricate ourselves from it. What we need at the United Nations Assembly is a restoration of Anglo-American leadership, and that means unity between the two governments. Things broke down last autumn in the United Nations largely because the British and Americans weren't working together. That division reflected itself in other delegations, too.

Mr. Shapiro: The reason why the British and the Americans didn't work together—isn't this as a result of a change in power relationships? Didn't the British feel, during the last summer, especially since the seizure of the Suez Canal, that America was now taking a new tack in international relationships, that its British and French allies had become regional allies rather than world partners in the old sense?

Mr. Pearson: I think that that had something to do with it. I think there was a growing feeling in London and Paris last autumn that United States policy was wavering and inconsistent and wasn't taking enough into consideration the needs and the interests of the Western European countries. But the immediate cause of the breakdown was the action that the United Kingdom and France

took and which got very little support at the United Nations. It
was British and French action in that sense that brought about the
immediate collapse of co-operation. But, that had been building
up, as you say, over the months. . . .

Mr. Fraser: Before we leave this question of the U.N., I'd like
to have your answer to the question a little more broadly. Do you
think it was the United Nations' vote, and not just the American
pressure and the Russian threat, that brought the cease-fire in
Egypt?

Mr. Pearson: There may have been other considerations.
Perhaps the things you have mentioned were very important, but
they couldn't have been worked out if the United Nations hadn't
been there, an international organization, to step in.

Mr. Fraser: You mean as a system of communications.

Mr. Pearson: As a system of international political communication, as machinery for the solution of disputes and as a forum for
the expression of world opinion. . . .

Mr. Shapiro: Last November 1 and 2, the United Nations
General Assembly extended debate on the issue. Suppose the
Anglo-French action had gone through to a military success. It
would be interesting to speculate on what the results would have
been.

Mr. Pearson: Well, I'll be glad to speculate on that because
speculation is an interesting historical pastime, sometimes an
interesting political pastime. My view—and historians will be
arguing about this a hundred years from now, if there are any
people left on the planet—my view is that if the fighting had gone
on, if the British and French military intervention had continued,
they would have had no difficulty, of course, in bringing about
military victory. That would have been simple, but they would not
have been able to keep control of the Canal without controlling and
occupying the whole of Egypt. Earlier the U.K. found it well-nigh
impossible to control and operate the Canal from their military
base on the Canal when the local population were bitterly opposed
to them. This would have meant that Great Britain, which is
having a pretty hard time economically and financially discharging
its present responsibilities, would have had the occupation of
Egypt on its hands. That is one result. Another result would have
been, I think, the deep and bitter and prolonged hostility of the
whole Arab Asian world. They wouldn't have fought, but they

would never have accepted that position. They would have been so
bitter and hostile that some of the Arab states would have been
tempted to call in Russia to help. As Sir Winston Churchill once
said: when you are really up against it you'll accept help from the
devil; and the Arab world would have been up against it then.
Now, that's two results. I will give you the others. I think the
strains and stresses on Asian members of the Commonwealth
would have been so great that they would not have been able to
withstand them. That's the third result. And then, the fourth
result would have been an even greater breach between Washington and London than that which actually existed.

Mr. Shapiro: If the Israelis had been left alone to beat Nasser
single-handed, as they could easily have done, then what would
have been the effect?

Mr. Pearson: The answer given to me by top people in London
to that question was that if the Israeli had been allowed to fight
alone the bitterness of the Arab world towards Israel would have
been so much greater than the bitterness which was aimed against
Britain and France that there would have been no question that
other Arab states would have invited the Russians in quickly. . . .

Mr. Shapiro: . . . I wondered whether you would comment on
our relations with Great Britain, especially since the events of
last November. I was referring particularly to two remarks made
in the debate in the Commons on the Suez question. One was by
the Prime Minister—the 'superman' remark; the other one was by
you—the 'colonial choreboy' remark. I was just wondering
whether this reflected a new attitude on our part toward very close
collaboration with Great Britain.

Mr. Pearson: My view on what our relationship with Great
Britain should be hasn't changed at all since the events of last
autumn. You mentioned the 'colonial choreboy' remark, but you
must recall that I wasn't suggesting that we were a colonial choreboy; this expression was thrown at me by somebody else as dero-
gatory about our relations with the United States, suggesting that
we were a choreboy of the United States. I said it is wrong to be a
choreboy of the United States or the United Kingdom or anybody
else. But we're not, of course, and our relations with the United
Kingdom remained close and friendly throughout all the difficult
days and hours of last autumn. Our delegations were in close touch
and even when we disagreed we talked things over. I have spent

hours with them trying to see how we could work things out together. Sometimes we couldn't, but that seems to me to be the kind of relationship that makes the Commonwealth worthwhile. It is easy to be on good terms with somebody when you always agree with them, but to work out your disagreements in a friendly way so that you won't disagree in the future—that's the test of good relationship. I think we met that test last autumn, and far from stabbing the British in the back we took the opposite course; we really tried to get closer to them.

Mr. Shapiro: I agree that you certainly tried to be very constructive, to build a bridge between the United States and Great Britain, and you were very friendly toward Great Britain. But what I mean is a subtle change in relationship. Let me give you a specific example. When the Prime Minister came back from Bermuda he was asked whether a common policy on the Suez had been agreed upon and he gave a very equivocal answer. The answer, I believe, was that if a satisfactory agreement is reached it will be fine for everybody who is satisfied—some very equivocal answer.

Mr. Pearson: It is not fair to base too many important conclusions on an observation that a man may make after a long plane trip.

Mr. Shapiro: No, the point I was trying to make was that before last November a Canadian prime minister coming back from a conference with a British prime minister over a question as far away from Canada as Suez and so close to Britain, as important to Britain as Suez is, there would have been no equivocation at all about it.

Mr. Pearson: I think there has been a change, but not in the manner that you indicate. At Bermuda we talked a great deal about the Middle East, but we did talk about it from the point of view of two governments trying to get together on a policy. There is no doubt that that was the approach, that was the atmosphere. But a difference in our relations seems to me to have developed in twenty years. In the Twenties and Thirties we were very preoccupied about our relations with the United Kingdom, partly because of economic considerations. Most of our trade was going there, but we were also developing our constitutional status. To some sensitive people it seemed that the British were getting in the way of Canadian national political development—and there was some basis for this—that our independence of action was being

prejudiced by British policy in Europe where the danger centres
were. If anything happened in Europe we thought we would be
dragged in again as we were in 1914. We weren't dragged in, of
course, but we did come in. We were very worried lest British
policy should bring us into a conflict in a way which would disturb
the unity of this country.

Now all these worries have gone. We don't have to worry about
our constitutional condition. Nobody is sensitive about that any
longer. Downing Street is a place where we go and have dinner
now, not a place where they are trying to decide what the Canadians will do. Moreover the British do not primarily determine now
the great forces of politics which lead to peace and war; they
influence those forces importantly. They have wisdom and experience, which perhaps we should use more. But they don't themselves determine events as much as the Americans do. Just as soon
as we realized in this country that the Americans were now the
people who might drag us into trouble again—again I use that
word 'drag' very loosely—we began to worry more about our
political relations with the United States and less about our
political relations with the United Kingdom. The latter country
changed from a rather formidable father to a kindly big brother to
whom we could go for comfort and encouragement! . . .

III. Australia

1. Extracts from speech by the Minister of External Affairs, Mr. R. G. Casey, 26 November 1956[1]

I have devoted much of my time here [in the United States]
to efforts to repair the breach, but I have not had any response to
speak of. I have found difficulty in even discussing the matter. . . .

You may agree or disagree with the action taken, but these
affairs of the last month have dealt a grievous blow at the relations
between Britain and your great country. British-American relations are very much more important to Britain than to the United
States, but if we are going to wrap ourselves in the Union Jack and
you in the Stars and Stripes, there is no future for the world. The
basis of democracy's hope for the future is good relations between
your country and Britain. The rift is deep and wide. I ask you to
do all you can to repair it. If it cannot be repaired, hope for the
world is very small indeed. Your country cannot continue to exist

[1] *Sydney Morning Herald*, 28 November 1956.

on its own, great and powerful as you are. You must have the co-operation of the other leaders of the West. If there is no such co-operation, the future indeed is a black one. . . .

2. Statement by the Prime Minister, Mr. R. G. Menzies, 30 November 1956[1]

If, from now on, constructive efforts are to be made to repair the breach which has occurred in Anglo-American relations, it seems quite clear that the breach should be that of realism and not that of theory.

The first thing is to get the facts right. I have doubts as to whether even now they are as clearly understood in the United States of America as they are in Australia. It seems to me that the stark facts of the present situation in the Middle East could be put in this way: Great Britain and France have been ordered out of Egypt by the General Assembly of the United Nations. Their forces are being replaced by a fragmentary United Nations force which is pretty clearly not designed to be a fighting body. The whole operation appears to be based on the consent of Colonel Nasser, and subject to whatever conditions he thinks fit to impose. This means that Egypt's military defeat having been arrested on the very threshold, Colonel Nasser remains in possession of the field, and appears to be dictating terms as if he were a victor.

The Soviet Union is notoriously supplying implements of war to both Egypt, Syria and, possibly, to other Arab States. Israel having been ordered out of territory which she captured so swiftly from the Egyptians, finds herself once more hemmed in by those who are hostile to her, and whose capacity to attack her is being steadily built up.

When the Soviet Union becomes a military benefactor of a small nation, that nation very soon becomes, in effect, a satellite. In the Communist mind, assistance once accepted involves domination. The humiliation of Great Britain and France in the Middle East, therefore, is contemporary with the first movements designed by the Soviet Union to give her access to the waters of the Mediterranean.

Meanwhile, Colonel Nasser, without any domination by the General Assembly, which appears to have been so eager to divert its attention from the massacre of Hungary that it has concentrated

---

[1] C.N.I.A., vol. 27, no. 11, pp. 748–50.

all its fury upon Great Britain and France, has sabotaged the
Canal; has directed the sinking of many vessels in it, and has put
it out of action for many months. In the meantime, the current oil
supplies of Great Britain and Western Europe have been terribly
reduced; petrol rationing has been re-introduced; and a great deal
of unemployment seems inevitable. We are thus witnessing a state
of affairs in which Russian morale has been elevated; the already
difficult situation of Great Britain aggravated; the prestige of
Great Britain and France in the Middle East swept aside; and the
basis of Western European defence, which includes the maintenance of defensive positions in the Middle East, gravely impaired.

These are all grim matters which must cause profound anxiety
amongst friends of freedom. I confess myself baffled by the activities of the General Assembly of the United Nations. The terms of
the Charter make it quite clear that what might be called the
Executive Body—the Security Council—was set up in the hope
that the peace of the world would be defended by co-operative
action primarily on the part of the Great Powers.

The new role of the General Assembly appears to reverse this
principle because not only are many small nations offering to give
instructions, but the Great Powers themselves have been excluded
specifically from any executive part in the operation. It may be
believed by theoretically minded people that when the British and
French forces leave under orders, the Suez Canal problem will be
back where it began. But we should not delude ourselves that this
is so. In my opinion all the entries in this disastrous accounting
will be on the debit side. Like most Australians I am tremendously
exercised as to what the effect of all this will be upon Turkey, the
most powerful Middle East member of NATO, and upon the
Baghdad Pact, which appeared to offer great promise of stability in
the Middle East. I will not attempt to elaborate these comments.
We can only hope that the world will learn some salutory lesson
from what has happened. But I do want to point out that the worst
result of all is that there is beyond question an output of angry
division of opinion among those who are, by tradition and practice,
bound to be friends. Unless the differences between Great Britain
and the United States can be speedily bridged, there will be a
great temptation for the Soviet Union to pursue more actively its
aggressive policies in order to take advantage of the confusion now
existing west of the Iron Curtain.

I am not attempting to exaggerate the significance of anything that I am saying, but I have many friends in both the United States of America and Great Britain, and at a slightly earlier stage I was made the principal spokesman for the Eighteen Nations which formed the majority in the London Conference. I would like to remind all the Eighteen Nations that the very essence of the proposals which my colleagues and I were sent to Cairo to present, was that the character of the Suez Canal as a free international waterway must be preserved by taking the operation of the Canal out of the political control of any one nation. It is to be hoped that the Eighteen Nations will not forget about this, and that it will continue to be regarded by them as their major task to bring this result about.

The real danger is that when the Canal has been cleared and the United Nations force withdrawn, it will be thought by many that the whole incident is closed. It can never be closed until the future of the Canal has been assured. It is certainly not to the interests of the American people, who are not only the greatest Power in the world, but have also done so much for the world since the Second World War, that there should be any discord between them and the people of Great Britain who after all have twice in this century been the first into the field in the defence of liberty, including the liberty of our American friends and allies. This is not the time to be standing on fine points of protocol. Statesmanship on both sides of the Atlantic must concentrate its attention on the rebuilding of the old bridges. It needs to be made clear once more that the defence of the free world depends essentially not upon relations of the General Assembly but on the closest and most friendly co-operation between the Great Powers of Western Europe and the United States.

It would be of inestimable benefit if these recent matters of difference being swiftly put aside, it can be made plain by some concerted statement on the part of the leaders of these great nations that their cohesion will remain and that under any and all circumstances they will work together and fight together to resist and defeat Communist aggression with goodwill and quick understanding, and the earliest possible meeting on the highest level the danger which indeed threatens the world may be successfully averted [*sic*]. If the present state of affairs is allowed to continue the United Nations itself may unhappily become for the world a discredited institution and the grand alliance which alone can put strength into it may be weakened so seriously that reckless action on the part of Communist Powers may be positively encouraged.

Nobody can be in any doubt of the tremendous goodwill and high idealism of the President of the United States nor can any-body have any doubt that the people of Great Britain and France, who would be the first objects of attack in the event of global nuclear war, have followed the course that they have in order to maintain the peace which is vital to their very existence. We have, therefore, vast reserves of peaceful friendship. It is to be hoped that they will be drawn upon by the world's leaders swiftly and effectively.

3. Statement by the Minister of External Affairs, Mr. R. G. Casey, 1 December 1956[1]

Most of my time overseas was concerned with the Middle East situation.

Many members of the United Nations have acted as though the fighting in the Suez Canal area took place out of a clear sky—that it 'just happened'. This is the opposite of the truth. It came about after many years of the most intense coldly calculated provocation by Nasser, both against Britain and against Israel. Even Nasser's seizure of the Suez Canal four months ago and Israel's kick-back against Egypt a month ago were by no means the beginning c i it all.

Egypt has in fact been conducting an undeclared war against Israel for years—and to a lesser extent against Britain in the Middle East. Egypt has blockaded Israel shipping through the Suez Canal in defiance of the United Nations and of international obligations to allow complete freedom of passage through the Canal. She has conducted continuous commando operations deep into the heart of Israel, and has killed large numbers of Israeli civilians over recent years. She has defied United Nations' resolu-tions against her. Egyptian wireless and other propaganda against Britain and France and against Israel has been intense. So far as Israel is concerned, the Egyptian Government has repeatedly stated that they intend to destroy the State of Israel. ('Egypt will fix the time and place of Israel's destruction.') The large scale arming of Egypt by Russia was directed to the destruction of Israel and to

[1] *C.N.I.A.*, vol. 27, no. 12, pp. 828–30.

the elimination of all legitimate Western interests in the Middle East.

All this reflected the state of undeclared and unrelenting war by Egypt against Israel. It was clear that the time of Egypt's assault on Israel was becoming ever closer. The setting-up of a joint Egyptian-Jordan-Syrian military command, under an Egyptian, was the final red light for Israel. In these circumstances, Israel moved against Egypt.

Britain came into the picture because the Suez Canal area looked like becoming a continual area of conflict between Egypt and Israel, and partly because if the fighting spread and involved other countries in the region, the Soviet Union would inevitably have intervened with all that entailed, including the major risk of a great war. France and Britain sought to place themselves between the two combatants and so to limit the armed conflict.

But all this chain of events was ignored by the majority of members of the United Nations. They ignored all the intense provocation of the past and charged Britain, France and Israel, in effect, with aggression, a grossly improper charge. They are now engaged in whitewashing Nasser and restoring the situation of a month ago, with all its tensions and frictions. There is little sign of the United Nations tackling the basic problems of Egyptian-Israeli relations and the future of the Suez Canal. Until these fundamental problems are solved the Middle East will remain explosive and highly dangerous.

The conception that seems to be prevailing in the United Nations is to restore the conditions which existed in the Middle East prior to a month ago. But bluntly, this means restoring the conditions of tension and friction that led to this outbreak. Something much more fundamental must be achieved, if the Middle East is not to be the scene of constantly recurring trouble.

In all this situation one cannot ignore the attitude of the United States. They have chosen to throw their voice and their vote with the great majority of members of the United Nations, who, in effect, charged the United Kingdom, France and Israel with aggression. This has produced the most grave rift between the United Kingdom and the United States in living memory.

In spite of hopeful words from high quarters, this rift remains. In the last three weeks, in Washington and New York, I did my

utmost in private and public to bring the United Kingdom and
United States back into high level discussions. Although I had
every evidence that Australia still remains on terms of the closest
intimacy with America and with Britain, what I had to say in this
regard on behalf of Australia did not produce any noticeable
response from the United States, although I was told on the highest
authority that as soon as the United Kingdom forces left Egypt,
the traditional intimacy between the United States and the United
Kingdom would be restored. But what is needed is that this should
be accompanied by a determination on the part of the United
States to throw the weight of their undoubted influence behind a
full-scale effort to remove the real causes of instability in the
Middle East—Egyptian-Israel relations—and the co-operative
control of the Suez Canal. The United States is just as convinced
as Britain and Australia that these are the fundamental essentials
for stability in the Middle East.

However, notwithstanding all this—the one thing that has to be
restored beyond all others is the intimate and confident relationship between the English-speaking peoples. That some anti-
American feeling should be expressed in Britain is understandable
and inevitable in the circumstances—but it is to be hoped that it
will pass quickly. The British have every reason to feel victimized
by the unfair treatment and lack of understanding to which they
have been subjected.

But let us make no mistake on the big question. Whatever
doubts some of us may have about the form or speed of American
actions in this present Middle East crisis, there is no doubt about
their firm determination to prevent the world being communized,
to back and sustain Western Europe, and to support democratic
nations everywhere in their efforts at collective defence.

How the United Nations will come out of this is another matter.
It is too early yet to attempt to form a proper judgment. The
increasing number of groups of countries that automatically vote
together irrespective of the merits of the case introduces the
equivalent of party politics into the international sphere. It is a
most dangerous manifestation that could bring the United Nations
down. Party politics is right enough in the national sphere, but it
should have no place internationally. The main function of the
United Nations is to stop war breaking out or to limit it or stop it
when it does break out. The counting of heads, irrespective of the

facts and merits of the situation, is no way to do this delicate and all-important task.

The United Nations is paying very much more attention to getting the British out of Egypt than to getting United Nations observers into Hungary. The United Nations' effort to bring pressure of organized world opinion to bear on Russia in Hungary has failed so far. Some more vigorous attempt to bring the spotlight of world opinion to bear on Russian savagery against helpless civilians in Hungary is long overdue—unless Russia is to be allowed to get away with the horror that she has perpetrated.

4. Extract from speech by the Minister of External Affairs, Mr. R. G. Casey, in the House of Representatives, 2 April 1957[1]

. . . I end this very short survey of the Middle East with some observations on the past and future role of the United Nations. The United Nations has had many of these questions before it for a number of years. Without wishing to be unduly critical, it can be said that it has reached no satisfactory solution to any of them. I believe this reflects the fact that we must recognise that the United Nations cannot always be counted upon to reach objective and fair and constructive conclusions on situations in which group pressures and the promotion of special interests have tended to weaken its effectiveness and its impartiality. This was all too evident in the Assembly's handling of the Israel-Egypt dispute. From this Egypt will no doubt draw considerable encouragement. For, as long as Egypt has the backing of a partisan group in the General Assembly, for so long will she be encouraged to pursue her own interests to the detriment of the international situation.

The Security Council was intended by the Charter of the United Nations to acknowledge the prime responsibilities of great powers in questions involving peace and security. It was not envisaged that these responsibilities should be submerged in the voting of 80 countries in the General Assembly. The temptation must be avoided of believing that it is a substantive foreign policy merely to put a question on the agenda of the United Nations and to invite discussion. There is, unfortunately, no basis for believing that the United Nations Assembly will automatically provide a just and effective solution for any and all problems that come

[1] Australia, H.R. Deb., 1957 (First Period), no. 3, p. 14.

before it. It is this Government's view that there is a compelling need, in the United Nations and outside it, for the great democratic powers to assert joint leadership directed towards peace and stability which is entirely consistent with the Charter of the United Nations. . . .

5. Extract from speech by the Leader of the Opposition, Dr. H. Evatt, in the House of Representatives, 2 April 1957[1]

. . . I say to the Government once more that it is making the greatest possible blunder by not adhering more firmly to the United Nations and its Charter. I have never contended that the Charter is a perfect instrument, but the United Nations organisation is the one wonderful body that can deal with these great issues. If a problem is not solved at the first attempt it should not be put away in despair and regarded as insoluble. Those who are seeking a solution should not act like a disappointed litigant and say, 'We are through with it.' Unfortunately, that is the attitude of this Government. The speech of the Minister for External Affairs [Document 4] reflects the attitude of his colleagues. He was bitterly annoyed with the United Nations organisation when it said, with regard to the Suez invasion, that the invading forces must leave, as, finally, they did. I regard that event, in substance, as a victory for the United Nations, but not as a defeat for the British people. Those who were responsible for that invasion were the members of the Government of Great Britain. The people had no say in it. So far as they were able to express their views, they were against it. I believe that the British people and the Australian people are strongly in favour of the United Nations, and the strength of that organisation will depend on the moral support given by its members, including Australia. I have always claimed that the Minister for External Affairs would stick up for the United Nations and at any time of the Suez crisis I felt that he was in favour of the United Nations' actions and that he differed very strongly from the course followed by this Government. . . .

[1] Australia, H.R. Deb., 1957 (First Period), no. 3, p. 423.

6. Extracts from speech by the Deputy Leader of the Opposition, Mr. A. A. Calwell, in the House of Representatives, 4 April 1957[1]

. . . The Minister for External Affairs could not say anything worthwhile because Cabinet would have prevented him from relating the course of events properly. Cabinet did not want him to tell the truth and discredit the Government. The Minister used words, not to express his views and give the facts, but to cloud the issues and hide their real meanings. Sir Anthony Eden at least had the decency to resign when he and the French Prime Minister, Mr. Mollet, outraged world opinion by attacking Egypt. . . . The Menzies Government not only supported the Eden-Mollet act of aggression but actually encouraged it. The Prime Minister of this country went to England and urged full-blooded economic sanctions. What did that mean? It meant trouble in the world. If he had been as wise as the Prime Minister of Canada, he would not have associated us with the Egyptian affair when the United Kingdom and France attacked Egypt. We did not attack Egypt. We should have followed the Canadian example. It would have been better for this country. . . .

7. Extracts from speech by the Prime Minister, Mr. R. G. Menzies, 8 July 1957[2]

I do not think that retrospective discussion of great events is of much help unless such discussion can illuminate the path for the future. Because I think that in this case it can, I think I ought to say one or two things about the Suez Canal and about the United Nations and its relationship to this great issue, an issue which has divided opinion in so many countries, and which has in the long run altered the balance of power and significance in the whole of the Middle East area.

The first thing I want to say to you is . . . that I am utterly unrepentant about anything that was done or said over this great issue on our side. Having said that, I will examine the matter further. . . .

. . . I believe that over this matter of the Canal the United Nations, through the General Assembly, acted with both haste and

---

[1] Australia, H.R. Deb., 1957 (First Period), no. 3, p. 584.
[2] *Speech is of Time: Selected Speeches and Writings by the Right Honourable Robert Gordon Menzies* (London, 1958), pp. 171–80.

injustice. . . . It dealt with it, of course, in a vast hurry. It said,
'This requires haste, this is an urgent matter.' The haste was so
great that back in my own country, though I had had some small
part in the matter, I hardly remember reading the terms of a
resolution in the Assembly until after I had read the division lists.
An odd thing it was to learn in my own country about what the
motion was or the amendment was only after it was all over.
Indeed somebody said to me the other day . . . that there have been
one or two occasions when the Assembly have voted first and
debated the amendment afterwards! And that may well be true,
because all I know is that they do not always appear to realize in
New York that the clock is different. The earth goes round the
sun—I am sure I am right in saying that. And because the earth
goes round the sun, when some profound fellow in New York is
writing out the second draft of an amendment, I, I am happy to
say, am sleeping the sleep of the just at six o'clock in the morning
in Canberra and the celebrated statesmen in [the United Kingdom]
are sleeping the sleep of the just, or the unjust, at some other hour.
But the Suez procedure was all done in a hurry, and because it
was done in a hurry there were some things that I believe were
never adequately put, or adequately understood; and I am enough
of a lawyer to know that unless the case is fully argued, the judgment may turn out to be wrong. Let the case be fully argued.

Now why do I say that? Well, take one example; I am sure that
in the General Assembly, as in one or two great organs of public
opinion in this country, it was taken for granted that if Colonel
Nasser wanted to nationalize the Canal, he had as much right to do
so as my friend Lord Attlee in his day would have to nationalize
an industry in Great Britain. The two things were utterly
different. . . .

. . . When the General Assembly, hastily convened, discussed
these matters, it discussed them on the footing that this was a
lawful act, and therefore, of course, that what Great Britain did,
and what France did, must be unlawful. Those nations were
treated as challenging, and, indeed, in the long run challenging by
arms, an act which had all the purity of an unchallengeable act, so
that if it was legal, they were illegal. That is the whole footing on
which this matter was considered, and that has given me a very
great deal of trouble. Not that we can go back and replough the
ground, reargue the case, secure different decisions, because, in

the long run, Great Britain and France, great law-abiding countries, having received the views of the General Assembly, went out;
and the Israelis, who had pursued the somewhat speedy Egyptians
when they crossed the frontier, had gone back. Then people say,
'Very well, it is all over.' But it is not all over. I am not advocating
some new military developments, but I do say this to you, gentlemen, that if the General Assembly of the United Nations is to
assume and exercise this moral and political authority in the world
—these eighty-one nations—then the time is more than overdue
for introducing procedures into that body which will enable both
sides to be effectively heard at the right time.

. . . If it is to work, then there are two observations to be made
about the General Assembly with its new powers, its new authority, its new prestige, which must be considered by everybody.
First of all, we are told, and repeatedly told, that the General
Assembly is the 'Parliament of the World'. Now I am an old
parliamentary 'sweat' myself, and I do not see many of the ear-
marks of a democratic Parliament in the Assembly. There are
eighty-one nations, each with one vote. A highly developed and
advanced country like the United States, with 170 million people,
one vote; and some country, perhaps not very advanced, perhaps
very backward, with a quarter of a million people, one vote. We
have some democracies fully developed; we have some aspiring
democracies (and how much this country has done to encourage
them and to develop them); and we have on the other hand some
dictatorships; we have the Soviet Union and the Soviet bloc, the
satellites. Here is indeed an odd gallery of nations, but each with
one vote. I do not suppose that is very easily altered, but I find it
hard to call the Assembly a 'Parliament' under those circumstances. However, if it cannot be altered, at least I would like to
say this to a lot of the new nations whose proper pride it is to be
admitted to the United Nations: 'Please understand that you will
have power in the Assembly; you will be able to vote in the Assembly; but the people with the great responsibilities for peace
are four or five or six great nations in the world. Now do not drive
them too hard by the power of the vote. Always remember that
though you have some power, your responsibility is small; you
must respect and treat with objectivity the arguments of the great
powers who, when it comes to the point, will accept and carry the
great responsibility for world peace.'

Now that is my first point about the General Assembly. The second is this: I think it is intolerable that an Assembly meeting in New York should be able at five hours' notice to vote on something when my Government cannot consider it, and perhaps your Government cannot consider it, and most of the countries around the world cannot consider it. If this is to be Parliament, then let it have parliamentary procedure; let there be notice given of meetings; let there be notice given of motions; let there be adequate debate; let there be time for everybody to cast an informed vote, and to precede it, if necessary, by an informed speech. That is the very essence, and unless it is done, we will have in the General Assembly snap votes, snap motions; we will have people of prudence and thought overruled by people of imprudence, in a hurry, and, in the long run, there will be a serious risk that the great nations, who are the custodians of the peace, will say, 'We cannot have this—we will withdraw.' And, if they withdraw, the whole structure falls to the ground. So let us have no more grouping for catch-votes to get majorities, or to frustrate majorities.

. . . There are some people in some countries in the world who appear to think that to take a matter to the United Nations is in itself an act of policy—'Our policy is to take this to the United Nations.' There could be no greater blunder. There have been symptoms of this blunder even among some of our traditional friends. It is not a policy to take a matter to the United Nations— it is just as if you signed an application for this thing to be referred to the United Nations, and at the same time you said: 'Well, I am in debt—I.O.U.—thank God that's paid.' It does not happen that way. My conclusion, much contrary to what I said in my anger a few months ago, is that the great and significant powers in the world must treat the United Nations more seriously; they must say, 'It is not a policy to go there, but it is a good procedure to go there with a policy.' . . . If that is done, great good sense may yet come to the world. . . .

IV. New Zealand

1. Extract from statement by the Minister of External Affairs, Mr. T. L. Macdonald, 1 December 1956[1]

. . . It is significant and disturbing that Colonel Nasser, in his negotiations with the Secretary-General of the United Nations,

[1] *E.A.R.*, vol. VI, no. 12, pp. 5–6.

has not been asked, or at any rate has not agreed, to abandon his previously announced objective of 'sweeping the Israelis into the sea'.

Nor has he abandoned his more recent objective of exercising full control of the Suez Canal. If his position is restored without clarification of these two essential points, then all will have been in vain, and this whole sorry business will be re-enacted in some form or other in future years.

Consideration of this question invariably raises the position of the United States' policy in respect of Suez, and the actions of Britain and France. It was most encouraging to note that editorial writers and columnists in leading New York and Washington newspapers were strongly critical of the line being followed by the State Department. . . .

V. SOUTH AFRICA

1. Extracts from statement by the Minister of External Affairs, Mr. Eric Louw, 16 December 1956[1]

The Suez Canal problem is still there to plague the Western nations, for whom the Canal is an economic lifeline.

There seems to be little doubt that pressure exerted by the United States through the medium of the United Nations led to the withdrawal of Anglo-French forces, and there would appear to be good grounds for those who say that the responsibility for maintaining peace in the Suez area, and indeed in the Middle East, now devolves upon the United States.

No responsible statesman will be so optimistic—or so naïve—as to believe that the small and mainly symbolic United Nations 'police force' is able to handle the Suez situation. . . .

The problems raised by Egypt's nationalization of the Suez Canal Company will not be solved by a thoroughly divided United Nations, in which the Bandung countries, generally acting in concert with the Communist States, are playing so important a role under the leadership of India. . . .

It has been suggested by some of the Press commentators and columnists that the United States is now looking to the East, rather than to the Western nations, for assistance in countering Communist expansion and aggression. Mr. Nehru's visit to the United States is cited in support of that suggestion. I am not informed in regard to probable changes in United States policy,

[1] *The Star* (Johannesburg), 18 December 1956.

and I will, therefore, say no more than that the Western nations run the risk of being sadly disillusioned, if in this matter they are relying on support from the East, either in their own national interests or for the purpose of resisting Communist expansion. Nehru, Chou En-lai and their Bandung associates have other fish to fry. . . .

If support is given, it will be at a price—and the price may be Africa. We in the southern half of the African continent have reason to believe it. Up to the present our warnings have not been heeded—often because it was not in accordance with the policy of a particular Western country, and thus of its Press, to direct attention to such warnings, which are therefore 'played down' as much as possible.

South Africa, which has so much at stake, will continue to warn.

2. Extract from speech by the Member for Green Point, Mr. P. Van der Byl, in the House of Assembly, 28 January 1957[1]

With regard to the Suez crisis . . . , Egypt is the front door through which Russia can infiltrate into the whole of Africa, and let Communism seep in till it permeates the 200,000,000 non-Europeans of this Continent—the most dangerous thing that could possibly happen to us. England and France, with all the urgent need for maintaining the goodwill of their non-European colonies (whether they acted wisely or not is a matter of opinion) had the courage to take a stand against an Egyptian dictator who obviously was in close association with Russia and who would at any time sell the pass leading to Africa. Here was a chance for our Government to rally to a cause which was our own, for once Russia is established in the Middle East, and stands in the gateway to Africa, our very existence is in jeopardy. Whether it was due to ineptitude, sheer folly or their personal dislike of Britain, I don't know. I don't know whether it was due to anti-Semitism—and here I look at the bench over there—from which we heard so much from 1939 to 1943 against the Jews when they thought Hitler was going to destroy all the Jews in Europe, or just stupidity. All I do know is that they informed an astonished world that the Suez crisis was not their affair. In short, they were neutral in a matter which vitally concerned South Africa. . . .

[1] South Africa, H.A. Deb., vol. 93, col. 54.

3. Extracts from speech by the Minister of External Affairs, Mr. Eric Louw, in the House of Assembly, 11 February 1957[1]

. . . The Opposition's attitude is that we should have acted in respect of something which could perhaps happen as a result of the nationalisation of the Suez Canal Company. Mr. Speaker, when we were faced with the position that the Government had to decide, we did not immediately jump in and issue a statement. We waited four or five days. The matter was fully discussed by the Cabinet and the statement about which hon. members are now complaining was then issued. The Government was concerned with only one matter, and that was the nationalisation of the Company, a company which is called a 'joint stock company', which is registered in Egypt, that is to say, an Egyptian company. We were not concerned with possibilities of what could perhaps happen in the future, should Egypt as a result of that step proceed to take some action or other. What was our attitude? The attitude of the Union Government was in the first place this: As far as we can ascertain there are no shares in the Company held here in South Africa; South Africa was not and is not to-day a user of the Canal. We do not have a single ship using the Canal. Furthermore the Canal is not and has never been a vitally important artery to South Africa. It is therefore clear that South Africa has no direct interest in the nationalisation of the Suez Canal Company, and that we therefore stated the position correctly, namely that it was a domestic affair of Egypt's whether it would nationalise the Company or not. That is also admitted today at U.N.O. That was also admitted later on here in South Africa. . . .

Mr. Speaker, the Opposition realise the weakness of their argument as regards the aspect of the nationalisation of the Canal Company, but they now say: 'Yes, but there was after all the International Convention of 1888.' Mr. Speaker, if hon. members will take the trouble to study my statement, they will see that we, and also the Prime Minister, said throughout that we were indeed interested in the honouring of international agreements. The fact remains, however, that since the nationalisation of the Canal Company, the Egyptian Government has not once done anything which could be regarded as an infringement of its undertakings

[1] South Africa, H.A. Deb., vol. 93, coll. 923–6.

under the 1888 Convention. The Egyptian Government was
especially careful not to do so. . . .

. . . Is it now the attitude of the United Party that we should also
have intervened and that we should also have suffered the same
consequences as [the United Kingdom and France]? Is their
attitude that South Africa should blindly have followed England's
example in this matter? Mr. Speaker, let me again state what
Canada's attitude was. I quote again from the speech by Mr.
Lester Pearson, the Minister of Foreign Affairs, as reported to me
by our High Commissioner. He says—

Referring to the recent Suez crisis, and Canada's *disapproval* of the
action taken by the United Kingdom, Mr. Pearson stated that on that
occasion, when the Commonwealth was so deeply split on the issue,
their relief was correspondingly great when the danger to the Commonwealth was removed by the Anglo-French decision to accept the cease
fire resolution of the United Nations.

I wonder what the Opposition and their Press would have said
had I made such a statement, had I stated in public that we did not
support the step taken by England and France. Up till now I have
not said one word about the British-French action, nor do I intend
doing so. My instructions to our delegation at U.N.O. when the
motion condemning Britain and France was passed, was to abstain
from voting. England is a fellow member of the Commonwealth;
we are on the best possible terms with the present British Government, and we are very good friends with France. Where they have
taken a step which in their opinion was in their own interests, it
was not for me or for South Africa to interfere. Mr. Speaker, you
will remember that according to the British statement the attack
was intended to protect the Canal. Just read Mr. Eden's speech in
the British House of Commons. And we have already explained
that the nationalisation of the Canal was a matter in which we
had no direct interest. . . .

Mr. Speaker, what is interesting is that even prior to the Anglo-
French attack, opinion here in South Africa had begun to veer.
Then already the *Star* and *Argus* amongst others pointed out the
dangers of the position and began to express doubts. The House
knows of the opinions expressed regarding that naïve scheme of
the Canal Users' Association. The United Party will gain very
little support for an attitude such as that. . . .

VI. INDIA

1. Extracts from speech by the Prime Minister, Shri Jawaharlal Nehru, in the Rajya Sabha, 7 December 1956[1]

. . . This resolution ['India's further continuation of her membership in the Commonwealth is inconsistent with the principles of Panch Sheel'] deals with a subject which has frequently come up in this House and in the other House. . . . I am prepared to admit that there are many people in this country who, for sentimental or other reasons, would advocate this or would approve of it if it takes place. I don't deny that I think also that those very people, or many of them, if we once explain the position to them, may change their views on the subject. It is very easy, in any issue of this type, to get people to agree to a sentimental approach to a problem but it would be a bad day when a Government's policy in such matters is governed by sentiment only and by the sudden passion of the moment. . . .

. . . Now, it is easy to break anything, it is much more difficult to join or build. You break in a moment. An artist or creator works hard at a piece of art; whether it be a beautiful vase, a potter's work; or a building . . . , enormous labour and creative instinct go to build it, but with a blow you can break that beautiful vase or with a bomb you can destroy that piece of architecture. It is easy to break. There is enough breakage in this world not to add to it, and our broad policy has been not to break with any country but to ever get closer to it. . . . It has been India's privilege and honour to help in joining countries, to be a bridge between them, and not to break the bridges that already exist. . . .

. . . India is not hostile to any country. Therefore, when any event occurs, we are not swept away by passions, as other countries are. That is no virtue in us. I do not say we are better. But because of geography and other causes—we get angry, we disapprove—we are not suddenly obsessed by that fever of fear and other things as other countries are. Therefore we can judge world events a little more objectively. We may judge wrongly. That is a different matter. I do not say we are cleverer or more intelligent, but we are not swept away by passions in these things so much and that too chiefly because of geography and Gandhiji and other things. We can judge these things more objectively, and therefore, we can

---

[1] Jawaharlal Nehru, *Speeches in Parliament, November 16–December 7, 1956* (New Delhi, 1957), pp. 72–73, 80.

often make suggestions which are helpful. We can see the other countries. We can act as a bridge to some extent and bring them together. . . .

In everything that we do we keep this in view: are we helping the cause of peace or not? I do not want to make a brave gesture—all right, so and so has done wrong and we condemn that country. I am quite sure that every kind of contact that we have got with such a country—whether it is the Commonwealth association, the Commonwealth group or whether it is any other contact—helps the cause of peace. It does not hinder my adopting any policy, as the House knows. It helps the cause of peace and brings us together and I am for everything that brings us together, without tying us up in any way. . . . The Commonwealth does not tie us in any way. All this is very different from the ties of the array of alliances, such as the Warsaw Treaty, N.A.T.O., Baghdad Pact, S.E.A.T.O. and the Balkan Treaty. The whole world is full of these knots which tie up and we want to unravel these knots, to open them out, so that people may live, as far as possible, without fear, and live their own lives. I submit that while normally I would have opposed such a resolution, as I have done in the past, in the present circumstances I would oppose it still more. . . .

2. Extracts from statement issued by the Prime Minister, Shri Jawaharlal Nehru, and the President of Syria, 22 January 1957[1]

. . . The prime need of the hour is that the passions and conflicts which have recently convulsed the world and threatened world peace should be allowed to subside. All nations should help in this process, and nothing should be done which would aggravate the tensions and conflicts in the Middle East. Progressive forces working for freedom and stability and for the realization of the national aspirations of the people in this area should be encouraged so that they may help in healing divisions and conflicts. The United Nations with its recent increased authority can assist in this process. A special responsibility lies on the big powers in this regard.

In reviewing the recent grave events in the Middle East, satisfaction was expressed during the talks at the clear and unequivocal stand taken by the United Nations in regard to the aggression against Egypt. The several resolutions adopted by the United

[1] *F.A.R.*, Vol. III, no. 1, January 1957, pp. 33–34.

Nations in this regard represented a triumph of those principles upon which are founded the faith and hope of countries which have lately emerged into full independence. It is a matter of gratification that the common loyalty of the two countries to these principles had led to a widening area of co-operation between them. . . .

The problems of the Middle East can only be solved if the countries in that area are able, in complete freedom and without domination by any foreign power, to develop in accordance with their genius and traditions, more particularly in the economic and social fields, in order to raise the standards of living of their peoples. A military approach to the problems of this area will only serve to create further disharmony and instability, besides contributing to the heightening of tension and endangering world peace. Intervention by the big powers in the form of military pacts and alliances is detrimental to peace and stability in the Middle East. The Baghdad Pact has caused bitter conflicts and divisions in the Arab world and has greatly increased international tension.

The two countries have subscribed to the declaration at Bandung that colonialism in all its manifestations is an evil which should be brought to an end. They reaffirm their support to the cause of freedom and independence of all people under foreign domination which constitutes a denial of fundamental human rights. . . .

3. Extracts from speech by the Prime Minister, Shri Jawaharlal Nehru, in the Lok Sabha, 25 March 1957[1]

. . . Whenever there is a debate on foreign affairs in the House, there are always some amendments dealing with India's association with the Commonwealth of Nations. . . . The question is an important one. And I can very well understand hon. Members, not only on the other side of the House, thinking about this matter much more now than they did previously, and enquiring from me, as they have done, sometimes in writing, sometimes orally, as to why in spite of all that has happened . . . we still think it is right to continue the Commonwealth association. . . .

. . . I have felt, and for the first time I felt, the first time in these many years, that it may some time or other require further consideration. But in this as in many other matters we are not going to

[1] Lok Sabha Deb., 1957, Part II, vol. I, coll. 668–70.

act in a huff or in a spirit of anger merely because we dislike something that has happened. I feel, as I said here, that in spite of these occurrences that have happened and that have distressed us, it is right for us to continue our association with the Commonwealth for a variety of reasons which I mentioned then, among them being primarily the fact that our policies, as is obvious, are in no way conditioned or deflected from their normal course by that association. . . .

Secondly, at this moment, when there are so many disruptive tendencies in the world, it is better to retain every kind of association, which is not positively harmful to us, than to break it. Breaking it itself is a disruptive thing. It does not add to that spirit of peaceful settlements and peaceful associations that we wish to develop in the world.

Therefore, after giving all this thought, I felt—and I felt clearly —in my mind that it would not be good to break up this association in spite of the painful shocks that all of us had experienced in these past few months.

But again, no decision that we can take in these or other matters for today can be said to be a permanent decision for ever. . . . One has to review these matters from time to time in view of changing conditions. . . .

4. Extracts from speech by the Member for Mysore, Shri M. S. Gurupadaswamy, in the Lok Sabha, 25 March 1957[1]

. . . I have been categorically stating that our association with the Commonwealth is not for our good. . . . Our Commonwealth association has not in any way influenced Great Britain while taking a decision on any matter. I feel that our membership has been misconstrued, has been viewed with suspicion by other countries. We are neutral, as we say. If we are neutral, how can we justify our membership in the Commonwealth? I think that by remaining in the Commonwealth we are only encouraging colonial tendencies of a few powers. We are used as mere scapegoats or instruments to further their ends. Indirectly, we are only feeding their ambitions, we are only helping them to commit more blunders against their colonial peoples. Therefore, I think this is

[1] Lok Sabha Deb., 1957, Part II, vol. I, coll. 690-1.

the time for review, whether our membership in the Commonwealth should continue or not. The Prime Minister said [Document 3] if the situation changes for bad, if the policy of the United Kingdom changes, then we may think of seceding from the Commonwealth. But I think this is the right occasion. The United Kingdom has committed a great blunder and I do not think there would be any benefit by waiting for a little while more. I feel that now itself we should take a decision, and it would be better, it would be fitting to our national prestige and honour that we should cease to be a member of the Commonwealth of Nations. . . .

5. Extracts from speech by the Member for Kendrapara, Shri Surendraneth Dwivedy, in the Lok Sabha, 23 July 1957[1]

. . . I wish to put forth a viewpoint which has been demanding our withdrawal from the Commonwealth. I feel that the pledge given to the nation in the year 1930 would be left unhonoured if we do not decide on our dissociation from the Commonwealth. There was a great expectation all over the country when, following the Anglo-French aggression on Egypt, the Prime Minister stated that the time had come to reconsider our association with the Commonwealth. When Britain conducted an unprovoked aggression against a small country like Egypt, India was not even informed but countries like the United States, which are outside the Commonwealth, were taken into confidence. India was neither consulted nor informed of the aggression. . . .

Therefore, it was expected when the recent Commonwealth Conference was meeting against this background, some reference to these two incidents would be made. But, I am sorry to say that the cryptic communiqué[2] that was issued gave no indication to these. . . . Sir, really it is a matter of surprise that the resentment that was shown all over the country, not only in India but even in

---

[1] Lok Sabha Deb., 1957, Second Series, vol. III, coll. 4764–5.
[2] Its sole reference to the Middle East crisis of the preceding year is as follows: 'The Commonwealth Ministers discussed the international problems of the Middle East. They agreed that, in the long term, economic and social progress must be the foundation for stability in the Middle East. They agreed, however, that in the short term the need is to work towards a relaxation of the tension arising from the dispute between the Arab States and Israel, the plight of the Arab refugees and the unresolved problems in connection with the Suez Canal. They considered that solutions of all these urgent questions should continue to be pursued by all practicable means.'

Asia and Africa, vanished into the sky when it reached the Commonwealth Conference and our Prime Minister remained calm. I think he owes an explanation to this country. Why is it that such a cowardly act of aggression was not taken into account? When he himself had condemned it in no uncertain terms, why is it that it was not included in the discussions? Will I be wrong, Sir, if I say that there was some sort of political horse trading in this Commonwealth Conference? . . .

VII. PAKISTAN

1. Extracts from speech by the Prime Minister, Mr. H. S. Suhrawardy, in the National Assembly, 22 February 1957[1]

. . . To preserve peace in the Middle East and to ensure stability, Pakistan acceded to the Turco-Iraqi Pact along with Iran and the United Kingdom. This is known as the Baghdad Pact. The Pact represents and offers to help the mutual defensive capacity of like-minded nations in the defence of the Middle East. Here, it may be asked: why did we feel it necessary to do so? There is little doubt that this was indicated by the instinct of self-preservation. . . .

. . . This pact . . . is of a purely defensive nature; it does not permit that any country which may have committed aggression would be helped by any of the members of the Pact and, therefore, the Pact was not invoked at the time when the United Kingdom considered it proper or expedient in pursuance of its own policy to give an ultimatum to Egypt and to Israel to draw back from the Suez Canal, otherwise it would intervene. The United Kingdom has from time to time placed before us the reasons why it considered it expedient to do so. The United Kingdom has tried to justify its action by saying that if it had not done so, the Middle East would have been involved in a conflagration, and it was possible seeing the advance that Israel was making that it would have advanced much further.

I am not here to defend the United Kingdom or to place before the House its policy. I am merely stating before the House the facts as have appeared before us. The Baghdad Pact, therefore, was not invoked to assist the United Kingdom. On the other hand the four Muslim members of the Baghdad Pact met together to condemn its conduct and to call upon the United Kingdom to withdraw from Egypt and to obey the call of the United Nations.

[1] Pakistan, N.A. Deb., vol. I, no. 13, pp. 916–26.

Happily it has obeyed that call. We cannot carry opposition and venom for all time. Things change, matters develop, new situations arise in the world and today we see no reason why the United Kingdom having obeyed the mandate of the United Nations should not sit with us in order to promote the security of the Middle East and also to strengthen ourselves. . . .

. . . Sir, I do not think that the review would be complete without referring to the Commonwealth. . . . We are fortunate, I maintain, in being part of this great comity of nations. The fact that we have adopted the Republican form of government in March, 1956, has not altered our position as a member of the Commonwealth. You may remember, Sir, that during the wave of anger which swept through Pakistan, there was a demand that we should secede from the Commonwealth. Personally I do not think it proper that we should cut off our nose to spite our face and I am glad to say that the moderation and the cautious attitude which we adopted at that time has proved correct. A similar wave of resentment went through India and it must be said to the credit of the Prime Minister of India that he too set his face against any severance of relations from the Commonwealth, as he too realised the advantages that it brings to a country to be a member of a body which comprises so many important nations. The role which the Commonwealth is playing may not be spectacular but it is constructive and definite. All the countries comprised within the Commonwealth [sic] are sovereign states, though they differ from each other in many respects. We have a unity of purpose and common belief in the democratic way of life and we are prepared to co-operate with each other for the peace, progress and security of mankind. If the member countries maintain their faith in the noble purpose underlying the concept of Commonwealth and we faithfully practise them, we are bound to exert strength, and moral influence, on international affairs. . . .

2. Extracts from speech by Mr. M. A. Khuhro (West Pakistan, Muslim League), in the National Assembly, 23 February 1957[1]

. . . I was really surprised to hear from the Prime Minister— after all he was not addressing children—when he said [Document 1] it is because of their [Baghdad Pact Muslim Power]

[1] Pakistan, N.A. Deb., vol. I, no. 14, pp. 963, 968.

conference that Britain and France pulled out their armies from Egyptian territories. Certainly not. If you want to have credit, it may be 5 per cent, it may be 1 per cent, but the fact of the matter is that the United States of America was determined that Britain must get out and France must get out. The fact of the matter is that the United Nations would not yield and succumb to the influence of Britain and France. The fact of the matter is that the U.S.S.R. threatened that as a counter-attack they will fight and defend this country. These are the reasons why they went out. . . .

. . . Now, Sir, about the Commonwealth the Prime Minister made certain remarks which were very unhappy ones. He says: Why should we leave the Commonwealth? . . . But the Prime Minister never elaborated and explained to us what are the real advantages that he gets from the Commonwealth. I wish he had elaborated on that point. . . . If you do not get any kind of benefit then the correct course would be to get out of it. . . . It is better to say 'good-bye'. . . .

3. Extracts from speech by Mr. Pir Ali Mohammad Rashdi (West Pakistan, Muslim League), in the National Assembly, 23 February 1957[1]

. . . The Muslim League's view is that we are not opposed to the Baghdad Pact but continued association of Britain in the context of the new conditions runs counter to the principle laid down in this Article[2]. . . .

The Muslim League also feels that under these changed circumstances it is not consistent with the interest of this country . . . to continue its association with the British Commonwealth. . . . By breaking that association it does not mean that we make war with Britain. Many friends of Britain are outside the Commonwealth. We are under obligation to Britain for many things and we are not an ungrateful people. The very tongue in which I am speaking has been supplied by them. I am cognisant of that. The foreign policy of free nations is a system of accounting with debit and credit sides, and in this light you have to act. Sir, we did not join the Commonwealth by an Article of the Constitution which we could

---

[1] Pakistan, N.A. Deb., vol. I, no. 14, p. 1012.
[2] Article 24 of the Constitution of the Republic of Pakistan (1956): 'The State shall endeavour to strengthen the bonds of unity among Muslim countries to promote international peace and stability.'

have done. We did it by a Resolution to keep our and Mr. Suhrawardy's hands free. . . . We were assured that Britain would not do
anything to imperil our prestige and position and relations with
our Muslim brethren in any other country. But what happened? . . .
They invaded Egypt and in such a manner that for the first time
the British nation was divided on international policy, and the
Prime Minister who was responsible for it was condemned and
not allowed to disfigure that high office occupied by him. They
brutally invaded Egypt and massacred the civilian population; it
was a war of vengeance and attrition. . . . Rivers of Muslim blood
were found flowing in the Suez, so much so that even Iran found
it difficult to sit with Britain at the same table. Now I leave it to
your judgment to determine whether it will be possible for Pakistan
to further its obligations under Article 24 of the Constitution. . . .

4. Extracts from speech by the Prime Minister, Mr. H. S. Suhrawardy, in the National Assembly, 25 February 1957[1]

. . . A great deal has been said for and against Britain. . . .
Britain has committed many acts, it is true, about which we can
complain but it is no use going over past history. It is no use
retelling all these tales for the purpose of assessing what should be
your attitude towards it today. I will not have Britain being beaten
on both sides, that we shall condemn it on this side and India shall
condemn it on the other for supporting us in the Kashmir dispute.
Let us be fair to our friends. If they are with us today, and they
have taken courage, even jeopardised their position in the Commonwealth—I am surprised that no one has noticed it.

One member even contradicted my statement that our friends
have jeopardised their interests in supporting us but in fact they
have done so. If they have done that, surely it is our duty to
recognise that fact. What is more—and this I consider in the
international plane to be a very great sacrifice—they have obeyed
the mandate of the United Nations and walked out of Egypt. We
have been told that they went out for this and the other reason;
they went out because the U.N. demanded they should get out.
Yes, they went out for that reason. It is said they went out because
Russia threatened that it will send its bombs. My answer is: no.
That threat meant nothing at all. If Russia had sent bombs across,
it would have received bombs as well. Threats need not be invoked

[1] Pakistan, N.A. Deb., vol. I, no. 15, pp. 1089–91.

to explain actions, when moral forces also dictate the same cause. The Muslim countries of the Baghdad Pact put forward a modest claim that Britain went out of Egypt because we requested her to do so. This has been derided; this has been ridiculed. I do not know why. I do not know why people are so cynical; I do not know why people consider that it is only the threat of enemies that counts and not the protests or pleadings of friends. . . .

Britain, Sir, is not here to reply to the charges which have been levelled against it. We all know its history, the past, as well as the present, and if you consider the future you must give credit to this country, that it is one by one shedding gradually its imperial connections, and giving self-government to the countries which had hitherto been under its control. And when you compare this to other countries, have you no word to say in its favour? When you compare it to those who have defied the United Nations, countries big and small, when you compare it to those who are members of the Security Council and in whose hands the honour and prestige of the United Nations rest, and they themselves defy it, have you nothing to say? Can you not give credit to Britain for its conduct? It could have stayed in Egypt; it need not have left. We have seen that the United Nations has no power whatsoever to enforce its will unless the nations affected accept its directions themselves. Had Britain and France desired to stay on in Egypt, there was no power in the hands of the United Nations that could have compelled them to get out. And today, when a small country like India says that it will not have a United Nations Force on its territory, because it considers Kashmir to be its territory, the United Nations have to temporise with it, because they know that they cannot send a force there, if India takes the offensive and objects to that force on its soil. It is only when it was agreed between the parties that the United Nations Force could be placed in Egypt, that it was possible for such a force to be placed there. . . .

VIII. Ceylon

1. Extracts from statement by the Prime Ministers of the Colombo Powers, 14 November 1956[1]

Upon the suggestion of the Prime Minister of Indonesia, a meeting of the Prime Ministers of the Colombo countries was held on the 12th, 13th and 14th November, 1956. The meeting took

[1] *C.N.I.A.*, vol. 27, no. 11, pp. 775–7.

place in Delhi at the invitation of the Prime Minister of India and was attended by the Prime Ministers of Burma, Ceylon, Indonesia and India. The Prime Minister of Pakistan was unable to participate owing to other preoccupations.

This meeting of the Prime Ministers was convened more especially to consider the grave situation that had arisen on account of the Israeli attack on Egypt and the military operations by the United Kingdom and France against Egypt. The Prime Ministers considered also the situation that had arisen in Hungary and the grave developments in the international situation which threatened the peace of the world. Each of them individually had already given expression to his concern at these developments and had expressed his strong disapproval and distress at the aggression of and intervention by great powers against weaker countries. . . .

In the course of the past year, many developments had taken place which indicated a relaxation of the fears and tensions that afflict the world. The Prime Ministers had hoped that the cooperative spirit of Bandung and the Five Principles of peaceful co-existence would spread and help in removing those fears and tensions. It has, therefore, come as a great shock to them that aggression can be committed and ruthless suppression take place in spite of the widespread desire of the peoples of the world for peace and freedom. Neither peace nor freedom can come if strong nations, trusting to their armed power, seek to compel weaker countries to obey their will. This reversal of a historic process has particular significance for the countries of Asia and Africa who have no great military strength and who have to rely on the justice of their cause and the firm determination of their peoples.

Many of the countries of Asia and Africa have recently become independent and emerged from a colonial or semi-colonial status, while some others are struggling to attain freedom. To all these countries, a revival of the spirit and methods of colonialism is a matter of great concern as it threatens their own freedom. Intervention in the affairs of another State, though meant to protect special interests, is sometimes justified on the plea of international co-operation. The Prime Ministers are opposed to any such intervention and are determined to resist any resurgence of colonialism, whatever form it may take. They have every confidence that the United Nations, in accordance with its Charter, will support the cause of freedom and oppose every attempt to revive or continue colonialism.

The Prime Ministers are firmly of opinion that world peace can only be assured on the basis of freedom and disarmament. Military pacts and alliances, in particular those intended to serve the interests of big powers, encourage fear and a race in armaments. Recent history has shown that these military pacts, instead of bringing security to any region, have brought apprehension, trouble and conflict. The Prime Ministers earnestly trust that this policy of military pacts and of stationing troops of one country in the territory of another will be abandoned. The real and urgent need is that the under-developed countries of the world should be helped to progress and to raise the standard of living of their peoples. . . .

2. Extracts from speech by the Prime Minister, Mr. S. W. R. D. Bandaranaike, 6 December 1956[1]

. . . I am sorry to say that the recent happenings were no source of pride to us. They brought the world to the verge of a third world war. Still, I am glad that even at this eleventh hour Britain has decided unconditionally to withdraw her forces from Egypt. I take it she must have done that in consultation with her French allies in this little venture. . . . As long as these forces remain on the Egyptian soil, the position will continue to be dangerous and capable of flaring up into a general war. I impressed that upon the British Prime Minister, Sir Anthony Eden. I said the same thing to President Eisenhower of the United States and Mr. St. Laurent, Prime Minister of Canada. On my return to London I said the same thing to Mr. Selwyn Lloyd. I tried once again to tell them what I felt impartially. I did not consider anybody to be particularly my own enemy. I am glad anyhow—for reasons of their own —they have decided to take that step. I appreciate the action of the British Government and I have no doubt that pressure was brought to bear on them by their own party. Be that as it may, they had the courage to do it. When a man commits a mistake it takes a lot of courage to admit it. Things like face-saving must not stand in one's way in dealing with these great issues. . . .

The United Nations has always been a vital instrument. It has been effective in some cases; it has dilly-dallied on others. It is in

[1] *Pakistan Horizon*, vol. IX, no. 4, pp. 182–9.

the nature of things that it should be so. The United Nations is not a super-state, with a force at its command, capable of overcoming and subduing any nation in the world. So why should the United Nations be blamed? We must build it up to that position. I think the United Nations has emerged with enhanced reputation and prestige out of the Egyptian and Hungarian issues but still some changes have to be made. I would like the United Nations Charter to be amended. It was framed ten years ago and it is due for reconsideration in the light of experience gained. I trust that will be done. I trust that they will set up a United Nations Police Force, however small it may be.

The big nations have realized that they cannot act on their own and in defiance of world opinion. It is not the time for us merely to spend our energy in mutual abuses and mutual recrimination. It serves no useful purpose. Let us learn from what has happened to improve the position for the future. There are things at stake and we have to act. There are certain valuable lessons which we have to learn. I trust that will be done. I hope that the clearing of the Suez Canal will be expedited. I hope that Egypt will permit, under the United Nations, the clearing to take place when the foreign forces are withdrawn from her territory. That should be followed up by sitting down, with as much friendship as they can muster, to discuss and solve the issue on the basis of the six principles to which England and Egypt are parties in the Security Council and which the Security Council endorsed. Why that was not done before this mad hare-brained military venture is a question to which I have not received a satisfactory answer from any source....

What we in Asia felt over the Egyptian affair was that it was in effect—and I have no doubt it was—an attempt at a resurgence of at least the spirit of that Imperialism or Colonialism under which we suffered for so many centuries and from which we rather hope we have now finally escaped. That is why we felt very strongly over the Egyptian question. We have felt for Hungary also, although Hungary does not touch us so closely as Egypt. . . .

. . . True, we are weak—judged by material standards. We have no large armies; we are not wealthy peoples; we are far behind the advanced peoples of the West. But today the people are beginning to understand that force alone cannot solve problems. The moral force of Asia is a potent thing, believe me. The Afro-Asian group in the United Nations consists of 26 nations. They have differences amongst themselves, I know, but in the United Nations they often act together. It is almost a decisive group. The entire membership of the United Nations, with 18 members admitted last year, is seventy-nine. You can well imagine that, with a bloc of twenty-six, we can do a great deal in the United Nations where the strength and power of a country is not the only measure by which its influence is judged. We are all equal there. However small or weak a country may be, it can exercise an important influence in the deliberations and decisions of the United Nations . . . and will thus have, in my opinion, an increasingly decisive voice in shaping the affairs of the world. I discovered in Washington that there is an acute realization of that position. . . .

3. Extracts from speech by the Prime Minister, Mr. S. W. R. D. Bandaranaike, in the House of Representatives, 11 December 1956[1]

Mr. Speaker, I thought that as soon after my return as possible the House would like to hear from me a fairly full and detailed statement regarding the trip that I made to attend the U.N.O., about the various important personalities with whom I had an opportunity of discussing matters of moment. . . .

The House will remember that the Colombo Powers held a meeting at New Delhi to discuss the position in Egypt and also the question of Hungary two weeks ago. . . . I do not wish to go into the details of that discussion as they were embodied in a joint communiqué which we issued from Delhi [Document 1]. . . . That was an important statement. . . . It had an influence on a good many countries in the world. Certainly it had an influence in the discussions of the United Nations. . . .

. . . In London, . . . I had the opportunity of meeting the British Prime Minister. . . . I am much obliged to Sir Anthony Eden for having given me an opportunity of explaining to him our point of view.

I pressed upon him our views, not only ours but of the Asian Powers as well, with whom I had discussions, that we did not consider their action to be justified. That is not a point of view with which I expected him to agree but I did impress upon him the fact that in our opinion it was vital and urgent that there should, in fact, be an unconditional withdrawal from Egyptian territory of

[1] Ceylon, H.R. Deb., 1956–7, vol. 27, coll. 793–805.

British, French and Israeli forces before, in fact, effectively any other question could have been taken up, even the question of the clearing of the Suez Canal, and that it was our firm belief that delays in such withdrawal would only result in continuing the danger of widespread hostility.

I shall tell you the reasons on which we based that view. We were aware that there was a certain build-up going on in Syria of Soviet forces coming in to assist the Syrians—I do not say that they were not entitled to do so—various people coming in as volunteers, volunteers within inverted commas, fighting personnel such as jet fighter pilots, pilots of tank forces, qualified personnel, etc., who might well have started hostilities there, personally, initially against the Israelis or maybe against some of these countries like Iran, on the footing of trying to impose the decision of the United Nations on those who were defying those decisions. The moment such a situation arose—there was the danger of such a situation arising—the chances of a world war could have been almost taken for granted at that stage.

I impressed that point of view upon Sir Anthony Eden. At that time he was obviously not quite prepared to accept that point of view. He was for a phased withdrawal, a conditional withdrawal. With reference to the future of the Suez Canal, or the future of Israel, I did my best to impress upon the British Prime Minister, Sir Anthony Eden, that that withdrawal should be unconditional and should be expedited to the fullest extent in view of the serious implications involved; in fact the time was passed for mere face-saving devices that might have set the whole world ablaze. After all, what a face to save! . . .

I think the House is aware that at the United Nations meeting I tried to put forward our point of view [Part IV, VIII, Document 4, *supra*], the point of view of some of our countries, but primarily, of course, I was speaking for our own policy. I explained to them the philosophy, the reasons and the grounds behind the policy that we follow of neutralism, of a refusal, in fact, to align ourselves with power blocs and the corollary to that attitude, the attitude of living and letting live, of being friends with all and having friendly relations with all countries irrespective of their political ideologies. . . .

I went to Canada at the invitation of the Government of Canada and received a cordial reception in Ottawa. May I say that from the conversations I had with the Canadian leaders I am very satisfied that Canada adopts a fairly impartial and objective attitude in dealing with these important problems. She wants to be friendly with all of us and, during a debate at which I happened to be present for a little time in the Canadian Parliament over Egypt and Hungary, Mr. Louis St. Laurent, Prime Minister, was so—may I say—annoyed by certain remarks of the Opposition that he threw his notes aside and made [II, Document 2, *supra*] a slashing attack upon big countries imposing their wishes upon small countries, particularly with reference to England and France. . . .

On my way back I stopped in London for three days. . . . I saw Mr. Selwyn Lloyd and had an opportunity of explaining to the Ministers again, as a result of my conversations in America and Canada, that I was confirmed in my view that they should unconditionally withdraw their forces from Egypt. . . .

I impressed upon British Ministers when I returned from New York that they must announce this. It was done. Indeed Mr. Selwyn Lloyd got at me the night before I left. . . . He wanted to see me—he was preparing his speech for the following day in the House of Commons—to find out what I really thought of the announcement. I said, 'Well, better late than never. You have done right. Please withdraw your forces. Personally I think that is the only thing to be done.' I think that process is proceeding now. . . .

On the question of the bases I have settled certain things with the British Government. They need further discussions. The bases of Trincomalee and Katunayake will be handed over to the Government of Ceylon in 1957 on a date to be decided between the two governments, and the flag of Ceylon will fly over the bases of Trincomalee and Katunayake in the course of a few months. . . .

Now there is the question of facilities. The facilities that Britain wants are the use of the arms dump in Trincomalee, the oil reserves that they have in Trincomalee, and landing facilities at Katunayake. Those are the main questions.

At this point may I proceed to deal with the question of the 'Superb' and the 'Newfoundland' [British warships]? The House knows and the people of this country know the position as these matters become public here. I was informed of it in London, I took it up, and I am satisfied with the explanation given by the highest authorities, namely that they were routine refuellings. The 'Superb' was not used in the Egyptian venture at all. As for the

'Newfoundland', it had a routine refuelling at Trincomalee about six weeks prior to the launching of the Egyptian incident when, presumably, the time was not yet even decided or thought of. The 'Newfoundland' refuelled somewhere else before it proceeded there. But in a way, may I say, I was glad that this incident arose in that manner because it enabled me to point out to the British Government that the indefinite continuation of such facilities, even after the bases have been handed over to us, creates a really impossible position in the light of the foreign policy of our Government. There are times when we will have to refuse them the use of these facilities; for instance, at a time such as the Egyptian incident, when we are not engaged with Britain in hostilities—when they are engaged in hostilities and we are not, in other words, what is the use of having facilities which are going to be denied to them; a denial which they will have to accept at a time when they probably need the facilities most? I suggested that these facilities should cease altogether and it has been now decided that in January —now they are somewhat preoccupied—we will discuss these details regarding the facilities: what facilities they precisely need, the duration for which those facilities are going to be permitted at all till they cease altogether, and while they continue, the conditions on which the Government is prepared to give them those facilities. . . .

4. Extracts from speech by the Leader of the Opposition, Mr. Leslie Goonewardena, in the House of Representatives, 18 December 1956[1]

. . . My party does not find itself in any disagreement worth mentioning with the broad lines of policy pursued by the Hon. Prime Minister in his discussions in New Delhi, in his discussions with the British Government or in his intervention in the United Nations General Assembly. . . .

With regard to the discussion with the British Government on the future of bases in Ceylon, I consider it a very welcome sign that the Hon. Prime Minister has evidently modified his position on those bases, at least in relation to the facilities which we were made to understand earlier this Government was prepared to grant to the British Government.

In the first discussions between the representatives of the Ceylon

[1] Ceylon, H.R. Deb., 1956–7, vol. 27, coll. 1058–62.

Government and the British Government on this question, as far as we were made to understand, although it was agreed that the bases were to be handed over to Ceylon, nevertheless, it was also agreed that the British would be permitted to continue to have certain facilities in relation to their navy and other armed forces.

The statement of the Hon. Prime Minister last week in regard to this matter [Document 3] has apparently modified that position. Apparently the experience at Suez, the Suez crisis, and the little incident where certain British warships are supposed to have re-fuelled in Trincomalee, have brought about a change. You will recollect that many of us on this side of the House were not satisfied with the earlier position that, although the bases were to be handed over to us, the British should, nevertheless, continue to enjoy facilities. There was a grave fear in our minds that the arrangement on the bases—handing them over to us—would be more nominal than real, and, particularly in a war situation, we would be exposed to the same dangers we would have been exposed to if the ownership of the bases had actually continued to remain in the hands of the British. We are therefore very glad to know that there is to be some modification as regards the position of the Government in relation to the granting of these facilities to the British in any agreement we come to with them in respect of the bases.

I do not think that the Hon. Prime Minister was very explicit on this point, and, although he did mention that it made it easier for him to raise the question with the British Government and that he was able to point out the practical example of what had happened during the Suez crisis, he was not able to draw attention to the fact that the granting of facilities would raise complications so long as the foreign policies of this Government and the British Government did not remain identical. I do not think we have received a categorical statement from the Hon. Prime Minister that in the discussions that are proposed early next year this Government will firmly take up the position that in whatever arrangement is made for the handing over of the bases, the granting of facilities to the British armed forces cannot be considered. We should be very glad if we can have such an assurance from the Prime Minister.

With regard to the Hon. Prime Minister's discussions in New Delhi, we must express our pleasure that the policy he followed there was one which was a discouragement to aggression wherever it took place. . . .

There is, however, one matter on which naturally we hold an opinion somewhat different and are not entirely satisfied; that is in relation to our connection with the Commonwealth. The position of the Government is, I believe, that while we would possibly at a not very distant date convert ourselves into an independent republic, we should nevertheless continue to be within the Commonwealth. I would in all seriousness suggest to the Hon. Prime Minister and to his Government that just as in the case of the facilities to be granted in relation to the bases experience is making them alter their position, so, in relation to our present position in the Commonwealth experience in the future might also lead them to a change of their position on that question too.

It does appear to me that our connection with the Commonwealth is somewhat unreal. It may have some reality for some of the other earlier Dominions which have not only similar historical raditions but also in the current world have a similarity of interests, particularly in relation to the foreign policies being pursued by those Dominions.

The present Suez crisis brought out the whole reality of this fact that one section of the Dominions, the white Dominions, if I might be permitted to call them so, were aligned completely or more or less completely, with Britain.

The Hon. S. W. R. D. Bandaranaike: Canada and South Africa also kept out.

Mr. Leslie Goonewardena: With the exception of Canada, which in fact and actual reality is more of a dominion of the United States. . . . In that case it may be somewhat different. But there is one fact that emerges from the Suez crisis, that whilst some of the Asian countries, erstwhile members of the British Empire, appear to see eye to eye on this question of the invasion of Suez, the other Dominions of today seem to fall into a different camp. That, I suggest, is a pointer to the fact that our continued membership in the Commonwealth is becoming an unreality. The foreign policies of the Asian members of the Commonwealth seem to be fundamentally different from the foreign policies of the other members of the Commonwealth. I would say this is one example, but as time goes on I do feel that this unreality will become more apparent and, rather than have a false relation with the Government of the United Kingdom it would be much healthier for us to have a relation which represents the real connection that should exist between our country and the United Kingdom.

I know that there are many people in this country who feel that the present connection that we have with the Commonwealth has more advantages than disadvantages, and, for that reason, prefer to continue to remain within the Commonwealth.

It is quite possible that for a big country like India, for example, the advantages of remaining within the Commonwealth may outweigh its disadvantages, but in relation to a small country, particularly like Ceylon, whose interventions in Commonwealth affairs cannot reasonably be expected to exert much influence, it is, I think, a matter that should be reviewed by the Government in the light of the fact that, unlike India, because we are a very small country, we are likely to get all the disadvantages and very few of the advantages as a result of our maintaining that connection in its present form. . . .

INDEX

Abadan, 110

Abu Suier airfield, attacked, 170

African National Congress, South Africa, statement on Suez Canal nationalization, 66

Afro-Asian countries: attitude to United Nations Emergency Force, 281; in United Nations, 464–5; reaction to Suez invasion, 191, 255, 259, 276; resolution on Anglo-French withdrawal, 279, 300

Age, The (Melbourne), 185

Al-Gomhouria (Cairo), 85

Almaza airfield, attacked, 170

Amery, Julian, on seizure of canal, 26

Amory, D. Heathcote, 174

Amritsar Bazar Patrika, 19

Anglo-French relations, Canadian view of, 416

Anglo-American relations, 171, 187, 235, 242, 350, 381, 389, 412, 414, 421, 424, 431, 434; Australian mediation, 440–1; Australian view of, 435–6; Bermuda meeting, 378, 410, 411

Arab world, Nasser's aims for, 35, 106–7

Arab-Asian countries: attitude to Suez, 432; resolution on ceasefire, 330, 331, 359, 370

Argus, The (Melbourne), 186

Aswan Dam, 4, 34, 35, 53

Attlee, Clement Richard, 1st Earl, 418, 445

Auckland Star, The, 16

Australia: aftermath of Suez, 388ff., 435–47; attitude to United Nations Emergency Force, 280, 289–90, 345–52; delay in instructions to U.N. representative, 227; Egypt breaks off diplomatic relations with, 183; foreign policy following Suez, 390–3; importance of Suez Canal to, 12–13, 53, 132, 237; Menzies states policy, 130–3; policy in General Assembly, 235–6; policy in Security Council, 183–4, 209, 225–9; objections to General Assembly's debate, 278–9; Press reaction to Suez invasion, 185–6; reaction to Israeli invasion, 225–6; reaction to seizure of Suez Canal, 51–58;

reaction to Suez invasion, 182–6, 209, 210, 225–41; relations with United Kingdom, 438; support for United Kingdom, 84, 300, 389; views on Commonwealth, 393; views on international control of Suez Canal, 53, 54, 127

Australian Labour Party: reaction to Suez invasion, 184, 233; views on use of force, 125, 133, 240

Azad, Maulana, 193

Baghdad Pact, 198, 453, 454, 457–8; meeting of Muslim members, 261, 267, 268, 457–8; statement of Muslim members, 265–6, 267, 268; British statement to Muslim members, 403–4

Balkan Treaty, 453

Bandaranaike, S. W. R. D.: answers questions on seizure of Suez Canal and London Conferences, 75–78; attitude towards seizure of canal, 23; message to General Assembly, 270; offers troops to United Nations Emergency Force, 293; on Commonwealth, 402

speeches: on Suez invasion, 463; on trip to United Nations, 465–8; on United Nations, 463–4; to General Assembly on United Nations Emergency Force, 372–3; to General Assembly on withdrawal of troops, 372

statements: offering troops, 373; on London Conferences, 162; on seizure of Suez Canal, 71, 73, 74; on Suez Canal Users' Association, 163; on Suez invasion, 195; on British use of bases, 165, 468–9

Bandung (Bandoeng) Conference, 22, 71, 248, 462

Bandung (Bandoeng) Declarations, 69

Barnett, Air Vice-Marshal Sir Denis, 170

Belgium: attitude towards withdrawal of troops, 299; resolution in General Assembly, 258

Beloff, Professor Max, 202

Ben-Gurion, David, 98

Bermuda meeting, 378, 410, 431